SOLAR AND STELLAR ACTIVITY: SIMILARITIES AND DIFFERENCES

COVER ILLUSTRATION:

BY THE KIND PERMISSION OF KLUWER, referenced in the current volume: Rutten et al., Fig 1, page 250, — Schematic vertical section through a super-granulation cell with the magnetic network at its borders.

A SERIES OF BOOKS ON RECENT DEVELOPMENTS IN ASTRONOMY AND ASTROPHYSICS

Printed by BookCrafters, Inc.

First published 1999

Library of Congress Catalog Card Number: 99-60528
ISBN 1-886733-78-3

Please contact proper address for information on:

PUBLISHING:
Managing Editor
PO Box 24463
211 KMB
Brigham Young University
Provo, UT 84602-4463 USA

Phone: 801-378-2298
Fax: 801-378-4049
E-mail: pasp@astro.byu.edu

ORDERING BOOKS:
Astronomical Society of the Pacific
CONFERENCE SERIES
390 Ashton Avenue
San Francisco, CA 94112 - 1722 USA
415-337-2624

Phone: 415-337-2624
Fax: 415-337-5205
E-mail: catalog@aspsky.org
Web Site: http://www.aspsky.org

A SERIES OF BOOKS ON RECENT DEVELOPMENTS IN ASTRONOMY AND ASTROPHYSICS

Vol. 1-Progress and Opportunities in Southern Hemisphere Optical Astronomy: CTIO 25th Anniversary Symposium
ed. V. M. Blanco and M. M. Phillips ISBN 0-937707-18-X

Vol. 2-Proceedings of a Workshop on Optical Surveys for Quasars
ed. P. S. Osmer, A. C. Porter, R. F. Green, and C. B. Foltz ISBN 0-937707-19-8

Vol. 3-Fiber Optics in Astronomy
ed. S. C. Barden ISBN 0-937707-20-1

Vol. 4-The Extragalactic Distance Scale: Proceedings of the ASP 100th Anniversary Symposium
ed. S. van den Bergh and C. J. Pritchet ISBN 0-937707-21-X

Vol. 5-The Minnesota Lectures on Clusters of Galaxies and Large-Scale Structure
ed. J. M. Dickey ISBN 0-937707-22-8

Vol. 6-Synthesis Imaging in Radio Astronomy: A Collection of Lectures from the Third NRAO Synthesis Imaging Summer School
ed. R. A. Perley, F. R. Schwab, and A. H. Bridle ISBN 0-937707-23-6

Vol. 7-Properties of Hot Luminous Stars: Boulder-Munich Workshop
ed. C. D. Garmany ISBN 0-937707-24-4

Vol. 8-CCDs in Astronomy
ed. G. H. Jacoby ISBN 0-937707-25-2

Vol. 9-Cool Stars, Stellar Systems, and the Sun. Sixth Cambridge Workshop
ed. G. Wallerstein ISBN 0-937707-27-9

Vol. 10-Evolution of the Universe of Galaxies: Edwin Hubble Centennial Symposium
ed. R. G. Kron ISBN 0-937707-28-7

Vol. 11-Confrontation Between Stellar Pulsation and Evolution
ed. C. Cacciari and G. Clementini ISBN 0-937707-30-9

Vol. 12-The Evolution of the Interstellar Medium
ed. L. Blitz ISBN 0-937707-31-7

Vol. 13-The Formation and Evolution of Star Clusters
ed. K. Janes ISBN 0-937707-32-5

Vol. 14-Astrophysics with Infrared Arrays
ed. R. Elston ISBN 0-937707-33-3

Vol. 15-Large-Scale Structures and Peculiar Motions in the Universe
ed. D. W. Latham and L. A. N. da Costa ISBN 0-937707-34-1

Vol. 16-Proceedings of the 3rd Haystack Observatory Conference on Atoms, Ions and Molecules: New Results in Spectral Line Astrophysics
ed. A. D. Haschick and P. T. P. Ho ISBN 0-937707-35-X

Vol. 17-Light Pollution, Radio Interference, and Space Debris
ed. D. L. Crawford ISBN 0-937707-36-8

Vol. 18-The Interpretation of Modern Synthesis Observations of Spiral Galaxies
ed. N. Duric and P. C. Crane ISBN 0-937707-37-6

Vol. 19-Radio Interferometry: Theory, Techniques, and Applications, IAU Colloquium 131
ed. T. J. Cornwell and R. A. Perley ISBN 0-937707-38-4

Vol. 20-Frontiers of Stellar Evolution: 50th Anniversary McDonald Observatory (1939-1989)
ed. D. L. Lambert ISBN 0-937707-39-2

Vol. 21-The Space Distribution of Quasars
ed. D. Crampton ISBN 0-937707-40-6

Vol. 22-Nonisotropic and Variable Outflows from Stars
ed. L. Drissen, C. Leitherer, and A. Nota ISBN 0-937707-41-4

Vol. 23-Astronomical CCD Observing and Reduction Techniques
ed. S. B. Howell ISBN 0-937707-42-4

Vol. 24-Cosmology and Large-Scale Structure in the Universe
ed. R. R. de Carvalho ISBN 0-937707-43-0

Vol. 25-Astronomical Data Analysis Software and Systems I
ed. D. M. Worrall, C. Biemesderfer, and J. Barnes ISBN 0-937707-44-9

Vol. 26-Cool Stars, Stellar Systems, and the Sun, Seventh Cambridge Workshop
ed. M. S. Giampapa and J. A. Bookbinder ISBN 0-937707-45-7

Vol. 27-The Solar Cycle: Proceedings of the National Solar Observatory/Sacramento Peak 12th Summer Workshop
ed. K. L. Harvey ISBN 0-937707-46-5

Vol. 28-Automated Telescopes for Photometry and Imaging
ed. S. J. Adelman, R. J. Dukes, Jr., and C. J. Adelman ISBN 0-937707-47-3

Vol. 29-Viña Del Mar Workshop on Catacysmic Variable Stars
ed. N. Vogt ISBN 0-937707-48-1

Vol. 30-Variable Stars and Galaxies
ed. B. Warner ISBN 0-937707-49-X

Vol. 31-Relationships Between Active Galactic Nuclei and Starburst Galaxies
ed. A. V. Filippenko ISBN 0-937707-50-3

Vol. 32-Complementary Approaches to Double and Multiple Star Research, IAU Collouquium 135
ed. H. A. McAlister and W. I. Hartkopf ISBN 0-937707-51-1

Vol. 33-Research Amateur Astronomy
ed. S. J. Edberg ISBN 0-937707-52-X

Vol. 34-Robotic Telescopes in the 1990s
ed. A. V. Filippenko ISBN 0-937707-53-8

Vol. 35-Massive Stars: Their Lives in the Interstellar Medium
ed. J. P. Cassinelli and E. B. Churchwell ISBN 0-937707-54-6

Vol. 36-Planets Around Pulsars
ed. J. A. Phillips, S. E. Thorsett, and S. R. Kulkarni ISBN 0-937707-55-4

Vol. 37-Fiber Optics in Astronomy II
ed. P. M. Gray ISBN 0-937707-56-2

Vol. 38-New Frontiers in Binary Star Research: Pacific Rim Colloquium
ed. K. C. Leung and I.-S. Nha ISBN 0-937707-57-0

Vol. 39-The Minnesota Lectures on the Structure and Dynamics of the Milky Way
ed. Roberta M. Humphreys ISBN 0-937707-58-9

Vol. 40-Inside the Stars, IAU Colloquium 137
ed. Werner W. Weiss and Annie Baglin ISBN 0-937707-59-7

Vol. 41-Astronomical Infrared Spectroscopy: Future Observational Directions
ed. Sun Kwok ISBN 0-937707-60-0

Vol. 42-GONG 1992: Seismic Investigation of the Sun and Stars
ed. Timothy M. Brown ISBN 0-937707-61-9

Vol. 43-Sky Surveys: Protostars to Protogalaxies
ed. B. T. Soifer ISBN 0-937707-62-7

Vol. 44-Peculiar Versus Normal Phenomena in A-Type and Related Stars, IAU Colloquium 138
ed. M. M. Dworetsky, F. Castelli, and R. Faraggiana ISBN 0-937707-63-5

Vol. 45-Luminous High-Latitude Stars
ed. D. D. Sasselov ISBN 0-937707-64-3

Vol. 46-The Magnetic and Velocity Fields of Solar Active Regions, IAU Colloquium 141
ed. H. Zirin, G. Ai, and H. Wang ISBN 0-937707-65-1

Vol. 47-Third Decennial US-USSR Conference on SETI
ed. G. Seth Shostak ISBN 0-937707-66-X

Vol. 48-The Globular Cluster-Galaxy Connection
ed. Graeme H. Smith and Jean P. Brodie ISBN 0-937707-67-8

Vol. 49-Galaxy Evolution: The Milky Way Perspective
ed. Steven R. Majewski ISBN 0-937707-68-6

Vol. 50-Structure and Dynamics of Globular Clusters
ed. S. G. Djorgovski and G. Meylan ISBN 0-937707-69-4

Vol. 51-Observational Cosmology
ed. G. Chincarini, A. Iovino, T. Maccacaro, and D. Maccagni ISBN 0-937707-70-8

Vol. 52-Astronomical Data Analysis Software and Systems II
ed. R. J. Hanisch, R. J. V. Brissenden, and Jeannette Barnes ISBN 0-937707-71-6

Vol. 53-Blue Stragglers
ed. Rex A. Saffer ISBN 0-937707-72-4

Vol. 54-The First Stromlo Symposium: The Physics of Active Galaxies
ed. Geoffrey V. Bicknell, Michael A. Dopita, and Peter J. Quinn ISBN 0-937707-73-2

Vol. 55-Optical Astronomy from the Earth and Moon
ed. Diane M. Pyper and Ronald J. Angione ISBN 0-937707-74-0

Vol. 56-Interacting Binary Stars
ed. Allen W. Shafter ISBN 0-937707-75-9

Vol. 57-Stellar and Circumstellar Astrophysics
ed. George Wallerstein and Alberto Noriega-Crespo ISBN 0-937707-76-7

Vol. 58-The First Symposium on the Infrared Cirrus and Diffuse Interstellar Clouds
ed. Roc M. Cutri and William B. Latter ISBN 0-937707-77-5

Vol. 59-Astronomy with Millimeter and Submillimeter Wave Interferometry, IAU Colloquium 140
ed. M. Ishiguro and Wm. J. Welch ISBN 0-937707-78-3

Vol. 60-The MK Process at 50 Years: A Powerful Tool for Astrophysical Insight: A Workshop of the Vatican Observatory
ed. C. J. Corbally, R. O. Gray, and R. F. Garrison ISBN 0-937707-79-1

Vol. 61-Astronomical Data Analysis Software and Systems III
ed. Dennis R. Crabtree, R. J. Hanisch, and Jeannette Barnes ISBN 0-937707-80-5

Vol. 62-The Nature and Evolutionary Status of Herbig Ae / Be Stars
ed. P. S. Thé, M. R. Pérez, and E. P. J. van den Heuvel ISBN 0-937707-81-3

Vol. 63-Seventy-Five Years of Hirayama Asteroid Families: The role of Collisions in the Solar System History
ed. R. P. Binzel, Y. Kozai, and T. Hirayama ISBN 0-937707-82-1

Vol. 64-Cool Stars, Stellar Systems, and the Sun, Eighth Cambridge Workshop
ed. Jean-Pierre Caillault ISBN 0-937707-83-X

Vol. 65-Clouds, Cores, and Low Mass Stars
ed. Dan P. Clemens and Richard Barvainis ISBN 0-937707-84-8

Vol. 66- Physics of the Gaseous and Stellar Disks of the Galaxy
ed. Ivan R. King ISBN 0-937707-85-6

Vol. 67-Unveiling Large-Scale Structures Behind the Milky Way
ed. C. Balkowski and R. C. Kraan-Korteweg ISBN 0-937707-86-4

Vol. 68-Solar Active Region Evolution: Comparing Models with Observations
ed. K. S. Balasubramaniam and George W. Simon ISBN 0-937707-87-2

Vol. 69-Reverberation Mapping of the Broad-Line Region in Active Galactic Nuclei
ed. P. M. Gondhalekar, K. Horne, and B. M. Peterson ISBN 0-937707-88-0

Vol. 70-Groups of Galaxies
ed. Otto G. Richter and Kirk Borne ISBN 0-937707-89-9

Vol. 71-Tridimensional Optical Spectroscopic Methods in Astrophysics, IAU Colloquium 149
ed. G. Comte and M. Marcelin ISBN 0-937707-90-2

Vol. 72-Millisecond Pulsars: A Decade of Surprise.
ed. A. A. Fruchter, M. Tavani, and D. C. Backer ISBN 0-937707-91-0

Vol. 73-Airborne Astronomy Symposium on the Galactic Ecosystem: From Gas to Stars to Dust
ed. M. R. Haas, J. A. Davidson, and E. F. Erickson ISBN 0-937707-92-9

Vol. 74-Progress in the Search for Extraterrestrial Life: 1993 Bioastronomy Symposium
ed. G. Seth Shostak ISBN 0-937707-93-7

Vol. 75-Multi-Feed Systems for Radio Telescopes
ed. D. T. Emerson and J. M. Payne ISBN 0-937707-94-5

Vol. 76-GONG '94: Helio- and Astero-Seismology from the Earth and Space
ed. Roger K. Ulrich, Edward J. Rhodes, Jr., and Werner Däppen ISBN 0-937707-95-3

Vol. 77-Astronomical Data Analysis Software and System IV
ed. R. A. Shaw, H. E. Payne, and J. J. E. Hayes ISBN 0-937707-96-1

Vol. 78-Astrophysical Applications of Powerful New Databases: Joint Discussion No. 16 of the 22nd General Assembly of the IAU
ed. S. J. Adelman and W. L. Wiese ISBN 0-937707-97-X

Vol. 79-Robotic Telescopes: Current Capabilities, Present Developments, and Future Prospects for Automated Astronomy
ed. Gregory W. Henry and Joel A. Eaton ISBN 0-937707-98-8

Vol. 80-The Physics of the Interstellar Medium and Intergalactic Medium
ed. A. Ferrara, C. F. McKee, C. Heiles, and P. R. Shapiro ISBN 0-937707-99-6

Vol. 81-Laboratory and Astronomical High Resolution Spectra
ed. A. J. Sauval, R. Blomme, and N. Grevesse ISBN 1-886733-01-5

Vol. 82-Very Long Baseline Interferometry and the VLBA
ed. J. A. Zensus, P. J. Diamond, and P. J. Napier ISBN 1-886733-02-3

Vol. 83-Astrophysical Applications of Stellar Pulsation. IAU Colloquium 155
ed. R. S. Stobie and P. A. Whitelock ISBN 1-886733-03-1

Vol. 84-The Future Utilisation of Schmidt Telescopes, IAU Colloquium 148
ed. Jessica Chapman, Russell Cannon, Sandra Harrison, and Bambang Hidayat ISBN 1-886733-05-8

Vol. 85-Cape Workshop on Magnetic Cataclysmic Variables
ed. D. A. H. Buckley and B. Warner ISBN 1-886733-06-6

Vol. 86-Fresh Views of Elliptical Galaxies
ed. Alberto Buzzoni, Alvio Renzini, and Alfonso Serrano ISBN 1-886733-07-4

Vol. 87-New Observing Modes for the Next Century
ed. Todd Boroson, John Davies, and Ian Robson ISBN 1-886733-08-2

Vol. 88- Clusters, Lensing, and the Future of the Universe
ed. Virginia Trimble and Andreas Reisenegger ISBN 1-886733-09-0

Vol. 89-Astronomy Education: Current Developments, Future Coordination
ed.John R. Percy ISBN 1-886733-10-4

Vol. 90-The Origins, Evolution, and Destinies of Binary Stars in Clusters
ed. E. F. Milone and J. C. Mermilliod ISBN 1-886733-11-2

Vol. 91-Barred Galaxies, IAU Colloquium 157
ed. R. Buta, D. A. Crocker, and B. G. Elmegreen ISBN 1-886733-12-0

Vol. 92-Formation of the Galactic Halo–Inside and Out
ed. H. L. Morrison and A. Sarajedini ISBN 1-886733-13-9

Vol. 93-Radio Emission from the Stars and the Sun
ed. A. R. Taylor and J. M. Paredes ISBN 1-886733-14-7

Vol. 94-Mapping, Measuring, and Modelling the Universe
ed. Peter Coles, Vincent Martinez, and Maria-Jesus Pons-Borderia ISBN 1-886733-15-5

Vol. 95-Solar Drivers of Interplanetary and Terrestrial Disturbances: Proceedings of 16th International Workshop, National Solar Observatory/Sacramento Peak
ed. K. S. Balasubramaniam, S. L. Keil, and R. N. Smartt ISBN 1-886733-16-3

Vol. 96- Hydrogen-Deficient Stars
ed. C. S. Jeffery and U. Heber ISBN 1-886733-17-1

Vol. 97-Polarimetry of the Interstellar Medium
ed.W. G. Roberge and D. C. B. Whittet ISBN 1-886733-18-X

Vol. 98-From Stars to Galaxies: The Impact of Stellar Physics on Galaxy Evolution
ed. Claus Leitherer, Uta Fritze-von Alvensleben, and John Huchra ISBN 1-886733-19-8

Vol. 99-Cosmic Abundances: Proceedings of the 6th Annual October Astrophysics Conference
ed. Stephen S. Holt and Geroge Sonneborn ISBN 1-886733-20-1

Vol. 100-Energy Transport in Radio Galaxies and Quasars
ed. P. E. Hardee, A. H. Bridle, and J. A. Zensus ISBN 1-886733-21-X

Vol. 101-Astronomical Data Analysis Software and Systems V
ed. George H. Jacoby and Jeannette Barnes ISSN 1080-7926

Vol. 102-The Galactic Center, 4th ESO/CTIO Workshop
ed. Roland Gredel ISBN 1-886733-22-8

Vol. 103-The Physics of Liners in View of Recent Observations
ed. M. Eracleous, A. Koratkar, C. Leitherer, and L. Ho ISBN 1-886733-23-6

Vol. 104-Physics, Chemistry, and Dynamics of Interplanetary Dust, IAU Colloquium 150
ed. Bo A. S. Gustafson and Martha S. Hanner ISBN 1-886733-24-4

Vol. 105-Pulsars: Problems and Progress, IAU Colloquium 160
ed. M. Bailes, S. Johnston, and M. A. Walker ISBN 1-886733-25-2

Vol. 106-Minnesota Lectures on Extragalactic Neutral Hydrogen
ed. Evan D. Skillman ISBN 1-886733-26-0

Vol. 107-Completing the Inventory of the Solar System: A Symposium held in conjuunction with the 106th Annual Meeting of the ASP
ed. Terrence W. Rettig and Joseph M. Hahn ISBN 1-886733-27-9

Vol. 108-M. A. S. S. Model Atmospheres and Spectrum Synthesis: 5th Vienna Workshop
ed. S. J. Adelman, F. Kupka, and W. W. Weiss ISBN 1-886733-28-7

Vol. 109-Cool Stars, Stellar Systems, and the Sun, Ninth Cambridge Workshop
ed. Roberto Pallavicini and Andrea K. Dupree ISBN 1-886733-29-5

Vol. 110-Blazar Continuum Variability
ed.H. R. Miller, J. R. Webb, and J. C. Noble ISBN 1-886733-30-9

Vol. 111-Magnetic Reconnection in the Solar Atmosphere: Proceedings of a Yohkoh Conference
ed. R. D. Bentley and J. T. Mariska ISBN 1-886733-31-7

Vol. 112-The History of the Milky Way and Its Satellite System
ed. A. Burkert, D. H. Hartmann, and S. R. Majewski ISBN 1-886733-32-5

Vol. 113-Emission Lines in Active Galaxies: New Methods and Techniques, IAU Colloquium 159
ed. B. M. Peterson, F. Z. Cheng, and A. S. Wilson ISBN 1-886733-33-3

Vol. 114-Young Galaxies and QSO Absorption-Line Systems
ed. Sueli M. Viegas, Ruth Gruenwald, and Reinaldo R. de Carvalho ISBN 1-886733-34-1

Vol. 115-Galactic and Cluster Cooling Flows
ed. Noam Soker ISBN 1-886733-35-X

Vol. 116-The Second Stromlo Symposium: The Nature of Elliptical Galaxies
ed. M. Arnaboldi, G. S. Da Costa, and P. Saha ISBN 1-886733-36-8

Vol. 117- Dark and Visible Matter in Galaxies
ed. Massimo Persic and Paolo Salucci ISBN 1-886733-37-6

Vol. 118-First Advances in Solar Physics Euroconference: Advances in the Physics of Sunspots
ed. B. Schmieder, J. C. del Toro Iniesta, and M. Vázquez ISBN 1-886733-38-4

Vol. 119-Planets Beyond the Solar System and the Next Generation of Space Missions
ed. David R. Soderblom ISBN 1-886733-39-2

Vol. 120-Luminous Blue Variables: Massive Stars in Transition
ed. Antonella Nota and Henny J. G. L. M. Lamers ISBN 1-886733-40-6

Vol. 121-Accretion Phenomena and Related Outflows, IAU Colloquium 163
ed. D. T. Wickramasinghe, G. V. Bicknell and L. Ferrario ISBN 1-886733-41-4

Vol. 122-From Stardust to Planetesimals: Symposium held as part of the 108th Annual Meeting of the ASP
ed. Yvonne J. Pendleton and A. G. G. M. Tielens ISBN 1-886733-42-2

Vol. 123-The 12th 'Kingston Meeting': Computational Astrophysics
ed. David A. Clarke and Michael J. West ISBN 1-886733-43-0

Vol. 124-Diffuse Infrared Radiation and the IRTS
ed. Haruyuki Okuda, Toshio Matsumoto, and Thomas L. Roellig ISBN 1-886733-44-9

Vol. 125- Astronomical Data Analysis Software and Systems VI
ed. Gareth Hunt and H. E. Payne ISBN 1-886733-45-7

Vol. 126-From Quantum Fluctuations to Cosmological Structures
ed. D. Vallis-Gabaud, M. A. Hendry, P. Molaro, and K. Chamcham ISBN 1-886733-46-5

Vol. 127-Proper Motions and Galactic Astronomy
ed. Roberta M. Humphreys ISBN 1-886733-47-3

Vol. 128- Mass Ejection from AGN (Active Galactic Nuclei)
ed. N. Arav, I. Shlosman, and R. J. Weymann ISBN 1-886733-48-1

Vol. 129-The George Gamow Symposium
ed. E. Harper, W. C. Parke, and G. D. Anderson ISBN 1-886733-49-X

Vol. 130-The Third Pacfic Rim Conference on Recent Development on Binary Star Research
ed. Kam-Ching Leung ISBN 1-886733-50-3

Vol. 131-Boulder-Munich II: Properties of Hot, Luminous Stars
ed. Ian D. Howarth ISBN 1-886733-51-1

Vol. 132-Star Formation with the Infrared Space Observatory (ISO)
ed. João L. Yun and René Liseau ISBN 1-886733-53-X

Vol. 133-Science with the NGST
ed. Eric P. Smith and Anuradha Koratkar ISBN 1-886733-53-8

Vol. 134-Brown Dwarfs and Extrasolar Planets
ed. Rafael Rebolo, Eduardo L. Martin,
and Maria Rosa Zapatero Osorio ISBN 1-886733-54-6

Vol. 135-A Half Century of Stellar Pulsation Interpretations: A Tribute to Arthur N. Cox
ed. P. A Bradley and J. A. Guzik ISBN 1-886733-55-4

Vol. 136- Galactic Halos: A UC Santa Cruz Workshop
ed. Dennis Zaritdky ISBN 1-886733-56-2

Vol. 137-Wild Stars in the Old West: Proceedings of the 13th North American Workshop
on Cataclysmic Variables and Related Objects
ed. S. Howell, E.Kuulkers, and C. Woodward ISBN 1-886733-57-0

Vol. 138-1997 Pacific Rim Conference on Stellar Astrophysics
ed. Kwing L. Chan, K. S. Cheng, and Harinder P. Singh ISBN 1-886733-58-9

Vol. 139-Preserving the Astronomical Windows, proceedings of Joint Discussion
No. 5 of the 23rd General Assembly of the IAU
ed. Syuzo Isobe and Tomohiro Hirayama ISBN 1-886733-59-7

Vol. 140-Synoptic Solar Physics – 18th NSO/Sacramento Peak Summer Workshop
ed. K. S. Balasubramaniam, J. W. Harvey, and D. M. Rabin ISBN 1-886733-60-0

Vol. 141-Astrophysics from Antarctica
ed. Giles Novak and Randall H. Landsberg ISBN 1-886733-61-9

Vol. 142-The Stellar Initial Mass Function, 38th Herstmonceux Conference
ed. Gerry Gilmore and Debbie Howell ISBN 1-886733-62-7

Vol. 143-The Scientific Impact of the Goddard High Resolution Spectrograph
ed. John C. Brandt, Thomas B. Ake III, and Carolyn Collins Petersen ISBN 1-886733-63-5

Vol. 144- Radio Emission from Galactic and Extragalactic Compact Sources,
IAU Colloquium 164
ed. J. Anton Zensus, G. B. Taylor, and J. M. Wrobel ISBN 1-886733-64-3

Vol. 145-Astronomical Data Analysis Software and Systems VII
ed. Rudolf Albrecht, Richard N. Hook, and Howard A. Bushouse ISBN 1-886733-65-1

Vol. 146-The Young Universe: Galaxy Formation and Evolution at
Intermediate and High Redshift
ed. S. D'Odorico, A. Fontana, and E. Giallongo ISBN 1-886733-66-X

Vol. 147-Abundance Profiles: Diagnostic Tools for Galaxy History
ed. Daniel Friedli, Mike Edmunds, Carmelle Robert,
and Laurent Drissen ISBN 1-886733-67-8

Vol. 148-Origins
ed. Charles E. Woodward, J. Michael Shull,
and Harley A. Thronson, Jr. ISBN 1-886733-68-6

Vol. 149-Solar System Formation and Evolution
ed. D. Lazzaro, R. Vieira Martins, S. Ferraz-Mello,
J. Fernández, and C. Beaugé ISBN 1-886733-69-4

Vol. 150-New Perspectives on Solar Prominences, IAU Colloquium 167
ed. David Webb, David Rust, and Brigitte Schmieder ISBN 1-886733-70-8

Vol. 151-Cosmic Microwave Background and Large Scale Structure of the Universe
ed. Yong-Ik Byun and Kin-Wang Ng ISBN 1-886733-71-6

Vol. 152-Fiber Optics in Astronomy III
ed. S. Arribas, E. Mediavilla, and F. Watson ISBN 1-886733-72-4

Vol. 153-Library and Information Services in Astronomy III, (LISA III)
ed. Uta Grothkopf, Heinz Andernach, Sarah Stevens-Rayburn,
and Monique Gomez ISBN 1-886733-73-2

Vol. 154-Cool Stars, Stellar Systems, and the Sun, Tenth Cambridge Workshop
ed. Robert A. Donahue and Jay A. Bookbinder ISBN 1-886733-74-0

Vol. 155-Second Advances in Solar Physics Euroconference:
Three-Dimensional Structure of Solar Active Regions
ed. Costas E. Alissandrakis and Brigitte Schmieder ISBN 1-886733-75-9

Vol. 156-Highly Redshifted Radio Lines
ed. C. L. Carilli, S. J. E. Radford, K. M. Menten and
G. I. Langston ISBN 1-886733-76-7

Vol. 157-Annapolis Workshop on Magnetic Cataclysmic Variables
ed. Coel Hellier and Koji Mukai ISBN 1-886733-77-5

Vol. 158-Solar and Stellar Activity: Similarities and Differences
ed. C. J. Butler and J. G. Doyle ISBN 1-886733-78-3

Book orders or inquiries concerning these volumes should be directed to the:
Astronomical Society of the Pacific Conference Series
390 Ashton Avenue
San Francisco, CA 94112-1722 USA

Phone:415-337-2126 catalog@aspsky.org
Fax: 415-337-5205 Web Site: http://www.aspsky.org

ASTRONOMICAL SOCIETY OF THE PACIFIC
CONFERENCE SERIES

Volume 158

SOLAR AND STELLAR ACTIVITY: SIMILARITIES AND DIFFERENCES

**Proceedings of a meeting held in
Armagh, N. Ireland
2-4 September 1998**

Edited by
C. J. Butler and J. G. Doyle

Table of Contents

Part 3. Physical Manifestations of Activity: Spots, Plages and Prominences

Part 4. Physical Manifestations of Activity: Flares

Part 5. Chromospheric Dynamics

Part 6. Coronal Dynamics

Part 7. Summary

Preface

On the morning of 16 September 1997, the astronomical community learned of the sudden death of Patrick Brendan Byrne, who had been taken ill the previous night whilst observing on the island of La Palma. His death at this time when at the peak of his career was a great loss, not only to Armagh Observatory where he had worked for nearly twenty years, but to his many colleagues overseas. Several of Brendan's friends felt the need for some tangible memorial to his work.

It was following the oral examination of one of his PhD students, that we met in Armagh one of Brendan's closest colleagues and friends, Professor Marcello Rodono. Together we formulated the proposal to hold a meeting in Armagh to commemorate the contribution of Brendan Byrne to astronomical research. The topic we chose, *solar and stellar activity*, was an area of research with which Brendan had been involved since coming to Armagh and one which held many still unanswered questions.

One such question, which particularly concerned Brendan was, to what extent we can use the observed characteristics of solar activity - spots, plages and flares - as paradigms for the variability we see in other stars ? Brendan, in his later years, was gradually moving away from the solar analogy towards a belief that some features of stellar activity are distinctly different from the solar picture, not only in their magnitude but also in their underlying physical nature. Others working in the field continue to treat the Sun as essentially the prototype of all magnetically active stars. However, whether or not the different manifestations of magnetic activity on the Sun are identical to those believed to exist on other stars, they continue to provide the background to our current understanding of stellar activity and are likely to do so for the foreseeable future. Thus the *similarities and differences* between solar and stellar activity are of fundamental concern to all of us working in the field of stellar activity and this is an opportune time for their exploration and debate. We hope that the proceedings of this meeting, will not only commemorate Brendan's work in this direction, but will continue and extend it for the benefit of us all.

In the preparation of these proceedings, the Editors would like to thank the contributors who provided such high quality and interesting papers and who made our life so much easier by sticking to the guidelines of the ASP Conference Series. In addition, we would like to acknowledge the kind assistance with the organization of this meeting in Armagh of Aileen Brannigan and Margaret Cherry, and Professor Mark Bailey for his encouragement and support. The photograph of participants, taken at the reception by the Mayor of Armagh, is reproduced with the kind permission of Vincent Loughran.

C.J. Butler
J.G. Doyle
Armagh, January 1999.

Conference Participants

Aarum, Vidar, Institute of Theoretical Astrophysics, University of Oslo PO Box 1029 Blindern, N-0315 Oslo, Norway (vidara@astro.uio.no)

Abbett, Bill, Space Sciences Lab., University of California, Berkeley, CA 94720-7450, USA (abbett@ssl.berkeley.edu)

Andretta, Vincenzo, NRC/NASA Goddard Space Flight Center, Code 682.0, Greenbelt MD 20771, USA (andretta@gsfc.nasa.gov)

Aznar, Regina, Armagh Observatory, College Hill, Armagh, BT61 9DG, N. Ireland (rea@star.arm.ac.uk)

Bailey, Mark, Armagh Observatory, College Hill, Armagh, BT61 9DG, N. Ireland (meb@star.arm.ac.uk)

Banerjee, Dipankar, Armagh Observatory, College Hill, Armagh, BT61 9DG, N. Ireland (dipu@star.arm.ac.uk)

Bromage, Barbara, Centre for Astrophysics, University of Central Lancashire, Preston, PR1 2HE, UK (b.j.i.bromage@uclan.ac.uk)

Bromage, Gordon, University of Central Lancashire, Preston, PR1 2HE, UK (g.e.bromage@uclan.ac.uk)

Butler, John, Armagh Observatory, College Hill, Armagh, BT61 9DG, N. Ireland (cjb@star.arm.ac.uk)

Collier-Cameron, Andrew, School of Physics and Astronomy, University of St Andrews, North Haugh, St Andrews, Fife, KY16 9SS, Scotland (acc4@st-and.ac.uk)

Cuntz, Manfred, Center for Space Plasma and Aeronomic Research, University of Alabama, Huntsville AL 35899, USA (cuntzm@cspar.uah.edu)

Donati, Jean-Francois, Observatoire Midi-Pyrenees, 14 Av. E. Berlin, F-31400 Toulouse, France (donati@obs-mip.fr)

Doyle, Gerry, Armagh Observatory, College Hill, Armagh, BT61 9DG, N. Ireland (jgd@star.arm.ac.uk)

Drake, Steve, Goddard Space Flight Center, Code 660.2, Greenbelt MD 20771, USA (drake@lheavx.gsfc.nasa.gov)

Foing, Bernard, ESTEC SCI-SO, Postbus 299, 2200 AG Noordwijk, The Netherlands (bfoing@estscn.estec.esa.nl)

Harra-Murnion, Loiuse, Mullard Space Science Laboratory, Holmbury St Mary, Dorking, Surrey, RH5 6NT, UK (lkhm@mssl.ucl.ac.uk)

Hawley, Suzanne, Space Sciences Lab., University of California, Berekeley, CA 94720-7450, USA (slh@ssl.berkeley.edu)

Jeffries, Rob, Department of Physics, Keele University, Staffordshire, ST5 5BG, UK (rdj@astro.keele.ac.uk)

Jevremovic, Darko, Armagh Observatory, College Hill, Armagh, BT61 9DG, N. Ireland (dcj@star.arm.ac.uk)

Kovari, Zsolt, Konkoly Observatory, 1525 Budapest, P.O. Box 67, Hungary (kovari@buda.konkoly.hu)

Lanza, Antonio, Osservatorio Astrofisico di Catania, Viale A. Doria 6, I 95125 Catania, Italy (nla@sunct.ct.astro.it)

Lanzafame, Alessandro, Instituto di Astronomia, Universita di Catania, Viale Andrea Doria 6, I-95125 Catania, Italy (acl@sunct.ct.astro.it)

Linsky, Jeff, JILA/University of Colorado, Campus Box 440, Boulder CO 80309-0440, USA (jlinsky@jila.colorado.edu)

Mason, Helen, Dept. of Applied Maths. and Theoretical Physics, Cambridge University, Silver Street, Cambridge, CB3 9EW, UK (h.e.mason@damtp.cam.ac.uk)

Mathioudakis, Mihalis, Dept. of Pure and Applied Physics, Queen's University, Belfast, BT7 1NN, N. Ireland (m.mathioudakis@queens-belfast.ac.uk)

Mitrou, Kassios, Section of Astrophysics, Astronomy and Mechanics, Physics Department, University of Athens, Panepistimiopolis 15783 Zogrofos, Athens, Greece (kam@rigel.da.uoa.gr)

Montes, David, Dept. of Astronomy and Astrophysics, Pennsylvania State University, 525 Davey Lab, University Park PA 16802, USA (dmg@astro.psu.edu)

O'Shea, Eoghan, Dept. of Pure and Applied Physics, Queen's University, Belfast, BT7 1NN, N. Ireland (e.oshea@queens-belfast.ac.uk)

Olah, Katalin, Konkoly Observatory, 1525 Budapest, P.O. Box 67, Hungary (olah@buda.konkoly.hu)

Oliveira, Joana, ESTEC/ESA, Solar System Division PB299, 2200 AG Noordwijk, The Netherlands (joana@so.estec.esa.nl)

Özeren, Ferhat, Armagh Observatory, College Hill, Armagh, BT61 9DG, N. Ierland (ffo@star.arm.ac.uk)

Pallavicini, Roberto, Osservatorio Astronomico di Palermo, Palazzo dei Normanni, Piazza del Palamento 1, I-90134 Palermo, Italy (pallavic@oapa.astropa.unipa.it)

Palle Enric Bago, Armagh Observatory, College Hill, Armagh, BT61 9DG, N. Ireland (epb@star.arm.ac.uk)

Peres, Giovanni, Intitudo ed Osservatorio Astronomico di Palermo, Piazza del Parlamento 1, 90134 Palermo, Italy (peres@oapa.astroa.unipa.it)

Perez, Elena, Armagh Observatory, College Hill, Armagh, BT61 9DG, N. Ireland (epp@star.arm.ac.uk)

Phillips, Ken, Rutherford Appleton Laboratory, Space Science Dept, Chilton, Didcot, Oxon, OX11 0QX, UK (k.j.h.phillips@rl.ac.uk)

Pres, Pawel, Rutherford Appleton Laboratory, Space Science Division, Chilton, Didcot, Oxon, OX11 0QX, UK (pres@solg2.bnsc.rl.ac.uk)

Priest, Eric, Mathematics and Computational Sciences Dept., St. Andrews University, St. Andrews, KY16 9SS, Scotland (eric@dcs.st-and.ac.uk)

Rodono, Marcello, Instituto di Astronomia, Universita di Catania, Viale A. Doria 6, I-95125 Catania, Italy (mrodono@alpha4.ct.astro.it)

Rolleston, Robert, Department of Pure and Applied Physics, Queen's University, Belfast, BT7 1NN, N. Ireland (r.rolleston@qub.ac.uk)

Ruediger, Guenther, Astronomical Institute Potsdam, An der Sternwarte 16, D 14467 Potsdam, Germany (gruediger@aip.de)

Rutten, Rob, Sterrekundig Instituut Utrecht, Postbus 80.000, NL-3508 TA, Utrecht, The Netherlands (r.j.rutten@fys.ruu.nl)

Schmieder, Brigette, Observatoire de Paris, Section Meudon, 92197 Meudon, Cedex Principal, France (schmiede@mesopo.obspm.fr)

Schrijver, Karel, Stanford-Lockheed Institute for Space Research, Dept. H1-12/Bldg. 252, 3251 Hanover Street, Palo Alto CA 94304, USA (schryver@lmsal.com)

Schussler, Manfred, Kiepenheuer-Institut, Schoneckstr 6, D-79104 Freiburg, Germany (msch@kis.uni-freiburg.de)

Short, Ian, University of Georgia, Department of Physics and Astronomy, Athens GA 30602-2451, USA (cis@calvin.physast.uga.edu)

Solanki, Sami, Institute of Astronomy, ETH-Zentrum, CH-8092 Zurich, Switzerland (solanki@astro.phys.ethz.ch)

Stern, Bob, Lockheed Martin Solar and Astrophysics Lab, Dept H1-12 Bldg 252, 3251 Hanover Street, Palo Alto CA 94304, USA (stern@sag.space.lockheed.com)

Teriaca, Luca, Armagh Observatory, College Hill, Armagh, BT61 9DG, N. Ireland (lte@star.arm.ac.uk)

Theissen, Armin, Armagh Observatory, College Hill, Armagh, BT61 9DG, N. Ireland (ath@star.arm.ac.uk)

Totten, Ed, Department of Physics, Keele University, Keele, Staffordshire, ST5 5BG, UK (ejt@astro.keele.ac.uk)

Ulmschneider, Peter, Institut fur Theoretische Astrophysik, Universitat Heidelberg, Turnerstr. 3, 69126, Heidelberg, Germany
(ulm@ita.uni-heidelberg.de)

van den Oord, Bert, Sterrekundig Instituut Utrecht, P.O. Box 80.000, 3508 TA Utrecht, The Netherlands (oord@fys.ruu.nl)

Walter, Fred, Department of Physics and Astronomy, State University of New York, Stony Brook, NY 11794-3800, USA (fwalter@astro.sunysb.edu)

Zboril, Milan, Astronomical Institute, Tatranska Lomnica, 05960 Slovakia
(zboril@ta3.sk)

In Honor Of

Patrick Brendan Byrne

P.B. Byrne at Dunsink Observatory

P. A. Wayman[1]

Glebe Cottage, Glebe Avenue, Wicklow, Ireland

The loss of a talented and able person when at the peak of his or her career is particularly poignant. We instinctively feel in the case of Brendan Byrne, who died so tragically and so suddenly in September 1997, that he still had many tasks in front of him, tasks that generally-speaking cannot be taken up by others.

This conference is called in memory of Brendan and many of the contributions will bear testimony to the effectiveness of his work in cooperation with others. I will not myself try to foresee the successes of Brendan's career that will receive special attention here. I wish to speak about Brendan's early years as a Scholar in the Astronomy Section of the School of Cosmic Physics; one of the three schools of the Dublin Institute for Advanced Studies (DIAS).

My first scholar to be appointed was John Butler, who has made his career here at Armagh; followed by two UCD physics graduates, Michael Norris and John Brady, neither of whom completed their degrees although they followed successful careers in computer programing in UCD and as a mathematics lecturer in London, respectively.

Thus when Brendan came to my notice in 1969 showing promise and ability I was hopeful of having a new active astronomy student. As I remember it, Brendan came to me, rather than waiting to be sought out.

Brendan was the eldest of a family of worthy industrious standing in Ringsend - his father was a baker as I recall, and Brendan concluded when he left school that he would have to earn some money to help him through four years of University. This he did by living economically at home and working hard at the Philips radio factory in Clonskeagh. He worked there for more than a year but I do not think he enjoyed it too well. I suspect he found they were doing it all wrong and some friction was set up, but I don't know this for a fact.

At Dunsink he accepted the terms offered and the course of his initial studies was discussed. I had at that time a suggestion by John V. Jelley, an independent research scientist at the Atomic Energy Research Establishment at Harwell who could pursue personal interests in cosmic ray science or astrophysics, provided it was understood that he would not be considered for further promotion. His wife, Joan Freeman Jelley, also worked there, and they did not have a family, so John was able to accept this limitation to worldly success.

The question was put to me by John as to whether I had any suggestion to make concerning the possibility of detecting brief flashes in the night sky -

[1]Professor P.A. Wayman 1927-1998, former director of Dunsink Observatory, died in Dublin on 21 December 1998, after a short illness

perhaps from the region of the galactic centre. Dunsink was part of the group of institutions with an informal controlling interest in the Harvard-Smithsonian Observatory at Bloemfontein. The Boyden Observatory - commonly regarded as the first international observatory - was considered seriously at one time as a suitable location for the proposed European Southern Observatory.

Although John Jelley and I both considered it unlikely that any optical or near infrared radiation from what have now been denoted 'gamma-ray bursters', detected in distant galaxies, could penetrate the intervening clouds of interstellar material, Boyden was situated in the latitude where the galactic centre passes through the zenith and crosses the meridian at midnight in the clear nights of the southern winter. Moreover there was, under-utilised, the *Armagh-Dunsink-Harvard (ADH)* telescope that could be adapted to hold two large photomultiplier tubes.

Thus between Brendan, myself and John Jelley, the construction of the required equipment at Harwell was planned and eventually completed in time for the winter observing season at Bloemfontein in 1971.

Brendan spent some time at Harwell working closely with John and his technical assistant and I was soon assured that the equipment would work well giving analog records on standard *AKAI* tape recorders, and so it was packed up and shipped to South Africa. This was the era of anti-apartheid legislation and care was taken to minimise the publicity.

Struggling with the limitations in the services offered at Boyden, Brendan carried out the work according to plan and timetable and returned by Union Castle line six or seven months later. He continued to keep up his schedule as regards rewards and interpretation into the writing up stage and presented his bound thesis volumes with a margin to spare, within four years of starting his DIAS scholarship.

Brendan's work as a scholar did not result in an outstanding discovery. There were a few single-channel flashes and by chance they occurred in the galactic centre channel rather than the comparison channel, but at 20-50 milliseconds in length I feel sure they came from reflection of sunlight on satellite panels. Still it was in many ways a model project using special equipment, needing many night hours with the telescope, being individualist and cooperative in spirit and above all being finished on time !

Anyway, it made a good start to a too-brief career and has been followed by keen scientific activity on Brendan's part as we all know. By 1975 it was clear in the atmosphere of S. African politics of apartheid that Boyden could not successfully continue. Efforts were made to ensure that practical work at telescopes would be possible from Ireland at the new observatory on La Palma. By the ironies of fate it was that beautiful Island that claimed Brendan's life in 1997, but he had achieved a great deal, and from my viewpoint I recall the words:

The pupil serves not his master well
Who does not his master soon excel.

P.B. Byrne at Armagh Observatory

The Editors

Brendan Byrne came to Armagh Observatory after a year as visiting lecturer at the University of Cape Town. Whilst in South Africa, he had kept in touch with his colleagues in Dunsink and Armagh and developed an interest in stellar activity, a topic which had already taken firm root in Armagh with the work of David Andrews and Dermot Mullan. When Brendan came to Armagh in 1978 to take up his position as a Research Astronomer, he fitted quickly into this small group of stellar astronomers which by then included John Butler and was shortly also to include the young solar astronomer, Gerry Doyle.

An application had already been submitted for time on the *International Ultraviolet Explorer Satellite (IUE)* to look at flare stars and in the resulting observational run, the Armagh group obtained the first ultraviolet spectrum of a flare on a dwarf star - Gliese 867A - a star which Brendan had previously observed optically. The scene was now set for a major push in multi-wavelength coverage of active stars, though at this time it was unlikely that anyone would have predicted how this would develop.

The need for continuous coverage of RS CVn and BY Dra stars with periods of a few days, in order to search for rotational modulation in the ultraviolet, was clear, and it was the requirement for collaboration between astronomers under the auspices of the three participating agencies, NASA, ESA and SERC, which lead to the formation of the Cool Star Consortium, involving initially JILA, Catania and Armagh astronomers, later to be joined by others from Paris, ESTEC and Rutherford Laboratory. The agreement to form the group and make a joint proposal to the *IUE* allocation committees took place at the Montreal General Assembly of the IAU in August 1979. This ushered in a period of dynamic research into active cool stars at Armagh with which Brendan was closely involved.

An ambitious programme of observation with 6 days of continuous *IUE* coverage of the five active stars: II Peg, HR 1099, AR Lac, AU Mic and BY Dra, took place in August 1980. These observations, together with many others that came later, provided some confirmation of the spatial association between dark photospheric spots and bright, ultraviolet emitting, plages similar to that which occurs on the Sun. Nevertheless, the results were not entirely conclusive and significant doubts remained.

After the first set of *IUE* observations, it was quickly realised that a multi-wavelength approach involving optical photometry and spectroscopy, simultaneously with the *IUE* observations, was required to disentangle the possible causes of the variable emission in the strong ultraviolet lines. Were the increases in line emission due to flares or plages, for instance ? Such a question, hopefully, could be settled by simultaneous optical work. Thus coordinated ground-based and satellite-based coverage became an integral part of the consortium's active star research. Ultimately, the two satellites, *IUE* and *EXOSAT* were employed simultaneously with ground-based optical and radio observatories. Brendan Byrne was closely involved in much of this work and chaired many of the annual meetings which followed as the pile of data got higher and the logistics of analysis and publication came to the fore. The experience he gained with *IUE* and his evident flair for the job eventually lead to his appointment as chairman of the ESA/UK IUE Time Allocation Panel. He was a good chairman; always sticking to the point and conscious of time and deadlines.

The period following Brendan's appointment at Armagh was one of gradually increasing research activity. When he arrived there was no in-house computer, only a modem connection to the Queens University Belfast Computer Centre. The need for greater access to a computing system, even one as modest as that currently at Dunsink, was plainly evident. At first Brendan was not overly interested in the *Starlink* Network, which had been set up in Great Britain, and favoured a *stand-alone* computer in Armagh. However, as time went on, it became increasingly evident that much duplication of effort could be avoided by access to *Starlink* and, once this became a feasible proposition in Northern Ireland, Brendan was soon converted and enthusiastically entered into the fray. It may be difficult for outsiders to realise the enormous boost the establishment of a *Starlink* node in Northern Ireland, based partly in Armagh and partly in Belfast, gave to astronomical research at Armagh Observatory. It was now possible to access and analyse data from various optical and space observatories using packages developed elsewhere in the Starlink Network and to have available equipment that was guaranteed to run the complex software required. As with *IUE*, Brendan went on to become a member of the Starlink Panel and to add his voice to those which applauded its philosophy of coordination and cooperation in the provision of astronomical software.

One of Brendan's greatest assets was his excellent command of the English language, both spoken and written. Who could forget Brendan's clearly expressed and often humorous remarks at colloquia and symposia and the ready wit, so characteristic of his native Dublin. Brendan was a popular speaker and one that never failed to catch the interest of his audience. No doubt his communication abilities, and his willingness to use them in the service of Armagh Observatory, lead to his appointment as assistant director from 1989 to 1994. During this period he contributed much to the reorganisation of this institution as it tried to come to terms with its expanding role in UK/Irish science.

Brendan's scientific pursuits also moved on and towards the last years of his life he developed an increasing interest in the evolution of angular momentum in late type stars and how cluster studies could help to track this evolution. This lead him to study rapidly rotating stars with their extended, prominence-like, structures. His work in this direction has prospered and the results of some of the projects he initiated are included in this volume. As Jeff Linsky so aptly put it at the closing banquet of this meeting,

Brendan, I am not sure where you are,
but you are still doing first-rate astrophysics

Part 1

MAGNETIC FIELDS IN THE SUN AND STARS: OBSERVATIONS

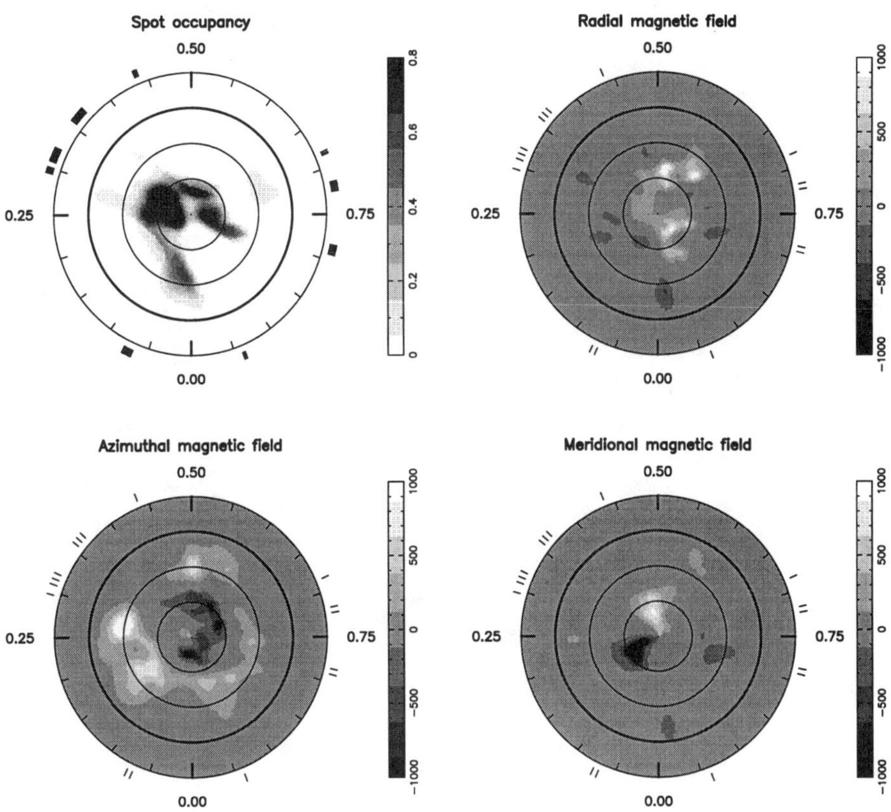

Flattened polar view of the brightness (upper left), radial field (upper right), azimuthal field (lower left) and meridional field (lower right) distributions at the surface of the K1 subgiant of HR 1099 at epoch 1995.94. Parallels are shown (as concentric circles) every 30° down to a latitude of -30° (the bold one featuring the equator). Magnetic fields are labelled in G. The radial ticks around each plot depict the rotational phases of observations.

Solar and Stellar Activity: Similarities and Differences
ASP Conference Series, Vol. 158, 1999
C.J. Butler and J.G. Doyle, eds.

Magnetic Fields in the Hertzsprung-Russell Diagram

Jeffrey L. Linsky

JILA/University of Colorado and NIST
Boulder, CO 80309-0440 USA

Abstract. Stellar activity consists of the phenomena that occur from the conversion of magnetic energy into heat, non-thermal particles, and kinetic energy after the energy in the magnetic field has been raised above its potential field value by mass motions. For stars located in different regions of the HR Diagram, the nature and magnitude of the mass motions responsible for stressing the field lines can be fundamentally different leading to a variety of activity phenomena. With this paradigm in mind, I lead a tour through the HR diagram, identifying those types of stars for which magnetic fields have been measured or inferred by reliable proxies. For each class of stars I identify the likely type of plasma flow that is stressing the magnetic fields to produce the observed active phenomena.

"Magnetic fields are to astrophysics what sex is to psychoanalysis."

Henk van de Hulst (1988)

1. Roles that Magnetic Fields Play in Solar and Stellar Atmospheres

During this conference, we will have the opportunity to learn about the latest observations and theory concerning the measurement of magnetic fields and magnetic field controlled phenomena in the atmospheres of the Sun and late-type stars. Unfortunately, it is easy to lose sight of the important questions. From my perspective there are four fundamental questions:

1. How are magnetic fields regenerated in different types of stars?

2. What are the properties (e.g., structure, field strengths, filling factors, geometry, energy content) of magnetic fields in different types of stars?

3. What types of plasma motions stress the magnetic fields, leading to conversion into other types of energy (e.g., heat, non-thermal particles, kinetic energy)?

4. What observable manifestations of these energy conversion processes can be studied as stellar active phenomena?

In this review I will concentrate on the last two questions and make some comments concerning the second. I anticipate that others will address the first and second questions in more detail. Since observers concentrate on the last question, I will start with it. It is difficult, however, to go back from the observed active phenomena to the physics of energy conversion (the third question), because different energy conversion scenarios could, in principle, lead to the same observable active phenomena. Theory can be an important guide in searching out the essential physics responsible for the observed phenomena. Let us begin by listing the various observable phenomena that magnetic fields can produce:

Heating Processes: MHD wave processes, rapid field annihilation processes

Flares: rapid enhancements of the emitted flux across the electromagnetic spectrum that result from rapid magnetic field annihilation events

Isolation: thermal isolation of adjacent plasma by suppression of conduction and convection

Geometry of Structures: magnetic flux loops and arcades; chromospheric network structures, spicules, coronal mass ejections, prominences

Convection: magneto-convection has different properties and cell sizes than normal convection

Wind Acceleration: Alfvén waves and other MHD wave acceleration mechanisms

Non-thermal Particle Acceleration: Maxwell's equations say that $\nabla \times H - (1/c)\partial D/\partial t = 4\pi J/c$. Therefore, twisted magnetic fields create electric currents that accelerate charged particles.

Turbulence: spectroscopic turbulence can be produced by bulk motions created by magnetic fields

Downflows: downflows observed as redshifts in strong field regions

Starspots: The photosphere in sunspots and starspots appears dark because of the suppression of convective flux by strong magnetic fields. The location of the visible photosphere is depressed (the so-called "Wilson Depression") because horizontal pressure balance requires lower gas pressures in spots, lower densities, and thus has less opacity.

Zeeman Broadening: Zeeman sensitive lines will be broader than non-sensitive lines and can increase line blanketing

Polarization: line and continuum polarization and depolarization processes

Radio Emission: gyrosynchroton emission and maser radio emission are much brighter than thermal free-free emission.

A fundamental problem: Since the gas pressure scale height $H = kT/\mu g$ is generally smaller than the magnetic field scale height, $\beta = 8\pi P_{gas}/B^2$ changes from $\beta > 1$ below the photosphere to $\beta < 1$ above the photosphere. Thus below the photosphere convective motions shake the field lines, whereas above the photosphere the field geometry controls the motion of the plasma. What are the consequences of this behavior in a dynamic atmosphere?

2. A Tour Through the HR Diagram in Search of Magnetic Fields

Before we begin our tour, I should first describe my model of the relation of magnetic fields to solar/stellar activity. This can help us to concentrate on what I believe are the critical questions concerning magnetic fields. I think of solar/stellar activity according to the paradigm: **Stellar activity consists of the phenomena that occur from the conversion of magnetic energy into heat, non-thermal particles, and kinetic energy after the energy in the magnetic field has been raised above its potential field value by mass motions. For stars located in different regions of the HR Diagram, the nature and magnitude of the mass motions responsible for stressing the field lines can be fundamentally different leading to a variety of activity phenomena.** Although this paradigm may not include all activity phenomena, probably most workers in the field would accept it. Given this paradigm, here are some important questions to ask during our tour:

- What direct or indirect techniques can measure magnetic field properties?
- Are magnetic fields detected or inferred from the data?
- What are typical magnetic field strengths and fluxes?
- What is the magnetic field geometry?
- Is there evidence for active regions and star spots?
- What is the nature of the mechanical stresses on the magnetic field?

2.1. O and Early-B Stars

We begin our tour in the upper left hand of the HR Diagram where the hottest and most luminous stars live. I have not found any direct measurements of magnetic fields on these stars and it is difficult to measure sub-kilogauss fields from Zeeman broadening or splitting of spectral lines as the these stars rotate rapidly and their spectral lines are very broad. However, magnetic fields may eventually be measured using the Hanle effect (cf. Ignace, Nordsieck & Cassinelli 1997). The Copernicus satellite observed O VI emission lines, which are now understood as produced by Auger ionization (O IV $\rightarrow$ O VI) by X-rays from the shocked hot plasma in in the radiation-driven winds of these stars. The presence of magnetic fields in some of these stars is inferred from the non-thermal radio emission spectrum seen in 24% of the O3–B2 luminous stars surveyed by Bieging, Abbott & Churchwell (1989).

White (1985) proposed a model in which Fermi acceleration by strong shocks in the wind accelerates electrons that emit by the synchrotron process. If this model is correct, then surface fields of a few Gauss are implied, and such fields could be primordial in these young stars. Also, Gagné *et al.* (1997) discovered periodic X-ray emission from θ^1 Ori C (O7 V). They argue that the X-ray periodicity is due to absorption by the wind in an extended magnetosphere. This star may be an example of a high mass oblique rotator star like the chemically peculiar B and A stars. Babel & Montmerle (1997) estimate a stellar surface magnetic field of 300 G for this star. I conclude that for these stars **the magnetic fields are stressed by bulk motions in the radiation-driven winds.**

2.2. Magnetic Chemically Peculiar B and A Stars

On and near the main sequence from about spectral type B2 down to about A5 lie several types of magnetic chemically peculiar stars. Magnetic fields with simple geometries (dipoles and quadrupoles) and mean surface field strengths up to 34 kG [Babcock's star = GL Lac (B8p SiHeW)] are easily measured. The hotter stars in this group are generally helium-strong and the cooler stars are helium-weak and Si-strong. For many years their atmospheres have been described with oblique rotator models in which the magnetic and rotational poles are well separated. Many of these stars are luminous non-thermal radio emitters as shown by their negative spectral indices (Linsky, Drake & Bastian 1992). Also a few have now been identified as intrinsic X-ray sources, that is the X-ray emission is not from an unknown companion star (Drake *et al.* 1994). Non-magnetic chemically peculiar stars are not radio sources (Drake, Linsky & Bookbinder 1994). Linsky *et al.* (1992) proposed a wind-fed magnetosphere model in which the radiation-driven wind emerging far from the magnetic poles is trapped in the magnetosphere. Gas pressure and centrifugal acceleration distend the field lines near the magnetic equator, forming an equatorial current sheet and leading to electron acceleration and gyro-synchrotron emission. The non-thermal radio radiation and perhaps the X-ray emission are thus a consequence of the **magnetic fields being stressed by the stellar wind and rotation.**

2.3. Normal B and A Stars

I know of no magnetic field detections for stars without chemical peculiarities in the spectral range B4–A5. Since many of these stars are slow rotators or more rapid rotators seen pole-on, Zeeman measurements of absorption lines in polarized light would have detected strong dipolar fields or in unpolarized light would have detected the presence of kG fields. I also know of no detections of non-thermal radio emission from main sequence stars in this spectral range. X-ray emission was detected from $< 3\%$ of the field stars in the B4–B9 spectral range in the Grillo *et al.* (1992) Einstein survey, so it is difficult to say that these stars constitute a class of intrinsic X-ray emitters. Schmitt *et al.* (1993) detected a few late-B stars in their *ROSAT* survey of wide binaries, but concluded that the cause of the emission is a "puzzle." Thus we have no direct or indirect information on the magnetic fields of these stars.

2.4. F–M Main Sequence Stars

If one were to observe the Sun as a star (i.e., as a point source), then the net magnetic flux would correspond to a uniform magnetic field strength of < 1 G because of the nearly complete cancellation of magnetic flux with opposite polarities. It would be extremely difficult to measure this average magnetic field with a Zeeman analyzer that compares the shift of line profiles measured in opposite circular polarizations. The failure of Zeeman polarization techniques to measure significant magnetic field strengths in main sequence stars with sub-photospheric convective zones is usually explained by these stars having very complex magnetic field geometries like the Sun. High resolution magnetograms reveal that most of the Sun has very weak photospheric fields, but in small clumps below the chromospheric network, typical field strengths of 1500 G are common. Thus kiloGauss fields with a variety of orientations are distributed across the solar photosphere and presumably other dwarf stars awaiting a sensitive technique for their measurement. Even though sunspots have stronger magnetic fields, their contribution to the magnetic field in integrated sunlight is negligible as sunspots are dark in the optical and cover $< 1\%$ of the solar surface.

Robinson, Worden & Harvey (1980) pioneered a very successful technique to measure the Zeeman broadening of optical lines and splitting of near-infrared lines. They analyzed high resolution spectra in unpolarized light for which there is no cancellation of oppositely polarized magnetic fields. Saar (1990) developed this method further and used it to measure the mean unsigned magnetic field strengths (B) and filling factors (f) in the active regions of many stars. Saar assumed either a two-component atmosphere $F_{obs} = fF_{mag}(B) + (1-f)F_{quiet}(B = 0)$ or a three-component atmosphere (including starspots), where F refers to the radiative flux observed from the magnetic or quiet components of the atmosphere. Uncertainty in the F_{mag}/F_{quiet} ratio leads to uncertainty primarily in the value of f. He then solved the radiative transfer equation in a Milne-Eddington atmosphere including the exact Zeeman patterns and magneto-optical effects in the line transfer. For each star he adopted published values for vsini and radial-tangential macro-turbulence. Saar (1995, 1996) has summarized this work which has lead to the following important conclusions concerning main sequence stars:

- Magnetic field strengths increase with B–V and decreasing T_{eff} in main sequence stars,
- The inferred magnetic field strengths are close to their equipartition values, i.e., $B \approx B_{eq} = (8\pi P_{gas})^{0.5}$, with P_{gas} evaluated where $T = T_{eff}$, except perhaps for the most active stars,
- Both the filling factor and the total magnetic flux (fB) increase as P_{rot} and the Rossby number decrease until saturation is reached at large values of f. (Note that the divergence of flux tubes with height should depend on the value of f.)
- Vilhu (1994) showed that most activity indicators reach a maximum value (saturate) at high angular velocity (Ω_{sat}), but when $\Omega > \Omega_{sat}$, B may increase beyond B_{eq} while f stays at its saturated value.
- X-ray and Ca II surface fluxes are correlated with fB.

Although very successful in measuring global properties of stellar magnetic fields, the Zeeman broadening method provides no information on the three-dimensional (3-D) structure of the magnetic field across the stellar surface. A very different technique described by Donati & Brown (1997) and by Donati et al. (1997) called Zeeman Doppler imaging (ZDI) can map the radial, meridional, and azimuthal components of the magnetic field for rapidly-rotating stars using profiles of a large number of spectral lines in circularly polarized light. Zeeman-Doppler images are now available for six active stars (including AB Dor and HR 1099) using the maximum entropy or optimal reconstruction techniques. Unlike the Sun, the field lines for these active stars are azimuthal in rings (3 for AB Dor). This technique now allows one to follow the magnetic field evolution in active stars and to monitor stellar magnetic cycles directly rather than through a proxy like the Ca II H+K flux. However, the ZDI technique likely misses most of the magnetic field because it does not see dark spots or weak field regions.

The existence of dark starspots, identified by photometric variability and Doppler images, is an important indirect indicator of stellar magnetic fields. Although sunspots cover less than 1% of the solar surface, starspots can cover as much as half of the observed hemisphere of stars in RS CVn systems like II Peg. Since starspots are very dark in the optical and near infrared, there are no measurements of starspot magnetic fields. Sunspot umbral fields are typically 3500 G, or about 2.3 times typical photospheric network fields. Whether or not starspot magnetic field strengths are similarly enhanced will require future Zeeman broadening measurements of molecular lines in the infrared.

Redshifted transition region lines (e.g., C II, Si IV, C IV) are one of several indirect indicators of magnetic fields in stellar atmospheres. Redshifts have been observed in solar spectra for a long time. Achour *et al.* (1995), for example, found that the redshift of the C IV 1548Å line is 6.2 km s^{-1} in quiet regions but is 13.0 km s^{-1} in active regions. Observations with *SoHO*/CDS and *SoHO*/SUMER show that redshifts are seen even in lines of Ne VIII formed at 650,000 K (Brynildsen *et al.* 1998) ... although see paper by Teriaca *et al.* in these proceedings. *IUE* spectra (e.g., Ayres, Jensen & Engvold 1988) and *HST*/GHRS spectra (e.g., Wood, Linsky & Ayres 1997) show redshifts in the transition region lines of late-type dwarfs and giants. Resonance and optically thin intersystem lines are both redshifted, confirming that the downflows are real and not an effect produced by optically thick lines formed in an atmosphere with a velocity gradient. Downflow velocities increase with line formation temperature between 3 10^4 and 1 10^5 K. Lower gravity stars (e.g., β Dra and Capella) show larger redshifts than main sequence stars (e.g., α Cen A, α Cen B, and ϵ Eri). Solar transition region models computed with a 2-D hydrodynamic code show that density perturbations in regions of strong density and temperature stratification cool radiatively to become condensations that flow downward by gravity (Reale *et al.* 1996). This effect is enhanced (as is observed on the Sun) in regions of vertical magnetic fields which constrain horizontal motions and conduction.

I conclude that for the F–M dwarf stars active phenomena result from the **stressing of magnetic fields by convective motions.**

2.5. Late-M Stars and Brown Dwarfs

Zeeman broadening analyses of near-infrared spectra have provided magnetic field measurements of M dwarfs as late as M4.5 V. Johns-Krull & Valenti (1996) measured $B = 3.8 \pm 0.5$ kG covering $50 \pm 13\%$ of EV Lac and $B = 2.6 \pm 0.3$ kG covering $50 \pm 13\%$ of Gl 729, but the line profiles indicate a distribution of field strengths across the surface or with depth. Strong magnetic fields in starspots may be contributing to the broadening of the observed line profiles. Field strengths of 4.0 kG and 4.3 kG have been measured for AD Leo (M4 Ve) and AU Mic (M2.5 Ve) by Saar (1994). Magnetic fields have not yet measured in the coolest M dwarfs and brown dwarfs, because these stars are too faint for high resolution spectroscopy. Nevertheless magnetic fields are almost certainly present as their internal structures are similar to the M4.5 Ve stars. I say this because there is no sharp change in the coronal heating efficiency as measured by

L_x/L_{bol} between stars with radiative cores and convective envelopes (spectral type M 5 and earlier) and stars that are fully convective (cooler stars with $M \leq 0.3 M_\odot$ and brown dwarfs) (Fleming, Schmitt & Giampapa 1995). Also, Jupiter, which is structurally similar to a low mass brown dwarf, has a strong magnetic field. Rüdiger (1998) argues that late M dwarfs with fully convective cores probably have α^2 dynamos because the stars do not show magnetic cycles.

One indirect indicator of strong magnetic fields on these stars is starspots that are identified by photometric variability. Doppler imaging is not yet feasible for these faint stars. Another indicator, flares, are readily observed at optical, UV, X-ray, and radio wavelengths. VB10 (M8 Ve) has been observed to flare in X-rays by the *ROSAT* HRI (Fleming 1998) with $L_x^{flare} \approx 10^{30}$ erg s^{-1}, and $L_x^{flare}/L_{bol}^{quiet} > 1.0$. Liebert *et al.* (1998) have observed an Hα flare on the M9.5 V star 2MASS J0149090+295613, which may be a brown dwarf. Outside of obvious flares, late M dwarfs are often bright X-ray sources with $L_x/L_{bol} \approx 10^{-3}$. In their *ROSAT* survey of all known K and M stars within 6 pc of the Sun, Schmitt, Fleming & Giampapa (1995) found that 94% are detected X-ray sources, the coolest detected source (by the Einstein satellite) being Gl 752B = VB 10. They argue that the smoothness of the L_x distribution function and the occurrence of flares are evidence for these coronae to have a heating mechanism in common with the Sun, and that this mechanism is magnetic in character. The correlation of increased spectral hardness (and therefore higher coronal temperatures) with increasing L_x is further evidence that the heating involves magnetic reconnection events. On the basis of X-ray surface fluxes for this sample of nearby stars, Mullan & Fleming (1996) argue that the coronae of at least the M stars with Hα emission (the so-called dMe stars) cannot be heated acoustically and therefore must be heated magnetically.

Typically brown dwarfs are rapid rotators. Basri (La Palma paper) reports that Kelu-1 (L2 star) has vsini = 40 km s^{-1}. Other brown dwarfs lie in the range 20–40 km s^{-1}. Neuhauser & Comeron (1998) have very recently reported the first detection of X-rays from a young brown dwarf in the Chamaeleon I star forming cloud with $L_X = 2.57\ 10^{28}$ erg s^{-1}, an X-ray luminosity typical of late-M dwarfs. There are as yet no observations of brown dwarfs in the radio or UV, but I anticipate detections at these wavelengths in the very near future. I conclude that for the late M dwarfs and brown dwarfs, like the warmer main sequence stars, active phenomena are a consequence of **magnetic fields being stressed by convective motions.**

2.6. RS CVn Systems

Magnetic fields have now been measured in several RS CVn systems using the Zeeman broadening technique. For example Bopp *et al.* (1989) reported $B = 2000 \pm 300$ G and $f = 0.66 \pm 0.14$ for the 17.4 day period VY Ari system (K3 III-IV + K3 III-IV). Using the Ti I 2.2μm line, Saar (1996) measured $B \approx 3000$ G and $f \approx 0.60$ for II Peg (K2-3 IV-V + ?), but the result may include a large spot contribution as the Ti I line is formed mainly in spots with $f_{spot} = 0.40 - 0.55$. Zeeman Doppler images, which are now published for six active stars including HR 1099 (Donati *et al.* 1990), show prominent azimuthal fields.

Starspots provide an important indirect indicator of strong magnetic fields. A recent analysis of 30 years of photometric monitoring of the 24 day period

HK Lac system (Oláh *et al.* 1997) provides information on the longevity and phase drifts of the large spots that together can cover up to 40% of the stellar surface. Doppler images of 13 RS CVn systems (Strassmeier 1996) also provide information on the location of large starspots. The high resolution *HST*/GHRS spectrum of the Fe XXI 1354Å line formed at $T = 1\ 10^7$ K in the Capella stars (G1 III + G8 III) shows that the high temperature coronal gas is stationary and therefore confined by strong magnetic fields (Linsky *et al.* 1998). For $n_e > 10^{12}$ cm^{-3}, the magnetic field must exceed 270 G to confine the hot gas.

In analogy with solar active regions (also called plages), areas of bright chromospheric line emission are likely regions of strong magnetic fields. Plages on RS CVn systems are identified by rotational modulation of transition region emission lines (e.g., II Peg studied by Rodonò *et al.* 1987) and bright features in Doppler images in the Mg II emission lines (e.g., Neff *et al.* 1989; Pagano *et al.* 1992). Plages have now been observed at the equator and at high latitudes (±50°). Plages cover 9% of the visible surface of AR Lac with emission line surface fluxes near the saturated heating rate (Linsky 1991).

There is a large literature concerning flares on RS CVn systems, but I call your attention to two very recent papers. In their analysis of the 1994 August 29 flare on UX Ari observed with *ASCA* and *EUVE*, Güdel *et al.* (1998) observed very hot plasma with $T \geq 100$ MK. Using a two-ribbon flare model, they find that $L_{peak} = 1.4\ 10^{32}$ erg s^{-1} with a loop length about $1R_{\star}$. A very interesting result is that the iron abundance increases from about 17% to 89% of solar during the flare, presumably due to the evaporation of solar abundance material from the lower atmosphere that fills the flaring loop. Osten & Brown (1998) studied four megaseconds of *EUVE* flare photometry on 16 RS CVn systems. In this unique data set with excellent statistics, they find that RS CVn systems flare 40% of the time, but short period systems like ER Vul (P=0.70 days, $R_1 \approx R_2 \approx R_{\odot}$, asini $\approx 3R_{\odot}$) flicker but do not show large flares. They conclude that the X-ray emitting regions are extended, because the duration of flares often exceeds the rotational period and eclipses are predicted but not seen. They find evidence that supports "quiescent" coronal heating by microflares in active stars: in the EUV $L_{flare} \propto L_{quiet}^{1.05}$. After they remove obvious flares, there is no obvious rotational modulation in the EUV flux in any of the 16 systems.

For the RS VCn systems, active phenomena result from the **magnetic fields being stressed by convective motions and the interactions of adjacent magnetospheres.**

2.7. Yellow and Red Giants

Linsky & Haisch (1979) proposed that a dividing line exists in the HR Diagram separating the yellow giants (spectral type K1 III and earlier), which have detected transition region emission lines, from the red giants (later spectral types), for which *IUE* spectra do not show detected emission lines. Ayres *et al.* (1981) found the same phenomenon in X-rays using Einstein data. What has happened to the so-called "Linsky-Haisch" dividing line, and is the boundary connected with some property of the magnetic field? The discovery of hybrid-chromosphere stars (e.g., Hartmann, Dupree & Raymond 1980) showed that the more luminous G Ib and K II stars have C IV and other emission lines, and the *ROSAT* survey of hybrid-chromosphere stars shows that many are also X-ray sources

(Reimers *et al.* 1996). While the X-ray dividing line is confirmed for giants on the basis of a *ROSAT* 25 pc survey (Hünsch & Schröder 1996), very sensitive *HST*/GHRS UV spectra (Ayres *et al.* 1997) show Si IV 1394 Å and in some cases C IV 1448 Å emission from both yellow and red giants with a lower plateau at $f_{Si\ \text{IV}}/f_{bol} \approx 10^{-8}$. Stellar evolution calculations show that the yellow giants and hybrid stars are more massive, younger stars with high rotation rates, whereas the red giants are old, slowly rotating lower mass stars ($\lesssim 1.5 M_\odot$). This suggests that the yellow giants and hybrid stars have magnetic fields and the red giants either have no magnetic fields (or submerged fields) with the Si IV emission ($T \approx 60,000$ K) due to purely acoustic heating, or the magnetic fields are very weak with the magnetic heating processes unable to heat the plasma to 10^6 K. Which conclusion is correct?

Measuring magnetic fields in yellow giants is difficult because the spectral lines are broad, but Hubrig *et al.* (1994) obtained $> 3\sigma$ detections for γ Tau (K0 IIIab), ϵ Tau (G9.5 III), ϵ Leo (G1 II), and ζ Her (G8 III) on at least one occasion each. They measured the shift in line centroids between opposite circular polarizations. Thus the measurements refer to the net flux and not to typical field strengths. Although these results should be tested with Zeeman broadening measurements, they and the X-ray and UV data clearly indicate that the yellow giants must have significant magnetic fields.

There are no direct measurements of magnetic fields in red giant photospheres, but Chapman & Cohen (1986) have measured the magnetic field strength of VX Sgr (M4 Ia) by analyzing the Zeeman splitting in the OH maser 1665 and 1667 MHz features. They identified a magnetic field strength of $\sim$ 2 mG at approximately 80 stellar radii, which implies a photospheric magnetic field strength of ~ 0.5 G if the field is spherically symmetric. Zeeman splitting for the 1612 MHz maser line (Szymczak & Cohen 1997) indicate a magnetic field strength of 1.1 mG where the line is formed. M giants are not known to be X-ray or non-thermal radio sources, however *HST*/FOS images of α Ori (M2 Iab) show a bright feature in near-UV light that could be a convective cell or possibly result from magnetic activity (Gilliland & Dupree 1996). The high rates of mass loss for non-pulsating luminous M stars apparently can only be explained by Alfvén-driven winds. Although the calculations are only approximate, Hartmann & MacGregor (1980) found acceptable mass loss rates for α Boo (K1.5 III) when a photospheric magnetic field strength of 10 Gauss was used. Alfvén-wave driven models for α Ori (M2 Iab) imply photospheric magnetic field strengths of 1–5 G (Hartmann & Avrett 1984; Airapetian *et al.* 1998). Thus there is a case for weak magnetic fields in red giants and **convective motions could stress these weak magnetic fields.**

2.8. Pre-Main Sequence Stars

Johns-Krull & Valenti (1998) have measured a total magnetic flux of 4.0 kG for the cTTS star BP Tau from their analysis of IRTF CSHELL spectra of the Ti I 2.22328 μm lines. I do not know any other direct measurements of magnetic fields in PMS stars, but strong fields are indicated by measurements of non-thermal radio emission, flares, and starspots on several of these stars. The youngest of the observable pre-main sequence stars, the so-called "embedded protostars," are bright X-ray sources with L_x/L_{bol} well above the level of saturated X-ray emission, $L_x/L_{bol} > 10^{-3}$. What can explain this result? Shu *et al.* (1997) propose that the X-ray emission from the embedded protostars consists of two components – a soft X-ray component produced in a magnetic corona like the somewhat older classical T Tauri stars and weak-lines T Tauri stars, and a second harder X-ray component produced where the magnetic fields of the star and the accretion disk interact. In the interaction region, field reconnection leads to electric currents that produce heating (and thus X-ray emission) and the acceleration of non-thermal particles. In their model stellar activity results from **the stressing of stellar magnetic fields by convective motions and by interactions with the magnetic field of the accretion disk.**

3. Conclusions

In this review of magnetic fields in the HR diagram, I have summarized the direct measurements of magnetic field properties in different types of stars and the indirect indicators (e.g., starspots, flares, non-thermal radio emission, and bright X-ray and UV line emission) of magnetic fields. Stellar activity phenomena are now thought to be the responses of a stellar atmosphere to the heating, non-thermal particle acceleration, and kinetic motions produced when stressed magnetic fields relax to lower energy states either by rapid field reconnection or by more gradual energy transfer. The critical point is that magnetic fields must be stressed (i.e., brought to a higher energy configuration) by some type of mechanical motion. For stars with convective zones, fluid motions below the photosphere where $\beta > 1$ jostle the field lines eventually, leading to reconnection above the photosphere where $\beta < 1$. However, this is not the only way of stressing the field lines. Shocks in the radiatively driven winds of O and early B stars, wind and centrifugal stretching of the magnetospheres of magnetic chemically peculiar stars, interactions with the fields of nearby stars in binary systems (e.g., RS CVn systems), and interactions with the fields of accretion discs for embedded protostars can also provide the stresses that lead to observable active phenomena. One should also look for other types of mechanical forces that can stress the fields leading to other active phenomena.

Acknowledgments. I thank NASA for grants to the University of Colorado under the *HST*, *AXAF*, and *FUSE* programs. I wish to thank Manfred Cuntz for his suggestions concerning the magnetic fields of red giants. I also thank for their hospitality the staff of Armagh Observatory, including Gerry Doyle, John Butler, and posthumously Brendan Byrne. We will long remember Brendan's scientific contributions and his good advice and companionship.

References

Achour, H., Brekke, P., Kjeldseth-Moe, O. & Maltby, P. 1995, ApJ, 435, 945

Airapetian, V., Ofman, L., Robinson, R.D., Carpenter, K.G. & Davila, J. 1998, in Cool Stars, Stellar Systems and the Sun, to appear

Ayres, T.R., Jensen, E. & Engvold, O. 1988, ApJS, 66, 51

Ayres, T.R., Linsky, J.L., Vaiana, G.S., Golub, L. & Rosner, L. 1981, ApJ, 250, 293

Ayres, T.R., Brown, A., Harper, G.M., Bennett, P.D., Linsky, J.L., Carpenter, K.G. & Robinson, R.D. 1997, ApJ, 491, 876

Babel, J. & Montmerle, T. 1997, ApJ, 485, L29

Beiging, J.H., Abbott, D.C. & Churchwell, E.B. 1989, ApJ, 340, 518

Bopp, B.W., Saar, S.H., Ambruster, C., Feldman, P., Dempsey, R., Allen, M. & Barden, S.C. 1989, ApJ, 339, 1059

Brynildsen, N., Brekke, P., Fredvik, T., Haugan, S.V.H., Kjeldseth-Moe, O., Maltby, P., Harrison, R.A. & Wilhelm, K. 1998, Solar Physics, 181, 23

Chapman, R.D. & Cohen, R.J. 1986, MNRAS, 220, 513

Dere, K.P., Bartoe, J.-D.F., Brueckner, G.E., Cook, J.W. & Socker, D.G. 1989, Solar Physics 123, 41

Donati, J.-F., Semel, M., Rees, D.E., Taylor, K. & Robinson, R.D. 1990, A&A, 232, L1

Donati, J.-F. & Brown, S.F. 1997, A&A, 326, 1135

Donati, J.-F., Semel, M., Carter, B.D., Rees, D.E. & Cameron, A.C. 1997, MNRAS, 291, 658

Drake, S.A., Linsky, J.L., Schmitt, J.H.M.M. & Rosso, C. 1994, ApJ, 420, 387

Drake, S.A., Linsky, J.L. & Bookbinder, J.A. 1994, AJ, 108, 2203

Fleming, T.A. 1998, private communication

Fleming, T.A., Schmitt, J.H.M.M. & Giampapa, M.S. 1995, ApJ, 450, 401

Gagné, M., Caillault, J.-P., Stauffer, J.R. & Linsky, J.L. 1997, ApJ, 478, L87

Gilliland, R.L. & Dupree, A.K. 1996, ApJ, 463, L29

Grillo, F., Sciortino, S., Micela, G., Vaiana, G.S. & Harnden, F.R., Jr. 1992, ApJS, 81, 795

Güdel, M., Linsky, J.L., Brown, A. & Nagase, F. 1998, ApJ, in press

Hartmann, L., Dupree, A.K. & Raymond, J.C. 1980, ApJ, 236, L143

Hartmann, L. & Avrett, E.H., 1984, ApJ, 284, 238

Hartmann, L. & MacGregor, K.B., 1980, ApJ, 242, 260

Hubrig, S., Plachinda, S.I., Hünsch, M. & Schröder, K.-P., 1994, A&A, 291, 890

Hünsch, M. & Schröder, K.-P. 1996, A&A, 309, L51

Ignace, R., Nordsieck, K.H. & Cassinelli, J.P. 1997, ApJ, 486, 550

Johns-Krull, C.M. & Valenti, J.A. 1996, ApJ, 459, L95

Johns-Krull, C.M. & Valenti, J.A. 1998, preprint

Liebert, J. *et al.* 1998, preprint

Linsky, J.L. 1991, in Mechanisms of Chromospheric and Coronal Heating, ed. P. Ulmschneider, E.R. Priest, & R. Rosner (Berlin: Springer-Verlag), p. 166

Linsky, J.L., Drake, S.R. & Bastian, T. 1992, ApJ, 393, 341

Linsky, J.L. & Haisch, B.M. 1979, ApJ, 229, L27

Linsky, J.L. & Wood, B.E. 1994, ApJ, 430, 342

Linsky, J.L., Wood, B.E., Brown, A. & Osten, R.A. 1998, ApJ, 492, 767

Mullan, D.J. & Fleming, T.A. 1996, ApJ, 464, 890

Neff, J.E., Walter, F.M., Rodonò, M. & Linsky, J.L. 1989, A&A, 215, 79

Neuhauser, R. & Comeron, F. 1998. Science, 282, 83

Oláh, K., Kövári, Zs., Bartus, J., Strassmeier, K.G., Hall, D.S. & Henry, G.W. 1997, A&A, 321, 811

Pagano, I., Rodonò, M. & Neff, J.E. 1992, in Cool Stars, Stellar Systems, and the Sun, ed. M.S. Giampapa & J.A. Bookbinder (San Francisco: Astronomical Soc. Pacific), p. 362

Reale, F., Peres, G. & Serio, S. 1996, A&A, 316, 215

Reimers, D., Hünsch, M., Schmitt, J.H.M.M. & Toussaint, F. 1996, A&A, 310, 813

Robinson, R.D., Worden, S.P. & Harvey, J.W. 1980, ApJ, 236, L155

Rodonò, M. *et al.* 1987, A&A, 176, 267

Rüdiger, G. 1998, in Very Low Mass Stars and Brown Dwarfs in Stellar Clusters and Associations, to appear

Saar, S.H. 1990, in IAU Symposium 138, The Solar Photosphere: Structure, Convection, and Magnetic Fields, ed. J.O. Stenflo, (Dordrecht: Kluwer), p. 427

Saar, S.H. 1994, in IAU Symposium 154, Infrared Solar Physics, ed. D.M. Rabin et al. (Dordrecht: Kluwer), p. 493

Saar, S.H. 1995, in Magnetodynamic Phenomena in the Solar Atmosphere - Prototypes of Stellar Magnetic Activity, ed. Y. Uchida, T. Kosugi, & H.S. Hudson (Dordrecht: Kluwer), p. 367

Saar, S.H. 1996, in Stellar Surface Structure, ed. K.G. Strassmeier & J.L. Linsky (Dordrecht: Kluwer), p. 237

Schmitt, J.H.M.M., Fleming, T.A. & Giampapa, M.S. 1995, ApJ, 450, 392

Schmitt, J.H.M.M., Zinnecker, H., Cruddace, R. & Harnden, F.R., Jr. 1993, ApJ, 402, L13

Shu, F.H., Shang, H., Glassgold, A.E. & Lee, T. 1997, Science, 277, 1475

Strassmeier, K.G. 1996, in Stellar Surface Structure, ed. K.G. Strassmeier & J.L. Linsky (Dordrecht: Kluwer), p. 289

Szymczak, M. & Cohen, R.J. 1997, MNRAS, 288, 945

Teriaca, L., Doyle, J.G. & Banerjee, D., 1999, these proceedings

Vilhu, O. 1994, A&A, 133, 117

White, R.L. 1985, ApJ, 289, 698

Wood, B.E., Linsky, J.L. & Ayres, T.R. 1997, ApJ, 478, 745

Solar and Stellar Activity: Similarities and Differences
ASP Conference Series, Vol. 158, 1999
C.J. Butler and J.G. Doyle, eds.

The Dynamic Nature of the Solar Magnetic Field

Carolus J. Schrijver & Alan M. Title

Stanford-Lockheed Institute for Space Research
H1-12/252, 3251 Hanover Street, Palo Alto CA 94304, USA

Abstract. Magnetic fields emerge onto the surface of the Sun as bipolar regions with a broad spectrum of sizes. The large–scale patterns formed by the dispersing fields of large bipolar regions survive for months. The individual concentrations forming these patterns merge and break up on a time scale of only hours to days. Any parcel of flux survives for only days to weeks, to be replaced over and over again primarily by flux emerging in small, ephemeral bipoles. The model that describes the details of field evolution on small scales complements the traditional diffusion model for large–scale flux dispersal. We discuss some implications for the study of activity and dynamos of cool stars in general.

1. Introduction

The solar photospheric magnetic field is incredibly dynamic and complex. Over the years the following scenario has been developed for the evolution of that field. Flux emerges in bipolar concentrations as buoyant flux bundles penetrate the surface. If the flux in an active region is large enough, the region manages to resist dispersive decay by super-granulation for days to weeks. During this phase, flux slowly disappears out of the region until the region loses coherence rather suddenly, and begins its ultimate decay. If the region contains a small amount of flux, the field appears to be passively advected by the flow after only an hour or so.

All flux is eventually subject to diffusive random–walk dispersal due to convective flows ranging from granular to super-granular scales. While flux is spreading, flux concentrations – each comprised of a number of intrinsically strong flux tubes – eventually collide with others. When they do, they either (partially) cancel against them if they are of opposite polarity, or temporarily form a larger concentration until the flows break that up again.

The multitude of small ephemeral regions that emerge within and around active regions continually replaces the flux. In the quietest regions of the photosphere, ephemeral regions replace all flux on a time scale $\hat{t}_{\rm r}$ of two days. In active regions that just begin their decay, $\hat{t}_{\rm r}$ reaches up to a few months, which exceeds the region's life time. But $\hat{t}_{\rm r}$ decreases quickly as the flux spreads out. Once the flux has spread into large areas of enhanced network of a single dominant polarity – the so–called unipolar areas – $\hat{t}_{\rm r}$ drops to at most a few weeks.

This replacement time scale is short relative to the time scale of some 6 months that differential rotation takes to shear patterns on large scales, as well

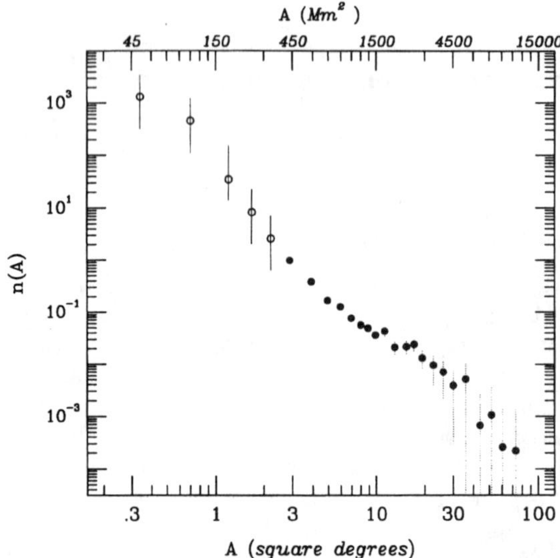

Figure 1. Size distribution $n(A)$ of bipolar regions measured at maximum development (each counted only once) for 978 active regions (dots) and 9492 ephemeral regions (circles) for 29 rotations during cycle 21 (from Zwaan & Harvey, 1994).

as relative to the 2 to 3 years it takes to transport the flux towards the poles from the zones of emergence. Consequently, every flux concentration in the large–scale patterns in magnetograms – such as the pole-ward streaks on synoptic charts that reflect the decay of the largest active regions – is repeatedly replaced before it reaches the polar cap. Only the net flux travels a long way, not the individual concentrations. Moreover, only a very small fraction ever reaches those high latitudes: the bulk of the flux canceling along the way.

The rapid replacement of flux on small scales does not affect the appearance or dispersal of flux on large scales. It does affect the small–scale field and the outer–atmospheric connectivity, potentially contributing significantly to field–line reconnection and braiding, and thus to coronal heating.

We review the processes involved in the above picture, and discuss some consequences for the study of stellar magnetic activity.

2. The Source Function of Photospheric Flux

Flux is injected into the solar photosphere on a wide range of scales, ranging from the detection limit to a size that matches the depth of the convective envelope. The source function $n(A)$ of the frequency of emerging bipolar regions as a function of area A is a smooth, monotonically decreasing function of A (Figure 1). For regions larger than about 500 Mm^2 (or 3.5 square degrees), the shape of $n(A)$ does not vary significantly with the phase of the cycle, $n(A)$ being modulated by a time–dependent constant only. For smaller bipoles the cycle amplitude decreases with decreasing size. Harvey (1993) found for the active regions matching the polarity of sunspot cycle 21 a cycle amplitude in the emergence spectrum $n(A)$ of a factor of 13.9 for regions with $A > 6.5$ square degrees, an amplitude of 8.8 for regions with $2.5 < A < 3.5$ square degrees, and substantially less for ephemeral regions. The cyclic variations are in phase for active regions in all size ranges, except for ephemeral regions for which the onset towards a new maximum occurs a year earlier than for larger regions, announcing the new cycle through high-latitude ephemeral regions.

Harvey (1993) found that all properties of bipolar regions change smoothly with the size of the regions and with the phase of the cycle. In other words, there are no unambiguous indications for well-defined classes of regions of distinctly different behavior. Specifically, ephemeral regions appear to be the smallest bipolar active regions, at one end of a broad spectrum of sizes. Other examples of smooth trends include the fact that whereas the latitude distributions peak precisely at the same latitude for active regions of all sizes, the distribution widths differ: for active regions larger than 1,000 Mm^2 the latitude distributions about the mean latitude are virtually identical, but for smaller regions the distribution widens progressively with decreasing size, particularly so on the pole-ward side. Another trend is found in the orientation of active regions: the average orientation of the dipole moment (as expressed in the Hale-Nicholson and Joy rules) is the same for all active regions, but the distribution around the mean increases steadily as the active–region size decreases. Whereas these gradual changes may indicate that all regions share the same origin, it is possible that some small fraction of the population of ephemeral regions, for example, is generated primarily by a turbulent dynamo, while the remaining fraction is associated with a global, deep–seated dynamo. Whereas recognizing this potential is important, it does not lead to deeper insight into the solar dynamo(s) until theoretical models provide us with more quantitative propositions to test.

Active regions show a marked tendency to cluster in nests: the probability that an active region emerges within an existing region is some 22 times larger than it is to emerge elsewhere, while more than 50% of active regions either emerge within an existing active region or at the location at which an active region emerged earlier. Such nests are surprisingly compact: the spread in emergence longitude for successive nest members is of the order of a single super-granular diameter! There are no convincing indications for a tendency of (nearly) simultaneous occurrences of more than one nest along the same longitude. Notions of simultaneous activity (nearly) 180° apart in longitude do not stand statistical tests either. Nests are also compact in the sense that magnetic flux emerging in a nest tends to remain confined within the nest perimeter during the nest's active lifetime; magnetic flux starts dispersing over the photospheric surface after the emergences stop (Gaizauskas *et al.*, 1983).

At the smallest scales of flux emergence is the intrinsically weak field (discovered by Livingston & Harvey, 1971). The weak field is found all over the solar disk, also in the otherwise empty cell interiors within the network of strong field – hence the weak field is often called the internetwork field. It consists of small fragments with polarities mixed at scales of a few thousand kilometers. Much of the internetwork field emerges in clusters of mixed polarities from small source areas, from where it streams outward, caught in the super-granular flow. Its properties suggest that the internetwork field is intrinsically weak, intersecting the photosphere at a variety of angles (see Zwaan, 1987).

Spruit *et al.* (1987) discuss various formation processes for internetwork field, including the formation of "U-loops" in between two Ω-loops which have emerged from the same toroidal bundle. Weak, small-scale fields are also produced by the aperiodic dynamo that is expected to operate in any region of turbulent convection (Durney, 1988). Note, by the way, that it is unlikely that the entire population of the larger ephemeral regions are the result of a purely turbulent dynamo, because of their preference to emerge at the mean sunspot latitude with a tendency for the proper orientation, and because the chromosphere over the quietest network fed by only ephemeral regions is still brighter than the chromosphere of the least active stars (see the review by Schrijver, 1995). Hence, the flux spectrum of a purely turbulent dynamo should drop rapidly somewhere below about 10^{19} Mx, although some overlap in the flux range is certainly allowed, as we already alluded to above.

The effect of this internetwork flux on the longer–lived, much larger concentrations discussed above appears small: in very quiet regions, at least 90% of the network concentrations originate in ephemeral regions, while the remaining 10% or less is caused by the piling up of internetwork patches (Martin, 1990, Wang & Zirin, 1988). The short life time of internetwork concentrations and their very small size are likely to limit their direct impact on the outer atmosphere. In order to study the flux spectrum for emerging active regions, the distribution $n(A)$ in Figure 1 can be used in combination with what is referred to as the plage state. The term "plage state" was introduced to describe the remarkable coherence and a well–defined outer perimeter at about 50 gauss for a resolution of a few thousand kilometers (see Figure 2) seen during the first few days to weeks following the emergence of relatively large active regions. The sharp transition in the mean flux density across the plage boundary is inconsistent with diffusive dispersal of flux: magnetic field in a plage appears constrained for some time. The flux appears to be distributed more or less evenly over the entire plage, with no clear preference towards the polarity–inversion line.

Within a plage perimeter, the average flux density – excluding spots – is characteristically 100 to 150 gauss for plage areas ranging from $4\ 10^8$ up to $5\ 10^{10}$ km^2 (Schrijver & Harvey, 1994). This plage state holds for newly emerged regions as well as for mature bipolar regions. Note that the constant mean flux density in active regions implies that $n(A)dA \propto n(\Phi)d\Phi$.

The cause of the plage state has not yet been identified. It is possible that the magnetic field is "anchored" at some depth until just before the ultimate disintegration of the corresponding active regions. Effective anchoring of the field could be the result of repeated emergence of convective blobs at roughly the same location; this could "freeze" the flux in nearly stationary down drafts of a flow pattern that evolves little (Simon *et al.*, 1995). This notion is supported by the exceptionally long–lived cells in regions of enhanced network, the puka's (Livingston & Orrall, 1974) that live for many days if not an entire disk passage. Another explanation focuses on the possibility that an abrupt change in the dominant scale of convection near a mean flux density of 50 gauss serves to contain flux (Schrijver, 1989). Not only would the flux dispersal within the plage be reduced because of the scale change ($D \propto \hat{\ell}^2/\hat{t}$), but it would also result in a rather sharp edge of the high–flux density plage area with a fairly homogeneous distribution of flux within the perimeter: in this model, the perimeter acts as a membrane with a preferential direction of transmission, caused by the difference in scales and the associated number density of paths leading away from the "membrane" on either side.

Associated with the plage state is a change in the observable properties of convection: openings on the scale of the super-granulation are rare, and the scale of the granulation is reduced while its pattern evolves more rapidly and less orderly (e.g., Spruit *et al.*, 1990). Yet there is no significant change in the convective energy transport, suggesting that the change in the pattern is either compensated for nearly perfectly as far as energy transport is concerned, or that it is a mere surface phenomenon in the convective overshoot.

3. The Quiet Photosphere and Ephemeral Regions

The shape of the flux input spectrum $n(\Phi)d\Phi$ is such that both the multitude of the smallest regions and the few largest regions contribute significantly to both the total flux budget (Schrijver & Harvey, 1989) and to the flux patterns on all length scales. That the largest regions are important has, of course, been recognized for a long time. The role of the ephemeral regions has only recently been appreciated, however. In order to illustrate this, we discuss some details of the so–called quiet photosphere, which covers much of the Sun and which contributes significantly to the outer–atmospheric emissions, particularly in chromospheric diagnostics.

Flux is located in a multitude of small concentrations evolving within the down flows of the super-granular convection. These down-flows form a network–like pattern with cell sizes that range from ≈10,000 km to ≈30,000 km (Spruit *et al.*, 1990). As flux moves along these lanes, it is subject to fragmentations and collisions. The latter lead either to (partial) cancellation or to merging of flux, depending on the polarities involved. Schrijver *et al.* (1997) developed a statistical description of flux collisions and fragmentations to model the distribution of fluxes in concentrations. Their model explains the distribution of fluxes in concentrations both in the mixed–polarity environment of the very quiet Sun as well as in the dense, unipolar environment of each polarity of active regions. They argue that the frequent collisions lead to rapid cancellation of flux, which locally balances the injection of new flux that emerges in ephemeral regions.

Ephemeral regions have fluxes ranging from $3\ 10^{18}$ Mx to $3\ 10^{19}$ Mx, averaging $1.3\ 10^{19}$ Mx. The initial rapid emergence phase lasts some 30 min., during which the poles separate up to about 7000 km, moving at $\approx 4\ km\ s^{-1}$. After this initial phase, the flux drifts with the super-granular flow, slowing down by a factor of about ten. Schrijver *et al.* (1998) measure an average emergence rate of ephemeral regions near disk center equivalent to ≈ 1 per area of 27,000 km squared per day. With this rate of emergence, it takes only ≈40 hrs for as much flux to surface in the quiet photospheric network as is present there.

Among the proposed mechanisms to heat the corona over the quiet photosphere, one leading contender invokes dissipation of currents driven by braiding of the coronal field lines that occurs because of the random shuffling of foot points caught in the photospheric turbulence. If we consider only this braiding,

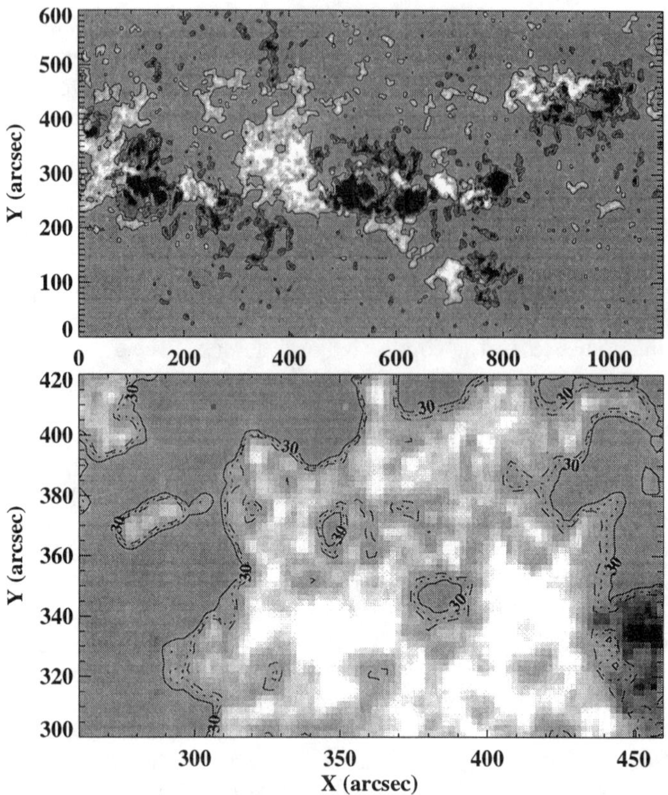

Figure 2. NSO/Kitt Peak magnetogram obtained on 11 September 1989, shown with 2 arcsec (1450 km) pixels. The gray scale saturates at ±400 gauss. The top panel shows a ±50 gauss contour, while the blow–up in the lower panel shows contours at ±30, ±50, and ±70 gauss.

the rate at which energy is supplied to the corona is given by a Poynting flux of $F \propto B_0^2 v^2 \hat{t}/L$, for an average photospheric flux density B_0, mean displacement velocity v, correlation time scale $\hat{t}$, and length L of the outer–atmospheric part of the field line involved (Parker, 1994, Van Ballegooijen, 1985). One consequence of flux replacement is effective braiding of the larger connections (Schrijver *et al.*, 1998): whenever flux from a newly emerged small bipole cancels against existing flux of opposite polarity, part of the flux of that opposite polarity flux is in effect "moved" to the location of the other pole of the bipole. The canceled flux has been replaced by the surviving opposite–polarity flux at another location (note that because there is no correlation between the frequency of the steps and the local concentration density, this merely introduces positional noise, not diffusive dispersal of the flux – see also the explanation given by Wang & Sheeley, 1991). This reconnection causes braiding of long loops, because one foot point is effectively moved, even though it is not being diffusively dispersed. The associated part of the Poynting flux for given B_0^2/L then scales with $v^2\hat{t}$. For

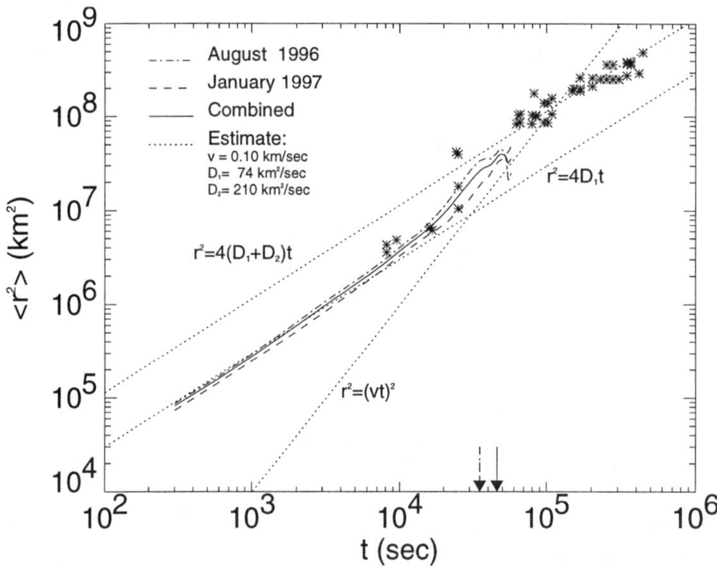

Figure 3. Mean-square displacements of magnetic concentrations for two sets of SOHO/MDI magnetograms (dashed and dashed-dotted), and for the combined set (solid). Arrows show the time span beyond which only 10 concentrations can be tracked. Data based on BBSO magnetograms of enhanced network are shown by asterisks. The line labeled '$r^2 = 4D_1t$' corresponds to diffusion on a granular scale with $D_1 = 74\ km^2\ s^{-1}$. The steeper dotted line labeled '$r^2 = (vt)^2$' corresponds to displacements caused by a slow super-granular drift for $v = 100$ m/s. The asymptotic case of dispersal on combined granular and super-granular scales is indicated by the line labeled '$r^2 = 4(D_1 + D_2)t$', with $D_2 = 210\ km^2\ s^{-1}$ (from Hagenaar et al., 1999).

braiding on the scale of granular convection (1000 km), the latter product is $v^2\hat{t} = 4D \approx 300$ km^2/s (e.g., Hagenaar *et al.*, 1999). The effective braiding associated with flux replacement involves a step length across a super-granule of 15,000 to 18,000 km on a time scale $\hat{t}$ of 4 to 6 hours. These braiding events occur relatively infrequently, because the total flux replacement time scale is some 40 hours. Correcting for this by the ratio of $\hat{t}$ to flux replacement time, the effective product $v^2\hat{t}$ reaches $v^2\hat{t} \times \hat{t}/40 \approx 2000$ km^2/s, which is much larger than that associated with granular braiding.

4. Flux Dispersal on Granular and Super-granular Scales

Leighton (1964) suggested that for the purpose of modeling the large-scale patterns in the photospheric magnetic field, the dispersal of photospheric magnetic field can be described as a random walk, in which the flux responds passively to evolving super-granular flows. The diffusion coefficient is related to the mean–square displacement by $D = \langle r^2 \rangle / 4t$. Lacking a detailed flow model, the value of D is to be estimated from observational data on displacements. The measurement of the diffusion coefficient characterizing the field dispersal is difficult because of a number of reasons, leaving us with a substantial ambiguity about the value of D and the validity of the model. One of the reasons is the superposition of the different length and time scales that contribute to the flux dispersal. Hagenaar *et al.* (1999) show that on time scale less than about 4-6 hours the granular dispersion dominates the dispersal (Figure 3). Thereafter the slow super-granular drift begins to contribute to the dispersal, but doesn't reach the complete randomization required for truly diffusive dispersal until more than half a day has passed.

For time scales of a few days, the dispersal of magnetic flux can in principle be studied by tracking concentrations of magnetic flux to determine their path as a function of time. A fundamental problem is that flux concentrations often fragment into smaller pieces, or coalesce with neighboring concentrations, which makes tracking over periods exceeding a day or two difficult. Relatively large flux concentrations in unipolar enhanced network can sometimes be tracked for up to several days (e.g., Schrijver & Martin, 1990), because in such an environment cancellation is less important, and the concentrations evolve to a larger average flux owing to the reduction in cancellation interactions. The resulting apparent diffusion coefficient $\langle D \rangle$ equals about 220 $km^2\ s^{-1}$ (see asterisks in Figure 3).

The diffusive flux–transport model on large scales requires $D \approx 600 \pm 200$ $km^2\ s^{-1}$. This high value allows a fraction of the flux in the leading polarities of the sources to reach the equator, where it can cancel with opposite flux originating from the other hemisphere. A substantially lower diffusion coefficient would allow the transport of comparable amounts of flux of both polarities to the poles. In that case, the polar fields would not reach observed values (see Wang *et al.*, 1989, and Sheeley, 1992). Perhaps this much larger value is caused by effects of giant cells or of correlations in evolution between neighboring cells, which would not be detectable until after many days, a time interval that tracking studies cannot cover (Hagenaar *et al.*, 1999). Another possible contribution to the apparent discrepancy between the inferred diffusion coefficients is that a large number of relatively small but more rapidly moving concentrations coexists with a smaller, but more conspicuous population of larger but slower concentrations, which could cause us to underestimate the dispersal rate by as much as a factor of two to three because of the bias to study primarily the larger concentrations (Schrijver *et al.*, 1996).

5. Long–term Flux Dispersal

The description of flux dispersal by a diffusion equation successfully reproduces observed global patterns of the field. This success is quite surprising, because this model ignores the inherent three–dimensionality of the magnetic and velocity field: the success of the diffusion model implies that once the field is released from the active–region plage, it is moved about completely passively by drag–coupling to the flows that are observed at the surface, ranging from granulation to meridional circulation.

The random–walk transport equation contains three distinct time scales (e.g., DeVore, 1987). In order to estimate them, let us approximate the differential rotation by $\Omega(\theta) = \Omega_{\rm eq} - \Omega_{\rm d} \sin^2\theta$, and the meridional flow by $v(\theta) = v_0 \sin 2\theta$, both as functions of latitude θ. The first of the three time scales is the shearing time scale $\hat{t}_{\rm s} = 2\pi/\Omega_{\rm d}$ associated with the differential rotation. For $\Omega_{\rm d} \approx 2°/{\rm day}$, one finds $\hat{t}_{\rm s} \approx 180$ days. The second time scale is that associated with the transport by the meridional flow: $\hat{t}_{\rm m} = R_\odot/v_0$, which for $v_0 \approx 10$ m/s equals $\hat{t}_{\rm m} \approx 800$ days. The global diffusion time scale is given by $\hat{t}_{\rm d} = R_\odot^2/D$. For a diffusion coefficient of 600 $km^2\ s^{-1}$ this yields $\hat{t}_{\rm d} \approx 9500$ days. If the meridional flow were reduced substantially, a pattern of pole-ward streaks of arched, unipolar areas containing flux of decaying large active regions (or complexes) would still form by the balance of diffusion and differential rotation. The hybrid time scale important for this pattern is given by $\hat{t}_{\rm h} = (\hat{t}_{\rm s}\hat{t}_{\rm d})^{1/2} \approx 1300$ days, i.e of the same order of magnitude as $\hat{t}_{\rm m}$. These time scales suggest that the large–scale patterns are primarily governed by $\hat{t}_{\rm s}$ and $\hat{t}_{\rm m}$, as most flux does not survive long enough to diffuse over a substantial fraction of the solar surface. This is confirmed by simulations. Sheeley (1992), for example, find that the time scale to reach the observed patterns after injecting the bipoles as observed, but starting with a field–free photosphere is approximately two years. Diffusion removes small scales and high gradients most rapidly. This is born out by the simulations by Sheeley (1992): only the largest 300 out of the total 2500 input active regions turn out to be important to the large-scale patterns, because only the flux in the largest regions has a reasonable chance of surviving long enough to be part of the large–scale pattern.

6. Surface Activity on Stars

In many ways the Sun is a typical cool star that we would like to use as an example to understand the dynamo action and outer–atmospheric radiative losses of other cool stars. In this review, we concentrate on two topics: what do we know about *(i)* the photospheric flux budget of cool stars, and *(ii)* the surface distribution of the field on their surface?

One of the hints towards addressing these questions lies in the remarkable tightness of the non–linear relationships between radiative losses from different temperature domains in stellar outer atmospheres. Schrijver & Harvey (1989) argued that data for the spatially resolved Sun throughout the cycle transform into relationships observed for stars because of how flux is distributed over the solar surface. That distribution depends on the details of the source function, for which disappointingly little is known for stars, and the dispersal process. As we demonstrated in the preceding sections, the latter depends on super-granular dispersal, meridional flow, and differential rotation.

Among these quantities, only the differential rotation has been established with some accuracy for other cool stars. Starspot records of some stars now span a period of over 80 years. The measured relative range in periods, $\delta P/P$, can be converted into a coefficient k of differential rotation, as defined by $\Omega(\theta)/\Omega(\theta = 0) = P(\theta = 0)/P(\theta) = 1 - k \cdot \sin^2(\theta)$, assuming that starspots cover the entire range of latitudes θ (Hall & Busby, 1990). Fortunately, the results are not very sensitive to the actual width of the belts, as long as these are relatively broad.

From data on a sample of 85 single and (close) binary stars covering a range of more than three orders of magnitude in rotation period P, Hall (1991) finds that $\log k = (-2.30 \pm 0.06) + (0.85 \pm 0.06) \log P(\theta = 0)$, for P in days, with an intrinsic range about the mean of an order of magnitude. This result suggests a small value of k and thus nearly rigid rotation *in relative terms* for rapid rotators. The time it takes to wrap a field line one full turn around a differentially rotating star, however, is the shearing time scale

$$\hat{t}_{\rm s} = \frac{P(\theta = 0)}{k} \approx 200 \cdot P^{0.15}(\theta = 0),$$

(for $\hat{t}_{\rm s}$ and P in days) which is remarkably insensitive to the stellar rotation period. A star with a rotation period of 1 day would wind up field lines in fact 1.7 times *faster* than a star with a rotation period of 30 days, despite the much smaller value of k. A similar study by Donahue *et al.* (1996) based on the spread δP in rotation periods observed in a sample of about 30 stars in the Mt. Wilson Ca II H+K program resulted in $\delta P/P \propto P^{0.3 \pm 0.1}$, implying a weaker dependence of k on rotation rate, and thus a somewhat stronger increase of the shearing time scale with P. It remains to be seen whether this difference is significant, and if so, whether it reflects different behavior between relatively slowly rotating single main–sequence stars that make up the sample of Donahue et al. and the RS CVn systems that dominate the sample of Hall.

The apparent insensitivity of the shearing time scale to the stellar rotation rate may be an ingredient of the explanation for the tightness of the relationships between stellar activity diagnostics. In order to obtain similar large–scale patterns of flux on stars other than the Sun, the four time scales that are important for the large–scale dispersal of magnetic flux over the solar surface (the time scales for shearing $\hat{t}_{\rm s}$, meridional flow $\hat{t}_{\rm m}$, random–walk diffusion $\hat{t}_{\rm d}$, and the hybrid time scale for shear and diffusion $\hat{t}_{\rm h}$) probably need to maintain the same order as on the Sun. Using the above result by Hall, this implies:

$$1 \lesssim \left(2.5 \frac{R_*}{R_\odot} \frac{v_{0,\odot}}{v_{0,*}} (\frac{P_\odot}{P_*})^{0.15}, 5.4 \frac{R_*}{R_\odot} (\frac{D_\odot}{D_*})^{\frac{1}{2}} (\frac{P_\odot}{P_*})^{0.075}\right) \lesssim 30 (\frac{R_*}{R_\odot})^2 \frac{D_\odot}{D_*} (\frac{P_\odot}{P_*})^{0.15},$$

for a meridional flow v_0 and a diffusion coefficient D for flux dispersal. These inequalities are rather insensitive to the value of the rotation period, at least where the value enters directly: the meridional flow may well depend on the rotation rate, while for the most rapid rotators even the super-granular velocity field might be affected. If we assume $v_{0,*}$ and D_* to be the same for all G2V stars like the Sun, these inequalities hold for rotation periods up to 500 $P_\odot$. Thus effectively these inequalities would hold for all truly Sun–like stars with any significant level of activity. Unfortunately, the meridional flow $v_{0,*}$ and the dispersal coefficient D_* are at present unknown for stars other than the Sun.

Little is known about the other important ingredient in the flux–flux relationships, namely the distribution of the flux over regions of different sizes. From the modulation of chromospheric signals, we know that stellar surfaces are inhomogeneously covered with evolving fields, sometimes exhibiting persistent zones of activity: active–region nests appear to survive for up to several months in cool stars, and active longitudes persist for decades in tidally interacting binaries – see the discussion in Schrijver & Zwaan (1999). But because we cannot resolve stellar surfaces, it is hard to even estimate the global flux budget, let alone derive the flux spectrum. The problem is that the flux present in the solar photosphere (which we can quantify fairly well, see the review by Schrijver, 1991) is only an indirect measure of the rate at which flux is being processed (Schrijver & Harvey, 1994): in a thoroughly mixed–polarity environment, collision – and therefore cancellation – rates depend on the square of the number density of flux concentrations, while at the interface of large–scale patterns this drops to a proportionality with number density. The global cancellation rate therefore depends on the geometry of the surface patterns. This cannot be observed for stars. So if measuring the dynamo strength involves establishing a flux emergence rate, we face a serious problem, because that quantity cannot be estimated in a straightforward way from the amount of flux present in the photosphere, but would involve knowledge of sources of new flux.

A complication in this is that whereas we may hope to "image" stellar surfaces well enough to resolve at least the relatively large active regions, we will still miss the smaller ones, particularly the ephemeral regions which – as we discussed – do have an important role in the formation of the quiet Sun and as a driver behind its non–thermal outer–atmospheric emission. At the coarse angular resolution on which the surface diffusion model applies, much of the total flux is missed, because the mixed–polarity field results in numerical cancellation of flux of opposite polarities within resolution elements. The difference can be substantial. In an inactive phase of the 11-year solar activity cycle most of the solar surface was covered by a mixed–polarity field in which fluxes in the two polarities balanced to about a tenth of a gauss at a resolution of about 50,000 km, whereas the average absolute flux density at the spatial resolution of $\approx$ 900 km, was $\approx$ 2 G, i.e., an order of magnitude greater! This big difference may prove important for our understanding of stellar dynamo processes, yet it remains unclear whether stellar observations contain sufficient indirect information to quantify the small end of the flux emergence spectrum. It seems that the extension of flux–flux relationships to stars even less active than the Sun requires substantial coverage of these stars by small active or ephemeral regions, much like the solar quiet network.

We do not yet know where the solar analogy breaks down, but the parameter domain for the ensemble of cool stars certainly exceeds that of the Sun. For example, some active stars have disk–integrated emissions well above the average emission from non–flaring solar active regions. And many of the most rapidly rotating stars show evidence of large starspots covering up to half of a hemisphere, or occurring at or near the rotational poles. Nevertheless, all of these stars obey the same non–linear relationships between radiative losses from different temperature domains in their outer atmosphere, suggestive of some similarity in the dynamics of the field and its distribution over the disk. Detailed solar studies are required in order to answer at least part of this unresolved riddle, charging the solar physicist in particular with the study of the small–scale magnetic fields, and the impact they have on outer–atmospheric emissions through processes of reconnection and field–line braiding.

References

DeVore, C.R.: 1987, Solar Phys., 112, 17

Donahue, R.A., Saar, S.H. & Baliunas, S.L.: 1996, ApJ, 466, 384
Durney, B. R.: 1988, A&A, 191, 374
Gaizauskas, V., Harvey, K.L., Harvey, J.W. & Zwaan, C.: 1983, ApJ, 265, 1056
Hagenaar, H.J., Schrijver, C.J., Title, A.M. & Shine, R.A.: 1999, ApJ, in press
Hall, D.S.: 1991, in I. Tuominen, D. Moss & G. Ruediger (Eds.), *The Sun and Cool Stars: activity, magnetism, dynamos*, IAU Coll. No. 130, Springer Verlag, Berlin, Germany, p. 352
Hall, D.S. & Busby, M.R.: 1990, in C. Ibanoglu (Ed.), *Active close binaries*, Kluwer, Dordrecht, The Netherlands, 377
Harvey, K. L.: 1993, *Ph.D. thesis*, Astronomical Institute, Utrecht University
Leighton, R. B.: 1964, ApJ, 140, 1547
Livingston, W.C. & Harvey, J.: 1971, in R. Howard (Ed.), *Solar Magnetic Fields*, IAU Symp. No. 43, Reidel, Dordrecht, 51
Livingston, W.C. & Orrall, F.Q.: 1974, Solar Phys., 39, 301
Martin, S.F.: 1990, in J.O. Stenflo (Ed.), *Solar photosphere: structure, convection, and magnetic fields*, IAU Symp. 138, Kluwer Academic Publishers, Dordrecht, The Netherlands, p. 129
Parker, E.N.: 1994, *Spontaneous Current Sheets in Magnetic Fields*, Oxford University Press, Oxford, U. K.
Schrijver, C.J.: 1989, Solar Phys., 122, 193
Schrijver, C.J.: 1991, in P. Ulmschneider, E. Priest & R. Rosner (Eds.), *Mechanisms of Chromospheric and Coronal Heating*, Springer–Verlag, Heidelberg, p. 257
Schrijver, C.J.: 1995, Astron. Astrophys. Rev., 6, 181
Schrijver, C.J. & Harvey, K.L.: 1989, ApJ, 343, 481
Schrijver, C.J. & Harvey, K.L.: 1994, Solar Phys., 150, 1
Schrijver, C.J. & Martin, S.F.: 1990, Solar Phys., 129, 95
Schrijver, C.J., Shine, R.A., Hagenaar, H.J., Hurlburt, N.E., Title, A.M., Strous, L.H., Jefferies, S.M., Jones, A.R., Harvey, J.W. & Duvall, Th.L: 1996, ApJ, 468, 921
Schrijver, C.J., Title, A.M., van Ballegooijen, A.A., Hagenaar, H.J. & Shine, R.A.: 1997, ApJ, 487, 424
Schrijver, C.J., Title, A.M., Harvey, K.L., Sheeley, N.R., Wang, Y.M., Van den Oord, G.H.J., Shine, R.A., Tarbell, T.D. & Hurlburt, N.E.: 1998, Nature, 394, 152
Schrijver, C.J. & Zwaan, C.: 1999, *Solar and Stellar Magnetic Activity*, Cambridge University Press, Cambridge, U.K.
Sheeley, N. R.: 1992, in K. L. Harvey (Ed.), *The solar cycle*, A.S.P. Conf. Series, Vol. 27, San Francisco, p. 1
Simon, G.W., Title, A.M. & Weiss, N.O.: 1995, ApJ, 442, 886
Spruit, H.C., Nordlund, Å. & Title, A.M.: 1990, ARA&A, 28, 263
Spruit, H.C., Title, A.M. & van Ballegooijen, A.A.: 1987, Solar Phys., 110, 115
van Ballegooijen, A.A.: 1985, ApJ, 298, 421
Wang, H. & Zirin, H.: 1988, Solar Phys., 115, 205
Wang, Y.-M., Nash, A.G. & Sheeley, N.R.: 1989, Science, 245, 712
Wang, Y.-M. & Sheeley, N.R.: 1991, ApJ, 375, 761
Zwaan, C.: 1987, ARA&A, 25, 83
Zwaan, C. & Harvey, K.L.: 1994, in M. Schüssler & W. Schmidt (Eds.), *Solar Magnetic Fields*, Cambridge University Press, Cambridge, UK, p. 27

Solar and Stellar Activity: Similarities and Differences
ASP Conference Series, Vol. 158, 1999
C.J. Butler and J.G. Doyle, eds.

Surface Magnetic Fields of Late-Type Stars

Jean-François Donati

Observatoire Midi-Pyrénées, 14 Av. E. Belin, F-31400 Toulouse, France

Abstract. In this paper, I first review the various existing techniques for measuring magnetic fields at the surface of cool stars. I will then describe the main results obtained to date with these different techniques, as well as the implications for our understanding of how dynamo processes operate in convective envelopes of cool stars. I will finally present some of the most promising directions of research to improve our knowledge of how magnetic fields affect the global evolution of late-type stars.

1. Introduction

The magnetic field is very often a crucial parameter in astrophysical problems. It indeed plays a key role basically everywhere in the universe, at all time and spatial scales, and in particular in stars, during their formation, their evolution and the last stages of their lives.

In the particular case of stars, the first concern is to understand the origin of large-scale magnetic fields, whether they are remnants (i.e. fossil fields) from an older evolutionary stage (like those of some chemically peculiar stars), or whether they are self-generated (i.e. dynamo) fields resulting from turbulent motions in a partly ionised plasma within the stellar envelope (such as those of the Sun and other active late-type stars). Studying large-scale magnetic structures can also inform us on the impact of these fields on various transport processes operating within or around stars (such as convection, turbulence, diffusion, large-scale circulation, internal rotation, accretion or mass loss) and therefore on stellar evolution as well. For cool stars, and the Sun in particular, a detailed and self-consistent picture of how large-scale fields are produced, evolve on both short and long timescales and affect the Sun and stars as well as their direct circumsolar/stellar environment is crucially lacking at the moment, even though of obvious interest for us.

The various existing techniques for detecting and modelling surface magnetic fields on cool non-degenerate stars are all based on the fact that magnetic fields, through the Zeeman effect (see e.g. Mathys 1989 for a detailed description of the atomic physics behind the Zeeman effect), affect the shape and polarisation of spectral line profiles. I will first recall the basic principles of these techniques (using either polarimetry, spectroscopy and spectropolarimetry), along with the main results obtained with each of them. Note that I will not mention the possibility of using the Hanlé effect (e.g. Faurobert-Scholl 1993) nor that based on observing cyclotron/synchrotron emission (e.g. Wickramasinghe & Ferrario 1988), as these methods have never been used to estimate magnetic fields at the surfaces of non-degenerate cool stars other than the Sun (as far as I know). I will outline in particular how these results help us understand how a dynamo operates in the convective envelopes of cool stars. I will then discuss some of the most promising direction of research to improve our knowledge of how magnetic fields affect the global evolution of late-type stars.

2. Conventional Polarimetric Methods

The very first methods that enabled one to diagnose the presence of surface magnetic fields on the Sun (Hale 1908) and stars (Babcock 1947) was aimed at detecting *wavelength shifts* between spectral lines recorded simultaneously in both orthogonal states of circular polarisation (Stokes V) with a high resolution spectrograph including a polarimetric module. When the field is not too strong, it may be shown (e.g. Mathys 1989) that this shift is approximately proportional to the line-of-sight component of the magnetic field vector averaged over the visible stellar disc (called longitudinal field). This shift being very small (about 1 $\mathrm{km\,s^{-1}}$ for a 1 kG longitudinal field on a spectral line at 600 nm with average magnetic sensitivity). This technique is usually only applicable to slow/moderate rotators. The method has progressively been replaced by a much simpler one, involving a narrow band photopolarimeter measuring *circular polarisation* in either wing of a broad spectral line (usually a Balmer line, Angel & Landstreet 1970). The observed circular polarisation is also translated into an estimate of the longitudinal field. Thanks to the large intrinsic width of the spectral line used, this technique (as opposed to the previous one) is just as efficient for slow and rapid rotators.

Another possibility consists of measuring *broadband linear polarisation* in stellar spectra (Leroy 1962; Tinbergen & Zwaan 1980). What one aims at detecting in this case is the net residual linear polarisation (Stokes Q and U) over the whole profile of one (or several) spectral lines, resulting from the *differential saturation* of the π and σ Zeeman components. The broadband filter is usually centred on a spectral domain which maximises the density of spectral lines for the stars of interest. The observed linear polarisation gives access to the disc-integrated component of the field vector perpendicular to the line of sight (called transverse field), as well as its *azimuth* with respect to the instrument angular reference (usually north-south), with an accuracy that does not depend on stellar rotation rate.

All attempts at detecting magnetic fields in solar-type stars other than the Sun with these methods have essentially failed (e.g. Brown & Landstreet 1981; Borra et al. 1984; Leroy & Leborgne 1989; Bedford et al. 1995) despite a few controversial claims (e.g. Huovelin et al. 1985). It is well known now that this failure is mostly attributable to the extremely intricate structure of magnetic fields on late-type stars, which feature multiple bipolar spot pairs cancelling their mutual effect in circular polarisation and yielding undetectable longitudinal fields (see § 4).

3. Stokes *I* Spectroscopy

Another option consists of extracting magnetic information from the *shape* of spectral lines in *unpolarised light* (Stokes I), recorded at high spectral resolution and high signal to noise ratio (Robinson 1980). With this technique, a measurement of the *differential broadening* between lines that are highly and weakly sensitive to magnetic fields yields an estimate of the disc area covered with magnetic fields (called filling factor), as well as an average field strength within these regions. The obvious advantage of this method is that magnetic field regions with opposite polarities at the surface of the star no longer cancel their mutual spectroscopic signatures, as they used to do with conventional polarimetric techniques. This method requires however the observed star to spin slowly enough (with $v \sin i < 8$ km s^{-1}) to avoid rotational drowning of the potential magnetic information contained in line profiles (Saar 1988). When the magnetic field is strong enough, spectral lines can even be split into individual Zeeman components, allowing the mean magnetic field *modulus* (called surface field) to be measured very accurately (Johns-Krull & Valenti 1996). In most cases though, magnetic field signatures in unpolarised light are fairly small and can easily be confused with small blends in line wings. For this reason, magnetic parameters inferred with this method are subject to large systematic errors (as much as a factor of three) as recently evidenced through the results of multiline techniques (e.g. Valenti et al. 1995; Rüedi et al. 1997), which are presumably less prone to systematic errors and thus more reliable.

Spectral lines are also subject to an increase in *equivalent width* in the presence of a magnetic field, whose magnitude depends on the actual Zeeman pattern of the selected line and orientation of field lines (Babcock 1949). However, the dependence of this effect on parameters such as field orientation for instance, is sufficiently complex that this method can only be used to indicate the presence of a field rather than to provide an accurate estimate of its strength.

Both methods have been very successful in detecting, or at least suggesting, the presence of magnetic fields in new types of stars, and in particular late-type stars (Saar 1988) and T Tauri stars (Basri et al. 1992). Altogether, magnetic fields have been detected with these methods in about 50 objects up to now. From a critical selection of all the newest measurements obtained with these methods, Saar (1996) confirms earlier claims that the area covered by magnetic regions scales up with stellar rotation rate, while the magnetic strength within these regions is roughly equal to the equipartition value (up to a unit filling factor at least). This result is taken as strong evidence that magnetic fields of active stars are indeed produced through dynamo processes.

The major drawback of these techniques is that they only provide extremely limited spatial information on the details of the actual magnetic topology. In an attempt to solve this problem, Saar et al. (1994) proposed to couple these techniques with a stellar surface imaging package and get magnetic maps of fast rotators. This new method is however still subject to caution as no simulation has been published to date to demonstrate unambiguously that the tiny magnetic broadening/amplification effects can indeed be reliably recovered in Doppler broadened line profiles. Moreover, none of these methods enables one to derive information on the *orientation* of field lines within magnetic regions.

4. Stokes Q, U, and V Spectropolarimetry of Rapid Rotators

Recently, Semel (1989) proposed to measure, in the particular case of rapidly rotating late-type stars, the full profiles of magnetic Zeeman signatures in circular and (whenever possible) linear polarisation, rather than their lowest moment as in conventional polarimetric techniques (see §. 2). For such stars, regions of opposite magnetic polarities located at different longitudes on the stellar surface (and therefore at different radial velocities with respect to the observer) contribute to the global Zeeman signature at different wavelengths and therefore no longer mutually cancel their effects in circular (or linear) polarisation.

Optimally, one would like to measure Zeeman signatures in individual line profiles, as one can do for solar magnetic regions for instance. However, in the particular case of late-type stars where Stokes V signatures rarely exceed 0.3% in relative peak-to-peak amplitude, one has to extract the desired polarisation information from thousands of spectral lines simultaneously, with the help of cross-correlation tools such as 'Least-Squares Deconvolution' (Donati et al. 1997). Magnetic fields have been detected in about 20 objects of various evolutionary stages up to now (e.g. Donati et al. 1992a, Donati et al. 1997) and mapped in six of them (e.g. Donati & Cameron 1997; Donati 1998). As an illustration, Figure 1 shows the rotationally modulated Stokes V signature (extracted from about 1,500 spectral lines with Least-Squares Deconvolution) for the rapidly rotating young K0 dwarf AB Doradus (Donati et al. 1998a).

Simulations have established that, with this method (often referred to as Zeeman-Doppler imaging or ZDI), one can convert, using a stellar surface imaging package, sets of rotationally modulated Zeeman signatures into surface maps of the vector magnetic field. Note that both the spatial distribution of magnetic regions and, to a certain extent, the *orientation* of field lines within them can in principle be recovered with this method (Brown et al. 1991; Donati & Brown 1997).

Results from ZDI (Donati et al. 1992b; Donati & Cameron 1997; Donati et al. 1998a; Donati 1998) reveals that magnetic images of rapidly rotating cool stars are indeed extremely complex (see for instance Figure 2 for the K1 subgiant HR 1099), explaining *a posteriori* the failure of conventional polarimetric methods on late-type stars. A particularly intriguing characteristic of these images is that they often include magnetic regions in which the field is mainly *azimuthal*, i.e. directed along lines of constant latitude. The polarity of these azimuthal field features is essentially longitude independent at a given latitude (e.g. clockwise and counterclockwise at high and low latitudes respectively on HR 1099, see Figure 2) while the latitude dependence is constant on a year-to-year basis (Donati 1998). These features are interpreted as the detection, at photospheric level, of the *toroidal component* of the mostly axisymmetric large-scale dynamo field. Observations also tell us that the latitudinal structure of this toroidal large-scale field component gets more complex for increasing rotation rates and convective depths, featuring for instance up to three rings of opposite polarities in the upper hemisphere of the ultra fast rotator AB Dor (Donati et al. 1998a).

In principle, such images could be used to study the temporal evolution of the magnetic field structure as the star evolves on its activity cycle, in a way very similar to what is done for the Sun. Data already suggest that toroidal

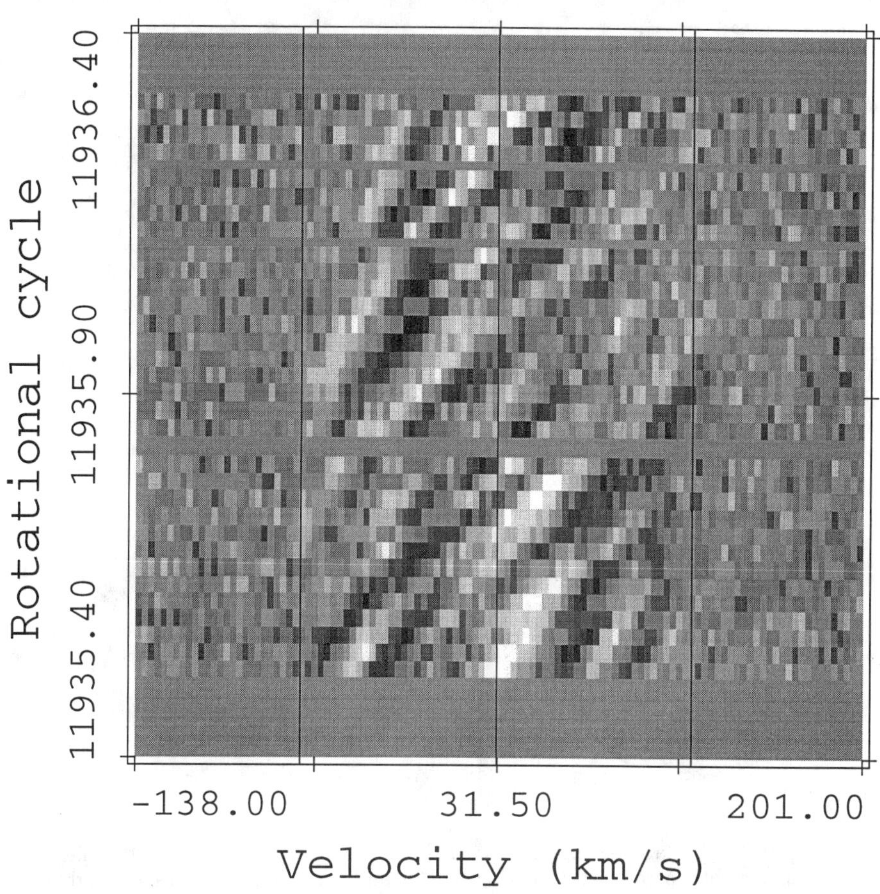

Figure 1. Rotationally modulated Least-Squares Deconvolved Stokes V signature of AB Dor in 1996 December. Black and white code relative circular polarisation levels of −0.06 and 0.06% respectively. The central and side vertical lines depict the radial and rotational velocities of AB Dor

components of large scale field structures do indeed evolve on a timescale of about a decade; the recent detection of a new azimuthal field polarity appearing at high latitudes on the young active star LQ Hydrae (confirmed with new observations in 1998 Jan., see Figure 3) may for instance be a forerunner of a global polarity switch in the magnetic field of this object (Donati 1998).

The detection at the photospheric level of the toroidal component of the large-scale dynamo field is interpreted as strong evidence that very active late-type stars trigger dynamo processes which are *distributed* throughout the whole convective envelope, rather than being confined to an interface layer with the radiative interior as in the Sun. An independent confirmation of this conclusion is that rising flux tubes from a deep-seated magnetic structure located at the base of the convective zone are expected to emerge at latitudes higher than 50° or so in rapidly rotating late-type stars (Schüssler et al. 1996), in strong

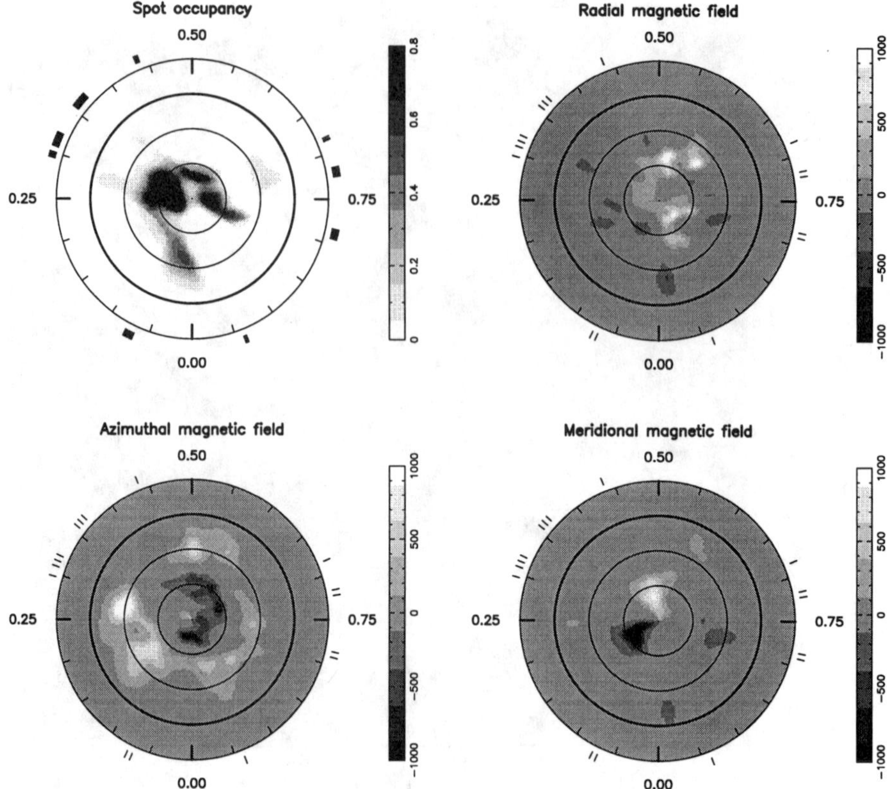

Figure 2. Flattened polar view of the brightness (upper left), radial field (upper right), azimuthal field (lower left) and meridional field (lower right) distributions at the surface of the K1 subgiant of HR 1099 at epoch 1995.94. Parallels are shown (as concentric circles) every 30° down to a latitude of −30° (the bold one featuring the equator). Magnetic fields are labelled in G. The radial ticks around each plot depict the rotational phases of observations.

contradiction with the brightness, radial and azimuthal field maps we reconstruct for these objects that clearly show features at latitudes lower than 30° (Donati & Cameron 1997; Donati et al. 1998a; Donati 1998). Furthermore, the orbital period fluctuations detected for RS CVn systems (and interpreted as evidence for changes in the quadrupole moment of the primary star, Applegate 1992) also argue in favour of this picture; generating the observed quadrupole moment changes indeed requires the associated change in the internal rotation of the primary star (and therefore also the underlying dynamo processes producing them) to be distributed in a large fraction of the convective zone (Lanza et al. 1998).

In addition to this, magnetic regions detected on late-type stars do not apparently correlate with photospheric brightness and are not confined within a restricted latitude range, making them more similar to the smallest scale magnetic structures on the Sun (i.e. intranetwork fields) than to the large scale

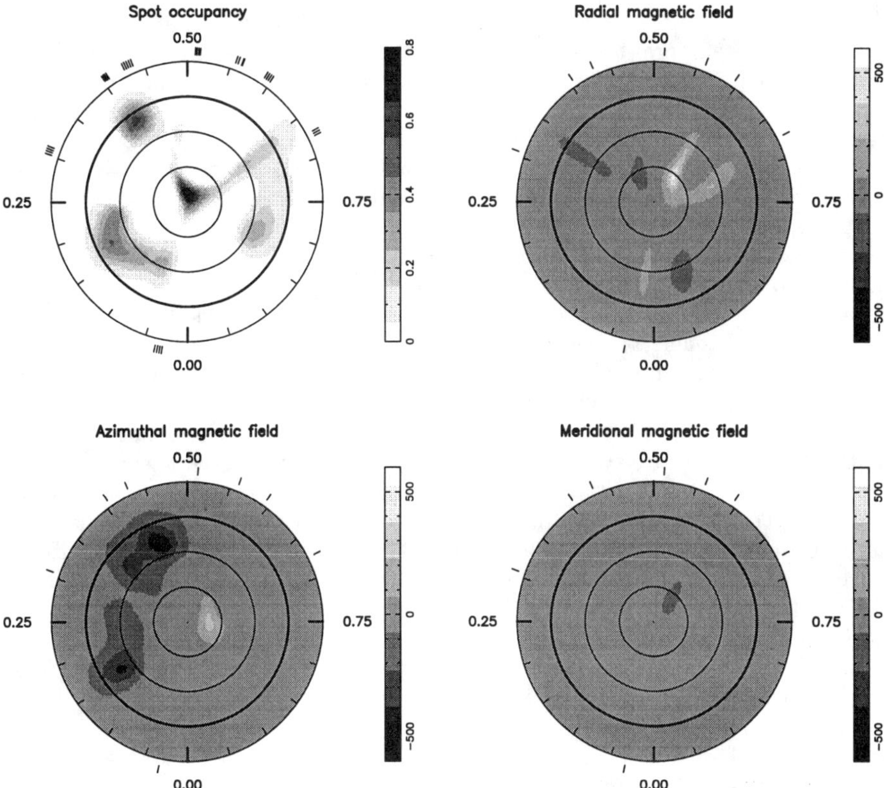

Figure 3. Same as Figure 2 for the young K0 dwarf LQ Hya at epoch 1998.03. Note the new azimuthal field polarity (showing up as a high latitude region of positive, i.e. counterclockwise, field) that recently appeared at the surface of this star.

active/ephemeral regions (Donati 1998). It may in turn indicate that these small-scale solar structures are solar analogues of the magnetic regions detected on rapidly rotating late-type stars, and therefore contribute to the solar magnetic cycle and to the large-scale field of the Sun (as suggested by Stenflo 1994) just as their scaled-up versions obviously do on rapidly rotating late-type stars. If confirmed, it may imply at the same time that a significant fraction of the large scale magnetic field of the Sun is also produced through a distributed dynamo.

5. Prospectives

Our knowledge of both the intensity and structure of the magnetic field of late-type stars other than the Sun has considerably improved in the last decade. The most recent results demonstrate in particular that we are now able to study the temporal evolution of the large scale magnetic structures of such stars as they evolve through their activity cycle, in a way very similar to that used in the particular case of the Sun. One of the obvious observational goals is thus to follow this temporal evolution for a small sample of carefully selected active stars (sampling in particular all evolutionary states from pre-main sequence to red-giant branch stage) and derive from such time sequences of magnetic images some detailed information on where and how dynamo processes operate within the convective layers of late-type stars. Simultaneous monitoring of orbital period fluctuations (Donati 1998) or surface differential rotation (see e.g. Figure 4, Donati & Cameron 1997; Donati et al. 1998a) can also give additional information on how angular rotation is distributed within convective zones and how this distribution is affected as the star evolves on its activity cycle. Altogether, we should learn how dynamo fields react to the internal velocity fields that generated them, and more generally, on how magnetic fields affect various transport processes in the interior of late-type stars (e.g. convection, turbulence, circulation) and thus the whole dynamics of convective zones.

Another goal is to understand the role and interaction of the magnetic field at the surface of a late-type star on the immediate circumstellar environment. One possibility consists in studying how dynamo field structures participate in producing, confining and ejecting coronal prominences around rapidly rotating stars (Cameron & Robinson 1989a, b; Donati et al. 1998a) where the centrifugal force plays a very important role (Jardine & Cameron 1991), and comparing the results to those obtained on the Sun. Another very promising direction of research would be to assess observationally the role of magnetic fields in the interaction of pre-main sequence stars (classical T Tauri stars, FU Ori objects) with their direct environment (circumstellar discs, jets, e.g. Ferreira & Pelletier 1993; Cameron & Campbell 1993) and thus on the global structure and rotation evolution of these stars. To detect magnetic fields of *accretion discs* (in which all ingredients for a vigorous dynamo are present, e.g. Horne & Saar 1991) is also an important issue, which should help to constrain the exotic shear-flow dynamo mechanisms invoked for these objects (e.g. Brandenburg et al. 1995).

To achieve these goals, we need to be able to observe new classes of objects, and in particular classical T Tauri stars or late-type stars in young open clusters, targets that are still hardly observable with the existing techniques and instruments. There is definitely a need for more efficient instruments, that can collect spectra in all polarisation states (circular and linear), with the widest possible spectral domain (possibly up to the infrared where the sensitivity to magnetic fields is stronger) and highest possible spectral resolution (up to 200,000 optimally). The new spectropolarimeter that CFHT has started building (called ESPaDOnS), which should yield full optical spectral coverage (i.e. 370 to 1,000 nm) at a spectral resolution of over 50,000 in a single exposure with 20% peak efficiency (atmosphere, telescope and detector included) is clearly a first step in this direction (Donati et al. 1998b). New generation instruments on 8 m telescopes will nonetheless be necessary to achieve our ultimate goals.

New magnetic imaging techniques must be explored as well. Computing spherical harmonic coefficients (up to orders 50 or so) from fitting sets of rotationally modulated Zeeman signatures is certainly an interesting possibility; in addition to providing an independent check for surface field maps obtained with the present imaging tools, it can also be used to derive an estimate of the 3D topology of the (assumed potential) field structure (as done in the particular

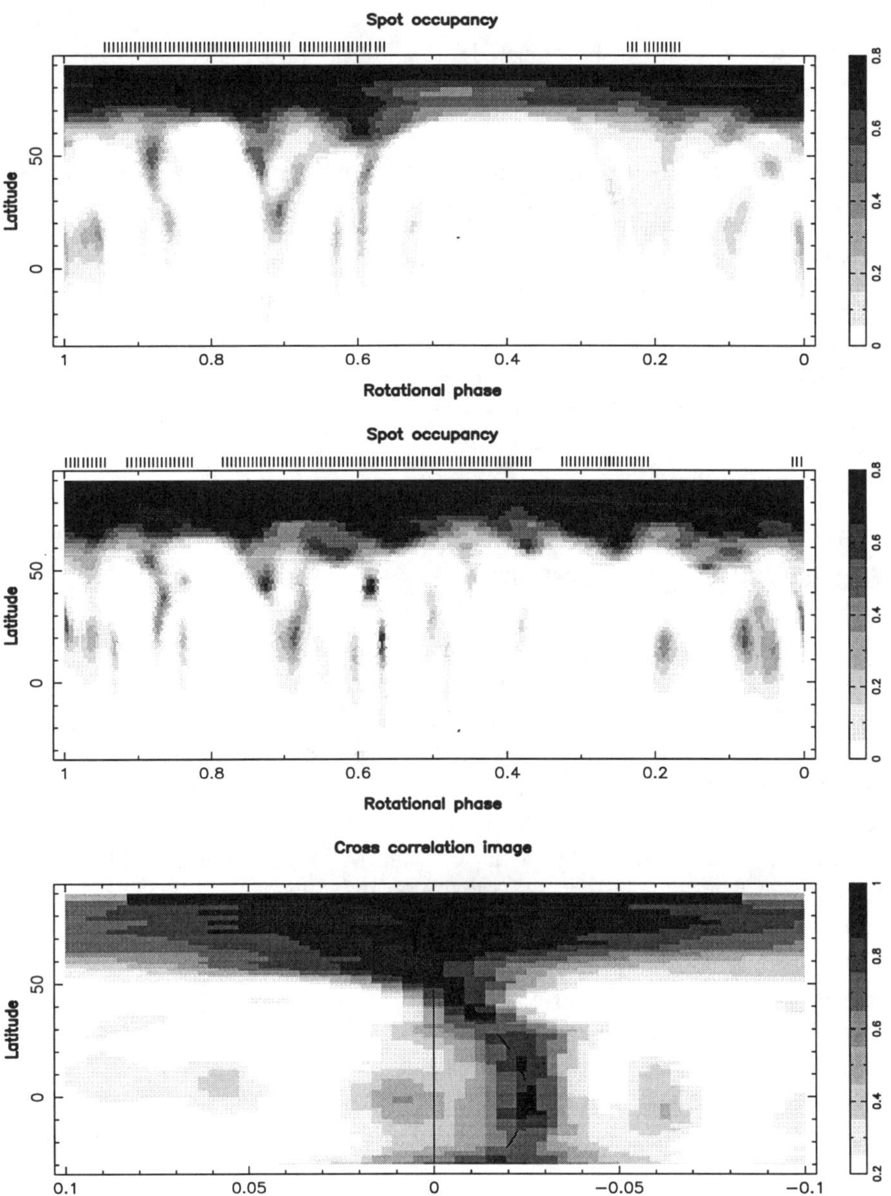

Figure 4. Cross-correlation image (lower panel) obtained by cross-correlating latitude belts of two brightness images (upper panels, rectangular projection) of AB Dor obtained 4 d (i.e. 8 rotation cycles) apart in 1995 Dec. Note that cross-correlation is performed on the fraction of the rotational cycles that was continuously monitored at both epochs, i.e. on phase interval [0.55,0.90] (phases of observations being indicated by vertical ticks above brightness images).

case of the Sun), and allow us to check at the same time (from the accuracy level at which data can be adjusted) how far the observed field departs from a potential configuration. These developments should be very interesting for studying the interaction of the large scale field structure with the immediate circumstellar environment.

All these results should improve our knowledge of how magnetic fields are produced and react on some of the most crucial transport processes operating within late type stars, and eventually bring us to the point where existing theories (on e.g. dynamo processes, prominence dynamics, coronal heating, accretion/ejection phenomena) that were specifically built to explain solar observations (the only available data at that time) will need to be reassessed and most likely revised in the broader context of both solar and stellar observations. Ultimately, we should be able to get a better insight on how magnetic fields influence the evolution of late-type stars.

References

Angel J.R.P. & Landstreet J.D., 1970, ApJ 160, L147

Applegate J.H., 1992, ApJ 385, 621

Babcock H.W., 1947, ApJ 105, 105

Babcock H.W., 1949, ApJ 110, 126

Basri G., Marcy G.W. & Valenti J.A., 1992, ApJ, 390, 622

Bedford D.K., Chaplin W.J. & Davies A.R., et al., 1995, A&A 293, 377

Borra E.F., Edwards G. & Mayor M., 1984, ApJ 284, 211

Brandenburg A., Nordlund Å., Stein R.F. & Torkelsson U., 1995, ApJ 446, 741

Brown D.N., Landstreet J.D., 1981, ApJ 246, 899

Brown S.F., Donati J.-F., Rees D.E. & Semel M., 1991, A&A 250, 463

Cameron A.C. & Campbell C.G., 1993, A&A 274, 309

Cameron A.C. & Robinson R.D., 1989a, MNRAS 236, 57

Cameron A.C. & Robinson R.D., 1989b, MNRAS 238, 657

Donati J.-F., 1998, MNRAS (in press)

Donati J.-F. & Brown S.F., 1997, A&A 326, 1135

Donati J.-F., Brown S.F., Semel M., et al., 1992b, A&A 265, 682

Donati J.-F. & Cameron A.C., 1997, MNRAS 291, 1

Donati J.-F., Cameron A.C., Hussain G.A.J. & Semel M., 1998a, MNRAS (in press)

Donati J.-F., Catala C. & Landstreet J.D., 1998b, in: Martin P., Rucinski S. (eds.), 'proceedings of the "fifth CFHT users' meeting" (in press)

Donati J.-F., Semel M., Carter B., Rees D. & Cameron A.C., 1997, MNRAS 291, 658

Donati J.-F., Semel M. & Rees D., 1992a, A&A 265, 669

Faurobert-Scholl M., 1993, A&A 268, 765

Ferreira J. & Pelletier G., 1993, A&A 276, 625

Hale G.E., 1908, ApJ 28, 315

Horne K. & Saar S.H., 1991, ApJ 374, L55

Huovelin J., Linnaluoto S., Piirola V., Tuominen I. & Virtanen H., 1985, A&A 152, 375

Jardine M. & Cameron A.C., 1992, Solar Phys. 131, 269

Johns-Krull C.M. & Valenti J.A., 1996, ApJ 459, L95

Lanza A.F., Rodonò M. & Rosner R., 1998, MNRAS 296, 893

Leroy J.L., 1962, Ann. Ap. 25, 127

Leroy J.L. & Leborgne J.F., 1989, A&A 223, 336

Mathys G., 1989, Fund. Cos. Phys. 13, 143

Robinson R.D., 1980, ApJ 239, 961

Rüedi I., Solanki S.K., Mathys G. & Saar S.H., 1997, A&A 318, 429

Saar S.H., 1988, ApJ 324, 441

Saar S.H., 1996, in: Strassmeier K.G. & Linsky J.L. (eds.), IAU Symp. 176, "Stellar Surface Structure". Kluwer Academic Publishers, Dordrecht, p. 237

Saar S.H., Piskunov N.E. & Tuominen I., 1994, in: Caillault J.-P. (ed.), 8th Cambridge Workshop on "Cool Stars, Stellar Systems and the Sun". ASP Conf. Series 64, p. 661

Schüssler M., Caligari P., Ferriz-Mas A., Solanki S.K. & Stix M., 1996, A&A 314, 503

Semel M., 1989, A&A 225, 456

Stenflo J.O., 1994, in: Ap&SS Lib. 189, "Solar Magnetic Fields: polarised radiation diagnostics". Kluwer, Dordrecht

Tinbergen J. & Zwaan C., 1980, A&A 101, 223

Valenti J.A., Marcy G.W. & Basri G., 1995, ApJ 439, 939

Wickramasinghe D. & Ferrario L., 1988, ApJ 334, 412

Solar and Stellar Activity: Similarities and Differences
ASP Conference Series, Vol. 158, 1999
C.J. Butler and J.G. Doyle, eds.

Magnetic Activity and Variability in Stellar Coronae (EUV)

M. Mathioudakis

Department of Pure and Applied Physics, Queen's University of Belfast, Belfast, BT7 1NN, N. Ireland

Abstract. The high time and spectral resolution achieved in the Extreme Ultraviolet in recent years, have opened a new era in the study of stellar coronae. Photometric observations show that coronal saturation occurs at much higher rotational velocities than chromospheric saturation. The high temperature tails ($T > 10^7$ K) that appear with the emission measure analysis, result from the simultaneous fit to lines and the non-zero "continuum" flux in the 70 – 130Å range. In the relative "cool" coronal sources such as Procyon and α Cen, the high temperature tails are not real and alternative interpretations have been proposed. The opacity has been the most controversial of these interpretations. The electron densities/pressures derived for the 10^7 K plasma, are much higher than those derived at 10^5 K, indicating that the transition region and coronal emission may arise in physically unconnected regions.

1. Introduction

Astrophysical research in the Extreme Ultraviolet (EUV 100 - 912Å), covers a wide range of objects and astrophysical problems from solar system sources, to late-type stars, to the interstellar medium and Active Galactic Nuclei. The strong attenuation of light at EUV wavelengths, limits the number of sources to those located nearby. Since its launch in June 1992, the *Extreme Ultraviolet Explorer* (*EUVE*) carried out extensive photometric and spectroscopic observations in the EUV (Bowyer & Malina 1996). The vast majority of the sources that has been observed are late-type stars. Spectroscopic observations in the EUV, have opened a new era in the study of cool stars. For the first time, we have achieved sufficiently high spectral resolution to resolve individual spectral lines formed in the $10^5 - 10^7$ K temperature range. The line to continuum ratios, elementals abundances, radiative losses and electron densities of the upper transition region and corona can now be determined. The analysis and interpretation of these observations has also led to a number of controversies. Here we review some of the results and controversies in the field of stellar coronae.

2. Coronal versus Chromospheric Saturation

One of the primary objectives of the *EUVE* mission was to carry out an all-sky photometric survey in four passbands covering the wavelength ranges 60

– 200Å (Lex/B), 160 – 240Å (Al/Ti/C), 345 – 605Å (Ti/Sb/Al) and 500 – 740Å (Sn/SiO). The all-sky survey was complemented by a deep survey (DS) in a $2^o \times 180^o$ strip of the sky along the ecliptic equator. The combination of higher effective area and longer exposure times, made the DS $\approx$ 20 times more sensitive than the all-sky survey. We point out that in 1990 – 1991, the Wide Field Camera (WFC) which was flown as part of the *ROSAT* mission, was used to survey the shorter EUV wavelengths 60 – 200Å (Pounds et al. 1993). The photometric observations have been especially prolific for the study of the EUV coronal emission as a function of fundamental stellar parameters. The ratio of rotational period (P_{rot}) to convective overturn time (τ_c) known as the Rossby number (R_O), was shown to be a better parameter for describing the levels of activity instead of the rotation period alone, when stars with different spectral types are considered in a single rotation-activity relation. This is because in F&G stars the convective turnover time has a steep dependence on color and is the parameter dominating R_O, whereas in later spectral types the rotation period dominates due to the weak dependence of τ_c on color (Noyes et al. 1984).

Mathioudakis et al. (1995) selected a group of main-sequence stars with known rotational periods and examined the activity-rotation relation in the EUV. A comparison with the chromospheric emission was also made (Figure 1). It is evident that the Mg II emission shows the effects of chromospheric saturation at log $R_O \approx 0$. This would correspond to $P_{rot} \sim 10$ days for an early G dwarf and $P_{rot} \sim 25$ days for a dM4 star. The Rossby diagram for the same group of stars shows no evidence for saturation in the *EUVE* Lex/B band. If coronal saturation exists in this sample, it occurs at considerably lower R_O/low P_{rot} than chromospheric saturation. The Rossby diagrams for open clusters have shown very clearly the effects of saturation and super-saturation in X-rays (see papers by R. Stern and R. Jeffries in these proceedings). There is a possibility that the EUV emission may behave differently from the X-rays. Houdebine et al. (1996) have shown that saturation in spectral lines may not necessarily mean saturation in the magnetic levels of activity, as the radiative losses in the E(UV) continuum are much larger than the radiative losses in the strongest chromospheric and transition region lines. Unfortunately, there is an insufficient number of fast rotating field dwarfs (Figure 1a) to answer this question. Rotational velocities of over 50 newly identified K & M stars with $L_{EUV}/L_{bol} \approx 10^{-3} - 10^{-4}$ will allow us to fully address this issue.

3. Low Activity Dwarfs - Brendan's favourites

Slowly rotating stars are found at the other end of the rotation activity relation. At the lowest levels of activity, the question arises: *Is there a point where chromospheric and/or coronal emission ceases to exist ?* The extensive surveys carried out with *IUE*, have shown that every late-type dwarf examined has chromospheric emission. The basal chromosphere was therefore suggested, which extends down to the latest spectral types (Schrijver 1987, Mathioudakis & Doyle 1992). A complete *ROSAT* X-ray survey of all K&M dwarfs within 6 pc has shown that 97% of the objects were detected, (63 out of 65, Schmitt et al. 1995), providing strong evidence that coronal formation is universal amongst late-type dwarfs. The PSPC spectra of low activity dwarfs are dominated by

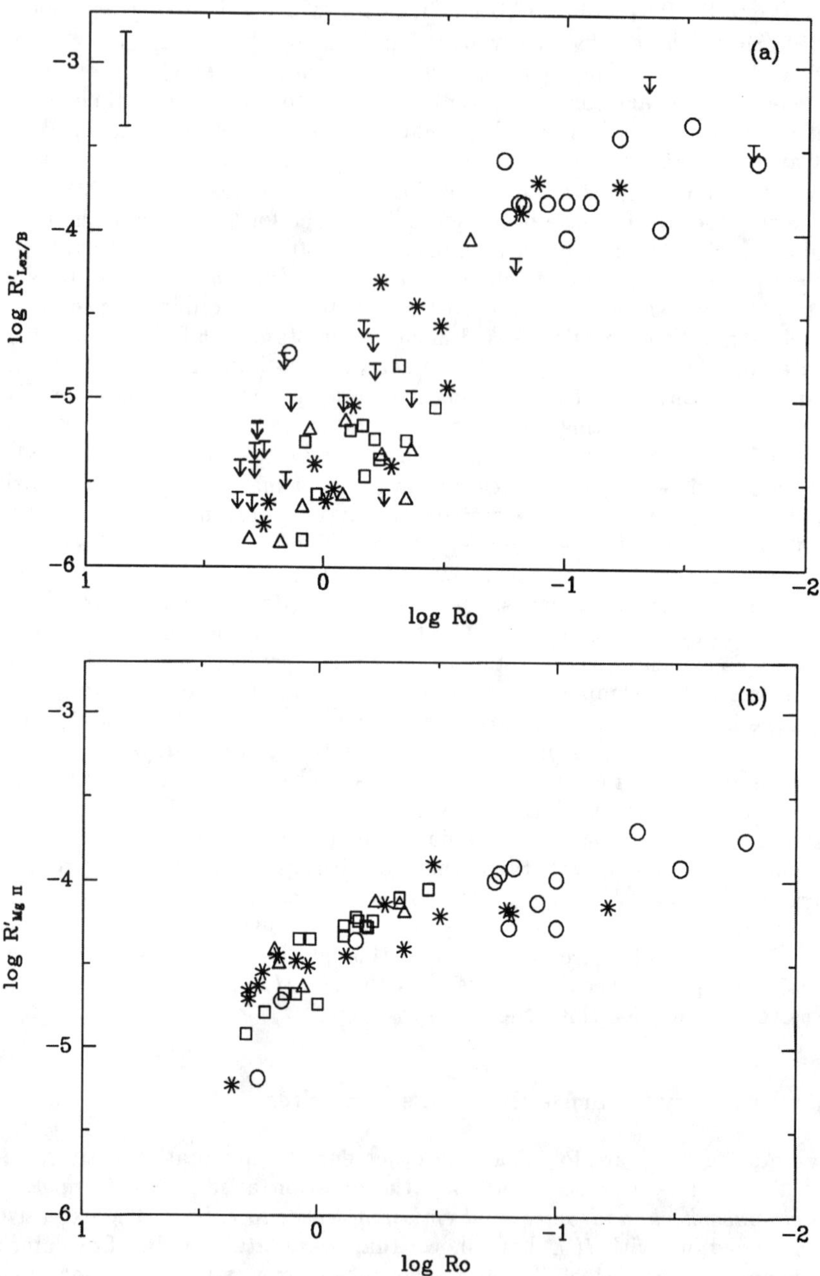

Figure 1. The Rossby number as a function of (a) $R'_{Lex/B}$ and (b) $R'_{Mg\ II}$, where R′ is the luminosity in these bands normalized to the bolometric luminosity. The symbols correspond to △ dF, □ dG, ∗ dK and ◯ dM stars.

a relatively soft component with a temperature of $\sim 10^6$ K (Giampapa et al. 1996). This cool component is also apparent in a simple comparison of EUV and X-ray filter ratios (Mathioudakis et al. 1995).

The study of low activity dwarfs was amongst Brendan's research interests. His favourite star was Gl 105B, a dM4 star with the weakest Mg II emission. A model atmosphere for Gl 105B indicates a T_{min} of less than 2650 K and a thin and steep chromosphere. Despite a 20ksec integration with the *ROSAT*/HRI instrument, Gl 105B was not detected (Doyle et al. 1998). Since the derived X-ray upper limit is higher than the weakest fluxes in the Schmitt et al. sample, the non-detection is most likely due to the reduced sensitivity of the HRI (compared to the PSPC) rather than the lack of a corona.

4. EUV Spectroscopy

The greatest capability of *EUVE* is through its spectroscopy which has allowed the observation of individual spectral lines with a resolution of $\frac{\lambda}{\Delta\lambda} \sim 150 - 350$ in the 60 – 700Å wavelength range; sample spectra are presented in Figure 2. The spectrum of Procyon is dominated by the lower ionization stages of iron (Fe IX - Fe XVI) indicating a maximum coronal temperature of 1.5 10^6 K. The EUV spectrum of the RS CVn binary σ Crb is dominated by the highest ionization stages of iron (Fe XV – Fe XXIV) indicating coronal temperatures in excess of 1.5 10^7 K. Procyon is a typical example of a *cool* corona whereas σ Crb is a *hot* coronal source. We emphasize that, Fe XXIV is the highest ionization stage of Fe detected in the EUV range; it is likely that σ Crb has a significant amount of material at even higher temperatures.

4.1. The reliability of High Temperature Tails

The large number of lines in the EUV range, require analysis techniques that account for the contribution of any particular temperature in the observed spectrum. A surprising feature in the differential emission measure distributions of several late-type stars, is a very strong component at temperatures in excess of 10^7 K (Mewe et al. 1995, Schrijver et al. 1995). The need for the high temperature tails, arises from an excess "continuum" flux in the 70 – 130Å wavelength range (Figure 2). These high temperature tails have been very controversial and a number of alternative interpretations have been proposed.

- If the high temperature tail is attributed to a real hot component, this would not be consistent with the X-ray fluxes of objects with "cool" coronae (e.g. Procyon, α Cen). The *ROSAT*/PSPC spectra are over-predicted by up to one order of magnitude (Schmitt et al. 1996a).

- Reduced iron abundances. The vast majority of lines detected in the EUV come from iron. The line flux is proportional to the elemental abundance whereas the continuum flux reflects the abundance of hydrogen. The lower line to continuum ratio could therefore imply that Fe is under-abundant in the corona of some late-type stars (Stern et al. 1995).

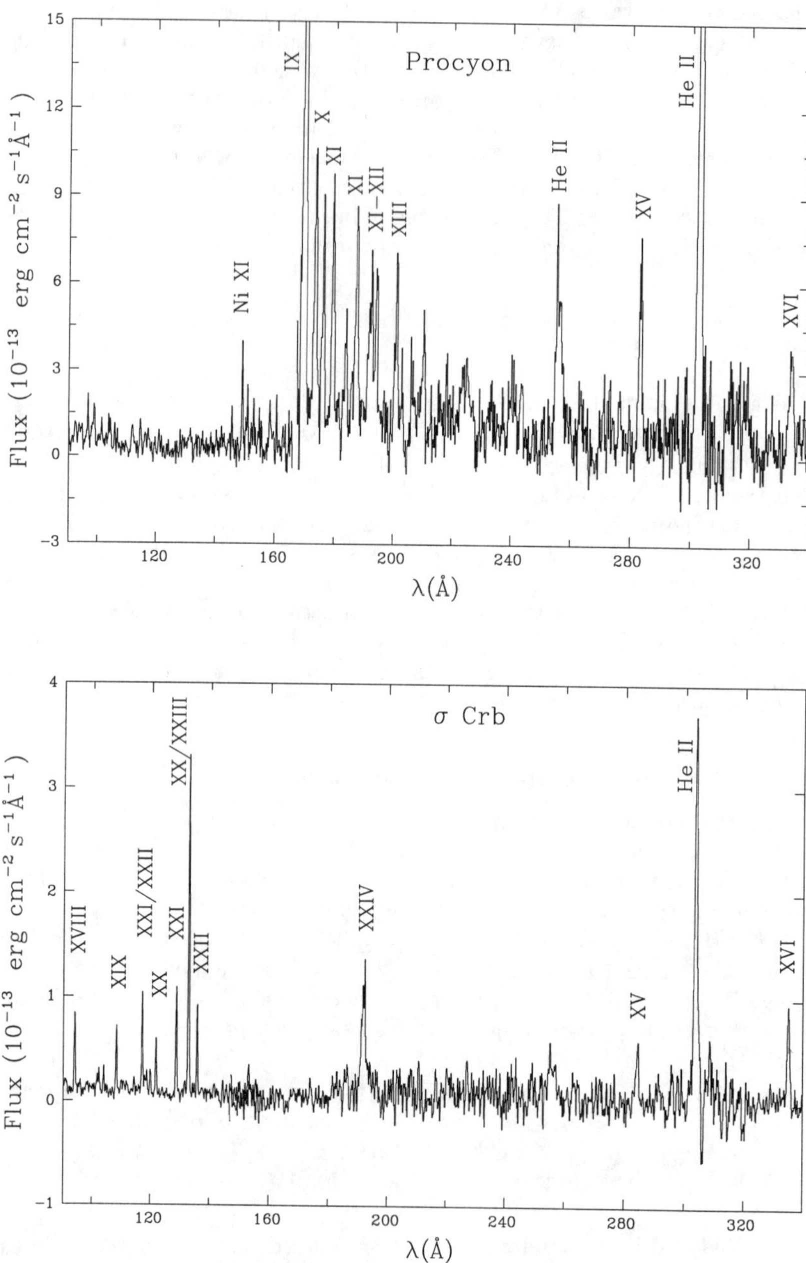

Figure 2. EUV spectra of Procyon and σ Crb. The strongest Fe lines are indicated by their ionization stage.

- The dependence of the total Gaunt factor on the square of the nuclear charge, led Drake (1998) to suggest that a low line to continuum ratio could be produced by an overabundance of He in active stars.

- The excess "continuum" flux could also be due to a number of missing lines that are not accounted for in the existing plasma codes (Jordan 1996). Given the limited spectral resolution of the instrument, these lines could merge together to create a *pseudo continuum* .. see paper by Phillips et al. (these proceedings).

- The increased continuum emission has also been attributed to opacity effects. We will outline the principles of this controversial suggestion in the following section.

4.2. Opacity effects in Stellar Coronae ?

Opacity effects in the solar corona have been known for several years. Acton & Catura (1976) provided the first convincing evidence that radiative transfer effects can be seen in X-rays. Schrijver et al. (1994) suggested that the EUV "continuum" in Procyon and α Cen could be attributed to resonant scattering. High optical depths arise from the fact that strong resonance lines have large oscillator strengths. For thermal broadening, the optical depth at line center is given by (Schrijver et al. 1994) :

$$\tau_0 = 1.2 \times 10^{-17} \frac{n_i}{n_{el}} A_Z \frac{n_H}{n_e} \lambda f \sqrt{\frac{M}{T}} n_e l \tag{1}$$

where $\frac{n_i}{n_{el}}$ is the ion fraction, A_Z the elemental abundance, $\frac{n_H}{n_e}$ the ratio of hydrogen to electron densities, f the line oscillator strength, M the atomic weight, T the temperature, l is the path length and n_e the electron density. In a homogeneous atmosphere, any photons scattered out of the line of sight will be compensated by photons scattered into the line of sight and no opacity effects will be detected. However, if the emission occurs in a localised volume of the atmosphere, such as a coronal active region, one can expect more photons to be scattered out of the line of sight when the active region is viewed along its radial extent, since the line of sight passes through the loop structures that make up a coronal active region. Thus, strong resonance lines can have significant optical depths and will appear weaker relative to the continuum. Strong arguments against this suggestion have been presented by Schmitt et al. (1996a) who have shown that significant opacity in EUV lines is not consistent with the PSPC spectra of Procyon.

In the solar case, the strong Fe XVII 15.01Å X-ray line ($f = 2.662$) is known to be resonantly scattered in quasi-stable active regions and flares (Phillips et al. 1996, Schmelz et al. 1997). In all case examined, including quiescent regions, the flux of this line is significantly reduced compared to the optically thin Fe XVII 16.78Å line ($f = 0.101$). With the launch of *AXAF* and *XMM* in 1999, we will be able to apply the solar techniques to stellar spectra and accurately estimate the importance of stellar coronal opacity. A 20ksec *AXAF* simulation of the dK7e star CC Eri in the Fe XVII wavelength region, is shown in Figure 3 .. see also paper by Drake (these proceedings).

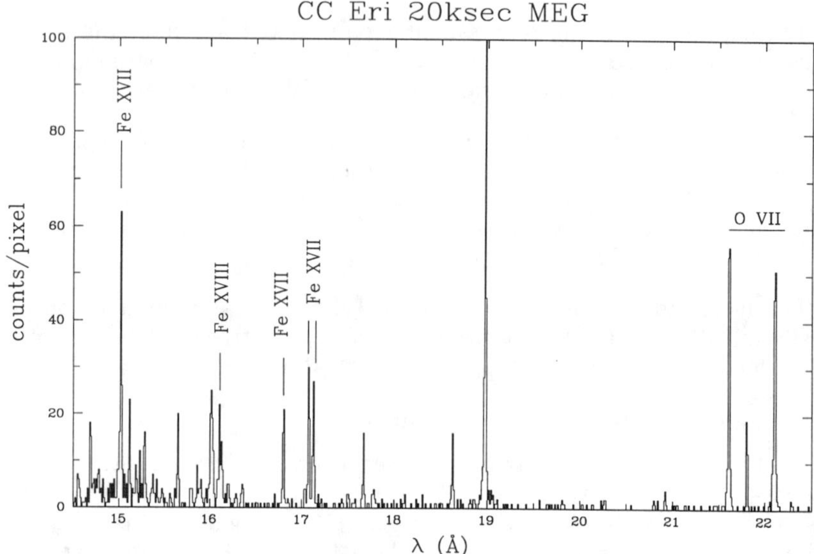

Figure 3. A 20ksec *AXAF* Medium Energy Grating simulation of CC Eri.

4.3. Coronal Electron Densities

The EUV spectral range offers several line ratios that can be used as density diagnostics (Keenan 1996). Line ratios of Fe IX – Fe XVI have been used on Procyon and α Cen and the derived electron densities are in the range of 10^9 - 10^{10} cm^{-3} (Schmitt et al. 1996b, Foster et al. 1996). In *hot* coronal sources, Fe XXI and Fe XXII ratios provide excellent density diagnostics for high density plasmas (10^{10} – 10^{14} cm^{-3}). The Fe XXI line ratios used in most cases are: R_1 = 145.65/128.74, R_2 = 142.15/128.74, R_3 = 102.22/128.74 while for Fe XXII R = 114.41/117.17. In an early *EUVE* observation of Capella, the Fe XXI ratios indicated densities as high as 3 10^{13} cm^{-3} (Dupree et al. 1993). For densities of this order of magnitude, the quasi-static approximation of coronal loops does not apply (van den Oord et al. 1997). However, in the case of Capella the values of n_e derived from the three Fe XXI line ratios were different by almost two orders of magnitude. It is hard to understand such a large discrepancy in the values of n_e derived from line ratios of the same ion. We have re-examined the *EUVE* short wavelength spectrum of Capella (Figure 4) and found that the only ratio that can provide a reliable measurement is R_3, with a corresponding value of n_e $\sim 5\ 10^{11} - 10^{12}$ cm^{-3}. The confinement of the gas therefore requires magnetic fields of $\approx$ 200 Gauss. The electron density combined with an emission measure of $\approx 10^{52}$ cm^{-3} at 10^7 K, implies that the volume of the emitting plasma is 10^{28} cm^{-3}. If we assume that the coronal plasma is distributed in semicircular loops with a cross-sectional diameter of 20% that of the footpoint separation, the radius R of the loop is given by

$$R = 3 \times 10^9 n^{-\frac{1}{3}} \tag{2}$$

where n is the number of loops and R is measured in cm. Electron densities of $10^{11} - 10^{12}$ cm^{-3} at temperatures of 10^7 K, have been reported in the quiescent state of II Peg, AR Lac and HR 1099. The transition region densities of these objects are of $10^{10} - 10^{11}$ cm^{-3}(Griffiths & Jordan 1998). The derived pressure in the corona is therefore much higher than in the transition region, indicating that the high and low temperature material originates in physically unconnected regions.

4.4. A final comment

There has been increasing evidence that the coronal abundances of some late-type stars may be lower than solar and a considerable number of papers have been dedicated to this issue (see Drake 1996). The new abundances have important implications for the radiative output of the corona. The radiative losses are computed as the product of the emission measure distribution with the radiative loss function of an optically thin plasma $\Lambda(T_e)$. Iron becomes an increasingly important contributor to the radiative loss function for temperatures above 10^5 K and dominates between $10^{5.7} - 10^{6.5}$ K (Cook et al. 1989). A reduced Fe abundance will increase the emission measure and decrease $\Lambda(T_e)$, causing little difference to the radiative losses in the temperature range where $\Lambda(T_e)$ is dominated by Fe. However, a significant increase in the radiative losses is found for $T > 10^{6.5}$ K where Fe is no longer the dominant radiative cooling agent (Mathioudakis & Mullan 1998). In these higher temperatures, radiative losses from bremsstrahlung radiation become important. The radiative losses provide a lower limit to the coronal heating requirements. The new Fe abundances must therefore be taken into account when considering the energy balance of the corona.

Acknowledgments. I will always be grateful to Brendan for his help, advice, encouragement and support in the early stages of my career.

References

Acton, L. & Catura, R.C., 1976, Phil. Trans. Roy. Soc. London, 281, 383.

Bowyer, S. & Malina, R.F., 1996, in Astrophysics in the Extreme Ultraviolet, Kluwer Academic Publishers

Cook, J.W., Cheng, C-C., Jacobs, V.L. & Antiochos, S.K., 1989 ApJ, 338, 1176

Doyle, J.G., Short, C.I., Byrne, P.B., & Amado, P.J., 1998 A&A, 329, 229

Drake, J.J., 1996, in Cool Stars Stellar Systems and the Sun, ASP Conference Series, R. Pallavicini & A.K. Dupree, Vol. 109, 203

Drake, J.J., 1998, ApJ, 496, L33

Dupree, A.K., Brickhouse N.S. & Doschek G.A., 1993, ApJ418, L41

Foster V.J., Mathioudakis M., Keenan F.P., et al. 1996, A&A, 473, 560

Houdebine, E.R., Mathioudakis, M., Doyle, J.G., & Foing, B.H. 1996, A&A, 305, 209

Giampapa, M. S., Rosner, R., Kashyap, V., et al. 1996, ApJ, 463, 707

Griffiths N.W., & Jordan C. 1998, ApJ, 497, 883

Jordan, C., 1996, in Astrophysics in the Extreme Ultraviolet, S. Bowyer & R.F. Malina, Kluwer Academic Publishers, 81

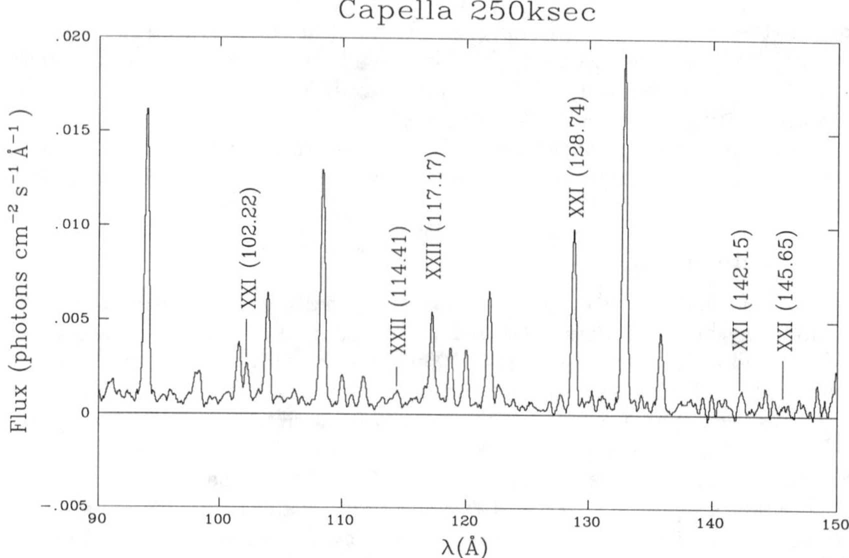

Figure 4. The *EUVE* short wavelength spectrum of Capella. The position of Fe XXI and Fe XXII lines that are used as density diagnostics is highlighted.

Keenan F.P., 1996, Space Sci Rev 75, 537
Mathioudakis, M. & Doyle J.G. 1992, A&A, 262, 523
Mathioudakis, M., Fruscione A., Drake J.J., et al. 1995, A&A, 300, 775
Mathioudakis, M. & Mullan, D.J., 1998, A&A, in press
Mewe, R., Kaastra, J.S., Schrijver, C.J., et al. 1995, A&A, 296, 477
Noyes, R.W., Hartmann, L.W. & Baliunas, S.L., 1984, ApJ, 279, 763
Phillips, K.J.H., Greer, C.J., Bhatia, A.K., & Keenan, F.P. 1996, ApJ, 469, L57
Pounds, K.A., Allan, D.J., Barber, C., et al. 1993, MNRAS, 260, 77
Schmelz, J.T., Saba, J.L.R., Chauvin, J.C. & Strong, K.T. 1997, ApJ, 477, 509
Schmitt, J.H.M.M., Fleming, T.A. & Giampapa, M.S. 1995, ApJ, 450, 392
Schmitt, J.H.M.M., Drake, J.J. & Stern, R.A. 1996a, ApJ, 465, L51
Schmitt, J.H.M.M., Drake J.J., Haisch B.H. & Stern R.A. 1996b, ApJ, 467, 841.
Schrijver, C.J. 1987, A&A, 172, 111
Schrijver, C.J., van den Oord G.H.J. & Mewe R., 1994, A&A, 289, L23
Schrijver, C.J., Mewe, R., van den Oord & G.H.J, Kaastra, J.S. 1995, A&A, 302, 438
Stern, R.A., Lemen, J.R., Schmitt, J.H.M.M. & Pye, J.P., 1995, ApJ, 444, L45
van den Oord, G.H.J., Schrijver, C.J., et al. 1997, A&A, 326, 1090

Solar and Stellar Activity: Similarities and Differences
ASP Conference Series, Vol. 158, 1999
C.J. Butler and J.G. Doyle, eds.

X-ray Magnetic Activity and Variability in Stellar Coronae: What are the Limits of the Standard Picture?

Robert A. Stern

Lockheed Martin Solar and Astrophysics Laboratory

Abstract.
Over the past 20 years, a "standard picture" of stellar coronal emission has developed from the results of both stellar and solar observations. Usually referred to as the age-rotation-activity connection, this picture is quite successful at qualitatively explaining the pattern of cool star stellar X-ray emission. However, some key puzzles remain, including the "supersaturation" phenomenon and the lack of significant long-term variability in active stars. Examination of such departures from the "standard picture" is likely to yield new insights into the physics of stellar coronae.

1. Introduction: the "Standard Picture"

It has now been more than 20 years since the study of extra-solar coronae began with the discovery of X-ray emission from Capella (Catura, Acton & Johnson, 1975; Mewe *et al.* 1975). Since that time, major advances in the understanding of stellar coronae have been achieved as a result of *Einstein*, *EXOSAT*, *ROSAT*, *EUVE* and other satellite observations. During the same period, solar X-ray observations from *SMM*, *Yohkoh*, *SOHO*, and the development of theoretical modeling tools for the solar corona have led to a clearer (yet hardly complete) picture of magnetic heating in the outer solar atmosphere.

As stellar observations have accumulated, it has become possible to synthesize what I will term the "standard picture" of stellar coronal activity. The term "picture" is used deliberately here: theories of stellar coronae are not sufficiently well developed that one could use the term "model." The "standard picture" is based upon using the solar analog, and dates from at least as far back as the work of Wilson (1966), Kraft (1967), and Skumanich (1972). By relating stellar *chromospheric* activity to stellar age and rotation, these authors coupled the concept of a self-regenerating stellar dynamo with magnetic heating, in a similar manner as was assumed for the Sun.

Perhaps the simplest way to summarize the "standard picture" is to examine the "Rossby relation", so aptly illustrated in Patten & Simon (1996), and reproduced as Figure 1 in Rob Jeffries' article on open cluster X-ray emission in these proceedings. In that figure, the ratio of stellar X-ray to bolometric luminosity versus Rossby number (ratio of rotation period to convective turnover time) is plotted for a variety of late type stars. The main features of this figure are that: (1) as the Rossby number decreases, $\frac{L_x}{L_{bol}}$ increases until it reaches a value of $\sim 10^{-3}$, whereupon it "saturates", and (2) for extremely small Rossby numbers, there is an indication of a turnover in this saturation curve, termed "supersaturation" (see §3 below). The rising portion of the curve is attributed to increased magnetic flux generation through the stellar dynamo as a star of a given convection zone depth rotates faster, with stars having deeper convection zones, hence longer convective turnover times, also having increased X-ray emission (see Noyes *et al.* 1984). The saturated regime is less well understood, though some plausible explanations have been given (§3).

The standard picture of enhanced coronal heating with stellar youth and rotation is also consistent with recent X-ray spectral results from, e.g., *ROSAT*. Güdel *et al.* (1997) have provided a compelling picture of "the Sun in time," demonstrating that the ratio of hotter to cooler coronal plasma in solar-like stars increases with increased $\frac{L_x}{L_{bol}}$ (see their Figure 1). This is entirely consistent with the concept of increased heating in magnetically confined plasma ("loops") resulting in increased coronal temperature (e.g., Rosner, Tucker & Vaiana 1978) An extreme version of this general trend is seen in stellar X-ray flares, where rapid heating occasionally produces thermal emission with plasma temperature up to 10^8 K (e.g. Pan *et al.* 1997, Tsuboi *et al.* 1998).

At this time, there is no quantitative theoretical model which can reproduce with authority the Rossby relation, or predict in detail the temperature of any given stellar corona. It is generally agreed that the basic concept of the activity-rotation connection must be correct, yet only parameterized fits to observations are available, no quantitatively falsifiable models. It should be pointed out, of course, that the detailed nature of *solar* coronal heating still eludes solar physics theoretical modelers.

Rather than try to develop better and better fits to observational data, it is instructive to examine the exceptions to the "standard picture", i.e. where the accepted wisdom apparently fails. No doubt some of these exceptions will later turn out to be either errors in the reduction or interpretation of observational data; however, it is likely that many "anomalous" results are telling us something important about the stellar coronal parameters. As such, they will be important guides to future theories of coronal heating mechanisms.

2. Observations that Challenge the Standard Picture/Solar Analogy

For the purpose of this review, I have selected a number of outstanding puzzles that challenge the standard model described above. They are listed below:

- Supersaturation (and saturation)
- The "Praesepe Puzzle"
- Wide binaries in the Hyades
- Paucity of solar-like cycles in active stars
- The concept of a static loop

One of the above subjects, the question of whether Praesepe and other clusters of similar age all have similar X-ray activity levels, I leave to Rob Jeffries, who discusses this issue in these proceedings. The last subject, the reality of so-called "static loops", I discuss only briefly below, since the beautiful *TRACE* movies shown during the meeting by Karel Schrijver make the case for a truly dynamic Sun better than any manuscript could.

3. Supersaturation and Saturation

Observations of the very youngest ($\lesssim$ 100 Myr) open clusters probe the "saturated" dynamo in stars with rotational velocities 10–100 times that of the Sun. It is now well known (e.g. Stauffer 1991, Prosser *et al.* 1995) that such young clusters have a population of cool, ultra-fast rotators (UFRs), stars with projected rotational velocities exceeding 20 $km\ s^{-1}$. In the youngest of these clusters (age $\sim$ 30–50 Myr) such as α Per (Randich *et al.* 1996), IC 2602 (Randich *et al.* 1995) and IC 2391 (Patten & Simon 1996), a population of G–K dwarfs with $v\ sin\ i > 50\ km\ s^{-1}$ has been found which exhibits a *decline* in activity level with increased rotation (Figure 1). This phenomenon is termed "supersaturation" (Randich 1997).

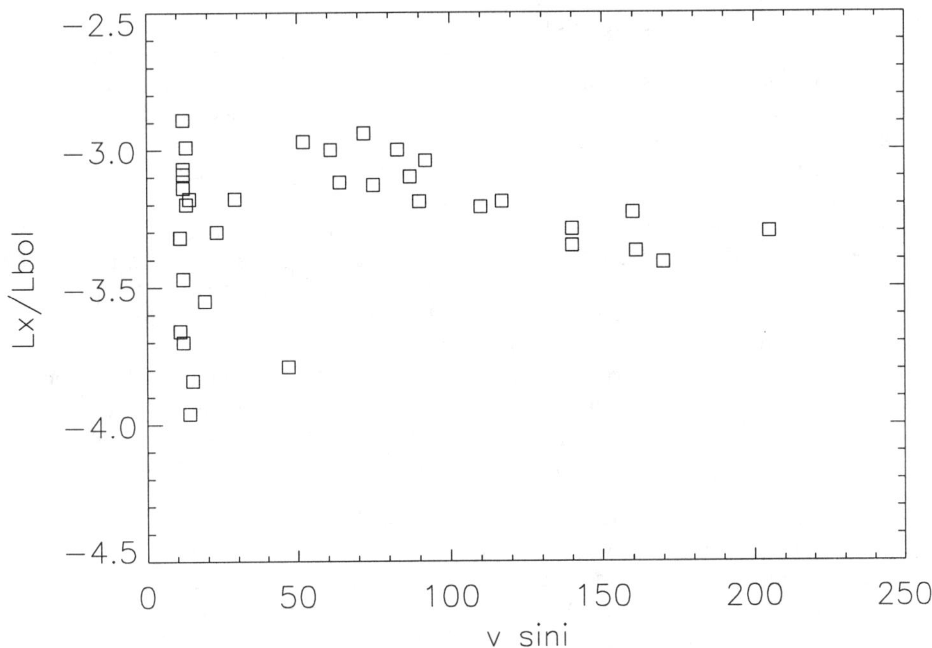

Figure 1. Log of X-ray/bolometric luminosity vs.$v\ sin\ i$ for α Per

Until the saturation phenomenon was observed, coronal activity was thought to increase monotonically with rotation (e.g. Pallavicini *et al.* 1981). Now, with the evidence for supersaturation, at first glance we are led to conclude that the efficiency of the stellar dynamo's heating may *decrease* for the most rapid rotators. However, these new results appear to fly in the face of previous suggested explanations of the saturation phenomenon. One popular hypothesis is that, given the typical X-ray surface flux of a solar active region, saturation occurs when a solar-type star is effectively 100% "covered" by active regions. A related idea is that saturation occurs when the magnetic flux generated by the dynamo reaches some limiting value due to the effect of Lorentz forces on convective zone flow patterns (Collier-Cameron & Li 1994). However, neither of these ideas can account for a *decline* in the magnetic heating rate as a star's rotational velocity, and presumably its rate of magnetic field generation, is increased beyond the saturation level. Thus, confirming that the overall dynamo heating rate or "efficiency" is truly decreasing would require theorists to step back and reconsider their preconceptions regarding the saturation phenomenon and the nature of the dynamo in the UFRs.

Supersaturation may have significant implications for the evolution of angular momentum in young stellar clusters. Recent results by Barnes & Sofia (1996) and Krishnamurthi et al. (1997) suggest that saturation in magnetic activity indicators is accompanied by a decrease in the magnetic braking efficiency, hence a change in the spindown timescale for rapidly rotating stars in young clusters such as α Persei. Barnes & Sofia suggest that such a decrease in magnetic braking efficiency may be caused either by an overall saturation in angular momentum loss rate or by a change in configuration of the magnetic field to more closely resemble a magnetic dipole. But would a decline of activity from the saturated condition for the UFRs imply an *increasing* angular momentum loss rate with increased rotational velocities? At the moment, we can only speculate about the answer, yet understanding the physics behind the supersaturation phenomenon could have a profound impact on models of stellar spindown from the pre-main sequence through the main sequence phases of stellar evolution.

We must, however, allow for another possibility: namely, the existence of a large population of "cool loops." Suppose that, as a result of the readjustment of a star's outer atmosphere to the increased angular velocity, the relative proportion of coronal heating producing higher (i.e. 5–10 MK) temperature structures compared to lower temperature (i.e. 100,000 K) structures changed as the apparent surface gravity of the star decreased due to enhanced rotation. At the stellar surface of a G5-K5 dwarf, increasing the rotational velocity to 100-200 km s^{-1} will decrease the apparent surface gravity by 5-20%. The effect will be proportionately greater away from the surface itself, decreasing gravity by up to 40% for loops of height 1 R_*. Decreasing the stellar surface gravity increases the gravitational scale height of coronal plasma: as Antiochos & Noci (1986) and Antiochos, Haisch & Stern (1986) pointed out, 100,000 K (and lower) temperature loops with heights smaller than the gravitational scale height at this temperature are stable. Thus with decreasing effective surface gravity, more such "cool loops" may exist, emitting UV and EUV radiation, but not X-rays. In fact, one of the principal avenues for radiative losses at this temperature is line emission from the C IV doublet at ≈ 1550 Å.

Thus, the decline in the ratio of X-ray to bolometric luminosity in these super-saturated UFRs need not be due to a decline in the heating efficiency of the dynamo, but, at least in one plausible scenario, to the heating of relatively more plasma at temperatures of 100,000 K or so. If this is true, then the extremely inefficient conductivity of plasma at these temperatures (compared to higher T coronal plasma), should result in directly observable increases in the relative radiative losses at 10^5 K, most notably in the C IV line. By comparing the C IV λ 1550 flux to the X-ray flux as a functional of rotational velocity, a direct test of such alternative hypotheses is possible, hence an indication of the true nature of the dynamo in the UFRs.

4. Wide Binaries in the Hyades Cluster

Pye *et al.* (1994) and Stern, Schmitt & Kahabka (1995) found that, in *ROSAT* pointed and all-sky survey observations of the Hyades cluster, there was a striking divergence between the binary/single X-ray luminosity functions for K stars, and possibly some difference for the M stars, unlike that for the F8-G5 stars in the cluster. In particular, Hyades K dwarf binaries of *any* type have a significantly different X-ray luminosity function than single stars, resulting in nearly an order-of-magnitude difference in the mean X-ray luminosity for the two samples (Figure 2, from Stern, Schmitt & Kahabka 1995).

Formal applications of statistical two-sample tests yield a probability $<10^{-4}$ that the binary and single K star samples come from the same distribution. This holds true for comparisons between single K stars and either spectroscopic binaries, or other (presumably wide) binaries. The M stars also exhibit this single/binary dichotomy, although in this case, the total number of confirmed binaries of all types amounts only to 9 objects. The formal results from the two-sample tests also yield an extremely low probabilities ($<10^{-4}$) that the M star single and spectroscopic binary populations are the same, and <0.02 that the samples for the single M stars and other M star binaries are the same. In contrast, with nearly equal numbers of single and binary stars, the F8-G5 single and binary X-ray luminosity functions are quite similar to each other: the probability that the single and spectroscopic binary G stars have the same distribution is $\approx$0.04-0.12, only marginally statistically significant, and P $\approx$ 0.15-0.20 for the single G and other G binaries, a null result. These formal quantitative results are qualitatively apparent in Figure 2.

This single/binary dichotomy is currently unexplained. It does not fit neatly into the picture of the age-rotation-activity connection, since the wide binaries, presumably unaffected by tidal spin-orbit coupling, should have the same rotational velocity distribution as single stars in the cluster. Perhaps this latter assumption is not valid, and the wide binaries do have different rotational histories in the pre-main sequence stage than do single stars in the cluster, resulting in a different set of initial conditions for subsequent rotational spin-down. In fact, as Hawley (1998) has pointed out in these proceedings, the time scale for spindown of field and cluster M dwarfs appears to be extremely long, essentially unmeasurable in clusters of Hyades age or younger. The shorter spindown time constant for the G stars would erase any memory of initial conditions, leading to the divergence of the X-ray distributions. This suggests that the spread in the K-M star X-ray luminosity function may be representative of the initial (ZAMS) conditions for the cluster. Clearly more work on rotational velocities of late type single and binary stars is needed.

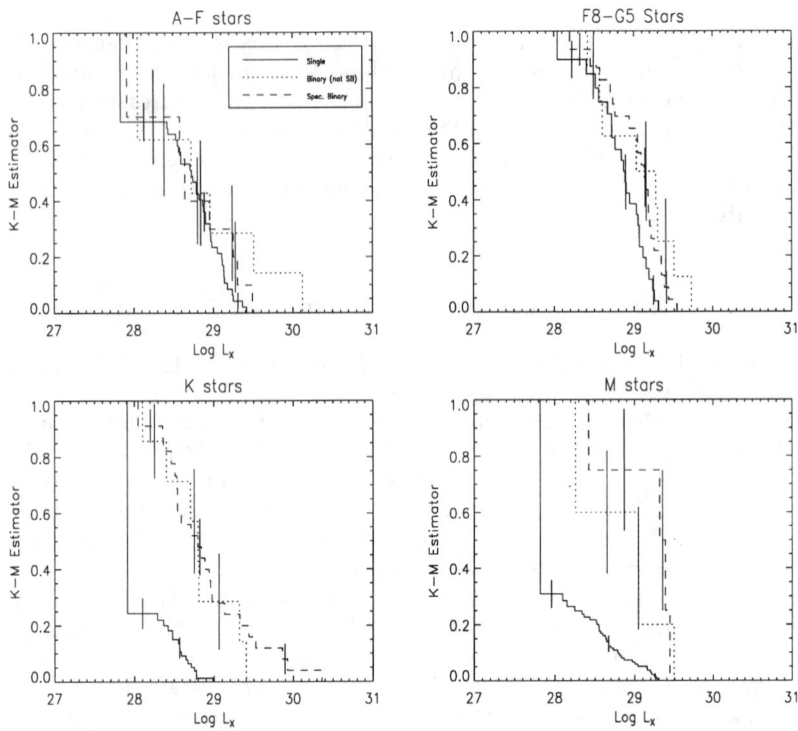

Figure 2. Cumulative X-ray luminosity functions for single (solid line), spectroscopic binary (dashed line) and other binaries (dotted line) in the Hyades sample grouped by A-early F stars, F8-G5 (solar type) stars, K stars, and M stars. Representative error estimates are shown.

5. The Paucity of Solar X-ray Activity Cycles in Active Stars

The magnetic activity cycle is one of the most well known aspects of solar coronal and chromospheric emission. Using *Yohkoh* Soft X-Ray Telescope (SXT) data the X-ray cycle has been observed as a waxing and waning of active regions, with a ratio of X-ray emission in the 0.5-2.0 keV band of at least a factor of 30 from maximum to minimum (Acton 1996). Even in the *ROSAT* band, which has a softer response than the *Yohkoh* SXT, a star exhibiting the same type of magnetic cycle as the Sun should change by a factor of 10 or more (Stern 1998, Peres *et al.* 1998).

But what of active stellar coronae? From purely anecdotal evidence, it is clear that active field stars or RS CVn binary systems *do not* change their X-ray luminosity by significant factors from observation to observation: otherwise successive observations of such stars and systems by the *Einstein*, *EXOSAT*, and *ROSAT* observatories over the last two decades would have found evidence for such variations. Yet typical measured X-ray fluxes for repeated observations of the brightest stellar coronae differ by no more than a factor of two or so

(Pallavicini 1993). Stern, Schmitt & Kahabka (1995) demonstrated that observations of the Hyades cluster taken a decade apart show little evidence of variability, possibly as much as a factor of two (Figure 3). As discussed in Stern (1998), observations of the Pleiades also suggest limited amounts of long-term variability, conceivably as great as a factor of two, but not much more. Why are there no order-of-magnitude X-ray activity cycle variations in these moderately to highly active stars?

Interestingly, some of the same phenomenology has been seen in long term studies of chromospheric Ca II reversals in the Mt. Wilson sample (Baliunas *et al.* 1995). These authors find that variability in this sample (which consists primarily of nearby, optically bright stars and a handful of cluster giants) seems to divide among three branches: (1) a "cyclic" branch of stars with similar Ca II activity as the Sun, which do exhibit long term (8–10 year) cycles, (2) a group of high-activity stars which show variability, but of a chaotic (not cyclic) nature, and (3) another group of "Maunder Minimum" stars of activity lower than the Sun, which seem to show little, if any variability. In a search to see if the "cyclic" activity group also demonstrated X-ray cycles, Hempelmann *et al.* (1996) undertook a statistical study of the residual variability of stars observed both in the Mt. Wilson program and the *ROSAT* All Sky Survey. These authors concluded that there was some marginal (3 σ) evidence for such cyclic variability for the sample as a whole: to date this is the only evidence of which I am aware for cyclic variability in stellar coronal X-ray emission.

What are we to make of these results? Is the stellar dynamo in rapidly rotating stars different in some way from that in slow rotators such as the Sun? Or are we somehow misinterpreting our limited data? Surprisingly, some of the uncertainty comes from the lack of recent "Sun-as-a-Star" long term X-ray flux data in a band comparable to the *ROSAT* bandpass. Certainly, we have extended data sets from the GOES satellites, and from *Yohkoh*, but these are peaked in the 0.5-10 keV region, where the variation in solar flux is due principally to variations in the number of active regions on the disk. We have to go back to the work of Kreplin (1970), using ionization chamber data, to see the solar cycle variation below 0.5 keV. Of course, it is here that *ROSAT* has some of its greatest sensitivity. Thus, although attempts at modeling the soft X-ray solar corona in the PSPC band have been made (e.g. Peres *et al.* 1998) , a direct measurement over several cycles is clearly desirable: this is where the new Solar X-Ray Imager (SXI) experiments on the next series of GOES satellites will be crucial (Bornmann *et al.* 1996).

Despite this incomplete data, the lack of variability in the more active stars is perfectly consistent with the Ca II monitoring cited earlier. From the results of the Hyades observations, Stern, Schmitt, and Kahabka (1995) suggested that this may be the result of a turbulent dynamo (as suggested by Durney *et al.* 1993) which dominates the X-ray flux in active cluster stars. If so, this could have profound implications for our understanding of how all stellar dynamos work.

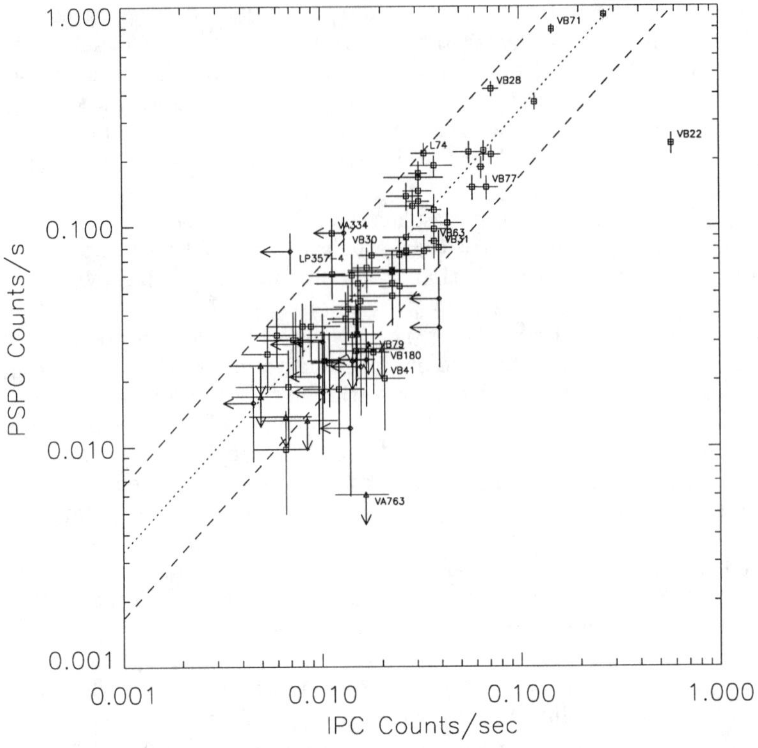

Figure 3. PSPC vs IPC count rates for stars in both Micela *et al.* (1988) study and Stern, Schmitt & Kahabka (1995). The central dotted line is *not* a fitted curve, but the best estimate of the relation between IPC and PSPC count rates for unabsorbed coronal sources. The dashed lines represent deviations from this relation by a factor of two.

6. The Concept of a Static Loop

Until the beginning of this decade, *Skylab* X-ray images and occasional sounding rocket images depicted a solar corona which could be conceived of as a collection of quasi-static loop structures, slowly evolving in time, with the notable exception of flares. Consequently, modeling of coronal structure was based upon the notion of a hydrostatic magnetic loop confining hot plasma, and indeed, such models were successfully applied to existing data (Rosner, Tucker & Vaiana 1978, Vesecky, Antiochos & Underwood 1979). In fact, Schmitt *et al.* (1985) and Stern, Antiochos & Harnden (1985) demonstrated that application of simple loop models to then-available crude stellar X-ray spectra was a viable alternative to single or "two-temperature" models. However, with the launch of *Yohkoh*, *SOHO*, and now *TRACE*, simply viewing the movies made by instruments on these remarkable observatories makes us question whether the solar corona, and by implication stellar coronae, can be thought of as static at all. Although these images can turn up examples of "classical" loops, the overwhelming impression is that many coronal structures are by their very nature dynamic. Since it is impossible to convey this picture through words or even with a single snapshot (as was demonstrated by Karel Schrijver and his astonishing *TRACE* movie), I recommend viewing various solar movies available on the WWW. One place to start would be our group's home page: *www.lmsal.com.*

7. Summary and Conclusions

The combined "solar analogy/standard picture" describes quite well the qualitative results of two decades of stellar X-ray observations. However, there are new observations suggesting that our *quantitative* understanding (using theoretical models) of stellar coronae is still very primitive. Even long-cherished pictures of the solar corona (e.g. static loops) are being overturned by new observations of coronal *dynamics*. With the upcoming launches of X-ray observatories having powerful spectroscopic capabilities (*AXAF*, *XMM*, and *ASTRO-E*), we expect to to gain further insight into the nature of stellar coronae as well as uncover additional puzzles.

Acknowledgments. This work was supported in part by the Lockheed Martin Independent Research Program.

References

Acton, L.W., 1996, in ASP Conf. Ser. 109, The Ninth Cambridge Workshop on Cool Stars, Stellar Systems and the Sun, eds. R. Pallavicini & A.K. Dupree (San Francisco: ASP), 45.

Antiochos, S., Haisch, B. & Stern, R., 1986, ApJ, 307, L55

Antiochos, S. & Noci, G., 1986, ApJ, 301, 440.

Baliunas, S., *et al.* , 1995, ApJ, 438, 269.

Barnes, S. & Sofia, S., 1996, ApJ, 462,746.

Bornmann,P., *et al.*, 1996, SPIE Proc. 2812, 309.

Catura, R.C., Acton, L.W. & Johnson, H.M., 1975, ApJ, 196, L47.

Collier-Cameron, A. & Li J., 1994, MNRAS, 269, 1099.

Durney, B.R., De Young, D.S. & Roxburgh, I. W., 1993, Solar Phys., 145, 207.

Güdel *et al.*, 1997, ApJ, 483, 947.

Hawley, S., 1998, these proceedings.

Hempelmann, A., Schmitt, J.H.M.M. & Stepien, K., 1996, A&A, 305, 284.

Jeffries, R.D., 1998, these proceedings.

Kraft, R.P., 1967, ApJ, 150, 551.

Kreplin, R.W., 1970, Ann. Geophys., 26, 567.

Krishnamurthi, S., Pinsonneault, M., Barnes, S. & Sofia, S., ApJ, 1997, 480, 303.

Mewe, R., *et al.*, 1975, ApJ, 202, L67.

Micela, G., Sciortino, S., Vaiana, G.S., Schmitt, J.H.M.M., Stern, R.A, Harnden, F.R., Jr. & Rosner, R., 1988, ApJ, 325, 798.

Noyes, R.W., Hartmann, L.W., Baliunas, S.L., Duncan, D.K. & Vaughan, A.H., 1984, ApJ, 279, 763.

Pallavicini, R., Golub, L., Rosner, R., Vaiana, G.S., Ayres, T.R., & Linsky, G.L., 1981, ApJ, 248, 279.

Pallavicini, R., 1993, in "Physics of Solar and Stellar Coronae,", J.F. Linsky & S. Serio, eds., Kluwer: Utrecht, 237.

Pan, H.C., *et al.*, 1997, MNRAS, 285, 735.

Patten, B.M. & Simon, T.S., 1996, ApJS, 106, 489.

Peres, S., Orlando, S., Reale, F., Rosner, R. & Hudson, H., 1998, these proceedings.

Prosser, C.F., *et al.*, 1995, PASP, 107, 211.

Pye, J.P., Hodgkin, S.T., Stern, R.A. & Stauffer, J.R., 1994, MNRAS, 266, 798.

Randich, S., Schmitt, J.H.M.M., Prosser, C.F. & Stauffer, J.R., 1995, Å, 300, 134.

Randich, S. & Schmitt, J.H.M.M., 1995, A&A, 298, 115.

Randich, S., Schmitt, J.H.M.M., Prosser, C.F. & Stauffer, J.R., 1996, A&A, 305, 785.

Randich, S., 1997, Memorie Societa Astron. Ital., 68, No. 4, 971.

Rosner, R., Tucker, W.H. & Vaiana, G.S., 1978, ApJ, 220, 643.

Schmitt, J.H.M.M., Harnden, F.R., Jr., Rosner, R., Peres, G., & Serio, S., 1985, ApJ, 288, 751.

Skumanich, A., 1972, ApJ, 171, 565.

Stauffer, J.R., 1991, in *Angular Momentum Evolution of Young Stars*, S. Catalano & J.R. Stauffer (eds.), p. 117.

Stern, R.A., 1998, in ASP Conf. Ser. 154, The Tenth Cambridge Workshop on Cool Stars, Stellar Systems and the Sun, eds. J.A. Bookbinder & R.A. Donahue (San Francisco: ASP), p. 223.

Stern, R.A., Antiochos, S.K. & Harnden, F.R., Jr., 1985, ApJ, 305, 417.

Stern, R.A., Schmitt, J.H.M.M. & Kahabka, P.T., 1995, ApJ, 448, 683.

Tsuboi, Y. *et al.*, 1998, ApJ, 503, 894.

Vesecky, J.F., Antiochos, S.K. & Underwood, J.H., 1979, ApJ, 233, 937.

Wilson, O.C., 1966, ApJ, 144, 695.

Solar and Stellar Activity: Similarities and Differences
ASP Conference Series, Vol. 158, 1999
C.J. Butler and J.G. Doyle, eds.

The Dutch Open Telescope

R.J. Rutten, R.H. Hammerschlag & F.C.M. Bettonvil

Sterrekundig Instituut, Utrecht, The Netherlands

Abstract. The Dutch Open Telescope is a new and novel optical solar telescope on La Palma. It aims at high resolution by combining an excellent site on La Palma with an open tower and an open telescope and leads the way to large-aperture high resolution telescopes. We briefly review the *DOT* principle, structure and goals. More information is found at the *DOT* website[1].

1. Introduction

The Dutch Open telescope (*DOT*) has grown out of the open telescope concept envisioned by C. Zwaan and R.H. Hammerschlag in the seventies. It has been pretty much a two-man project (Hammerschlag and technician P.W. Hoogendoorn) through the years. Its progress accelerated from a substantial grant by the Dutch technology foundation (STW) covering the completion and installation at La Palma – where the *DOT* now stands close to the *Swedish Solar Vacuum Telescope* (*SVST*), on top of its stilts as if a Martian invader has descended straight out of H.G. Wells' "The War of the Worlds". Its first images demonstrated its potential for high-resolution solar physics (Figure 1).

Recently, between the meeting and the writeup of this poster presentation, funding has been allocated for a three-year "science validation" period in which we shall get a chance to demonstrate the *DOT* capabilities for optical solar physics. To a large extent, the future of solar physics in The Netherlands rides on this demonstration.

2. *DOT* Principle

The *DOT* is a reflector with a parabolic mirror that sits out in the open at a height of 17 m (Figure 1). The open design departs radically from existing solar telescopes. All current high-resolution telescopes rely on internal evacuation to avoid internal turbulence. Examples are reflectors such as the NSO 76 cm Dunn Telescope at Sacramento Peak, the German 70 cm VTT at Tenerife, the French-Italian 90 cm THEMIS at Tenerife, and refractors such as the Swedish 47 cm *SVST* on La Palma. The vacuum window (reflectors) or the objective lens (refractors) sets a restrictive size limit of about 100 cm which does not apply to

[1] http://www.astro.uu.nl/~rutten/dot

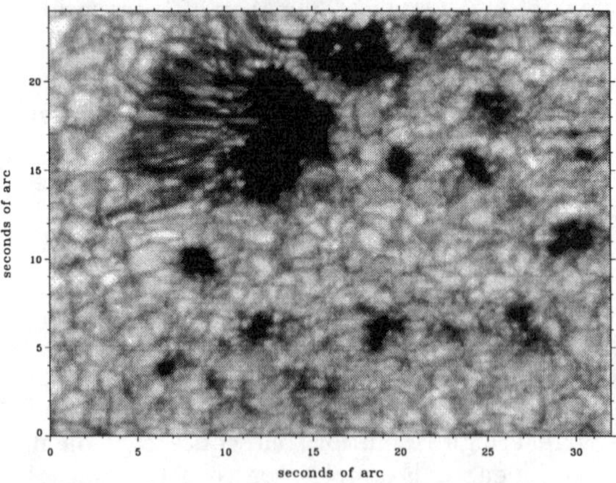

Figure 1. Top: the DOT on La Palma, December 5, 1997. North is to the right. Bottom: wide-band continuum image of an active region taken with the DOT on December 5, 1997.

open telescopes. Thus, the *DOT* represents an important test for large solar telescope concepts. The current *DOT* aperture is 45 cm, but the mechanical structure accepts a 76 cm mirror without change, a 100 cm mirror with minor adaptation or a yet larger mirror with major modification.

The telescope and the support tower are both open. There is no dome, only a fold-away bad-weather canopy. This novel design exploits the often excellent La Palma conditions through posing minimal obstruction to the strong trade winds that bring the best seeing. At La Palma, the best daytime seeing tends to occur when strong winds (5–10 m/s) from Northern directions suppress the convective plumes that arise from local ground heating. When the wind is sufficiently strong it confines the boundary convection to a thin layer of only about 10 m, well below the *DOT* telescope height of 17 m. This extraordinary circumstance may last all day and makes La Palma intrinsically better than all US mountain sites, where good seeing occurs just briefly after sunrise, before the convection starts and produces turbulence in a layer many tens of meters thick. (Comparable quality is reached at Pic du Midi after snowfall in summer, not a frequent event.)

Thus, the DOT's principle is to minimize obstruction to the local air flow. It also relies on the same strong winds, blowing right through the telescope, to inhibit convective turbulence within the telescope and in its immediate surroundings.

In addition, the simple optical scheme of the *DOT* guarantees optimum performance by avoiding the severe alignment problems that come with more complex arrangements such as Gregorian designs. The extraordinary mechanical stability gives high pointing precision even in strong winds. The fold-away canopy survives even the severe La Palma winter storms and ice loads.

3. The *DOT* Structure

The primary mirror (Cervit, 45 cm diameter, focal length 200 cm) focuses the incoming beam onto a water-cooled diaphragm that reflects most of the solar image out of the telescope and transmits only a 2 by 2 arcmin subfield.

The mirror is mounted deformation-free with nine-point axial and three-point radial support in a parallactic telescope structure that is considerably over-dimensioned as well as unbalanced in order to obtain extreme pointing stability. Brushless pairs of servo motors in push-pull preload configuration without backlash drive four-step gear trains achieve a 1:75,000 reduction with self-aligning gears. The extreme reduction serves to drive the telescope at very low dissipation (only about 20 W) in order to avoid local heat sources.

The 15 m *DOT* support tower puts the telescope above the turbulent boundary layer, especially when the strong trade wind blows uphill from Northern directions in the best-seeing weather pattern. The tower consists of eight triangularly arranged steel-tube stilts. The configuration permits only lateral motion of the platform, inhibiting any tilt with respect to the incoming wavefront, so that the telescope maintains precise angular tracking even in strong wind buffeting. At 13 tons the tower is considerably lighter than the combined platform (5 tons) and telescope (17 tons) which it supports. Nevertheless, it is designed to withstand large ice loads and (simultaneous) wind pressure.

The bad-weather enclosure opens clam-like and folds away to the sides. It is made of heavy polyester fabric mounted on heavy steel ribs and may be closed in winds up to 30 m/s (or opened, less likely). When closed it should withstand the 70 m/s (Bf 12) winds that might hit Roque de los Muchachos in the harsh La Palma winter storms. The coated surface tends to remain ice-free.

4. *DOT* Goals

The *DOT* is intended for imaging the solar photosphere and chromosphere with high spatial resolution. Resolution is the key frontier in optical solar physics. A major quest is to locate, diagnose and follow the basic elements of the solar magnetic field.

The three-year initial science program aims to obtain simultaneous imaging in three diagnostics:

- G band: field topology in the photosphere;
- Ca II K: field topology in the lower chromosphere;
- Hα: field topology in the upper chromosphere.

At present, simple secondary optics are mounted along the telescope axis in a tube behind prime focus (Figure 1). A G-band camera will be placed there, while filtergrams in Ca II K and Hα will be taken with cameras mounted besides the incoming beam.

These three imagers will serve as "proxy-magnetographs" in support of studies with space-based instrumentation (*SOHO* and *TRACE*). The program does not only follow techniques pioneered by G. Scharmer at the *SVST*, but it will actually rely on *SVST* hardware and software and on *DOT* operation from the *SVST* building. This will make it easy to operate the two telescopes in tandem. The program aims to chart and track the magnetic field patterns in the solar photosphere and chromosphere. In the meantime, we hope to develop instrumentation for future Stokes vector magnetometry and to implement advanced image processing to improve on the remaining seeing. These efforts are part of the European Solar Magnetometry Network[2].

Acknowledgments The *DOT* was built by the Sterrekundig Instituut Utrecht, the Physics Instrumentation Group of Utrecht University, and the Central Workshop of Delft Technical University. The *DOT* completion and installation at La Palma were funded by the Stichting Technische Wetenschappen (STW) of the Netherlands Organization for Scientific Research (NWO), and proceeded under the hospitable umbrella of the Instituto de Astrofísica de Canarias. The three-year science validation phase is funded by the Netherlands Organization for Scientific Research (NWO), the Netherlands Graduate School for Astronomy (NOVA), Utrecht University, the Sterrekundig Instituut Utrecht, and the European Commission via a TMR grant to the European Solar Magnetometry Network. The *DOT* team enjoys splendid hospitality of the Royal Swedish Academy of Sciences at the *SVST* with much support from Paco Armas, Göran Hosinsky, Rolf Kever and Göran Scharmer.

---

[2] http://www.astro.uu.nl/~rutten/tmr

Part 2

MAGNETIC FIELDS IN THE SUN AND STARS: EVOLUTION

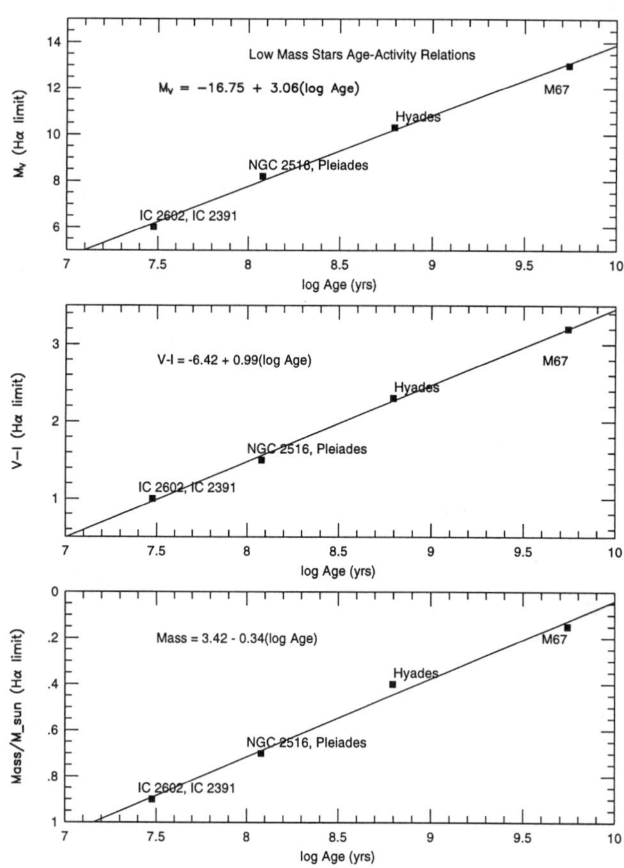

Age-activity relations for M dwarfs. The panels show linear relations with log age for M_V, V-I and stellar mass.

Solar and Stellar Activity: Similarities and Differences
ASP Conference Series, Vol. 158, 1999
C.J. Butler and J.G. Doyle, eds.

Chromospheric Activity in Low Mass Stars: Observational Results from Clusters and the Field

Suzanne L. Hawley[1,2]

Department of Physics and Astronomy, Michigan State University, East Lansing, MI 48824, USA

I. Neill Reid

Palomar Observatory, Caltech, Pasadena, CA 91125, USA

John E. Gizis

Astronomy Program, University of Massachusetts, Amherst, MA 01003, USA

P. Brendan Byrne

Armagh Observatory, College Hill, Armagh, BT61 9DG, N. Ireland

Abstract. Observations of low mass stars in the field and in open clusters are providing clues to the production and evolution of magnetic activity, many of which we still do not understand. Some molecular bands and colors respond much more strongly to the presence of activity than others. The observation that the dMe stars lie above and/or redder than the dM star main sequence in an M_V vs. V-I color magnitude diagram can be reconciled if the depth of the CaOH molecular band at 6230Å is used for the temperature indicator. Several clusters now have enough data to identify the "Hα limit" mass (color, absolute magnitude) where chromospheric activity becomes prevalent. There is a good correlation between the Hα limit mass (color, absolute magnitude) and the age of the cluster. This age-activity relation for M dwarfs is quite different from previous age-activity-rotation relations found for earlier type stars. The activity strength in Hα and in X-rays is shown for a range of M dwarf masses in several clusters and in the field, and two unexplained results are described. A brief discussion of an ongoing investigation of binarity and activity in NGC 2420 is given.

---

[1]NSF Young Investigator

[2]Current address: Space Sciences Laboratory, University of California, Berkeley, CA 94720

1. Introduction

We are carrying out a large survey of the low mass stellar populations in open clusters, with the twin aims of exploring chromospheric activity as a function of mass/age and investigating the initial mass function and subsequent dynamical evolution of the cluster. Previously we have carried out a similar investigation of field M dwarfs, which has been reported in Reid, Hawley & Gizis (1995, PMSU1) and Hawley, Gizis & Reid (1996, PMSU2). In this talk I will concentrate on the chromospheric activity results from these field and cluster surveys, which were of particular interest to Brendan. He and I were also involved in a project to observe several younger clusters to investigate the connection between magnetic activity and angular momentum evolution. It was our eventual hope to tie all of this work together into a coherent picture. With Brendan's students and postdocs who have worked on this project, including Robert Rolleston, Duncan Foster, Vincenzo Andretta, Alistair Gunn and Armin Theissen, this hope may still be realized, despite Brendan's untimely passing.

2. Colors, Magnitudes and Band-strengths

Figure 1 shows spectra of a star from our NGC 2516 cluster survey (Hawley, Tourtellot & Reid 1998) and the nearby dMe star AD Leo (no paper is complete without at least one AD Leo observation), illustrating the various molecular bands and atomic lines that we measure in the wavelength region near Hα. Hα is, of course, used to look for emission and thus the presence of magnetic (chromospheric) activity. In PMSU2, we showed that there were several molecular bands that responded strongly to the presence of magnetic activity in low mass M dwarfs. The active (dMe) stars had deeper TiO2 and shallower TiO4 bands than inactive dM stars at the same V-I color or TiO5 band-strength, which we use as a spectral type indicator. A second point that we noted in PMSU2 was that the dMe stars tend to lie above (brighter) and to the red of the main sequence defined by the dM stars in an M_V vs. TiO5 HR diagram. The effect is also present in the M_V vs. V-I HR diagram (see Figure 3a below). A possible explanation is that the dMe stars are all in binary systems, but there is evidence that at least some of the field dMe stars are single (e.g. AD Leo).

Using cluster data from IC 2602, NGC 2516 and the Hyades, we have further investigated these effects by studying the CaOH 6230Å molecular band. In Figure 2 we show that CaOH is shallower in dMe stars than in dM stars both a) at a given V-I color and b) at a given TiO5 band-strength. Figure 3a shows the M_V vs. V-I HR diagram with the active stars clearly offset redder and brighter than the main sequence. Figure 3b shows that if CaOH is used instead of V-I on the temperature axis, the active and inactive stars are brought into agreement. Thus it appears that CaOH is a better measure of temperature than either TiO5 or V-I in active stars. Note that these results could not be explained by the active stars being binaries.

These data will be useful to constrain atmospheric models of active stars. In an effort to provide more diagnostics at deeper layers, we have also investigated some strong atomic lines. Figure 4 shows Na D, Ca I and Fe I absorption line equivalent widths vs. V-I color for stars in the NGC 2516 cluster. The lines

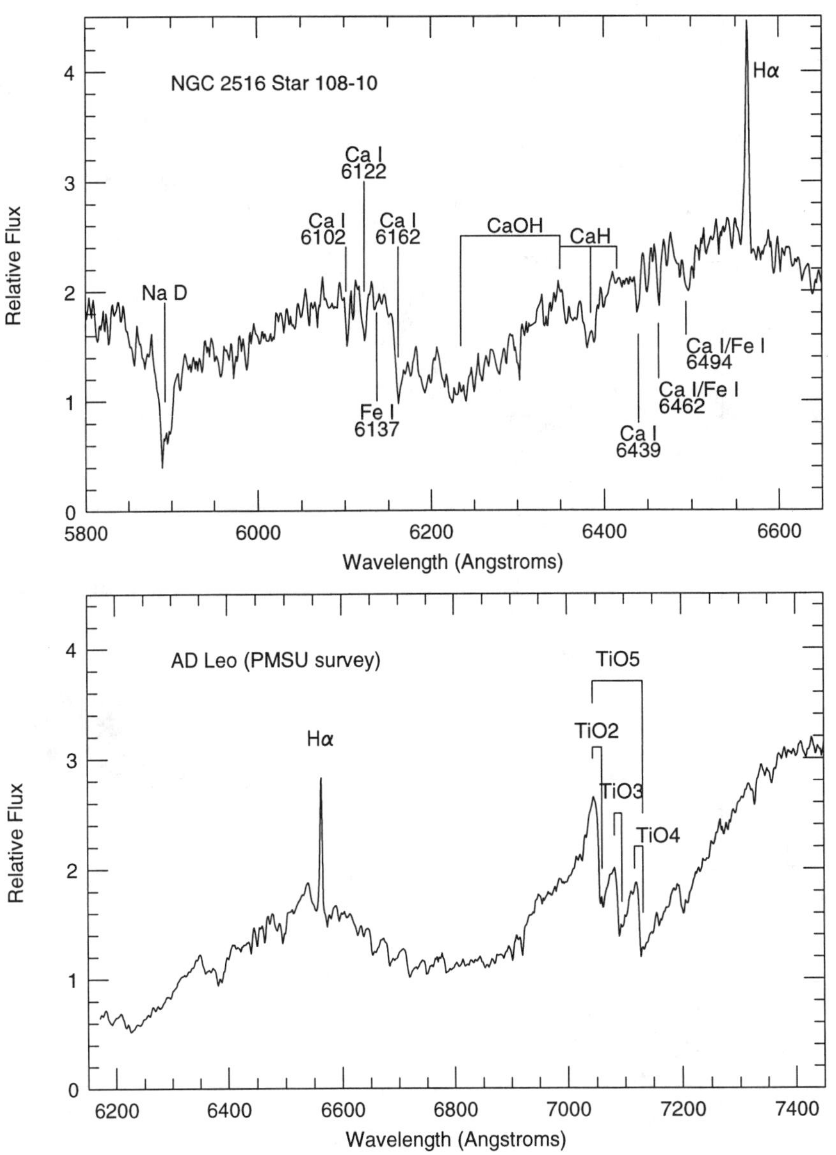

Figure 1. Spectra of a dMe star in the young cluster NGC 2516 and the nearby dMe star AD Leo. The molecular bands and atomic lines that we measure are marked.

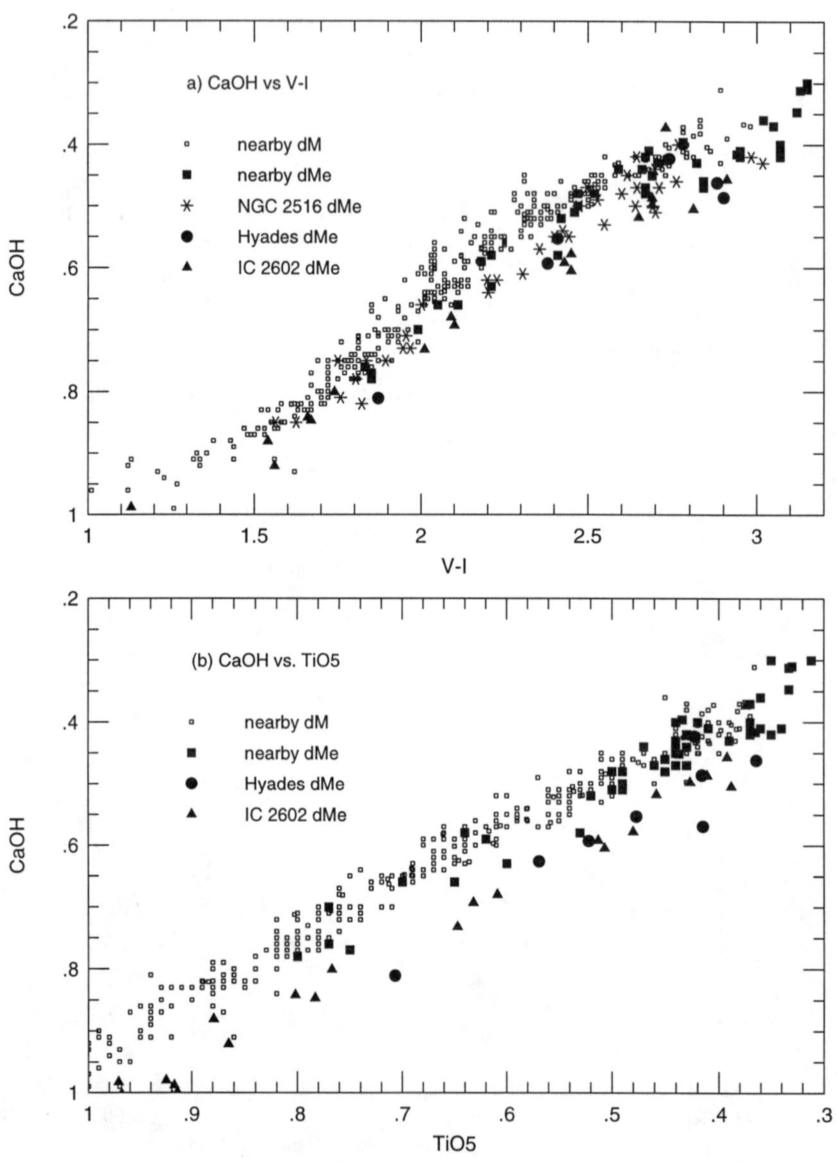

Figure 2. The CaOH molecular band-strength is plotted against a) V-I color and b) TiO5 molecular band-strength. The active (dMe) stars have shallower CaOH band-strength at a given V-I or TiO5.

show different behavior with decreasing mass; some increase in strength while others decrease. However, there is no obvious difference in the behavior of the dMe and the dM stars for these lines.

3. An Age-Activity Relationship for M dwarfs

It is well known that the relative number of active stars increases as you examine later spectral type (lower mass) stars on the main sequence (Joy & Abt 1974). By spectral type M5, more than half of all stars are dMe stars, compared with $< 10\%$ at spectral types earlier than M3 (see results in PMSU2). We showed in PMSU2 that this was not a selection effect due to the lower continuum flux at Hα in later type stars, which would make weaker Hα lines visible. In fact, in later type stars the Hα line generally has much larger equivalent width, so that the fraction of the bolometric luminosity emitted in Hα is relatively constant at least down to spectral type M6. (This activity strength indicator is discussed in more detail in the next section.)

Our kinematic analysis of the field stars showed that the dMe stars as a group had younger kinematic properties (smaller asymmetric drift, smaller velocity dispersions) confirming previous studies (Wielen 1975). A more intriguing result was that the early type dM stars appeared to have somewhat younger kinematics than the later type dM stars. In other words, the early type dM stars lost their magnetic activity sooner than the later type dM stars. Assuming that star formation has not occurred at a preferential mass in any given epoch, this result would then explain the prevalence of activity among the later type stars as an age effect – the later type stars remain active longer, hence a greater fraction will be active when we observe them.

A way to explore and quantify this hypothesis is to look at active stars in clusters of known age. Stauffer et al. 1991 showed that the color (mass) at which Hα emission became prevalent was bluer in the Pleiades compared to the Hyades. To investigate this effect, we have determined this "Hα limit" color, mass and absolute magnitude using data for several clusters including IC 2391 and IC 2602 (Stauffer et al. 1998), the Pleiades (Stauffer et al. 1994), the Hyades (Reid, Hawley & Mateo 1995, RHM), NGC 2516 (Hawley, Tourtellot & Reid 1998) and M67 (Hawley, Reid & Tourtellot 1998). We plot the results as a function of the log of the cluster age in Figure 5. Note that the M67 age is somewhat controversial, with recent determinations ranging from 4-7 Gyrs (e.g. Dinescu et al. 1995, Anthony-Twarog et al. 1991); we have chosen to use an intermediate value of 5.5 Gyrs. Finding the position of the Hα limit in mass, color and absolute magnitude is a subjective exercise, so our values and the resulting fits shown on Figure 5 should be taken as suggestive rather than definitive. What we do illustrate with some confidence is that the Hα emission lasts longer in lower mass stars, and the Hα limiting color/mass/absolute magnitude varies approximately linearly with log age.

This age-activity relation for the M dwarfs is in sharp contrast to the typical age-activity-rotation relations found for earlier type F, G and K stars (e.g. Skumanich 1972), where the **strength** of the emission decays with age. In the M dwarfs the emission remains strong, but disappears at different spectral types according to age. Earlier type M dwarfs lose their activity quickly, while later

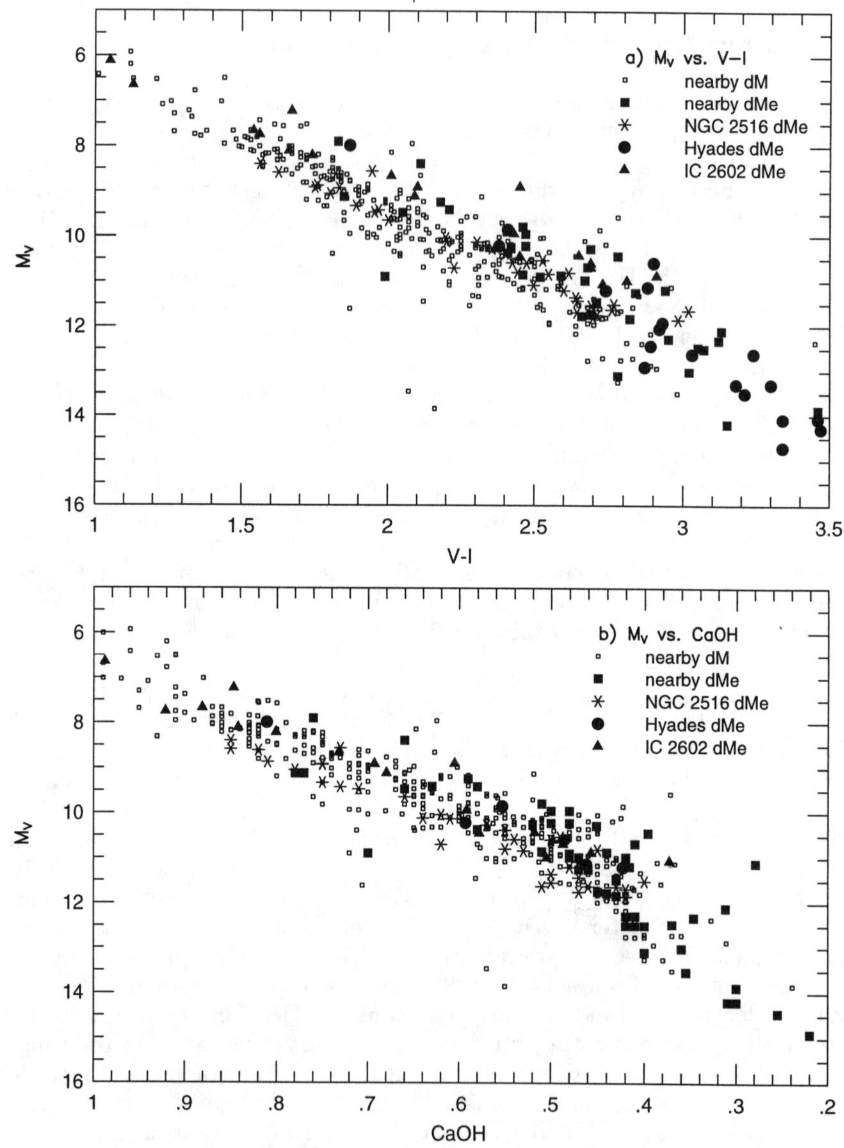

Figure 3. The absolute magnitude M_V is plotted against a) V-I color and b) CaOH molecular band-strength. The active (dMe) stars lie above/to the red of the dM star main sequence in the V-I plot, but are in agreement with the dM stars in the CaOH plot.

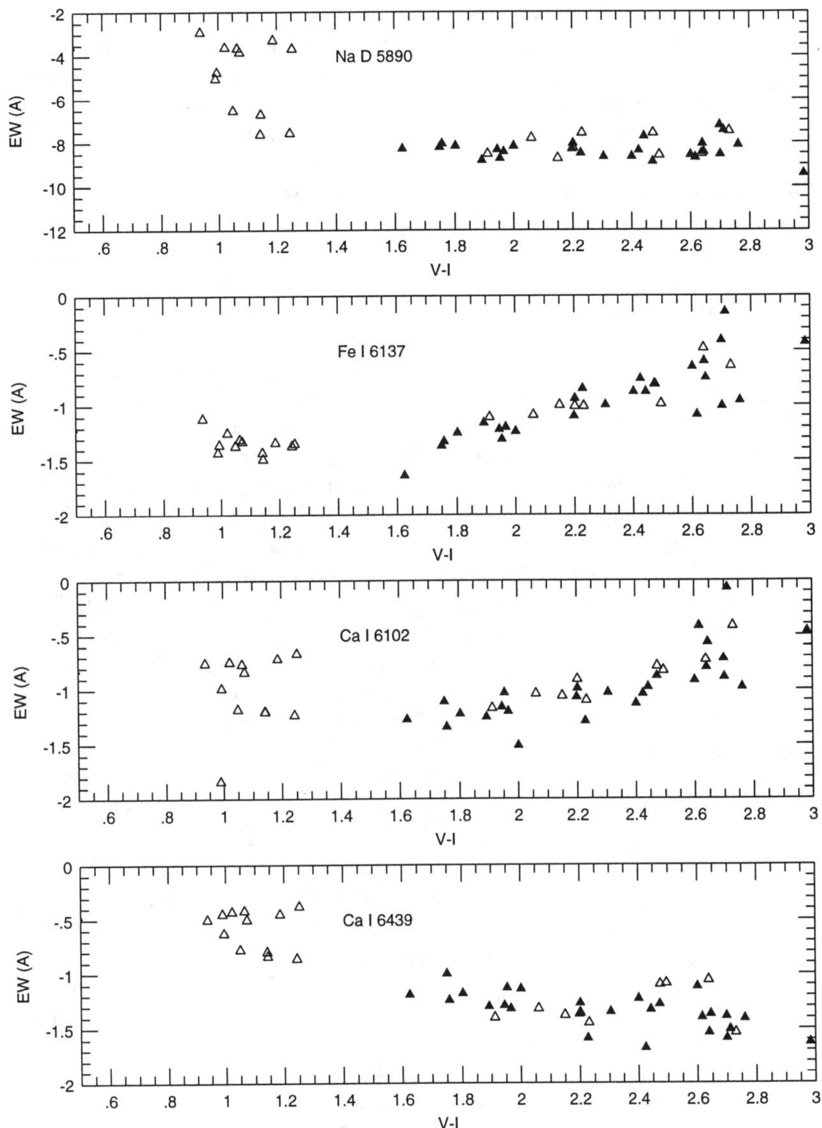

Figure 4. The equivalent widths in Angstroms of several atomic absorption lines are shown as a function of V-I color for stars in NGC 2516. The lines exhibit varying behavior with color, but the dMe stars (closed triangles) are similar to the dM stars (open triangles).

type stars retain it for much longer. The relationship between rotation and age which provides the physical basis for the Skumanich relation is no longer clearly applicable to the M dwarfs, since there is not an obvious connection between the slowing of rotation and the mass (spectral type) of the star. Other conflicting information about rotation in these late type stars also exists: in PMSU2 we showed that stars with slow and fast rotation rates had identical activity strength, while Basri & Marcy (1995) have pointed out the existence of very low mass stars with high rotation rates that have apparently no activity. All of these point to an additional physical mechanism affecting the production of magnetic activity, unrelated to rotation but probably connected to the stellar interior structure, which becomes important somewhere in the M dwarf mass range.

4. Activity Strength

The ratio of the Hα luminosity to the bolometric luminosity, $L_{H\alpha}/L_{bol}$, is a measure of the chromospheric activity strength which is independent of the continuum flux at Hα and hence can be used to compare activity strength in stars of different spectral type (bolometric magnitude). In PMSU2 we found that no changes occurred in the field stars in the range $6 < M_{bol} < 12$, and that there was some evidence for a decrease in activity strength among the faintest stars in our sample. Figure 6c shows the field Hα results. The intermediate age ($\sim$ 700 Myr) Hyades cluster also shows no sign of a decrease in activity strength with decreasing mass to $M_{bol} = 12$ (see RHM) as shown in Figure 6b. There are no known Hyads at lower mass, although we recently investigated some very low mass (VLM) candidates (Reid & Hawley 1998). Several of the stars turned out to be background pre-main sequence brown dwarfs, while others were foreground objects. Only one star remains as a possible Hyad, shown in Figure 6b as a closed circle.

Stauffer et al. (1994) have previously shown that there are a small number of VLM stars in the Pleiades which have very low activity strength. We combine their Pleiades data with our data for the young clusters NGC 2516 and IC 2602 in Figure 6a. Although the two more distant clusters do not have data for the lowest mass stars, there are a few NGC 2516 stars with quite low activity even at brighter magnitudes. Another remarkable feature of this plot is that all of the young clusters show an **increase** in chromospheric activity strength with decreasing mass between $4 < M_{bol} < 10$. This is in accord with the X-ray activity strength results for both the Pleiades and the Hyades shown in Figures 6d and 6e (from RHM and references therein). The field X-ray data (Figure 6f) does not show this effect, possibly because of the inhomogeneous nature of the field star sample (large spread in age, metallicity, distance). However, the field stars do show slight evidence for a decrease in X-ray activity strength in the lowest mass stars; the cluster samples do not extend to this faint a bolometric magnitude as yet.

There are thus two interesting and unexplained characteristics of the activity strength data:

1) the chromospheric activity decreases for stars with $M_{bol} > 12$ (mass $< 0.1 M_{\odot}$), at least in the Pleiades and the field. The young cluster NGC 2516 has

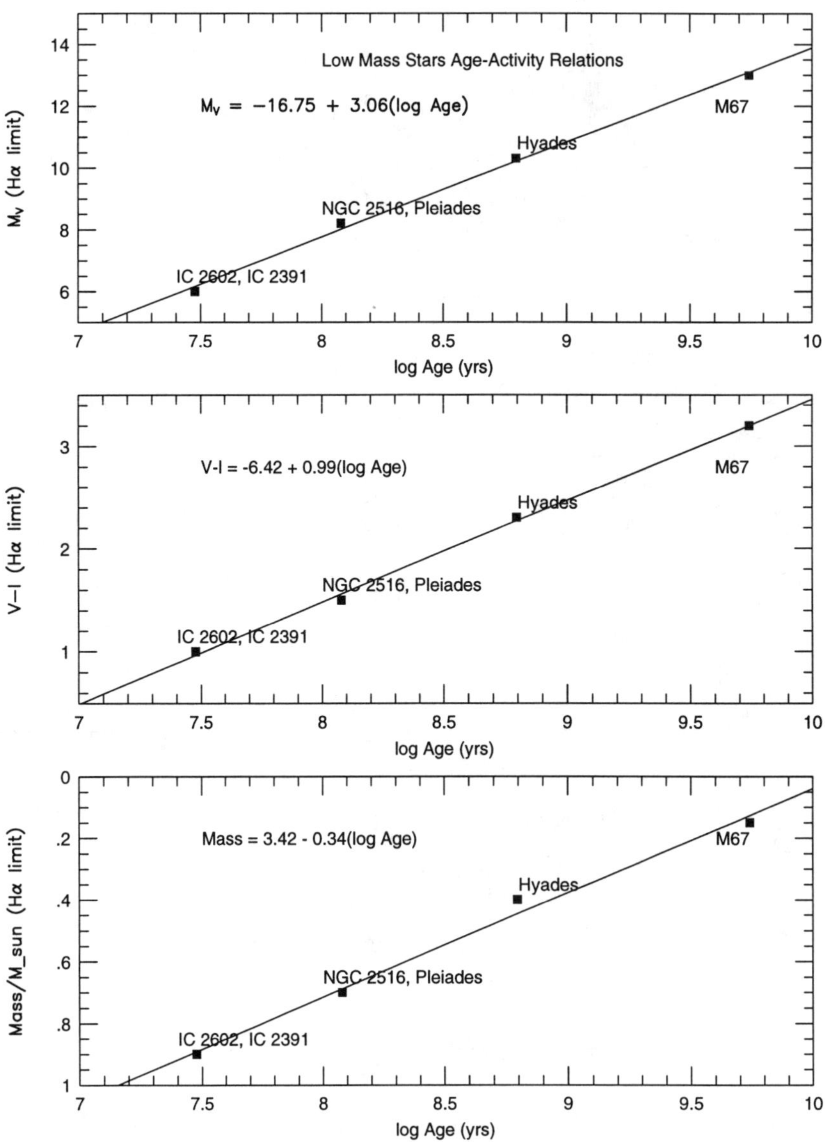

Figure 5. Age-activity relations for M dwarfs. The panels show linear relations with log age for M_V, V-I and stellar mass.

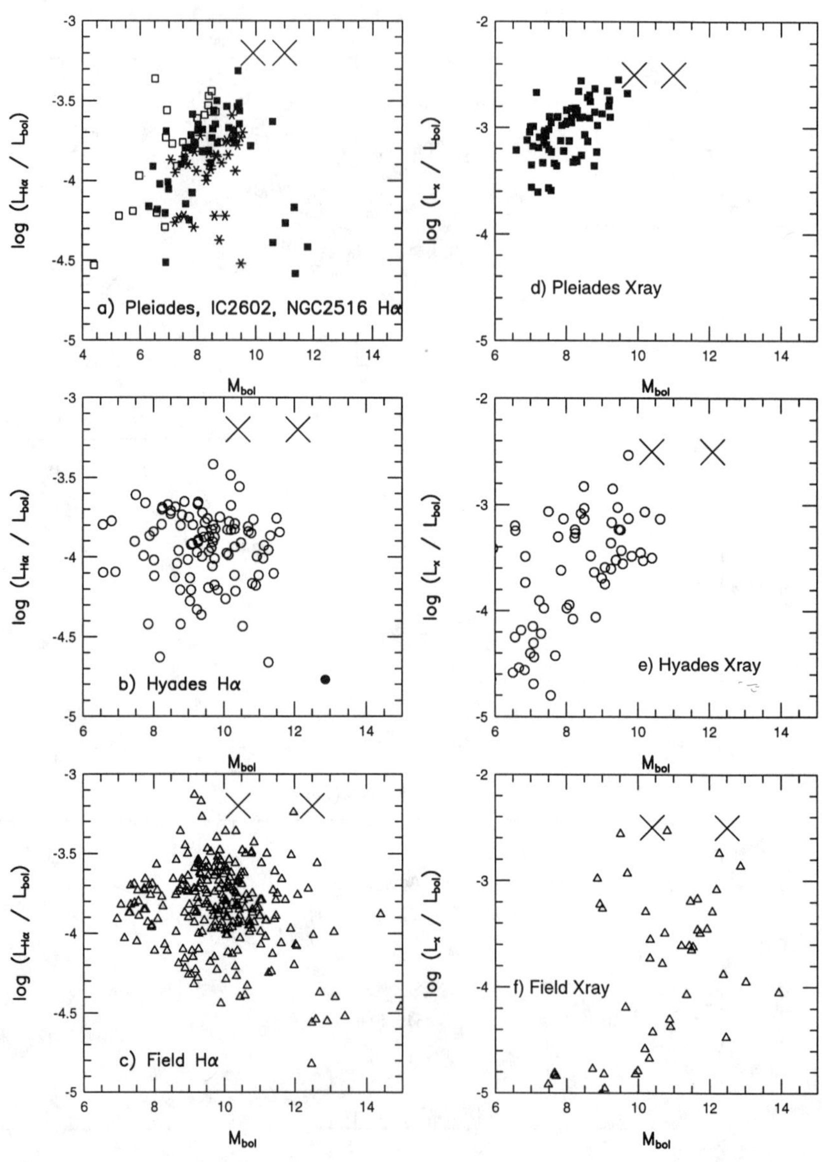

Figure 6. The Hα and Xray activity strengths are shown as a function of the bolometric magnitude for the Pleiades, Hyades and a sample of field stars. Figure 6a also includes Hα data for IC 2602 (open squares) and NGC 2516 (stars). The large X's at the top of each panel show the bolometric magnitudes for stars of 0.2 $M_\odot$ (left) and 0.1 $M_\odot$ (right).

a few earlier type stars with surprisingly low activity. X-ray activity may also be decreasing in the VLM field stars. We need new X-ray data for VLM cluster stars to determine whether the decrease also occurs there.

2) the chromospheric activity in young clusters (< 150 Myr) and the X-ray activity even in intermediate age clusters (< 700 Myr) increases with decreasing mass in the range $4 < M_{bol} < 10$.

Understanding the underlying physics that produces these effects will be instrumental in understanding the nature of magnetic activity in M dwarfs.

5. Brief comment on Binarity and Activity

Binary stars that are in close systems, presumably inducing fast surface rotation, can maintain high levels of activity even after the usual age-activity relations would suggest they should have become inactive. RS CVn stars are good examples of this phenomenon, as is the star Gliese 781 (seen at V-I=2, $M_V = 11$ in Figure 3a), a nearby subdwarf system consisting of an M dwarf and a white dwarf (Gizis 1998). Being a subdwarf, we assume the M dwarf is old, yet it shows strong Hα emission. Another example is that most of the K dwarfs in the Hyades with X-ray emission are in binaries (RHM). We are investigating whether most (all?) binary systems are active, and as a consequence the importance of rotation in promoting activity at various masses. As a preliminary test case, we observed a few stars in the fairly old cluster NGC 2420 (age $\sim$ 3–4 Gyrs) which shows a well-populated and quite pronounced binary sequence offset from the main sequence. We chose four stars with V-I = 1.6, three on the main sequence and one on the binary sequence. This color was the faintest we could observe with Keck, but is still not red enough to probe stars that should still have activity at this age, according to our relations shown in Figure 5. As expected, the three stars on the main sequence did not show activity, and the one star on the binary sequence also had no Hα emission. These results are by no means conclusive, and we intend to follow up with observations in other clusters to further investigate the connection of rotation with activity in low mass stars.

6. Summary

Our investigation of chromospheric activity in field stars and open clusters has led to several intriguing puzzles, and as yet no definitive answers to our questions regarding the connection between rotation and activity in these low mass stars. This presents a challenge that Brendan would have greatly enjoyed. We will miss his enthusiasm and insight in our further studies in this field. We close with a favorite Irish saying of his: "go n-eiri an bothar leat", which means "may the road rise to meet you". Hopefully our understanding of magnetic activity will find a smooth road in the future.

Acknowledgments. SLH was partially supported by NSF (NYI) grant AST 94-57455. Many of the observations referred to here were made at Cerro Tololo Inter-American Observatory, Palomar Observatory and the Keck Observatory.

References

Anthony-Twarog, B.J., Heim, E.A., Twarog, B.A. & Caldwell, N. 1991, AJ 102, 1056

Basri, G. & Marcy, G.W. 1995, AJ 109, 762

Dinescu, D.I., Demarque, P., Guenther, D.B. & Pinsonneault, M.H. 1995, AJ 109, 2090

Gizis, J.E. 1998, AJ 115, 2053

Hawley, S. L., Gizis, J. E. & Reid I. N. 1996, AJ 112, 2799 (PMSU2)

Hawley, S.L., Tourtellot, J.G. & Reid, I.N. 1998, AJ, submitted.

Hawley, S. L., Reid, I.N. & Tourtellot, J.G. 1998, Proceedings of the La Palma conference on Very Low Mass Stars and Brown Dwarfs in Stellar Clusters and Associations, ed. R. Rebolo, (Cambridge University Press), in press

Joy, A.H. & Abt, H.A. 1974, ApJS 28, 1

Reid, I.N., Hawley, S.L. & Gizis, J.E. 1995, AJ 110, 1838 (PMSU1)

Reid, I.N., Hawley, S.L. & Mateo, M. 1995, MNRAS 272, 828 (RHM)

Reid, I.N. & Hawley, S.L. 1998, AJ in press.

Skumanich, A. 1972, ApJ 171, 565

Stauffer, J.R. et al. 1997 ApJ 479, 776

Stauffer, J.R., Liebert, J., Giampapa, M., Macintosh, B., Reid, N. & Hamilton, D. 1994, AJ 108, 160

Stauffer, J.R., Giampapa, M.S., Herbst, W., Vincent, J.M., Hartmann, L.W. & Stern, R.A. 1991 ApJ 374, 142

Wielen, R. 1975, Highlights in Astronomy ed. G. Contopolous,(D. Reidel - Dordrecht-Holland), 3, 395

Solar and Stellar Activity: Similarities and Differences
ASP Conference Series, Vol. 158, 1999
C.J. Butler and J.G. Doyle, eds.

X-rays from Open Clusters

R.D. Jeffries

Department of Physics, Keele University, Keele, Staffordshire ST5 5BG, UK

Abstract. The present state of X-ray observations of cool stars in coeval open clusters is reviewed. Concentrating on *ROSAT* results for solar-type stars, the available observational dataset is summarized along with those details of the evolution of X-ray activity of low mass stars which have been *firmly* established as a result. Observational questions which are as yet unresolved are then addressed, including the origin of "super-saturation" and whether observations of one cluster can represent the X-ray properties of all clusters at the same age. The role of high spatial resolution X-ray imaging as a tool for identifying cluster members is highlighted and the prospects for future developments with *AXAF* and *XMM* are discussed.

1. Introduction

Open clusters are perhaps the best laboratories in which to study the evolution of stellar properties. One might hope they contain stars with a variety of masses, but with roughly the same age, distance and composition. It is not surprising then that much X-ray satellite time has been devoted to the study of open clusters in an attempt to elucidate the roles that rotation, age, composition, binarity, mass and initial conditions play in determining the levels of X-ray emission from a star, if indeed such a deterministic approach is possible.

A major achievement of the *Einstein* observatory in the 1980s was to show that solar analogues in young clusters could exhibit coronal X-ray activity orders of magnitude greater than our own Sun and that this activity declined as the fast rotating young stars were magnetically braked and lost their initial angular momentum - the *age-rotation-activity paradigm* (ARAP). A clear goal for the stellar X-ray astronomer would be to describe the history of coronal activity in our own solar system, but as we shall see, this is not yet possible in detail. Furthermore, a comprehensive empirical understanding of X-ray luminosity may provide a route to studying the age distributions of other stellar populations. In the era of the *ROSAT* satellite the ARAP has been largely confirmed, although various details remain puzzling. Confidence in the ARAP means that X-ray observations can now be used as a tool to find and study the low mass members of open clusters, in circumstances where optical methods of membership selection are difficult or nearly useless.

2. The ROSAT Era

The *Einstein* Observatory Imaging Proportional Counter (IPC) showed that stars throughout the H-R diagram emit X-rays, particularly those in close binary systems (RS CVns) and the low mass stars of young open clusters like the Pleiades and Hyades. X-ray luminosity in cool stars with convective envelopes clearly declined with age (Micela et al. 1990), but it had become clear that the primary determinant of X-ray activity was the rotation rate. This was interpreted as a natural consequence of the dynamo mechanism whereby the magnetic fields which confine and heat the coronae, are amplified by rotation and convection. Indeed, X-ray activity, expressed as the ratio of L_X to bolometric luminosity (L_X/L_{bol}), was found to be even better correlated with Rossby number (N_R), the ratio of rotation period to convective turnover time (Dobson & Radick 1989). The decline of X-ray activity with age was then simply explained in terms of the angular momentum loss (AML) suffered by young, single stars, with tidally locked, short period binary stars remaining a high L_X polluting factor due to their continued fast rotation. The discovery of spreads in rotation rate among low mass stars at the same age (Stauffer 1991) could then also explain why the X-ray luminosity functions (XLFs) of the Pleiades and Hyades showed some overlap. X-ray emission from higher mass stars was either interpreted as due to a shallow convective layer (early F stars), a lower mass binary companion (A stars) or intrinsic emission from a shocked, radiatively driven wind (early B stars).

The *Einstein* observations of open clusters were only partially satisfactory. The relatively low sensitivity meant that many cluster members (especially K and M stars) were undetected and much of the analysis relied on statistical treatments of upper limits, with consequently uncertain XLFs. Many questions were left hanging, such as: If the ARAP operates in clusters, why is there such a small range of L_X in the Pleiades where there is more than an order of magnitude spread in rotation rate? What happens to the X-ray activity of stars younger than the Pleiades ($\sim$ 100 Myr) and older than the Hyades ($\sim$ 700 Myr)? How much of the spread in X-ray activity at a given age can be attributed to short timescale variability or magnetic activity cycles? Is the activity of one cluster necessarily representative of all clusters at the same age? The launch of *ROSAT* in 1990 offered the opportunity to answer these questions. Its Position Sensitive Proportional Counter (PSPC) and High Resolution Imager (HRI) had greater sensitivity, higher spatial resolution and in the case of the PSPC, more spectral resolution than the IPC.

Table 1 is an update from the reviews of Caillault (1996) and Randich (1997), which summarizes the *ROSAT* dataset on clusters (older than 10 Myr). References to published results are given or if unpublished, the PI on the observation is indicated. Ages and distances are adopted from the Lyngå (1987) catalogue and should be treated with caution! There are now deep, reasonably consistently calibrated soft X-ray observations of more than 25 open clusters. This massive database has answered most of the questions posed by *Einstein*, but yielded a number of new mysteries. In particular the XLFs of F G and K stars are now well determined, with few upper limits, in several open clusters at ages from 30 Myr to 600 Myr.

Table 1. Open clusters observed with *ROSAT* (ages and distances from Lyngå 1987).

Cluster	Log Age (yr)	Distance (pc)	*ROSAT* Instrument[a]	References
IC 2602	7.00	155	PSPC (R)	Randich et al. 1995, A&A, 300, 134
NGC 2232	7.35	400	HRI (P)	PI Prosser
Col 140	7.35	300	HRI (P)	PI Prosser; PI Theissen
IC 2391	7.56	140	PSPC (P)	Patten & Simon 1993, ApJ, 415, L123 + Patten & Simon 1996, ApJS, 106, 489
			HRI (P)	Simon & Patten 1998, PASP, 501, 624
IC 4665	7.56	430	HRI (P)	Giampapa et al. 1998, ApJ, 501, 624
			HRI (P)	PI Giampapa
NGC 2451	7.56	220	HRI (P)	PI Schmitt; PI Huensch
Blanco 1	7.70	190	HRI (P)	Micela et al. 1999, A&A in press
Alpha Per	7.71	170	PSPC (R)	Randich et al. 1996, A&A, 305, 785
			PSPC (P)	Prosser et al. 1996, AJ, 112, 1570
NGC 2547	7.76	400	HRI (P)	Jeffries & Tolley 1998, MNRAS, 300, 331
NGC 2422	7.89	480	PSPC/HRI	Barbera et al. 1996, CSSS9, p.355
Pleiades	7.89	125	PSPC (P)	Stauffer et al. 1994, ApJS, 91, 625 + Gagné et al. 1995, ApJ, 450, 217
			PSPC (P)	Micela et al. 1996, ApJS, 102, 75
			PSPC (S)	Schmitt et al. 1993, A&A, 277, 114
			HRI (P)	Harnden et al. 1996, CSSS9, p.359
			HRI (P)	PI Prosser
Stock 2	8.00	320	HRI (P)	PI Sciortino
NGC 2516	8.03	440	PSPC (P)	Dachs & Hummel 1996, A&A, 312, 818
			PSPC (P)	Jeffries et al. 1997, MNRAS, 287, 350
			HRI (P)	PI Micela
NGC 1039	8.29	440	HRI (P)	PI Simon
NGC 6475	8.35	240	PSPC (P)	Prosser et al. 1995, AJ, 110, 1229
			PSPC (P)	James & Jeffries 1997, MNRAS, 292, 252
			HRI (P)	PI James
NGC 7092	8.43	270	HRI (P)	PI Favata; PI Micela
NGC 3532	8.54	500	HRI (P)	PI Simon
Coma Ber	8.60	86	PSPC (P)	Randich et al. 1996, A&A, 313, 815
IC 4756	8.76	400	HRI (P)	Randich et al. 1998, A&A, 337, 372
NGC 6633	8.82	320	PSPC (P)	Totten et al. these proceedings
Hyades	8.82	48	PSPC (S)	Stern et al. 1992, ApJ, 399, L159; Stern et al. 1995, ApJ, 448, 683
			PSPC (P)	Pye et al. 1994, MNRAS, 266, 798
			PSPC (P)	Stern et al. 1994, ApJ, 427, 808
			PSPC (P)	Reid et al. 1995, MNRAS, 272, 828
			HRI (P)	PI Walter
Praesepe	8.82	180	PSPC (R)	Randich & Schmitt 1995, A&A, 298, 115 + Barrado et al. 1998, ApJ in press
NGC 6940	9.04	800	PSPC (P)	Belloni & Tagliaferri 1997, A&A, 326, 608
NGC 752	9.04	400	PSPC (P)	Belloni & Verbunt 1996, A&A, 305, 806
NGC 3680	9.26	800	HRI (P)	PI Tagliaferri
IC 4651	9.38	710	PSPC (P)	Belloni & Tagliaferri 1998, A&A, 335, 517
M67	9.60	720	PSPC (P)	Belloni et al. 1993, A&A, 269, 175
NGC 188	9.70	1550	PSPC (P)	Belloni 1997, MemSAIt, 68, 993

[a] P - pointed observation, R - raster scan observation, S - all-sky survey observation

2.1. The Age-Rotation-Activity Paradigm

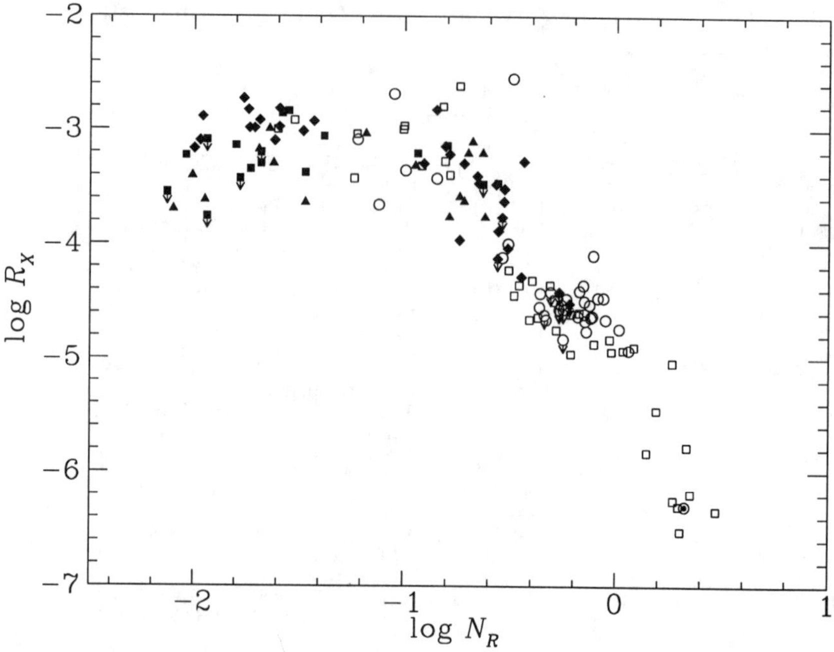

Figure 1. X-ray activity (L_X/L_{bol}) as a function of N_R for late type stars (F5-M5). Data are for IC2391 (filled triangles), α Per (filled squares), Pleiades (filled diamonds), Hyades (open circles) and field stars (open squares). (From Patten & Simon 1996)

At the same time as the emergence of new *ROSAT* results, observations of rotational broadening ($v \sin i$– see Stauffer 1991) or photometrically determined rotation periods (*e.g.* Prosser et al. 1995 and refs. therein) have allowed the evolution of stellar rotation with age to be studied in detail. The prevailing interpretation of these data is that young PMS stars contract and spin-up as they approach the ZAMS after magnetically uncoupling from any circumstellar disk. A variety of disk coupling lifetimes leads to more than an order of magnitude spread in rotation rates at the ZAMS (Bouvier et al. 1997). Meanwhile, angular momentum (AM) is lost via a magnetized stellar wind and in order to maintain a big spread in rotation rates, saturation of the AML rate is needed in the fastest rotators. Once on the ZAMS, with a constant moment of inertia, stars lose AM at a mass-dependent rate. Spindown timescales vary from a few tens of Myr for G stars to a few hundred Myr and longer for K and M stars. To reproduce this in models requires that the rotation rate above which AML saturation takes place is such that saturation occurs at *roughly* constant N_R (Krishnamurthi et al. 1997).

Coronal activity depends on rotation through the dynamo process, hence a correlation between X-ray activity (usually expressed as a fraction of bolometric luminosity - L_X/L_{bol}) and rotation is expected. This was well demonstrated by Stauffer et al. (1994) in the Pleiades, Randich et al. (1996a) in the α Per cluster and Stauffer et al. (1997) in IC 2391/2602. If one looks at G stars in the Pleiades for instance, stars with $v \sin i < 15 \mathrm{km\,s^{-1}}$ show an order of magnitude spread in L_X/L_{bol}, with the fastest rotators having the highest activity. This is even more clearly seen when using the more precise $v \sin i$ measurements of Queloz et al. (1998). However, the surprising result was that above a $v \sin i$ of $\sim 15 \mathrm{km\,s^{-1}}$, L_X/L_{bol} seems to reach a saturated plateau value of 10^{-3}. The reason for this saturation is still unknown. It may be caused by filling of the available coronal volume with plasma or it may be due to a negative feedback in the dynamo mechanism itself. Saturation explains why there is a limited range in L_X among Pleiads of a given mass, even though rotation rates vary by more than a factor 10.

There is a considerable scatter around the rotation-activity correlation when stars with higher or lower masses are added. Following earlier work on magnetic activity by Dobson & Radick (1989), it has been found that including a parameter describing the convective cell turnover time, significantly improves correlations and unifies the view of dynamo generated coronal activity. The new parameter of choice is N_R. Figure 1 (from Patten & Simon 1996) shows data from IC 2391, α Per, the Pleiades, Hyades and main sequence field stars. There is still some scatter around this relation, which may be due to time variability/activity cycles, but there appears to be a flattening of activity at $\log N_R = -0.8 \pm 0.2$, *irrespective of spectral type*, which compares favourably with similar analyses by Stauffer et al. (1997) and Queloz et al. (1998). Because convection zones get deeper and convective turnover times longer in lower mass stars, a uniform saturation N_R indicates a rotation period for saturation that gets longer at lower masses. It is tempting to speculate (see Krishnamurthi et al. 1998) that the coronal saturation coincides with the mass dependent saturation of AML that seems to be required in rotational evolution models (Barnes & Sofia 1996, Krishnamurthi et al. 1997). As yet, the details of (for instance) internal differential rotation are too uncertain for this speculation to be confirmed.

The age dependence of X-ray emission is thought to arise solely as a consequence of the rotation dependence. A clear demonstration of the effect is given by Caillault (1996 - Figure 1). The XLFs of young clusters have a spread *approximately consistent* (see below) with the rotation-activity relation (and saturation) and have peak and median values of L_X that decrease with age at a mass-dependent rate. A telling confirmation of the ARAP is that the spectral type of stars at which the saturation value of L_X/L_{bol} is achieved gets cooler in older clusters as the higher mass stars spindown below their saturation threshold. Unfortunately, the convergence of rotation rates in solar-type stars by the time they reach the Hyades age, means that at the moment we cannot say much about the history of our own Sun's activity prior to this epoch. However, the dispersion in rotation and hence coronal emission remains for somewhat longer in lower mass stars (Stern et al. 1995; see also Hawley in these proceedings), because of their longer spindown timescales.

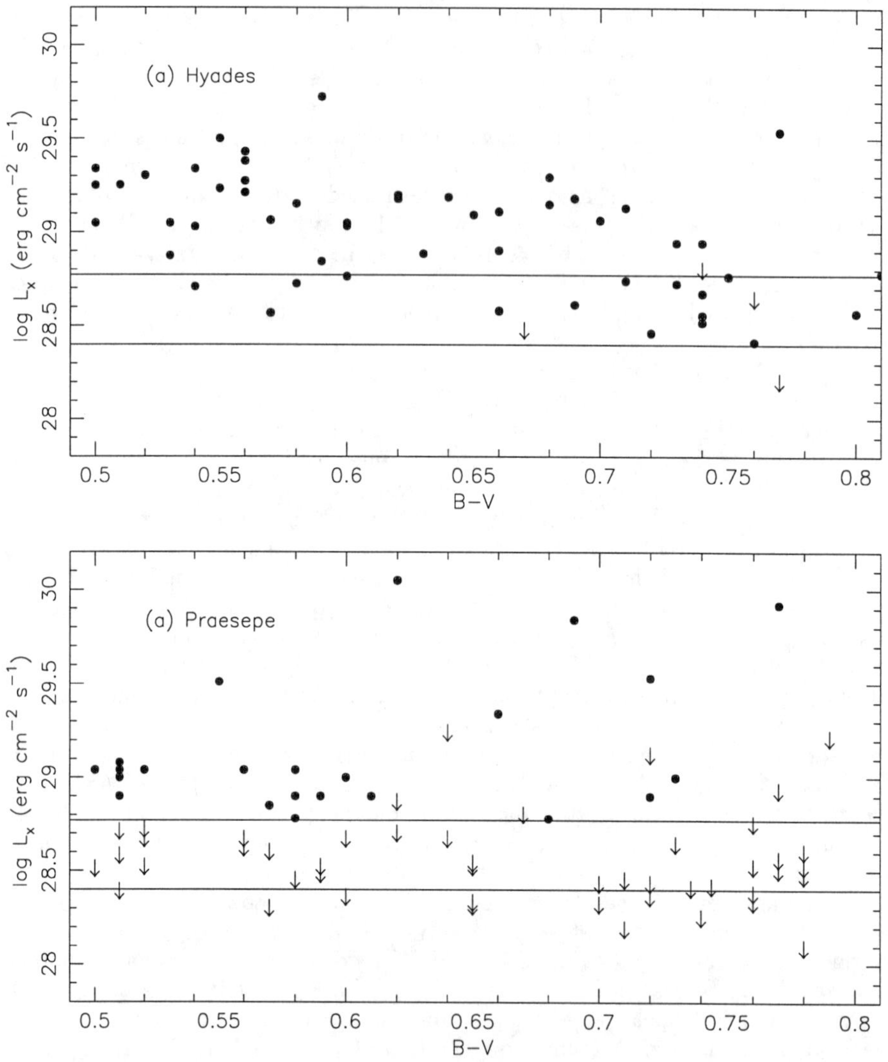

Figure 2. The L_X distribution for solar-type stars in the (a) Hyades and (b) Praesepe. The two horizontal lines indicate the approximate sensitivity limits of the *ROSAT* surveys for the Hyades (lower line) and Praesepe (upper line) The average level of X-ray emission in Praesepe is significantly lower than the Hyades (adapted from Barrado et al. 1998).

2.2. Time variability

Prior to *ROSAT* measurements, little was known about coronal variability on timescales of months or years and whether it might be responsible for the spreads in L_X seen in young clusters. Coronal variability is reviewed in these proceedings by Stern, so only a brief summary is given here. Solar coronal variability is $>$ a factor of 10 in soft X-rays during its activity cycle (Peres et al.– these proceedings), but how do younger stars behave? The Hyades and Pleiades now have multiple epoch X-ray observations. Both Gagné et al. (1995) and Micela et al. (1996) show that variability in Pleiades stars is smaller than solar variations on timescales of 6 months to ten years. Perhaps 20% to 40% are variable by as much as a factor 2. This might be thought due to the ceiling on X-ray emission provided by saturation, but Stern et al. (1995) show that similar results hold for G and K stars in the Hyades, which are not rotating fast enough for saturation. Stern et al. suggest that the difference in the behaviour of the Sun and younger stars might result from the action of a "turbulent" dynamo in their convection zones, which does not exhibit the cyclical behaviour seen in the Sun. In any case it could be considered fortunate that time variability is not sufficient to cause the spreads in activity seen in young clusters and cannot disguise the ARAP. Unfortunately the L_X of single solar-type stars in older clusters is too low for *ROSAT* – so we still do not know at what age solar-type variability sets in, although there are numerous examples of short-period binary systems which have been detected at the expected levels of emission (*e.g.* Belloni 1997 and refs. therein).

3. Unsolved Mysteries

Despite the success of the ARAP in describing most X-ray observations of open clusters, there remain a number of problems that either require a rethink or extension of the paradigm. Among these are the possibility of a "third parameter problem", the phenomenon of "supersaturation" and whether one cluster at a given age has X-ray properties representative of all similar clusters.

That rotation and spectral-type (or mass) alone might be insufficient to determine X-ray activity was postulated by Micela et al. (1996). They claim that the spread in X-ray activity among slowly rotating G/K Pleiades stars is too large to be accounted for by uncertainties in flux, inclination angle or variability. Similarly, Figure 10 in Stern et al. (1995) shows more than an order of magnitude spread in L_X for F8-G5 Hyades stars, even though their rotation rates are thought to be almost uniform and their variability is demonstrated to be small in the same paper. Micela et al. suggest that the internal rotation profile may be the ingredient that is missing from the ARAP. I believe that this problem may yet be due to a combination of errors in $v \sin i$ measurements (in the case of K stars) and grouping stars with a range of convective turnover times (for late F and G stars). Using more accurate $v \sin i$ measurements, Queloz et al. (1998) show that $L_X/L_{\rm bol}$ is well correlated with $N_R/\sin i$, with only about a factor of 3 spread, which is perfectly consistent with uncertain inclination angles, X-ray flux variability and errors. The key to resolving this issue definitively is to obtain accurate rotation periods for many more stars, rather than new X-ray observations.

"Supersaturation" is the observed phenomenon that at very fast rotation rates, L_X/L_{bol} appears to decrease by a factor of 3-5 below the canonical saturation limit of 10^{-3} (Prosser et al. 1996; Randich 1998). Because it affects only the fastest rotating late-type stars ($v \sin i >$ 100km s^{-1}) and there are only a few ($\sim$ 10) of these in the very young α Per and IC 2391/2602 clusters, it is still not clear whether supersaturation sets in at a particular rotation rate or a particular N_R. The latter is not favoured by observations of dMe stars (see James et al. in these proceedings). An explanation for supersaturation is not obvious. Randich (1998) suggests that it may represent a fall-off of the dynamo mechanism itself or perhaps a shift in the distribution of the radiative losses out of the *ROSAT* band to either hotter or cooler temperatures. The latter explanation may draw some support from the few X-ray spectra available for stars in the Pleiades (Gagné et al. 1995) which indicate that the fastest rotating G stars have hotter coronae than the slow rotators. Unfortunately there are almost no nearby field stars that rotate fast enough to exhibit supersaturation *and* have accurate X-ray spectra that could test this hypothesis. Perhaps a more likely explanation is centrifugal shrinkage of the available coronal volume at very high rotation rates.

Most of the early interpretation of X-ray observations relied on the assumption that one cluster was representative of all clusters at the same age. In the *ROSAT* era, this assumption can be tested (see Table 1). One of the most important results of this decade, illustrated in Figure 2, was that the coronal activity of low mass stars (especially solar-type) in Praesepe was significantly lower on average than those in the Hyades at the same age, although peak L_X values were similar (Randich & Schmitt 1995). Explanations range from contamination by non-members in the Praesepe sample, differing binary fractions or orbital distributions, differing initial conditions (specifically the AM distribution) or rotational evolution (perhaps influenced by composition differences – Jeffries et al. 1997). Mermilliod (1997) shows that the rotation rate distributions in the two clusters are similar and Barrado y Navascués et al. (1998) find that contamination with non-members is unlikely. Although much more work needs to be done on identifying and parameterizing binaries in Praesepe, a genuine explanation remains elusive. Observationally, the situation has now become more confusing. Randich et al. (1996b) find that the pattern of X-ray emission in the F/G stars of the similarly aged Coma Berenices cluster resembles the Hyades rather than Praesepe, whereas the X-ray activity of F/G stars in NGC 6633 and F stars in IC 4756 is less than the Hyades (Totten et al. these proceedings, Randich et al. 1998). Deeper observations of Praesepe, NGC 6633 and IC4756 to remove the upper limits will certainly assist interpretation, as will careful optical work to identify spectroscopic binaries. AXAF offers the opportunity to extend these tests to more distant open clusters with a range of ages, composition and richness.

4. High Spatial Resolution

High spatial resolution is useful in several distinct ways in studies of open clusters. First, as one moves towards more distant open clusters, the angular separation between cluster X-ray sources becomes small and accurate analysis becomes

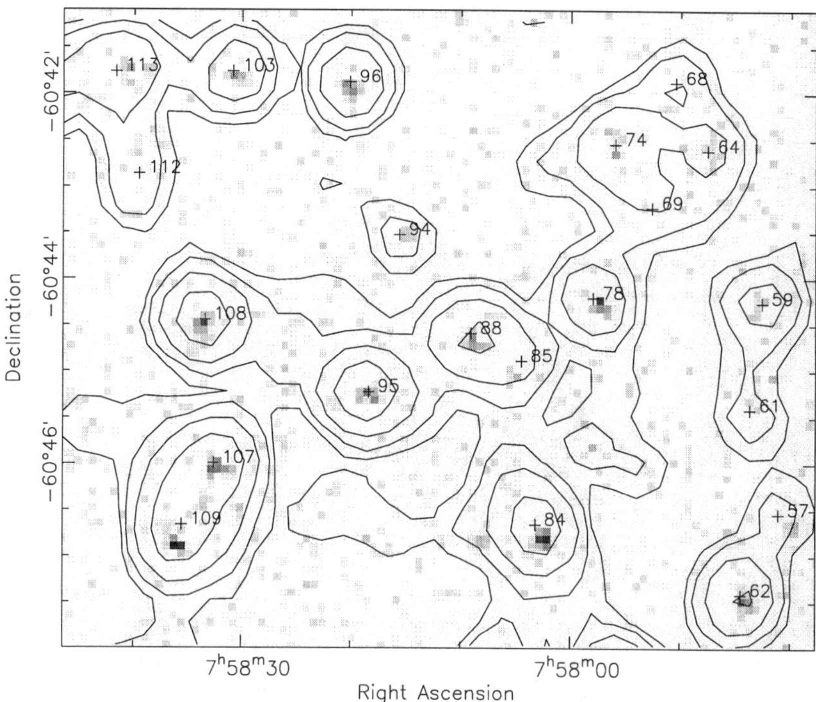

Figure 3. The core of NGC 2516 as seen by the PSPC (contours) and by the HRI (greyscale). The numbered sources are from Jeffries et al. 1997.

difficult as PSFs start to overlap. A good example is NGC 2516 (Jeffries et al. 1997), where 159 X-ray sources were found within 20$'$ of the centre of a PSPC pointing, and multiple PSF fitting (analogous to DAOPHOT for optical photometry) was required. This cluster has been re-observed at about 8 times the spatial resolution (but less sensitivity) with the HRI (FWHM$\sim 3''$) and the results are shown in Figure 3. Essentially all the sources found by the HRI are also extracted from the PSPC data at the right positions. At this sensitivity NGC 2516 represents the most distant cluster (for its age) which could be studied with the spatial resolution of the PSPC. As nowhere near all the cluster members were detected by the PSPC one would like to go deeper, but it is clear that HRI-like spatial resolution will be required to do this. The problem will be more acute for more distant clusters with a similar "richness".

A different problem is the identification of X-ray sources with their optical counterparts. In clusters close to the Galactic plane (and more distant clusters will tend to be so), there may be several candidate stars within each X-ray error circle. Clearly, improvements in spatial resolution concomitantly reduce the number of possible optical counterparts. For most nearby clusters the increase in resolution from PSPC to HRI is not too important from this perspective (*e.g.* Simon & Patten 1998), but for others near the Galactic plane or for deeper

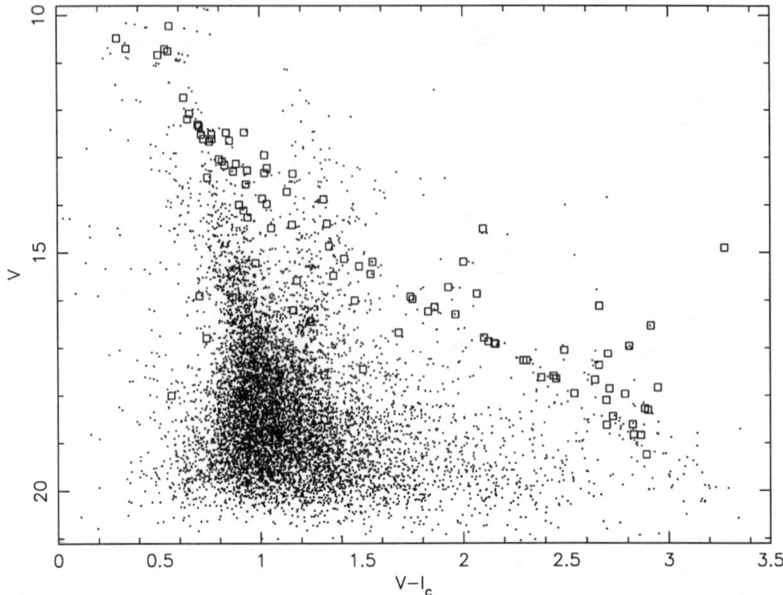

Figure 4. A colour-magnitude diagram for NGC 2547 together with those stars residing inside HRI X-ray error circles (squares).

studies with fainter possible optical counterparts, higher spatial resolution than offered by the PSPC is absolutely essential (*e.g.* Giampapa et al. 1998).

A related problem that plagues studies (not just in X-rays) of low mass stars in open clusters is a lack of firm membership lists for faint stars. Proper motions and photometric selection can be useful among nearby clusters, but as one moves to more distant clusters, which tend to be projected against the Galactic plane, both background contamination and small proper motions become problematic. An alternative approach, which relies on the ARAP, is to find low mass members by X-ray selection, because the contrast in X-ray flux between young, low mass cluster stars and a general field population is large. This approach has now been used in a number of young clusters where optically selected membership catalogues are either absent or very uncertain (*e.g.* Prosser et al. 1994; Patten & Simon 1996). A combination of photometric selection plus small X-ray error circles can be especially powerful and if the X-ray survey is deep enough then one can be reasonably sure that the selected members are a complete sample, rather than biased towards high activity levels. A recent example is shown in Figure 4. NGC 2547 is a young cluster observed with the HRI (Jeffries & Tolley 1998). The ~ 6 arcsec error circles contain one star on average and the majority of these form a sequence along the expected locus of the cluster in a colour-magnitude diagram. Less than one correlation is expected by chance in this part of the CMD, so these stars must be genuine cluster members. Furthermore, the X-ray survey is sufficiently deep that the X-ray selected G and K stars are expected to be a complete sample from which other investigations can be based.

5. AXAF and XMM

High resolution spatial imaging, moderate spectral resolution and increased effective area over *ROSAT* are the key attributes of the *AXAF* satellite. The ACIS instrument is capable of 1 arcsec resolution with spectral resolution of ~ 20 at ~ 2 keV over a 16 arcmin field. The ACIS sensitivity threshold will be a factor of 3-5 better than the PSPC/HRI for the same exposure time. The HRC instrument has even better spatial resolution over a larger 31 arcmin field but little spectral resolution. *XMM* has a larger overall effective area and the capability of providing spectral resolution of several hundred for individual targets. Together, the capabilities of the two instruments will make significant advances in our understanding of the X-ray evolution of cool stars in open clusters.

- For nearby clusters (the Pleiades and NGC 2516 in AO-1), X-ray emission can be probed further down the main sequence to see if the nature of the dynamo changes in fully convective stars or even brown dwarfs.

- The spectral resolution will allow detailed modelling of the X-ray spectra. At present, a single, crude spectral model is normally assumed to convert from count rate to flux. We will be able to see if coronal temperatures continue to increase in the most rapidly rotating stars and whether a shift of flux to higher energies causes supersaturation in the *ROSAT* pass band. *XMM* will be able to provide detailed, high resolution spectroscopic studies of even low activity stars in both the Hyades and Pleiades at L_X thresholds of roughly 10^{28} and 10^{29} erg s^{-1} respectively, in reasonable exposure times.

- For the first time, the detection of X-ray emission from main sequence stars in clusters older than the Hyades will be possible. The *AXAF* mission time should be long enough to also get second epoch observations to study variability in these older clusters.

- *AXAF* should be able to study clusters out to distances of 1 kpc and the high spatial resolution means that even clusters with low b will be accessible. Deeper observations of clusters at the same age as the Hyades are required to remove the remaining ambiguities posed by *ROSAT* upper limits. Several clusters at the same age as the Pleiades (~ 100 Myr) and NGC 6475 (~ 300 Myr) can be studied.

- More distant, compact open clusters can be examined to find more of the rare open clusters with ages between 10 and 40 Myr, which are so important in understanding the history of AML and circumstellar disks.

Acknowledgments. I would like to thank Giusi Micela, David Barrado y Navascués and Brian Patten for providing data and figures for this review.

References

Barnes, S. & Sofia, S. 1996, ApJ, 462, 746

Barrado y Navascués, D., Stauffer, J. R. & Randich, S. 1998, ApJ in press

Belloni, T. 1997, Mem.S.A.It, 68, 993

Bouvier, J., Forestini, M. & Allain, S. 1997, A&A, 326, 1023

Caillault, J. -P. 1996, in Ninth Cambridge Workshop on Cool Stars, Stellar Systems and the Sun, R. Pallavicini & A. K. Dupree, San Francisco: ASP Conf. Ser. Vol. 109, p. 325

Dobson, A. K. & Radick, R. R. 1989, ApJ, 344, 907

Gagné, M., Caillault, J. -P. & Stauffer, J. R. 1995, ApJ, 450, 217

Giampapa, M. S., Prosser, C. F. & Fleming, T. A. 1998, ApJ, 501, 624

Jeffries, R. D., Thurston, M. R. & Pye, J. P. 1997, MNRAS, 287, 350

Jeffries, R. D. & Tolley, A. J. 1998, MNRAS, 300, 331

Krishnamurthi, A. et al. 1998, ApJ, 493, 914

Krishnamurthi, A., Pinsonneault, M. H., Barnes, S. & Sofia, S. 1997, ApJ, 480, 303

Lyngå, G. 1987, Catalogue of open cluster data, 5th ed., Lund Obs.

Mermilliod, J. -C. 1997, Mem.S.A.It, 68, 853

Micela, G., Sciortino, S., Vaiana, G. S., Harnden, F. R., Rosner, R. & Schmitt, J. H. H. M. 1990, ApJ, 348, 557

Micela, G., Sciortino, S., Kashyap, V., Harnden, F. R. & Rosner, R. 1996, ApJS, 102, 75

Patten, B. M. & Simon, T. 1996, ApJS, 106, 489

Prosser, C. F. et al. 1995, PASP, 107, 211

Prosser, C. F., Randich, S., Stauffer, J. R., Schmitt, J. H. H. M. & Simon, T. 1996, AJ, 112, 1570

Prosser, C. F., Stauffer, J. R., Caillault, J. -P., Balachandran, S., Stern, R. A. & Randich, S. 1994, AJ, 110, 1229

Queloz, D., Allain, S., Mermilliod, J. -C., Bouvier, J. & Mayor, M. 1998, A&A, 335, 183

Randich, S. 1998, in Tenth Cambridge Workshop on Cool Stars, Stellar Systems and the Sun, R. A. Donahue, J. A. Bookbinder, San Francisco: ASP Conf. Ser. Vol. 154, p. 501

Randich, S. 1997, Mem.S.A.It, 68, 971

Randich, S. & Schmitt, J. H. H. M. 1995, A&A, 298, 115

Randich, S., Schmitt, J. H. M. M., Prosser, C. F. & Stauffer, J. R. 1996a, A&A, 305, 785

Randich, S., Schmitt, J. H. H. M. & Prosser, C. F. 1996b, A&A, 313, 815

Randich, S., Singh, K. P., Simon, T., Drake, S. A. & Schmitt, J. H. H. M. 1998, A&A, 337, 372

Simon, T. & Patten, B. M. 1998, PASP, 110, 283

Stauffer, J. R. et al. 1997, ApJ, 479, 776

Stauffer, J. R., Caillault, J. -P., Gagné, M., Prosser, C. F. & Hartmann, L. W. 1994, ApJS, 91, 625

Stauffer, J. R. 1991, in Angular momentum evolution of cool stars, S. Catalano & J. R. Stauffer, Dordrecht: Kluwer, p. 117

Stern, R. A., Schmitt, J. H. M. M. & Kahabka, P. T. 1995, ApJ, 448, 683

Solar and Stellar Activity: Similarities and Differences
ASP Conference Series, Vol. 158, 1999
C.J. Butler and J.G. Doyle, eds.

Activity in Pre-main Sequence and Very Active Stars

Frederick M. Walter

Dept. of Physics and Astronomy, SUNY, Stony Brook NY USA

Abstract. The "activity" characteristic of the youngest, and the most rapidly rotating cool stars, is difficult to explain solely in terms of a scaled-up Sun. I present an overview of the observed characteristics of low mass pre-main sequence stars and very active cool stars. To understand the evolution of the activity, I begin with the pre-main sequence stars and extrapolate forward in time, rather than beginning with the Sun and working backwards. This new paradigm for stellar activity is based on the decay of large-scale organized magnetic fields.

1. Introduction

Stellar magnetic activity manifests itself in the chromospheric and coronal emissions ubiquitous among the late-type stars. It is best observed on the Sun, by virtue of the Sun's proximity, and therefore the Sun has become our paradigm for stellar magnetic activity. Attempts to describe the activity seen in other stars in terms of solar-like structures (the *solar analogy*), by varying filling factors, densities, and scale heights, have been reasonably successful (e.g., Schrijver 1996). However, some characteristics of the most active stars are not easily explicable in terms of the solar analogy. This led Walter & Byrne (1998) to posit a non-solar-like magnetic morphology in the most active stars.

Here I review the activity seen in the most active late-type stars, and discuss its implications for stellar magnetic activity in general.

2. The Pre-main Sequence Stars

2.1. Classical T Tauri Stars

The classical T Tauri (cTT) stars (Herbig 1962) are low mass ($\lesssim$2.5 $M_\odot$) pre-main sequence (PMS) stars. At ages of less than a few million years (Myr), they are fully convective, and are contracting on a Kelvin-Helmholtz timescale. Recent reviews of the cTT stars and their evolution include those by Basri & Bertout (1993) and Stahler & Walter (1993).

Among their most striking characteristics are their strong emission lines. Most cTT stars were discovered in objective prism Hα surveys. Their spectra show a wide range in properties, from stars that look like "normal" late-type stars to objects which exhibit bizzare emission lines on a featureless blue continuum [see the Cohen & Kuhi (1979) atlas]. The Hα emission equivalent widths exceed about 10Å; for all but the coolest stars this implies far more emission flux than is seen in any "normal" chromosphere.

The low excitation emission lines (Ca II H&K; Mg II h&k; Na D) do not appear chromospheric in origin: they tend to be broad, with P-Cygni absorptions to the blue indicating significant stellar winds. Reipurth *et al.* (1996) provide an atlas of Hα emission profiles. Figure 1 shows Mg II h&k lines of 6 cTT stars observed with the Hubble Space Telescope.

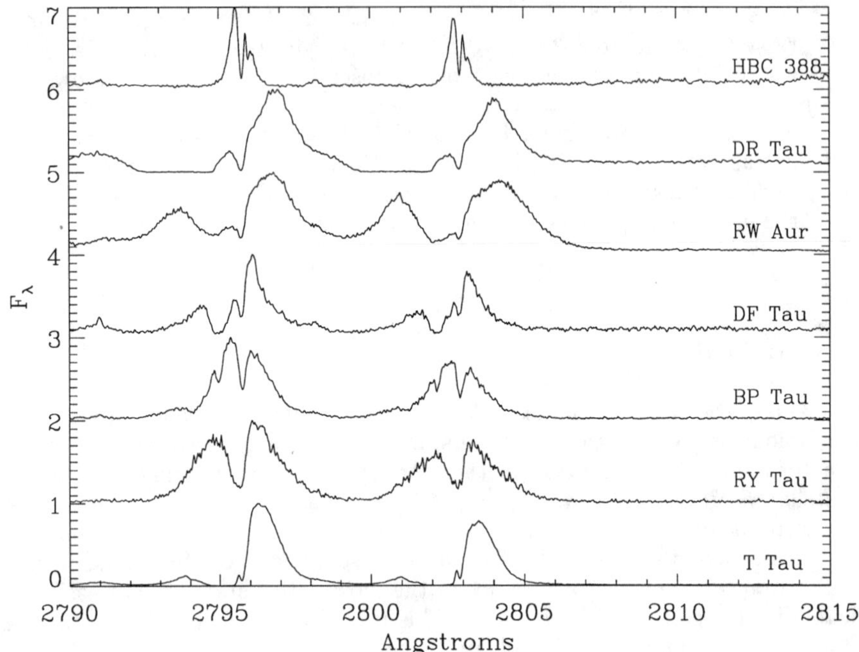

Figure 1. Mg II h&k line profiles for 6 cTT stars and one nTT star (HBC 388). The fluxes are normalized to the peak of the Mg II emission. HBC 388 has a fairly narrow chromospheric line profile, with an interstellar absorption line and an apparent redshifted absorption suggestive of infall. The cTT stars uniformly show very broad lines with blue-shifted absorption indicative of mass loss. The subordinate Mg II lines are visible in the spectra of DF Tau and DR Tau at about 2791 and 2798Å. The P-Cygni absorption in DR Tau really does go to zero.

Far-ultraviolet spectra of cTT stars (e.g., Figure 2) reveal a typical set of chromospheric and transition region lines.

Cram *et al.* (1980) and Brown *et al.* (1982) discussed the emission measure distribution of T Tauri. It is similar to a solar-like chromosphere/transition region, but elevated by 4-5 orders of magnitude. However, the emission measures of the Sun and solar-like stars reach a minimum at the top of the transition region, and then increase into the corona as temperature $T^{\frac{3}{2}}$ (e.g., Bohm-Vitense 1987). This increasing emission measure is not seen in the cTT stars (other-

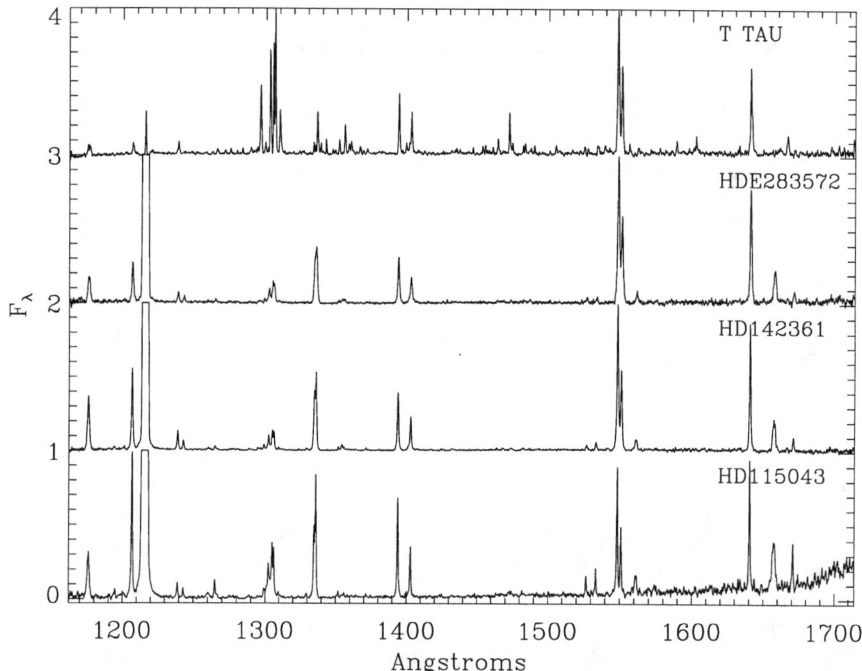

Figure 2. Low dispersion HST/GHRS spectra of 3 PMS stars. The fluxes scales are relative. **Upper panel:** T Tauri. Geocoronal Lyman α and O I emission is negligable in this small aperture observation. In addition to the strong transition region lines (Si IV, C IV), note the strong low temperature lines of O I. Most of the weak emission is H_2. **Middle panels:** The nTT stars HDE 283572 and HD 142361. Lyman α is geocoronal. These appear to be normal cool star spectra, without any H_2 emission. Relative to the hot (C IV, Si IV) lines, the intermediate temperature lines (C II, C III, Si III) are more prominent in the nTT stars. **Lower Panel:** For comparison, the main sequence G0 star HD 115043, a member of the UMa cluster, with an age of about 2.5×10^8 years. The strength of the continuum reflects the relative weakness of the emission lines.

wise they would be among the brightest X-ray sources in the sky); the coronal emission measures lie well below the emission measures in the transition region.

Neuhäuser *et al.* (1995) summarize the X-ray properties of a complete sample of cTT stars in Tau-Aur. While the X-ray fluxes are low in comparison to the UV fluxes, they are not low in absolute terms. The mean X-ray luminosity (taking into account the non-detections) is about 10^{29} erg s^{-1}, and ranges up to about 10^{30} erg s^{-1}, which is comparable to the X-ray luminosity of the most active main sequence stars. The low resolution X-ray spectra show a characteristic coronal temperature of about 1 keV, which is similar to that derived for the most active main sequence and post-main sequence stars.

Detailed spectra of cTT stars with ASCA (e.g., Skinner *et al.* 1997, Skinner & Walter 1998) show the same basic characteristics seen in active late-type stars in general: the data require multiple thermal components, and the inferred coronal abundances are often (but not always) sub-solar.

2.2. The Magnetospheric Accretion Model

The outer atmospheres of the cTT stars have confounded theorists and modelers for decades, because of the great variety of line profiles. While thermal winds cannot explain the Hα line profiles, spherical winds driven by Alfven waves (e.g, Hartmann & Kenyon 1990) showed some promise. However, it now appears that the cTT lines are formed in accreting material. The magnetospheric accretion model (Shu *et al.* 1994), motivated by the observed correlation between slow rotation of the cTT stars and the presence of disks (Edwards *et al.* 1993), posits a large-scale magnetic field threading the protoplanetary disk and enforcing co-rotation of the star with the inner edge of the disk. While there is still a significant stellar wind, the "chromospheric" emission arises in the infalling matter. Rather than spiraling in along the equator to form a boundary layer, the material is lifted centrifugally along the magnetic field lines, and then falls ballistically onto the star at high latitude. The emission arises in these accretion columns. Hartmann *et al.* (1994) show that the variety of Hα line profiles arises from the various inclinations of the stars to the line of sight.

The gravitational potential of a cTT star is not large; gravitational infall can result in temperatures up to only a few $\times 10^5$K. The X-ray emission cannot be produced by the infall. The similarity of the X-ray emission in the cTT stars and the active late-type stars, including the RS CVn systems, which occupy similar positions in the H-R diagram, suggests a common source.

2.3. Naked T Tauri Stars

The naked T Tauri (nTT) stars (e.g., Walter 1986) are contemporaries of the cTT stars. They have the same range of masses, ages, and radii. They have the colors of "normal" stars, and lack the continuum excesses of the cTT stars. Although the mean soft X-ray luminosities are larger, and the mean extinctions are lower (e.g., Neuhäuser *et al.* 1995), the X-ray temperatures and surface fluxes are comparable to those of the cTT stars and active late-type stars. These stars lack optically-thick circumstellar disks, and rotate faster than do the cTT stars[1]. Rotation periods range from 12 hours to about 5 days, with the more massive stars generally having shorter periods.

Their UV spectra (Figures 1 and 2) appear normal. The Hα emission is at the chromospheric level (Figure 3); Hα line profiles (see, e.g., Walter *et al.* 1988, 1994, 1997) are narrow, aside from rotational broadening, with dMe-like line profiles (see Figure 5).

---

[1]Stars spin up from conservation of angular momentum as they evolve down their Hayashi tracks. The cTT stars are relatively slow rotators, with mean periods near 8 days, because their rotation is braked by the interactions of the magnetic field and the disk. Once the disk and the magnetic field decouple, the star is free to spin up. This is thought to happen when the inner disk becomes optically thin.

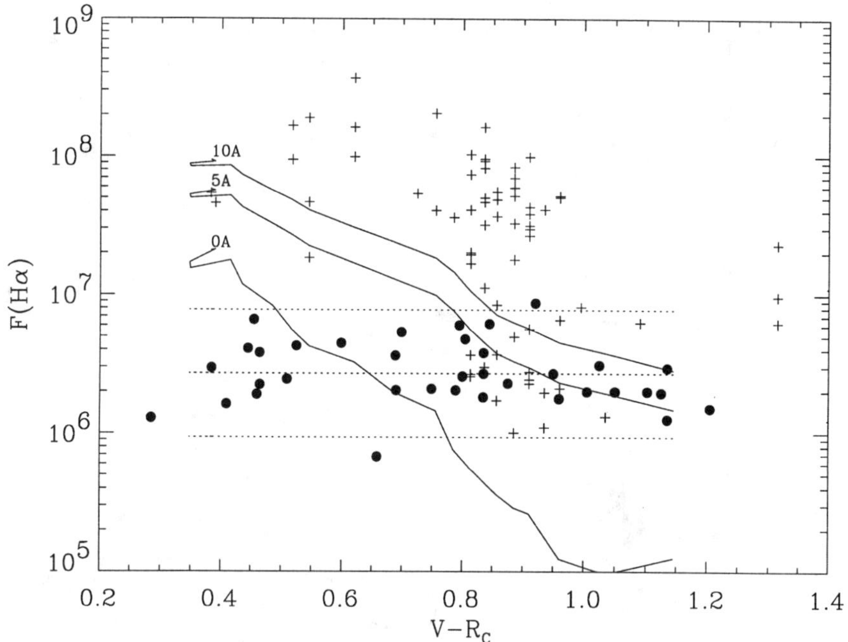

Figure 3. The Hα surface fluxes (erg cm^{-2} s^{-1}) for a sample of pre-main sequence stars in the Tau-Aur star forming region, plotted as a function of their unreddened Cousins V-R color. The plus signs indicate the cTT stars; filled circles are nTT stars (Walter *et al.* 1988). Lines of constant Hα equivalent width are overplotted. The Hα fluxes of the nTT stars are similar to those of stars with highly active chromospheres (e.g., dMe stars and RS CVn systems); the horizontal lines mark the mean and the $\pm 2\sigma$ ranges. Some of the cooler cTT stars, selected by Hα equivalent width, may actually be diskless nTT stars.

The nTT and cTT coronae are similar. The spectrum of HDE 283572 (Favata, Micela, & Sciortino 1998) is similar to that of SU Aur (Skinner & Walter 1998). This supports the idea that the cTT corona is not driven by the mass accretion process. Various diagnostics of the X-ray emission are shown in Figure 4. $\frac{F_X}{F_{bol}}$ is saturated, with the mean $\log(\frac{F_X}{F_{bol}})$=-3.25±0.11, somewhat lower than the mean value seen among active stars (e.g., Stauffer *et al.* 1994).

There is no strong evidence for any direct rotation-activity relation; rather it appears that the X-ray surface flux scales with mass, and the X-ray luminosity increases as a stronger function of mass[2]. Any apparent rotation-activity relation may be due to the inverse correlation between mass and rotation period.

[2]because the radius is an increasing function of mass at this age.

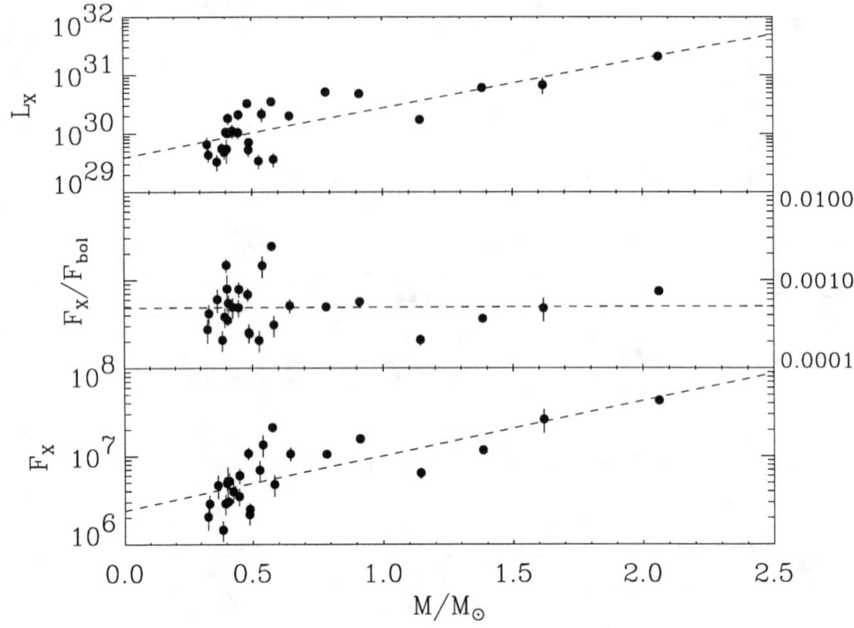

Figure 4. Variation of the X-ray luminosity L_X, $\frac{F_X}{F_{bol}}$, and the surface flux F_X with mass for the nTT stars of the Upper Scorpius OB association (Walter *et al.* 1994). The dashed lines represent least-squares fits to the data. The $\frac{F_X}{F_{bol}}$ ratio is flat; F_X increases as $M^{0.6\pm0.1}$ and L_X increases as $M^{0.8\pm0.2}$. These relations are not universal, and apply only at the ~1 Myr age of the Upper Scorpius association, because the radii and luminosities of PMS stars evolve.

The few Doppler images of nTT stars produced to date (Joncour *et al.* 1994; Strassmeier *et al.* 1994) show evidence of dark polar caps or crowns, similar to those seen in the RS CVn systems and in the ultra-fast rotators.

The Hα profiles of the nTT stars often show evidence for mass infall, either as sporadic episodes of infall (e.g., Wolk & Walter 1996) or as continuous infall. Given that the nTT stars recently had circumstellar disks, residual circumstellar material and accretion is not unexpected. However, very similar characteristics are seen in more evolved active stars (§3.2), and we will return to this.

3. Active Stars

The most active stars include both young and old stars united by a common characteristic: rapid rotation. These rapid rotators include close binaries, where tides force synchronous rotational and orbital periods (RS CVn and BY Dra systems), as well as single stars that have not yet spun down, or have been spun up again (ultra-fast rotators; dMe stars; FK Com stars). There is no clear observational distinction between active main sequence and active post-main sequence stars in terms of activity levels. The characteristics of the most active stars include strong X-ray emission, with $\frac{f_X}{f_{bol}} \sim 10^{-3}$, and strong UV emission lines. In some cases Hα is seen in emission above the continuum. The most active stars exhibit the non-solar-like characteristics described below.

3.1. Large Prominences and Extended Atmospheres

A peculiar characteristic of many of these very active stars is the absorption dips seen (primarily) in the Hα line profile. First reported in AB Dor (Collier Cameron *et al.* 1990), similar absorption dips have now been seen in the dMe stars HK Aqr (Byrne, Eibe, & Rolleston 1996) and RE1816+541 (Eibe 1998). These are thought to be due to cool absorbing clouds or prominences co-rotating with the star at heights of a few stellar radii above the photosphere. van den Oord *et al.* (1998) conclude that the neutral hydrogen clouds seen in this type of star are held in a large-scale magnetic field in the equatorial plane and are dynamically stable. However, they must have significant meridional extent to be seen projected on the stellar disks from these heights. Recently we have found similar absorption dips in AB Dor in the light of Mg II h&k and C IV (Walter *et al.*, in preparation).

Similar prominence-like structures may have been seen in the RS CVn system AR Lac. Neff *et al.* (1989) identified discrete structures with velocity amplitudes in excess of $v \sin i$, which suggests that these structures are suspended *above* the surface but in rigid rotation with the photosphere. A pronounced dark hemisphere on the G star in AR Lac in the 1987 Mg II Doppler image (Neff *et al.* 1989) is likely due to absorption of the chromospheric Mg II emission by an extended cool prominence with a size of order 3 $R_\odot$. X-ray light curves of AR Lac show that primary eclipse is often longer than expected for a pure geometric eclipse, which could suggest the presence of extended structures above the limb of the foreground star[3]. The 1984 light curve shows a slow egress (White *et al.* 1990). The 1993 light curve (White *et al.* 1994 shows a slow ingress, but egress is consistent with a geometric occultation of the background star. Four months later, the EUVE DS/S light curve showed an ingress consistent with geometrical obscuration, but the egress was slow. A simple model (Walter 1995) requires obscuring material out to about 6 $R_\odot$, with electron densities of 10^{11} to 10^{12} cm^{-3}, which is at the upper end of the range observed in solar prominences. In addition, Heuenemorder *et al.* (1989) noted that the $\frac{H\alpha}{H\beta}$ ratio in UX Ari is similar to that in solar prominences, and Hall & Ramsey (1994) modeled the Hα profiles of 4 RS CVn systems as prominence-like structures with scale heights of the order of the stellar radius.

3.2. Infalling Material

A signature of infalling material is redshifted absorption in the Hα emission profile. Discrete infall events are seen as transient redshifted absorption features

---

[3]Eclipse light curves suggest that most of the coronal X-rays do arise in compact structures of approximately solar scale in AR Lac (Walter, Gibson, & Basri 1983).

in Hα. Continuous or permanent infall can cause a noticeable asymmetry in the Hα emission line profile.

We noted above that many of the nTT stars exhibit evidence of infall. There is growing evidence of similar infall in a wide variety of active stars. While the nTT stars may not be completely naked, there is little expectation that the post-main sequence RS CVn systems, or the older ultra-fast rotators, should be accreting. Nevertheless, Byrne (1987) noted a transient absorption feature in the RS CVn system II Peg with an asymptotic velocity close to the free fall velocity, suggesting infall from a great height. Hall & Ramsey (1992) discussed similar events in other RS CVn systems.

An assortment of Hα line profiles showing evidence for continuous infall in ultra-fast rotators, RS CVns, and nTT stars, is shown in Figure 5.

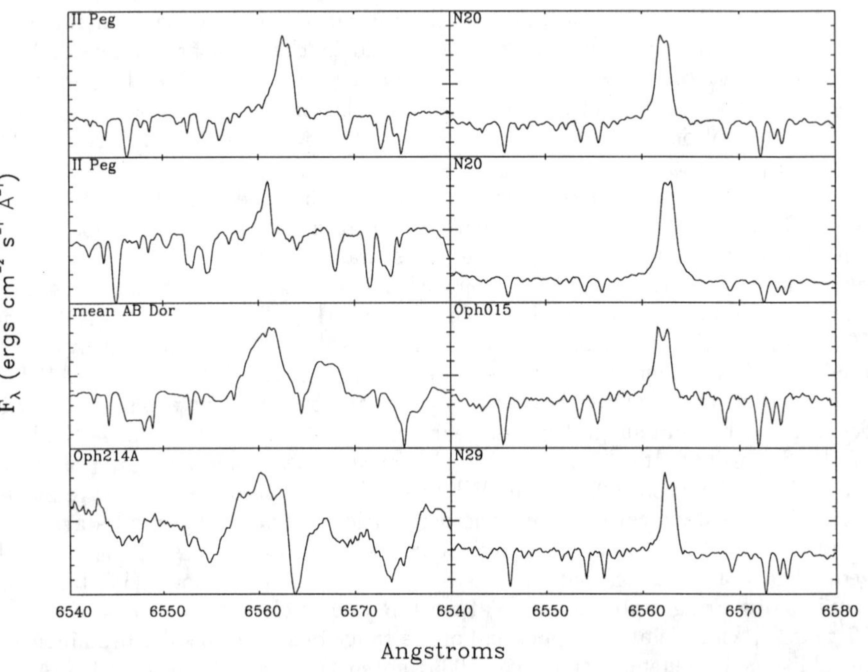

Figure 5. Representative Hα line profiles. The line is expected to be symmetric, but the ratio of the peak blue flux to the peak red flux generally exceeds unity. The pairs of profiles of II Peg and the naked T Tauri star Lk Ca 7 (N20) show variability in a 24 hour interval. The upper spectrum of N20 shows a smooth blue wing, a cut-off red wing, and V>R in the emission peaks, while the lower spectrum is more symmetric. In AB Dor and the nTT star Oph214A (Walter *et al.* 1994) the narrow features are telluric, while the broad central reversals are at the rest wavelength of the star. The dMe-like Hα lines of the other two nTTs have V>R and a sharp cutoff of the red wing (the telluric line does not contribute significantly to the cutoff).

The "quiescent" Hα profile of the quintessential ultra-fast rotator AB Dor shows an absorbed red wing. Assuming an intrinsically symmetric Hα line profile, this requires the presence of material falling at close to the escape velocity. The rapidly rotating late-type dwarf BD+22°4409 (Jeffries *et al.* 1994) also shows clear evidence of permanent inflow in Hα (Eibe, Byrne & Robb, in preparation).

3.3. Extended Magnetic Fields

The most direct evidence for large-scale organized magnetic fields in active stars arises from radio observations. VLBI observations of RS CVn systems show magnetospheres with sizes of a few stellar radii, (e.g., Lestrade *et al.* 1984; Mutel *et al.* 1985). The observed sizes, and the circular polarization, require large-scale, quasi-dipolar magnetic field configurations. Alef *et al.* (1996) show that the radio source associated with YY Gem has an extent of about 2 stellar diameters.

4. A Unified Model

The current interpretation of the cTT stars involves a large-scale organized magnetic field. This can be envisioned as a dipole (though it is probably far more complex), with magnetic lines of force emanating at high latitudes and threading the equatorial disk at heights of 5-10 stellar radii. This magnetic field is a *stellar* field; as the disk dissipates (the dust accumulates into large particles which are effectively invisible), one does not expect the large-scale field to be affected. Since the cTT and nTT stars have similar ages, we may expect them to possess similar large-scale magnetic fields.

There are two sets of observations which may support this picture.

- From equipartition arguments, one expects the photosphere to be cooler (starspots) at the footprints of the magnetic field. A dipolar field emanating at high latitude may be the cause of the polar caps or polar crowns observed in nTT stars.

- A large-scale magnetic field confines a large volume. The nTT stars are luminous radio sources (Brown *et al.* 1996). By analogy to the RS CVn systems, this suggests a large magnetospheric volume.

Then by analogy, since the active stars and the nTT stars have similar polar caps or crowns, large magnetospheres, coronal X-ray characteristics, and evidence for infall, we hypothesize that the large-scale magnetic field persists in these stars.

One can explain the observed infall seen in many active stars as due to material trapped in the large-scale field. This material need not be accreted, but may be supplied by the star itself. All active stars have large X-ray flares, and large flares are often accompanied by coronal mass ejections (CMEs). In the Sun, the CME joins the general flow of the solar wind. However, if a CME occurs under the canopy of a strong quasi-dipolar magnetic field, and if the energy density of the field is high enough, the field could trap the ionized material. This material will then drain back to the surface along meridional magnetic field lines, impacting at high latitudes, but a component may remain trapped in the equatorial plane. Stars inclined to the line of sight by about the latitude of the magnetic footpoints (such as AB Dor) will show accretion velocities close to the free-fall velocity; in systems with lower inclinations the absorption will be at smaller (though still redshifted) velocities (such as BD+22°4409).

5. Evolution of the Activity

Stellar activity can be divided into two components: the solar-like component, which is dominated by the small-scale magnetic structures, and the primordial component, which is dominated by the large-scale organized magnetic field.

Güdel, Guinan, & Skinner (1997) demonstrated that the evolution of the coronae of solar-like stars consists mainly of a rapid fading of a hot (~1 keV) component, and a slow decay of a cooler (~0.2 keV) component. In the active rapid rotator EK Dra the ratio of emission measures of the hot and cool components is about 1:1. Skinner & Walter (1998) and Favata, Micela, & Sciortino (1998) show that the emission measure distribution of PMS stars continue this trend, with about 80% of the coronal emission measure in the hotter component.

The evolution of the coronal X-rays appears smooth from the youngest cTT stars through the Sun. The implication is that the coronal structures responsible for the 0.2 keV plasma are similar on the Sun and stars, while the structures responsible for the 1 keV plasma decay, and are largely absent on the Sun. It is possible to speculate that this component is composed of a superposition of flares, with the flare rate being large enough in the active stars to yield an apparently quiescent hot component, or that the hot plasma is associated with longer, lower density magnetic loops. In static loop models, the maximum temperature scales as the product of the loop length and pressure $(pL)^{\frac{1}{3}}$ (Rosner, Tucker, & Vaiana 1978), so high pressure loops with lengths of order the stellar diameter can account for the higher coronal temperatures.

The large-scale organized field of the cTT stars clearly has weakened considerably by the age of the Sun. However, that component is not absent: the Sun does have a quasi-dipolar large-scale field, as reflected in the shape of the white-light corona, albeit strongly distorted by the solar wind. It is tempting to speculate that the active stars turn into solar-like stars when the stellar wind begins to dominate over large-scale field. This seems to occur in dwarfs with rotation periods of a few days, and in subgiants with periods of about 10 days.

References

Alef, W., Benz, A.O., & Güdel, M. 1997, A&A, 317, 707

Basri, G. & Bertout, C. 1993, in *Protostars and Planets III*, eds. E.H. Levy and J.I. Lunine (Univ. of Arizona Press), p. 543

Bohm-Vitense, E. 1987, ApJ, 317, 750

Brown, A. Ferraz, M.C., & Jordan, C. 1984, MNRAS, 207, 831

Brown, A., Walter, F.M., Ambruster, C., Stewart, R.T., & Jeffries, R. 1996, in *Radio Emission From the Sun and Stars*, eds. A.R. Taylor and J.M. Paredes (San Francisco: ASP), p. 294

Byrne, P.B. 1987, in *Cool Stars, Stellar Systems and the Sun*, eds. J.L. Linsky and R.E. Stencel, (Berlin:Springer-Verlag), p. 491

Byrne, P.B., Eibe, M.T., & Rolleston, W.R.J. 1996, A&A, 311, 651

Cohen, M., & Kuhi, L.V. 1979, ApJS, 41, 743

Collier Cameron, A., *et al.* 1990, MNRAS, 247, 415

Cram, L.E., Giampapa, M.S., & Imhoff, C.L. 1980, ApJ, 238, 905

Edwards, S., *et al.* 1993, AJ, 106, 372

Eibe, M.T. 1998, A&A, 337, 757

Favata, F., Micela, G., & Sciortino, S. 1998, A&A, in press

Güdel, M., Guinan, E.F., & Skinner, S.L. 1997, ApJ, 483, 947

Hall, J.C. & Ramsey, L.W. 1992, AJ, 104, 1942

Hall, J.C. & Ramsey, L.W. 1994, AJ, 107, 1149

Hartmann, L., Hewett, R., & Calvet, N. 1994, ApJ, 426, 669

Hartmann, L. & Kenyon, S.J. 1990, ApJ, 261, 279

Herbig, G.H. 1962, Adv. Astr. Astrophys., 1, 47

Heuenemoerder, D.P., Buzasi, D.L., & Ramsey, L.W. 1989, AJ, 98, 1398

Jeffries, R.D., Byrne, P.B., Doyle, J.G., Anders, G.J., James, D.J., & Lanzafame, A.C. 1994, MNRAS, 270, 153

Joncour, I, Bertout, C., & Bouvier, J. 1994, A&A, 291, L19

Lestrade, J-F., Mutel, R.L., Phillips, R.B., Webber, J.C., Niell, A.E., & Preston, R.A. 1984, ApJL, 282, L23

Mutel, R.L., Lestrade, J-F., Preston, R.A., & Phillips, R.B. 1985, ApJ, 289, 262

Neff, J.E., Walter, F.M., Rodonò, M., & Linsky, J.L. 1989, A&A, 215, 79

Neuhäuser, R., Sterzik, M.F., Schmitt, J.H.M.M., Wichmann, R., Krautter, J. 1995, A&A, 297, 391

Reipurth, B., Pedrosa, A., & Lago, M.T.V.T. 1996, A&AS, 120, 229

Rosner, R., Tucker, W.H., & Vaiana, G.S. 1978, ApJ, 220, 643

Schrijver, C.J. 1996 in *Stellar Surface Structure*, eds. K.G. Strassmeier and J.L. Linsky, (Dordrecht: Kluwer), p. 1

Shu, F., Najita, J., Ostriker, E., Wilkin, F., Ruden, S., & Lizano, S. 1994, ApJ, 429, 781

Skinner, S.L., Güdel, M. Koyama, K., & Yamauchi, S. 1997, ApJ 486, 886

Skinner, S.L. & Walter, F.M. 1998, ApJ, in press

Stahler, S.W. & Walter, F.M. 1993, in *Protostars and Planets III*, eds. E.H. Levy and J.I. Lunine (Univ. of Arizona Press), p. 405

Stauffer, J.R., Caillault, J.-P., Gagne, M., Prosser, C.F., & Hartmann, L.W. 1994, ApJS, 91, 625

Strassmeier, K.G., Welty, A.D., & Rice, J.B. 1994, A&A, 285, L17

van den Oord, G.H.J., Eibe, M.T., & Byrne, P.B. 1998, A&A, submitted.

Walter, F.M. 1986, ApJ 306, 573

Walter, F.M. 1995, in *Astrophysics in the EUV*, eds. R. Malina and S. Bowyer, (Dordrecht:Kluwer) p. 129

Walter, F.M., Brown, A., Mathieu, R.D., Myers, P.C., & Vrba, F.J. 1988, AJ, 97, 297

Walter, F.M. & Byrne, P.B. 1998, in *Cool Stars, Stellar Systems, and the Sun, X*, ASP Conf. Ser 154, eds. R.A. Donahue & J.A. Bookbinder, p. 1458

Walter, F.M., Gibson, D.M., & Basri, G.S. 1983, ApJ, 267, 665

Walter, F.M., Vrba F.J., Mathieu R.D., Brown A., & Myers P.C. 1994, AJ, 107 692

Walter, F.M., Vrba, F.J., Wolk, S.J., Mathieu, R.D., & Neuhäuser, R. 1997, AJ, 114, 1544.

Wolk, S.J. & Walter, F.M. AJ, 111, 2066

White, N.E, Shafer, R.A., Horne, K., Parmar, A.N., & Culhane, J.L. 1990, ApJ, 350, 776

White, N.E., *et al.* 1994, PASJ, 46, L97

Solar and Stellar Activity: Similarities and Differences
ASP Conference Series, Vol. 158, 1999
C.J. Butler and J.G. Doyle, eds.

Magnetic Λ-quenching and Grand Activity Minima

G. Rüdiger, R. Arlt & M. Küker

Astrophysikalisches Institut Potsdam,
An der Sternwarte 16, D-14482 Potsdam, Germany

Abstract. The nonlinear interaction between dynamo-induced magnetic fields and differential rotation in stellar convection zones has been found to significantly control the long-term activity behavior of solar-type stars. The magnetic back reaction is threefold: Lorentz force, α-quenching and Λ-quenching. The first feedback may provide an interesting non-periodic behavior which, however, disappears when α quenching is included. Yet if a strong Λ-quenching is allowed to modify the differential rotation, the dynamo resumes its exotic performance with periodically changing cycle periods and fluctuating magnetic parities. The magnetic Prandtl number must be smaller than unity.

1. Introduction

The period of the solar cycle and its amplitude vary considerably around their averages. Sunspots were even almost entirely absent during the Maunder minimum between 1666 and 1715 (Spörer 1889). Similar lulls of activity, which we will henceforth call grand minima, were detected after the Maunder minimum and in earlier medieval times. The activity cycle of the Sun is not exceptional: The observation of chromospheric Ca-emission of solar-type stars yielded activity periods between 3 and 20 yr (Noyes et al. 1984, Baliunas & Vaughan 1985, Saar & Baliunas 1993). A few of these stars do not show any significant activity. This suggests that even the existence of grand minima is a typical property of cool main-sequence stars like the Sun.

2. The Dynamo and Differential Rotation Model

The model is based on magnetic feedback to the internal solar rotation (Weiss et al. 1984, Jennings & Weiss 1991). Kitchatinov et al. (1994) and Tobias (1996, 1997) even introduced the conservation law of angular momentum in the turbulent convection zone (CZ) including magnetic feedback in order to produce grand minima of the dynamo cycle. In the present paper, we present a 2D mean-field theory in spherical polar coordinates based on a solar overshoot dynamo model (Rüdiger & Brandenburg 1995) together with a theory of differential rotation based on the Λ-effect concept (Küker et al. 1993).

We assume axial symmetry of the hydro-magnetic state of the star and ignore meridional flows. The field equations for the CZ include the effects of diffusion, α-effect, toroidal field production by differential rotation, and the Lorentz force. The dimensionless form of the equations used is

$$\begin{aligned}\frac{\partial\Omega}{\partial t} &= \frac{\mathrm{Pm}}{r^4}\frac{\partial}{\partial r}\left(r^3\left(r\frac{\partial\Omega}{\partial r}-V^{(0)}\Omega\right)\right)+\frac{\mathrm{Pm}}{r^2\sin^3\theta}\frac{\partial}{\partial\theta}\left(\sin^3\theta\frac{\partial\Omega}{\partial\theta}\right)+\\ &+\frac{\mathrm{E}}{r^2\sin\theta}\left(\frac{1}{r}\frac{\partial A}{\partial\theta}\frac{\partial(rB)}{\partial r}-\frac{1}{\sin\theta}\frac{\partial A}{\partial r}\frac{\partial}{\partial\theta}(\sin\theta B)\right),\end{aligned} \tag{1}$$

$$\frac{\partial A}{\partial t}=\frac{\partial^2 A}{\partial r^2}+\frac{\sin\theta}{r^2}\frac{\partial}{\partial\theta}\left(\frac{1}{\sin\theta}\frac{\partial A}{\partial\theta}\right)+r\sin\theta\, C_\alpha\,\alpha B, \tag{2}$$

$$\begin{aligned}\frac{\partial B}{\partial t} &= \frac{1}{r}\frac{\partial^2}{\partial r^2}(rB)+\frac{1}{r^2}\frac{\partial}{\partial\theta}\left(\frac{1}{\sin\theta}\frac{\partial}{\partial\theta}(\sin\theta B)\right)+\frac{C_\Omega}{r}\frac{\partial\Omega}{\partial r}\frac{\partial A}{\partial\theta}\\ &+\frac{C_\Omega}{r}\frac{\partial\Omega}{\theta}\frac{\partial A}{\partial r}-\frac{C_\alpha}{r\sin\theta}\frac{\partial}{\partial r}\left(\alpha\frac{\partial A}{\partial r}\right)-\frac{C_\alpha}{r^3}\frac{\partial}{\partial\theta}\left(\frac{\alpha}{\sin\theta}\frac{\partial A}{\partial\theta}\right),\end{aligned} \tag{3}$$

with the poloidal-field potential A, toroidal field B and angular velocity Ω. We used the normalizations $r=R\tilde{r}$, $t=\tilde{t}R^2/\eta_\mathrm{T}$, $\Omega=\Omega_0\tilde{\Omega}$, and $A=R^2B_\mathrm{eq}\tilde{A}$, $B=B_\mathrm{eq}\tilde{B}$ with the equipartition field $B_\mathrm{eq}=(\mu_0\rho\langle u'^2\rangle)^{1/2}$. For the quantities ν_T and η_T as the turbulent viscosity and magnetic diffusivity, only the ratio $\mathrm{Pm}=\nu_\mathrm{T}/\eta_\mathrm{T}$ is relevant. The one-point correlation tensor of the velocity fluctuations, Q_{ij}, is expressed by

$$Q_{r\phi}=\nu_\mathrm{T}\sin\theta\left(-r\frac{\partial\Omega}{\partial r}+V^{(0)}\Omega\right),\qquad Q_{\theta\phi}=-\nu_\mathrm{T}\sin\theta\frac{\partial\Omega}{\partial\theta}. \tag{4}$$

$V^{(0)}$ determines the rotation law without magnetic field. The computation domain consists of three layers: the top of the CZ with the Λ-effect, the bottom of the CZ with both Λ-effect and α-effect working, and the top of the core with 100 times lower viscosity and diffusivity. The boundary conditions are specified as $Q_{r\phi}=\partial A/\partial r=B=0$ at at $r=R$ and $Q_{r\phi}=A=B=0$ at the inner boundary.

The model is defined by five dimensionless numbers, namely the magnetic Reynolds numbers of differential rotation and α-effect, the magnetic Prandtl number, Elsasser number and the strength of the Λ-effect:

$$C_\Omega=\frac{\Omega_0R^2}{\eta_\mathrm{T}},\quad C_\alpha=\frac{\alpha_0R}{\eta_\mathrm{T}},\quad \mathrm{Pm}=\frac{\nu_\mathrm{T}}{\eta_\mathrm{T}},\quad \mathrm{E}=\frac{B_\mathrm{eq}^2}{\mu_0\rho\eta_\mathrm{T}\Omega_0},\quad \text{and } V^{(0)}. \tag{5}$$

With the eddy diffusivity, $\eta_\mathrm{T}=c_\eta\langle u'^2\rangle\,\tau_\mathrm{corr}$, and the Coriolis number $\Omega^*=2\tau_\mathrm{corr}\Omega_0$, the Elsasser number reads $\mathrm{E}=2/(c_\eta\Omega^*)$. In the α-effect, α_0 is the dynamo-alpha amplitude and in the spatial distribution, $\alpha=\alpha_0\cos\theta\sin^2\theta$, the factor $\sin^2\theta$ has been introduced to restrict magnetic activity to low latitudes and $\alpha_0\simeq l_\mathrm{corr}\Omega_0$, so that

$$|C_\alpha|\simeq\frac{\Omega_0\,R}{c_\eta\,u'}. \tag{6}$$

Similarly we find $C_\Omega/|C_\alpha|\simeq R/l_\mathrm{corr}$, whence C_Ω generally exceeds C_α ('$\alpha\Omega$ dynamo'); here we used $C_\alpha=-10$ and $C_\Omega=10^5$. $V^{(0)}$ is positive in order to produce the required super-rotation, its amplitude is 0.37.

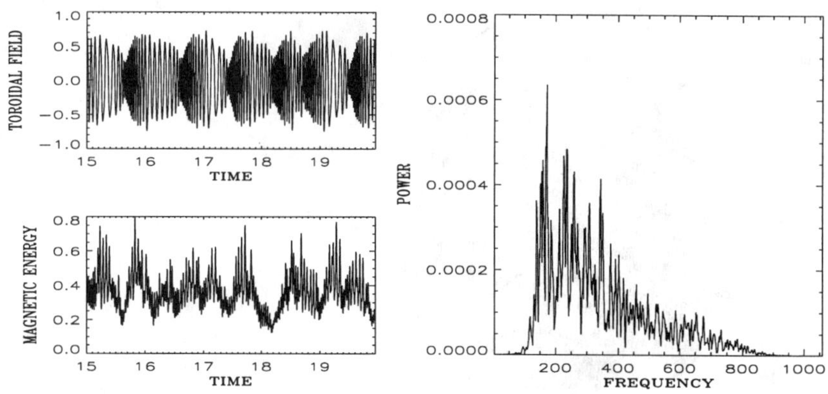

Figure 1. Time series (LEFT) and spectrum (RIGHT) of the full model with Λ-quenching ($\lambda = 25$). TOP-LEFT: Toroidal magnetic field, BOTTOM-LEFT: magnetic energy

3. Results and Discussion

When only considering large-scale Lorentz forces on differential rotation (Malkus & Proctor 1975), one gets irregular grand minima with strong variations in the cycle period (Tobias 1996). If the suppression of dynamo action by α-quenching, such as

$$\alpha \propto \frac{1}{1 + (B_{\rm tot}/B_{\rm eq})^2}, \tag{7}$$

is included, the dynamo returns to oscillations with one frequency. If a strong feedback of small-scale flows on the generation of Reynolds stress (Λ-quenching) as expressed by

$$V^{(0)} \propto \frac{1}{1 + \lambda (B_{\rm tot}/B_{\rm eq})^2}, \tag{8}$$

is added, grand minima occur at a reasonable rate between 10 and 20 cycle times if $\lambda \geq 20$. The cycle period varies by a factor of 6–8. Figure 1 gives the toroidal field at a fixed point, total magnetic energy and the power spectrum for $\lambda = 25$. All times and periods are given in units of a diffusion time $R^2/\eta_{\rm T}$. Field strengths are measured in units of $B_{\rm eq}$. The spectral graph of Figure 1 contains a set of frequencies with the highest peaks near the activity cycle frequency.

The correlation between the dipolar component $E_{\rm A}$ of the magnetic field, the quadrupolar component $E_{\rm S}$, and the differential rotation in terms of $(\partial\Omega/\partial r)^2$, averaged over the latitude θ, is shown in Figure 2. The dots in the interior of the diagram represent the actual time series; projections of the trajectory are given at the sides. The phase graph is similar to those given in Knobloch et al. (1998). The trajectory resides at strong differential rotation during normal cyclic activity and oscillates in a wide range of energies. Oscillation amplitudes diminish as the field starts to suppress the differential rotation. The actual grand minimum is found at very low dipolar energies; the quadrupolar component, however,

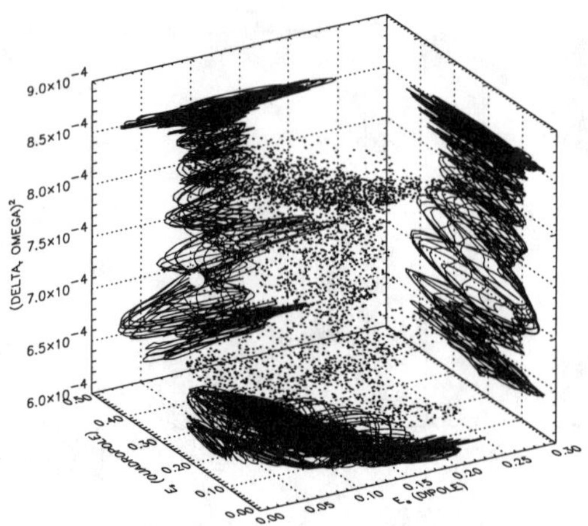

Figure 2. Correlations between magnetic field energies for both parities (dipolar and quadrupolar) and a measure of differential rotation

remains present throughout the minimum. The differential-rotation measure is already growing at that time.

The northern and southern hemispheres differ slightly in their temporal behavior. This is a general characteristic of mixed-modes dynamo explanations of grand minima. In all cases studied, the magnetic Prandtl number is Pm = 0.1. For Pm = 1.0, however, grand minima *do not appear* in models as Fig. 1. The magnetic Prandtl number controls the intermittency of the activity cycle; for smaller Pm, grand minima occur less and less frequent.

References

Baliunas, S.L. & Vaughan, A.H., 1985, Ann. Rev. Astron. & Astroph. 23, 379

Jennings R.L. & Weiss N.O., 1991, MNRAS 252, 249

Kitchatinov L.L., Rüdiger G. & Küker M., 1994, A&A 292, 125

Knobloch E., Tobias S.M. & Weiss N.O., 1998, MNRAS 297, 1123

Küker M., Rüdiger G. & Kitchatinov L. L., 1993, A&A 279, 1

Malkus W.V.R. & Proctor M.R.E., 1975, J. Fluid Mech. 67, 417

Noyes R.W., Weiss N.O. & Vaughan A.H., 1984, ApJ 287, 769

Rüdiger G. & Brandenburg A., 1995, A&A 296, 557

Saar S.H. & Baliunas S.L., 1993, Recent advances in stellar cycle research. In: Harvey K.L. (ed.) The Solar Cycle Workshop. ASP Conf. Ser. 27, p. 150

Spörer G., 1889, Verhandl. Kgl. Leopold.-Carol. Deutsch. Akad. Naturf. 53, 283

Tobias S.M., 1996, A&A 307, L21

Tobias S.M., 1997, A&A 322, 1007

Weiss N.O., Cattaneo F. & Jones C.A., 1984, Geoph. Astroph. Fl. Dyn. 30, 305

Solar and Stellar Activity: Similarities and Differences
ASP Conference Series, Vol. 158, 1999
C.J. Butler and J.G. Doyle, eds.

X-rays from NGC 6633

E.J. Totten, R.D. Jeffries & S. Harmer.

Department of Physics, Keele University, Keele, Staffordshire, ST5 5BG, UK.

Abstract. NGC 6633 is a cluster which has a similar age ($\sim 600\,\mathrm{Myr}$) to the Hyades and Praesepe, but probably a lower metallicity. We present the results of multi-colour photometry and high spatial resolution X-ray imaging in order to investigate whether the assumption, that one cluster has X-ray properties representative of all clusters at the same age, is justified.

1. Introduction.

NGC 6633 is a young open cluster of a similar age to the Hyades, Praesepe, IC4756 and Coma Berenices (500 Myr – 800 Myr). Such a sample of similarly aged clusters allows us to investigate whether and to what extent, properties such as rotation and coronal activity are a function of age alone or whether factors such as composition or initial conditions are important.

Randich & Schmitt (1995) have reported that barely 30% of Praesepe F and G stars are detected in a survey with similar detection limits to one that detected almost all Hyades F and G stars (Stern *et al.* 1995), suggesting that stars in Praesepe are less X-ray active on average than the Hyades. A similar analysis of the open cluster in Coma Berenices (Randich *et al.* 1996) shows that X-ray properties of late F and G stars in Coma Berenices are similar to those in the Hyades. Recently, Randich *et al.* (1998) analysed HRI observations of IC 4756 and found that while their G-type results are inconclusive, for F-type stars at least, IC 4756 is less X-ray active than the Hyades, but consistent with Praesepe.

The difference in X-ray activity between these two groups, the Hyades and Coma Berenices on one hand, and Praesepe and IC4756 on the other, has been investigated but no definitive explanation has been offered - the distribution of rotation rates (and hence average rotational properties) between Praesepe and the Hyades is similar (Mermilliod 1997), contamination by background sources has been ruled out (Barrado y Navascues *et al.* 1998) and levels of chromospheric activity (in M dwarfs) are similar in both clusters.

We present here, an analysis of X-ray observations of NGC 6633 and compare it with the Hyades and Praesepe, in the hope of lending more weight to the argument that X-ray properties of clusters are not solely dependent on cluster age alone.

2. Observations.

The X-Ray observations were obtained with the *ROSAT* High Resolution Imager (HRI) in September 1995. The nominal exposure time was 119 ks and the data were binned into an image with 1 arcsec pixels using pulse height channels 3-9 inclusive. For the purposes of this paper we obtained lists of potential cluster members from several sources. We used proper-motion selected objects (>85% probability) from Sanders (1973) and obtained photoelectric photometry for most of these bright stars from Hiltner *et al.* (1958). For fainter objects we used photometrically and radial velocity selected stars from Jeffries (1997) along with a dozen objects from a further radial velocity study (Jeffries in preparation). Photometry for the fainter stars is from our own CCD photometric survey (Jeffries 1997). Our optical catalogue contains 81 stars in total.

The X-ray data were searched *at the optical positions* (after bore-sight correction) to obtain the maximum likelihood source flux, utilising a model instrument PSF from David *et al.* (1996). For sources detected at less than 3σ significance we present 98% confidence upper limits. The source fluxes were converted to on-axis count-rates and then to 0.1 – 2.4 keV luminosities using a conversion factor of $3.8\ 10^{-11}\ erg\ cm^{-2}\ s^{-1}$ per HRI ct s^{-1} and a distance of 350 pc. The approximate luminosity threshold of the observation was $\approx 5.5\ 10^{28}\ erg\ s^{-1}$. Assumptions about the X-ray spectrum and cluster distance cannot plausibly alter our conclusions.

3. Results.

Cluster members were detected across a colour (B-V) range equivalent to spectral type late A through G. The average detection limit is $8.5\ 10^{28}\ erg\ s^{-1}$.

Figure 1a shows a comparison plot for NGC 6633 and the Hyades, with X-ray luminosity plotted against intrinsic (B-V) colour for F- and G-type stars – known spectroscopic binaries have been removed.

Figure 1b shows a comparison plot for NGC 6633 and Praesepe, with X-ray luminosity plotted against intrinsic (B-V) colour for F- and G-type stars – known spectroscopic binaries have been removed.

We have calculated the mean X-ray luminosity for the Hyades, Praesepe and NGC 6633 (for those stars shown in Figure 1a and Figure 1b) to be $(11.0 \pm 0.3)\ 10^{28}\ erg\ s^{-1}$, $(5.2 \pm 1.5)\ 10^{28}\ erg\ s^{-1}$, $(5.5 \pm 0.8)\ 10^{28}\ erg\ s^{-1}$ respectively. For the Hyades and Praesepe non-detections, we have assumed the X-ray flux to be equal to either the upper limit or zero, in order to calculate the range of possible mean values. For NGC 6633 we have used the maximum likelihood source flux, and associated error, at each optical position.

This suggests that the X-ray properties of NGC 6633 put it into a group of clusters along with Praesepe and IC4756, showing that it is less X-ray active than the Hyades and Coma Berenices.

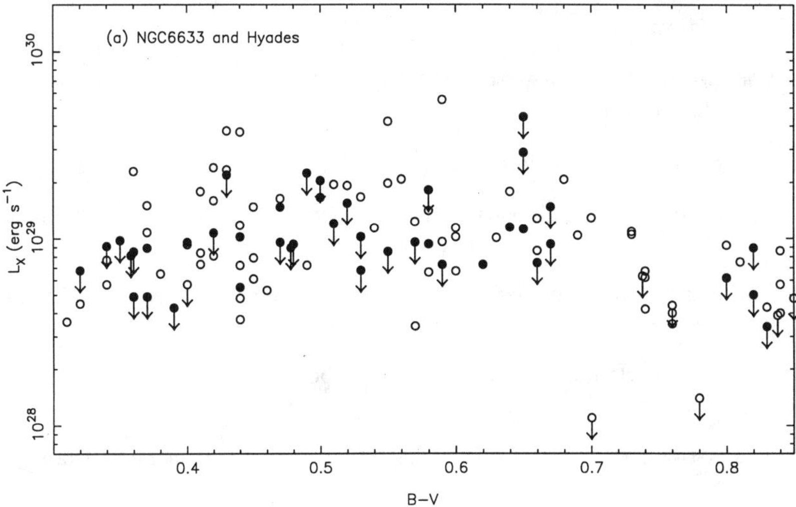

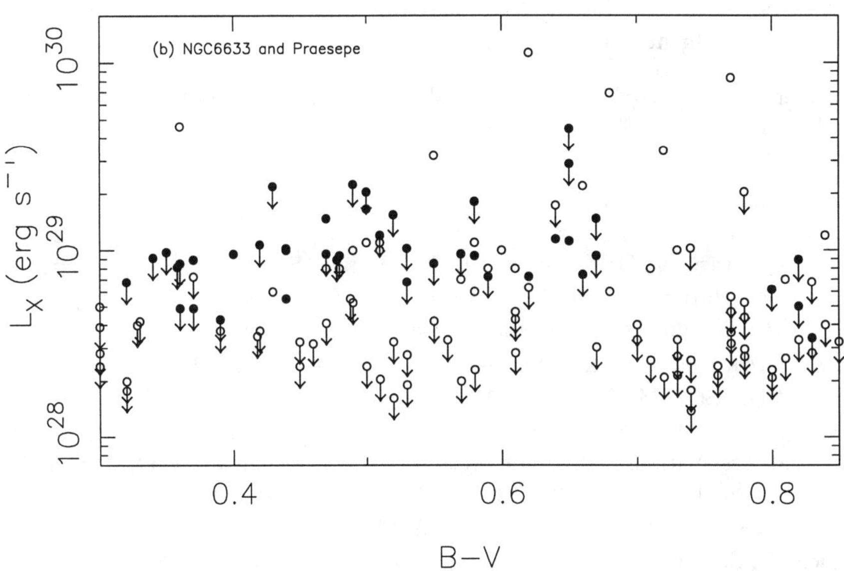

Figure 1. (a) L_X against intrinsic (B-V) for F-G type members of NGC 6633 and the Hyades. NGC 6633 stars are shown as filled circles, Hyades stars are open circles. (b) L_X against intrinsic (B-V) for F-G type members of NGC 6633 and Praesepe. NGC 6633 stars are shown as filled circles, Praesepe stars are open circles.

4. Conclusions.

- We detected X-ray members of NGC 6633 covering the spectral range late A- to late G-type.
- 21% of G-type stars were detected in X-rays.
- The mean X-ray luminosity of F- and G-type stars in NGC 6633 is about half that calculated for similar stars in the Hyades.
- The mean X-ray luminosity of F- and G-type stars in NGC 6633 is comparable to that of similar stars in Praesepe.
- Neither the Hyades nor Praesepe can be considered as the "definitive" open cluster.
- The X-ray properties of open clusters must depend on more than simply cluster age.
- Deeper X-ray observations and further optical spectroscopy are needed to shed light on the apparent difference in X-ray properties of similarly aged cluster groups.

5. Acknowledgments.

S. Harmer was supported by an undergraduate research bursary from the Nuffield Foundation (NUF-URB98).

References

Barrado y Navascues, D., Stauffer, J.R. & Randich, S. 1998 , ApJ, in press.

David, L.P., Harnden, F.R. Jr, Kearns, K.E. & Zombeck, M.V., 1996 The ROSAT High Resolution Imager Calibration Report.

Hiltner, W.A., Iriarte, B. & Johnson, H.L., 1958, ApJ, 127, 539.

Jeffries, R.D., 1997, MNRAS, 292, 177.

Mermilliod, J.C., 1997 Proceedings of the Workshop on Cool Stars in Clusters and Associations: Magnetic Activity and Age Indicators, G. Micela, R. Pallavicini & S. Sciortino (eds.), Mem. SAIt, **68**, 859.

Randich, S. & Schmitt, J.H.M.M., 1995, A&A, 298, 115.

Randich, S., Schmitt, J.H.M.M. & Prosser, C.F., 1996, A&A, 305, 785

Randich, S., Singh, K.P., Simon, T., Drake, S.A. & Schmitt, J.H.M.M., 1998, A&A, 337, 372.

Sanders, W.L., 1973, A&AS, 9, 213.

Stern, R.A., Schmitt, J.H.M.M. & Kahabka, P.T., ApJ, 448, 683.

Part 3

PHYSICAL MANIFESTATIONS OF ACTIVITY: SPOTS, PLAGES AND PROMINENCES

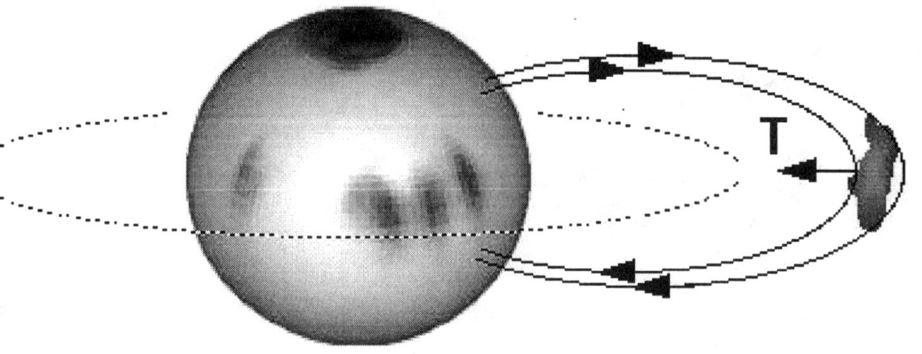

Schematic illustration of magnetic confinement of a coronal condensation outside the co-rotation radius, by a simple dipole-like magnetic field.

Solar and Stellar Activity: Similarities and Differences
ASP Conference Series, Vol. 158, 1999
C.J. Butler and J.G. Doyle, eds.

Spots and Plages: The Solar Perspective

Sami K. Solanki

Institute of Astronomy, ETH-Zentrum, CH-8092 Zurich, Switzerland

Abstract. Spots and plages are prominent markers of magnetic fields on the Sun and on other cool stars. Here a brief overview is given of their observed properties, with similarities and differences emphasized. The distribution of sunspot sizes is discussed in greater detail. An extrapolation from the solar to the stellar case is made. For active stars it provides a natural explanation for the larger spot areas that are deduced from the strength of molecular spectral lines than from Doppler imaging.

1. Introduction

The Sun is the only star whose surface can be resolved at high resolution. Solar observations thus provide a unique insight into the fine-scale structure of the magnetic field on cool stars and its manifestations. On the other hand, the Sun covers only a restricted parameter range compared to the sum of other cool stars. The importance of this interplay between the study of the Sun and stars was well understood by Brendan Byrne, as witnessed by his keen and typically critical, but always good humoured, interest in solar physics.

Sunspots and plages are manifestations of the solar magnetic field which, in the solar photosphere, is filamented into flux tubes. These possess diameters ranging from below the best current spatial resolution of roughly 200 km to over 50000 km. Sunspots and plages are brightness signatures of magnetic flux tubes, with sunspots being the largest and plages being composed of the smallest (called magnetic elements). The small tubes do not fill the whole solar surface in a plage, but rather only some fraction, typically 5–25%.

The brightness of individual flux tubes is a strong function of their size, as can be seen from Figure 1. Small flux tubes are bright (with a possible decrease of excess brightness for the very smallest), while larger ones are dark. Plages, being conglomerates of many small and bright flux tubes, are consequently bright. The dark flux tubes are subdivided into sunspots, which are composed of an umbra and a penumbra, and pores, naked umbrae which, however, are considerably brighter than the umbrae of mature sunspots (Sütterlin 1998). In general, pores are smaller, with diameters of 500–5000 km, whereas sunspots have diameters larger than approximately 4000 km. When discussing the brightness of sunspots we need to distinguish between the average over the whole spot (solid curve in Figure 1) and the umbral brightness (dot-dashed curve).

The second basic parameter of flux tubes is their field strength. In contrast to their brightness, however, the field strength inside the flux tubes (when averaged over their cross-section) is remarkably constant over the whole range of

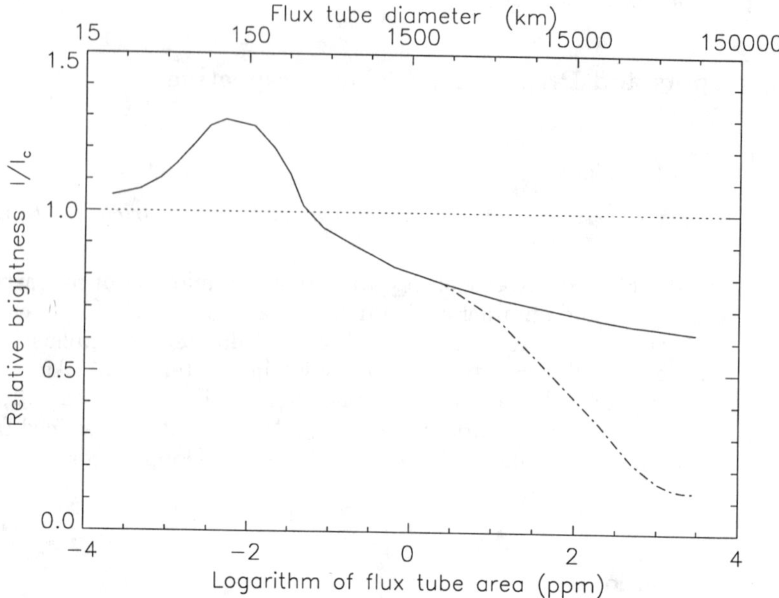

Figure 1. White-light brightness of magnetic features relative to 'quiet' Sun brightness vs. the (logarithmic) area of magnetic features in 10^{-6} times the solar hemispheric area (lower axis) and their diameter (upper axis). The solid line represents the brightness averaged over the whole flux tube, i.e., over both umbra and penumbra for sunspots. The dot-dashed lines represents the brightness of the umbra only.

their sizes, as can be seen from Figure 2. Only the very smallest flux tubes show signs of a strongly reduced field strength (at the left edge of Figure 2, cf. § 5).

In the case of sunspots the field strength varies strongly over their cross-section, so that it is once again necessary to distinguish between the field strength averaged over the whole sunspot (solid curves in Fig. 2) and the *maximum* field strength in the umbra (dot-dashed curves).

2. Size Distribution of Solar Flux Tubes

Consider the question of the relative abundance of flux tubes of different sizes. Basically, the number of flux tubes rapidly increases with decreasing size, just as the number of bipolar magnetic regions rapidly increases with decreasing size or magnetic flux (Schrijver & Title 1999).

For sunspots this size distribution can be deduced from direct observations. Bogdan et al. (1988) considered the Mt Wilson sunspot areas measured between 1921 and 1982 and found that the distribution of umbral areas A larger than $A_{\min} = 1.5 \times 10^{-6} A_{1/2\odot}$ (where $A_{1/2\odot} = 2\pi R_\odot^2$ is the surface area of the visible solar hemisphere) is roughly described by a log-normal distribution, i.e., the number of sunspots N with umbral area A is given by

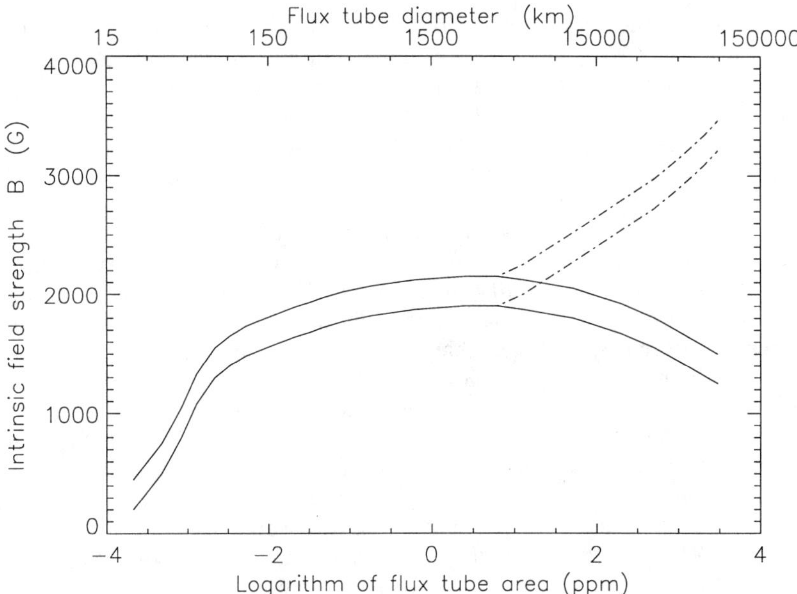

Figure 2. Intrinsic field strength B of magnetic features vs. the (logarithmic) area of magnetic features (lower axis, see Fig. 1) and their diameter (upper axis). The solid lines roughly enclose the observed range of values of the field strength averaged over the whole flux tube, including over both umbra and penumbra for sunspots. The dot-dashed lines represents the maximum field strength in the umbra.

$$\frac{dN}{dA} = \left(\frac{dN}{dA}\right)_{\max} \exp\left(-\frac{(\ln A - \ln \langle A \rangle)^2}{2 \ln \sigma_A}\right). \tag{1}$$

The distribution is characterized by three free parameters determined from the observations: $(dN/dA)_{\max}$ is the maximum value reached by the distribution (for the Sun it is 9.2 in units of $10^{-6} A_{1/2\odot}$), $\langle A \rangle$ is the mean sunspot area ($= 0.62$ in the same units) and σ_A is a measure of the width of the distribution ($= 3.8$). The range of measured umbral diameters is approximately 1200–11500 km.

Bogdan et al. (1988) display the measured sunspot distributions for the three years around solar activity minimum and maximum (in their Figure 3), together with log-normal fits to them. The log-normal curves describing the sunspot distributions at solar maximum and minimum differ only in their $(dN/dA)_{\max}$ values.

How does the distribution of flux-tube sizes continue to flux tubes with smaller diameters than those considered by Bogdan et al. (1988)? This question cannot be answered in a direct manner since area measurements become more and more unreliable as the flux-tube size decreases. Thus, no data sets of the distribution of the area of even pores exist. Hence, the sizes and numbers are most uncertain for the smallest and most common magnetic elements.

Nevertheless, qualitative estimates can be made. Although individual magnetic elements (i.e., the small flux tubes composing plages) cannot be counted, the total plage area in active regions can easily be determined (note, however, that this includes the intervening space between the magnetic elements). It turns out that the ratio between the areas coverd by plages and sunspots increases from approximately 12 at the maximum of the solar cycle to over 25 at solar activity minimum (Chapman et al. 1997). Hence, solar active-region plage covers *at least* an order of magnitude more of the surface area than sunspots. The above numbers are probably lower limits, since weaker, less bright plage regions are more easily missed than sunspots. Furthermore, magnetic elements also form the network, which is found on all parts of the 'quiet' Sun. If the network area is included in the above ratio then it probably varies between values of 20–30 and 50–60 over the solar cycle. Since only 10–20% of the surface area is covered by fields in a plage the amount of magnetic flux carried by plages and sunspots is on the same order of magnitude around activity maximum.

Indirect evidence that the surface area covered by magnetic elements is comparable to or larger than that covered by sunspots is also provided by the fact that the Sun seen as a star is brighter at times of high magnetic activity, i.e. at times when the number of sunspots on the solar surface is largest (e.g., Willson & Hudson 1988, 1991, Fröhlich & Lean 1998). Although the total irradiance (i.e., the wavelength-integrated flux as seen at the location of the Earth) varies by only 0.1–0.2% over the solar cycle, the measurements carried out from space are sufficiently accurate to detect it with ease.

Models which assume that only surface magnetism is the cause have been highly successful in reproducing the irradiance record (e.g., Foukal & Lean 1990, Chapman et al. 1996, Lean et al. 1998), and also a variety of other related diagnostics (Solanki & Unruh 1998, Fligge et al. 1998, Unruh et al. 1999). Hence, the brightening of the Sun at activity maximum is a direct result of the increased number of small flux tubes relative to large ones.

A particularly intriguing observation in this context is that whereas the brightness correlates directly with activity on solar-type stars with low activity levels (less than approximately three times the mean solar value), there is an anticorrelation for more active stars (Radick et al. 1989, Lockwood et al. 1992). This suggests that as the activity level rises, i.e. as the magnetic flux increases, the average size of magnetic flux tubes also increases. This is equivalent to saying that the number of larger flux tubes increases relative to smaller ones at higher activity levels.

In the following section I explore the consequences of a possible similar trend toward larger sunspots with increasing activity.

3. Extrapolation to more Active Stars

First consider the two functions dN/dA (given by Eq. 1) and $A\,dN/dA$, the latter of which describes the contribution of spots with area A to the total area covered by (sun)spots. The solar dN/dA and $A\,dN/dA$ distributions are represented by the thick lines in Figure 3. Both functions have been normalized to their values at $A = 10$. We evaluate numerically the integrals over both these distributions between $A_{\rm min}$ and A. The second of these reads:

$$\int_{A_{\rm min}}^{A} A' \frac{dN}{dA'} dA' = \left(\frac{dN}{dA}\right)_{\rm max} \int_{A_{\rm min}}^{A} A' \exp\left(-\frac{(\ln A' - \ln \langle A \rangle)^2}{2 \ln \sigma_A}\right) dA'. \quad (2)$$

These integrals, computed using the values for σ_A and $\langle A \rangle$ taken from Bogdan et al. (1988), are represented by the thick curves in Fig. 4. Plotted are $S_N(A)$ and $S_A(A)$, i.e. the integrals normalized to their maximum values:

$$S_N(A) = \int_{A_{\rm min}}^{A} \frac{dN}{dA'} dA' \Big/ \int_{A_{\rm min}}^{A_{\rm max}} \frac{dN}{dA'} dA' = \int_{A_{\rm min}}^{A} dN / N_{\rm tot} \, , \quad (3)$$

$$S_A(A) = \int_{A_{\rm min}}^{A} A' \frac{dN}{dA'} dA' \Big/ \int_{A_{\rm min}}^{A_{\rm max}} A' \frac{dN}{dA'} dA' = \int_{A_{\rm min}}^{A} A' dN / A_{\rm tot} \, . \quad (4)$$

Figure 4 suggests that at least for the Sun small sunspots cover a far larger portion of the solar surface than larger sunspots. According to Figure 3 of Bogdan et al. (1988) the distribution of sunspot sizes is the same at activity maximum as at activity minimum. If on more active stars only $dN/dA_{\rm max}$ would differ then the thick curve in Figure 4 would be valid for them as well, so that spots well below the resolution achievable with Doppler imaging would completely dominate the area coverage. In this case all spots on Doppler images would be (tight) clusters of many smaller spots.

A careful look at Figure 3 of Bogdan et al. (1988) does indicate that the distribution of the data points is somewhat flatter at activity maximum than at minimum. A flatter dN/dA curve can be achieved by increasing σ_A.

A flatter dN/dA curve implies that the $A_{\rm tot}/N_{\rm tot}$ ratio is larger, since the increase in the number of larger sunspots produces a disproportionately larger increase in the spot area coverage. However, the change in shape between solar activity minimum and maximum is so small that we can expect:

$$\left(\frac{A_{\rm tot}}{N_{\rm tot}}\right)_{\rm solar\ max} \lesssim 1.5 \left(\frac{A_{\rm tot}}{N_{\rm tot}}\right)_{\rm solar\ min} . \quad (5)$$

Over the same period the total area covered by sunspots increases by a factor of almost 10. We then extrapolate to even larger activity levels by continuing to increase σ_A with increasing activity (or, equivalently, with increasing total sunspot area). The σ_A and total sunspot area coverage $A_{\rm tot}$ connected with the individual curves in Figures 3 and 4 are listed in Table 1 ($A_{\rm tot}$ is fixed such that $A_{\rm tot} \approx 3000$ around solar maximum, i.e. 0.3% of the visible hemisphere is covered by sunspots, in good agreement with observations published by, e.g., the Royal Greenwich Observatory). We have assumed that σ_A changes linearly with $(dN/dA)_{\rm max}$. Note that on the most active star in the table the spots cover

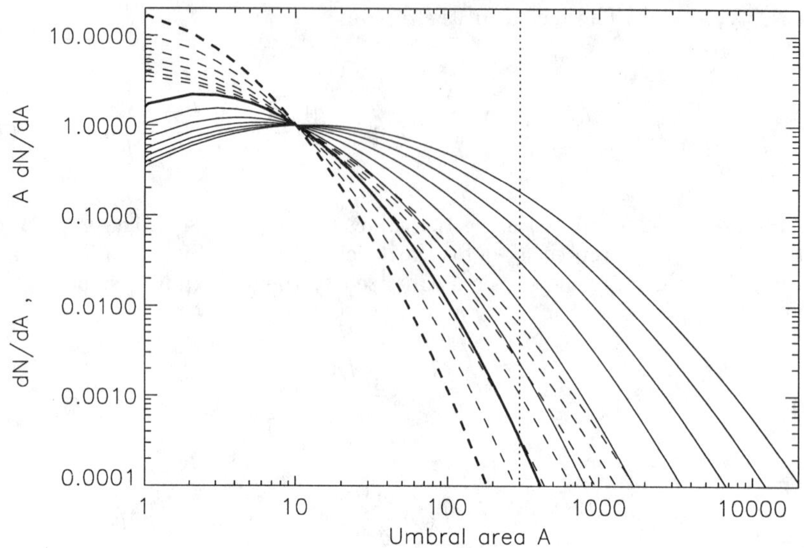

Figure 3. The distributions dN/dA (dashed curves) and $A\,dN/dA$ (solid) vs. spot umbral area A. All distributions have been normalized to their values at $A = 10$. Thick curves represent solar activity minimum. The curves next to them correspond roughly to solar activity maximum. From left to right the plotted curves describe increased spottedness, i.e. increased stellar activity.

the whole stellar surface in this model (this corresponds to the rightmost curves in Figures 3 and 4).

Table 1: Parameters of Sun- and starspot distributions

solar:	min	max					
σ_A	3.8	5.0	6.8	9.2	12.2	15.8	20.0
$(dN/dA)_{\max}$	5	25	65	125	205	305	425
A_{tot}	80	760	4000	15170	46410	121110	250000
$A_{\text{spot,tot}}$	320	3000	16000	61000	186000	484000	10^6
$A_{\text{tot}}/N_{\text{tot}}$	5.0	7.4	11.6	18.0	27.3	39.7	55.7
$A_{\text{DI}}/A_{\text{tot}}$	0.0002	0.003	0.02	0.06	0.14	0.23	0.33

Accepting that good Doppler images for rapidly rotating stars can resolve spots 5° in longitude and assuming that spots are roughly round, we find that spot areas larger than 1200 in units of 10^{-6} can be resolved (unless smaller spots appear in tight clumps). Assuming further that the ratio of umbral to penumbral area of 1:3 is independent of sunspot size means that spots with umbrae having

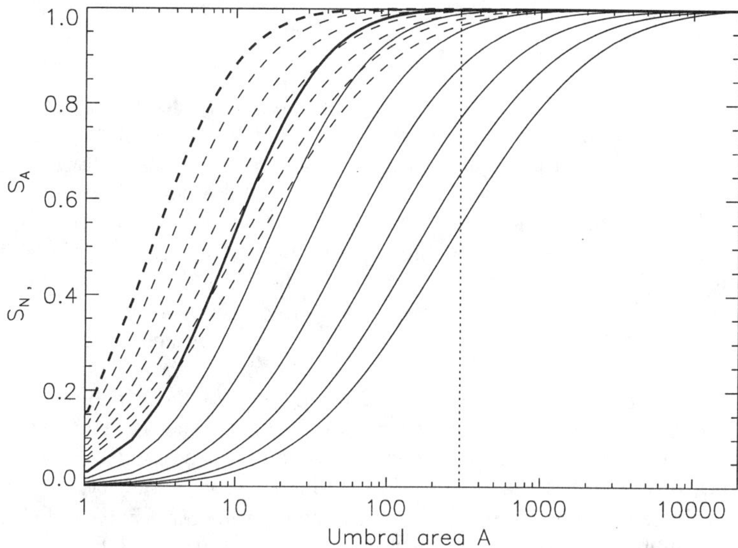

Figure 4. The integral functions S_N (dashed curves) and S_A (solid) vs. umbral area A. S_N and S_A are defined in the text.

areas above 300 can be resolved. This rough limit of resolvability is marked by the vertical dotted line in Figures 3 and 4.

The largest polar (or other) spots seen on Doppler Images of some very active stars cover up to 5–10% of the stellar hemisphere on dwarfs (e.g., Barnes et al. 1998, Strassmeier & Rice 1998) and on RS CVn subgiants (e.g., Vogt & Penrod 1983, Donati et al. 1992, Strassmeier 1997). This corresponds to $A \approx 12000$. Clearly, for the chosen form of the extrapolation (and the other assumptions of the current analysis) these large stellar spots are not single spots, but more likely are tight groups of smaller starspots.

If one now recalls that even the largest *sun*spots would not be resolved by even the highest resolution Doppler images this suggests that very likely in addition to the large spots seen on Doppler images of rapidly rotating stars a large number of smaller unresolved spots are present, which may well be more evenly distributed on the stellar surface and hence would not contribute to the Doppler image. This may provide an explanation for the fact that the spot filling factor derived from Doppler images (10–15%) is smaller than that obtained from the depths of molecular lines (30–50%; Neff et al. 1995, O'Neal et al. 1996, 1998).

The above conclusions depend on the various assumptions underlying the analysis. It is necessary to redo this analysis using other assumptions, which could influence the results.

4. Properties of Sunspots

The large-scale magnetic structure of sunspots is straightforward, but on a small scale it has turned out to be rather complex. Basically, a sunspot as seen in white light is the intersection between a magnetic flux tube and the solar surface. The magnetic field is strongest and most vertical in the central, darkest part of the spot (maximum field strengths range from 2000 to 3500 G). It weakens and becomes more horizontal towards the outer boundary, where it is 700–1000 G and is inclined by 70–80° to the vertical. Beyond the visible edge of the sunspot the magnetic field forms a so-called canopy, composed of almost horizontal field lines overlying mainly field-free gas. The base of the canopy lies in the photosphere not too far from the sunspot boundary. In the penumbra the field is filamented, with most of the magnetic flux being in the form of field inclined by 40–70° to the vertical. However, a smaller proportion is in the form of thin, almost horizontal flux tubes. There is evidence that at least some of these horizontal flux tubes return into the solar interior at the sunspot boundary.

The Evershed effect, i.e. shifts of photospheric spectral lines showing a horizontal outflow of matter, is connected with these horizontal flux tubes. This material also returns to the solar interior at the sunspot boundary. Outflow velocities are typically a few km s^{-1}. In the chromosphere and transition region a rapid inflow is observed (known as the inverse Evershed effect), which appears to follow the inclined magnetic field lines. Above the umbra it is seen mainly as a downflow that can be supersonic at transition-region temperatures (velocities vary between 5 km s^{-1} and 100 km s^{-1}; Kjeldseth-Moe et al. 1988, 1993). Nonetheless, the mass flux carried by the inverse Evershed flow is over an order of magnitude smaller than in the photospheric Evershed flow.

The magnetic field strength of sunspots is closely linked with their temperature (Kopp & Rabin 1992, Solanki et al. 1993, Martínez Pillet & Vázquez 1993). The relation between the two is, for relatively simple and unstressed sunspots, universal (with a small amount of scatter). Consequently, there is also an almost universal relation between the maximum field strength B_0 and the minimum brightness of different sunspots. Both these quantities scale roughly linearly with the sunspot diameter (larger sunspots have darker umbrae; Kopp and Rabin 1992, Collados et al. 1994, Solanki 1997b). It has also been suggested that the intrinsic brightness of sunspots changes over the solar cycle (Albregtsen & Maltby 1978), with sunspot umbrae being approximately 17 % darker at the beginning of the solar cycle than at its end (in total flux, i.e. integrated over all wavelengths).

The temperature stratification of sunspot umbrae as compared to the 'quiet' Sun reveals that umbrae are significantly cooler only in photospheric and low chromospheric layers (e.g., Maltby et al. 1986; the comparison is made at equal optical depth or column mass. Due to the Wilson depression of 400–800 km a direct comparison at equal geometrical depth is problematic). In the upper chromosphere and lower transition region sunspots are distinctly hotter. The late chromospheric temperature rise and the higher chromospheric temperature give sunspots the distinctive Ca II H&K line profiles, with narrow, single peaked, but strong core reversals (Linsky & Avrett 1970).

In the transition region and corona the important quantity is the electron density at a given temperature, since most of the radiation is optically thin. The low umbral temperature in the photosphere implies that although the transition region above umbrae lies deeper than in the 'quiet' Sun (even after the Wilson depression is accounted for), the electron density at a given temperature is almost the same as in the 'quiet' Sun, as may be seen by comparing the various models of Maltby et al. (1986) with each other. This is the property directly responsible for the low umbral emissivity of X-rays and of transition-region line radiation (e.g., Harmon et al. 1993, Gurman 1993). Finally, however, this low emissivity implies that the heating per unit mass is not very different from the 'quiet' Sun value and certainly far lower than in plages.

5. Plages

Plages are the chromospheric and transition-region signature of particularly dense collections of small flux tubes. The corresponding brightenings in the photosphere are called faculae. The contrast relative to the 'quiet' Sun at wavelengths emanating in photospheric layers increases strongly towards the solar limb. The underlying flux tubes have diameters below approximately 500 km, with the smaller ones being bright at most wavelengths. The largest of the flux tubes forming plages are slightly dark in the continuum, but appear brighter (or neutral) in filtergrams taken in line cores or near the limb (cf. Sütterlin 1998, for a similar behaviour seen in pores).

Roughly 90% of the magnetic flux in plages is concentrated into these small flux tubes, which possess field strengths of 1500–1700 G at the solar surface (Rüedi et al. 1992, cf. Stenflo 1973). When expressed in terms of the plasma $\beta = 8\pi p/B^2$, where B is the field strength and p is the gas pressure within the flux tube, this implies that 90% of the magnetic flux has plasma $\beta < 1$. Hence, inside these flux tubes the magnetic field dominates energetically over the gas.

In the 'quiet' Sun not more than approximately half of the magnetic flux is in strong-field form (corresponding to $\beta < 1$, Meunier et al. 1998). A large fraction of the "weak field" is also in the form of flux tubes (field strength of 200–1000 G at the solar surface, Solanki et al. 1996). Hanle-effect measurements indicate that there is an even larger amount of flux hidden in the form of a turbulent field of 5–30 G (Faurobert-Scholl 1993; Faurobert-Scholl et al. 1995), i.e. a field that is tangled up on a small scale.

The thermal stratification of magnetic elements is largely responsible for the brightenings that characterize faculae and plages. Magnetic elements are hotter than the 'quiet' Sun at practically all heights in the solar atmosphere. The temperature contrast increases with height in the upper photosphere and higher layers, so that the brightness contrast increases rapidly towards shorter wavelength in the UV and is larger in spectral lines (the larger sensitivity of the Planck function to temperature at short wavelengths also plays a role). Two geometrical effects also help to determine the brightness of magnetic elements. Firstly, the magnetic elements expand with height, so that the *intensity* contrast in higher layers is larger even if the *temperature* contrast is not greater there. Secondly, magnetic features are partially evacuated, so that one sees deeper inside them (Wilson depression). The resulting hot walls are best visible away from disc centre. This leads to an enhancement of photospheric emission from faculae near the limb of the sun (see Schüssler 1992 for a review).

In the upper atmosphere the brightness of plages is due to additional, non-radiative heating. Depending on the level in the atmosphere the dissipation of MHD waves (in the case of the chromosphere, Ulmschneider 1999) or magnetic reconnection (for the corona, Priest 1999) are the currently preferred modes of energy input. In either case the heating is driven by the dynamics of the magnetic flux tubes.

Relatively little is known of the internal dynamics of magnetic elements. The presence of stationary flows within the magnetic features is controversial. Evidence, both against steady flows (Solanki 1986, Stenflo & Harvey 1985, Frutiger & Solanki 1998) and for downflows (Bellot Rubio et al. 1997), has been presented. I favour the absence of such flows since then there is no need to worry about conserving mass inside the flux tube (a steady downflow would rapidly drain the corona due to the exponential decrease of the gas density with height).

Oscillations and waves in flux tubes have been intensively studied theoretically (e.g., Roberts & Ulmschneider 1997), but are difficult to observe (e.g., Solanki 1997a). There is evidence for low amplitude magneto-acoustic waves both below and above their cutoff frequency, which determines whether they propagate or not. There exists also indirect evidence for much larger amplitude non-stationary velocities (from, e.g., line widths). Magneto-acoustic and kink waves are the most likely sources of heating of the chromospheric layers of flux tubes (and hence of plages).

Finally, flux tubes are moved around as a whole due to the everchanging pattern of convective cells (the granulation). Thus each flux tube carries out a random walk on the solar surface. Since each flux tube is the footpoint of a bundle of field lines reaching up into the corona, such a random walk leads to the intertwining of field lines at greater heights. This in turn produces a build-up of magnetic energy and the formation of spontaneous current sheets and tangential discontinuities (Parker 1994). A part of this excess energy can be released through reconnection (Priest 1999, Schrijver & Title 1999).

Large flux tubes such as sunspots and pores are far less affected by the buffeting due to granulation in the sense that a smaller wave flux is excited. Random motions leading to braidings of field lines are also reduced. Consequently, the heating per unit mass of the chromosphere and coronae over larger flux tubes is reduced. This is the probable reason for the relatively small X-ray intensity exhibited by sunspot umbrae and pores.

More details on solar magnetic elements are given by Solanki (1993) and Steiner (1994).

Acknowledgments. It is a pleasure to thank Gerry Doyle and John Butler for the organization of this conference. Yvonne Unruh clarified a number of my questions concerning Doppler imaging and made many useful comments on the manuscript, which I gratefully acknowledge.

References

Albregtsen F. & Maltby P., 1978, Nature, 274, 41

Barnes J.R., Collier Cameron A., Unruh Y.C., Donati J.F. & Hussain G.A.J., 1998, , *Monthly Notices Royal Astron. Soc.* , 299, 904

Bellot Rubio L.R., Ruiz Cobo B. & Collados M., 1997, ApJ, 478, L45

Bogdan T.J., Gilman P.A., Lerche I. & Howard R., 1988, ApJ, 327, 451

Chapman G.A, Cookson A.M. & Dobias J.J., 1996, , *J. Geophys. Res.* **101**, 13541

Chapman G.A., Cookson A.M. & Dobias J.J., 1997, ApJ, 482, 541

Collados M., Martínez Pillet V., Ruiz Cobo B., del Toro Iniesta J.C. & Vázquez M., 1994, A&A, 291, 622

Donati J.-F., Brown S.F., Semel M., Rees D.E., Dempsey R.C., Mathews J.M., Henry G.W. & Hall D.S., 1992, A&A, 265, 682

Faurobert-Scholl M., 1993, A&A, 268, 765

Faurobert-Scholl M., Feautrier N., Machefert F., Petrovay K. & Spielfiedel A., 1995, A&A, 298, 289

Fligge M., Solanki S.K., Unruh Y.C., Fröhlich C. & Wehrli Ch., 1998, A&A, 335, 709

Foukal P. & Lean J., 1990, Science, 247, 556

Fröhlich C. & Lean J., 1998, in *New Eyes to see inside the Sun and Stars*, F.-L. Deubner, J. Christensen-Dalsgaard, D. Kurtz (Eds.), Kluwer Academic Publ., Dordrecht, Netherlands, *IAU Symp.* **185**, 89

Frutiger C. & Solanki S.K., 1998, A&A, 336, L65

Gurman J.B., 1993, ApJ 412, 865

Harmon R., Rosner R., Zirin H., Spiller E. & Golub L., 1993, ApJ, 417, L83

Kjeldseth-Moe O., Brynildsen N., Brekke P., Engvold O., Maltby P., Bartoe J.-D.F., Brueckner G.E., Cook J.W., Dere K.P. & Soker D.G., 1988, ApJ, 334, 1066

Kjeldseth-Moe O., Brynildsen N., Brekke P., Maltby P. & Brueckner G.E., 1993, Solar Phys., 145, 257

Kopp G. & Rabin D., 1992, Solar Phys., 141, 253

Lean J.L., Cook J., Marquette W., Johannesson A., 1998, ApJ, 492, 390

Linsky J.L. & Avrett E.H., 1970, PASP, 82, 169

Lockwood G.W., Skiff B.A., Baliunas S.L. & Radick R.R., 1992, Nature, 360, 653

Maltby P., Avrett E.H., Carlsson M., Kjeldseth-Moe O., Kurucz R.L. & Loeser R., 1986, ApJ, 306, 284

Martínez Pillet V. & Vázquez M., 1993, A&A, 270, 494

Meunier N., Solanki S.K. & Livingston W., 1998, A&A, 331, 771

Neff J.E., O'Neal D. & Saar S.H., 1995, ApJ, 452, 879

O'Neal D., Saar S.H. & Neff J.E., 1996, ApJ, 463, 766

O'Neal D., Saar S.H. & Neff J.E., 1998, ApJ, 501, L73

Parker E.N., 1994, *Spontaneous current sheets in magnetic fields : with applications to stellar x-rays*, Oxford University Press, New York

Priest E.R., 1999, these proceedings

Radick R.R., Lockwood G.W. & Baliunas S.J., 1989, Science, 247, 39

Roberts B. & Ulmschneider P., 1997, in *Solar and Heliospheric Plasma Physics*, C.E. Alissandrakis, G. Simnett & L. Vlahos (Eds.), Proc. 8th European Meeting on Solar Physics, Springer, Berlin, p. 75

Rüedi I., Solanki S.K., Livingston W. & Stenflo, J.O., 1992, A&A, 263, 323

Schrijver C.J. & Title, A.M., 1999, these proceedings

Schüssler M., 1992, in *The Sun — a Laboratory for Astrophysics*, J.T. Schmelz, J.C. Brown (Eds.), Kluwer, Dordrecht, p. 191

Solanki S.K., 1993, , *Space Sci. Rev.* , 61, 1

Solanki S.K., 1986, A&A, 168, 311

Solanki S.K., 1997a, in *Solar and Heliospheric Plasma Physics*, C.E. Alissandrakis, G. Simnett & L. Vlahos (Eds.), Springer, Berlin, p.49

Solanki S.K., 1997b, in *Advances in the Physics of Sunspots*, B. Schmieder, J.C. del Toro Iniesta & M. Vázquez (Eds.), Astron. Soc. Pacific Conf. Ser., Vol. 118, p. 178

Solanki S.K. & Unruh Y., 1998, A&A, 329, 747

Solanki S.K., Walther U. & Livingston W., 1993, A&A, 277, 639

Solanki S.K., Zuffrey D., Lin H., Rüedi I. & Kuhn J., 1996,A&A, 310, L33

Steiner O., 1994, in *Infrared Solar Physics*, D. Rabin, J. Jefferies & C.A. Lindsey (Eds.), Kluwer, Dordrecht *IAU Symp.* , 154, 407

Stenflo J.O., 1973, Solar Phys., 32, 41

Stenflo J.O. & Harvey J.W., 1985, Solar Phys., 95, 99

Strassmeier K.G. & Rice J.B., 1998, A&A, 330, 685

Strassmeier K.G., 1997, A&A, 319, 535

Sütterlin P., 1998 A&A, 333, 305

Ulmschneider P., 1999, these proceedings.

Unruh Y.C., Solanki S.K. & Fligge M., 1999, A&A, submitted

Vogt S.S. & Penrod G.D., 1983, PASP, 95, 565

Willson R.C. & Hudson H.S., 1988, Nature, 332, 810

Willson R.C. & Hudson H.S., 1991, Nature, 351, 42

Solar and Stellar Activity: Similarities and Differences
ASP Conference Series, Vol. 158, 1999
C.J. Butler and J.G. Doyle, eds.

Observations and Modelling of Starspots: a Tool to Understand Stellar Dynamos

A.F. Lanza[1] & M. Rodonò[1,2]

[1] *Osservatorio Astrofisico di Catania, Viale A. Doria 6, I 95125, Catania, Italy*

[2] *Istituto di Astronomia dell'Università degli Studi di Catania, Viale A. Doria, 6 - I 95125 Catania, Italy*

Abstract. We briefly review the relevant features of starspot activity in very active stars, which are characterized by a sizeable modulation of the optical flux, by focusing on the problems of the surface distribution of spots, stellar differential rotation and activity cycles. Hydromagnetic dynamo models, proposed to account for the observed spot behaviour are also discussed and the advantages and drawbacks of the suggested approaches are pointed out. The possibility that the dynamo in rapidly rotating stars is qualitatively different from the solar dynamo is also addressed by taking into account recent observational results.

1. Introduction

Sunspot activity provides several clues to the processes which produce the solar magnetic field, i.e., the solar dynamo (e.g., Moffatt 1978). The quasi-periodic variation of the number and area of the sunspot groups, the latitudinal migration of the belts where they appear and their polarity properties (Hale's law) indicate the existence of a large scale toroidal field inside the Sun. This field moves in latitude and is modulated during the 11-yr cycle with a reversal of its sign at the beginning of each new spot cycle. The observation of surface magnetic fields indicates that their sign change is related to the inversion of the poloidal field near the previous maximum of the sunspot cycle (Stix 1976, Moffatt 1978).

The theoretical picture proposed for the interpretation on these observations assumes that an $\alpha - \Omega$ dynamo is at work at the base of the convection zone or in the interface layer between the radiative interior and the convection zone (cf., e.g., Rosner & Weiss 1992, Brandenburg & Rüdiger 1995). The existence of a shear layer at the interface between the two zones, as indicated by helioseismological studies (Kosovichev 1996), naturally provides an amplification mechanism for the toroidal field based on the presence of differential rotation, the so-called Ω-effect. The poloidal field component is in turn regenerated from the toroidal field by helical convective motions and by the effect of the Coriolis force on the rising and expanding flux loops in the convection zone, the so-called α-effect (Moffatt 1978, Krause & Rädler 1980).

The possibility of testing dynamo models with solar observation alone is rather limited because we can not vary any of the global parameters of the star.

However, in recent years it has become possible to study also stellar dynamos through observations of atmospheric activity phenomena produced by magnetic fields and assumed activity and rotation tracers. In the near future it will become possible to measure directly long-term changes of large scale fields in stellar atmospheres through the Zeeman Doppler Imaging technique (Donati 1999), but at present we are limited to the information derived from indirect field indicators, especially photospheric starspots, which can be regarded as the stellar analogous of sunspots, and chromospheric plages, which produce variations of the flux in the cores of chromospheric lines such as Ca II H&K and Mg II h&k. The latter approach was pioneered by O. Wilson at Mt. Wilson Observatory and has now led to the detection of solar-like activity cycles in several late-type dwarfs (Baliunas & Jastrow 1990, Baliunas et al. 1995, Donahue 1996). The objects included in the Mt. Wilson monitoring program are mostly single main sequence stars with a very small or undetected optical flux modulation (Lockwood et al. 1997, Radick et al. 1998) and may possibly behave differently from the most active stars which show spot-induced optical flux modulations with amplitudes ranging from ~ 0.05 up to 0.5 mag. Our discussion of starspot activity and its connection with stellar dynamos will be limited to such very active stars, because only for them do we have a suitably extended database. They are characterized by sizeable convection zones and fast rotation (see § 3). Their main properties are reviewed in, e.g., Byrne & Rodonò (1983), Rodonò (1992), Byrne & Mullan (1992), Strassmeier & Linsky (1996).

The observed properties and the behaviour of the sunspots provide us with a general paradigma to address the study of stellar dynamos through the observations of starspots, but qualitative and quantitative differences are of course expected. Even in the solar case the dynamo mechanism is understood only in a very approximate way with current models relying on a mean field approximation, with heavy parametrization of the small-scale physics. Stellar studies may provide additional and important constraints. Therefore, in the following discussion of the general properties of starspot activity, we shall emphasize the aspects which suggest a non-solar behaviour and those which can not be explained by current dynamo models.

2. Mapping Techniques

The mapping of the starspot distribution on the photosphere of active stars is performed using indirect methods because we can not resolve their disks, as for the Sun. All mapping techniques involve the solution of an ill-posed problem and in order to single out a unique and stable solution it is actually necessary to add some kind of a priori information to the problem. The simplest models which are capable of fitting wide-band light curves assume that the starspot pattern consists of a few uniform spherical caps and derive their radius and location on the stellar surface by a simple χ^2 minimization (*few-spot models,* cf. Budding 1996). When a sequence of light curves is available, it is possible to evaluate the change of the geometric parameters of the starspots vs. time performing a simple *time series spot modelling,* which may allow us to derive information on stellar differential rotation, activity cycles and possibly spot lifetimes (e.g., Strassmeier et al. 1994, Henry et al. 1995, Oláh et al. 1997). The reliability of the results obtained by this approach has been recently discussed by, e.g., Eaton et al. (1996). The temperature of the spots may also be estimated from the rotational modulation of the $(V - I)_{KC}$ color index (Byrne et al. 1995) or from the observations of the TiO λ7055 Å and λ8860 Å bands (O'Neal et al. 1996).

More detailed information on the spot pattern can be obtained in the case of eclipsing binaries by the *eclipse mapping technique.* This method exploits the scanning of the occulted active star by the occulting companion disk during eclipses. The method has been applied by, e.g., Rodonò et al. (1995), Cameron & Hilditch (1997) and Lanza et al. (1998a) to RS CVn systems. However, the technique which provides us with the most detailed information is based on the modelling of photospheric line profiles vs. rotation phase in fast rotating stars ($20 \leq v \sin i \leq 80$ km s^{-1}), the so-called *Doppler imaging,* as discussed by, e.g., Vogt et al. (1987), Rice et al. (1989), Piskunov et al. (1990), Vincent et al. (1993). The same method can be applied also to spectropolarimetric data to derive maps of the surface magnetic field intensity and orientation (Donati 1999).

It is important to notice that maps based on photometric data alone can provide rather accurate information only on the distribution of spot longitudes and on the variation of the total spotted area, whereas Doppler imaging allows us to determine also their latitudinal distribution, thanks to the radial velocity information provided by narrow features moving over the line profiles.

3. Main Characteristics of Starspot Activity

The presence of sizeable (≥ 0.01 mag) optical flux modulations in late-type stars appears to be related with the Rossby number, which is defined as the ratio between the rotation period (P_{rot}) and the convective turnover time at the base of the convection zone (τ_c) estimated in a semi-empirical way (see Hall 1991a). More precisely only the stars having $Ro \equiv P_{rot}/\tau_c \leq 0.65$ show a significant modulation (Hall 1991a, 1994). The Rossby number is indeed related to the dynamo number, thus indicating that a new kind of dynamo or a different regime of the solar-like dynamo may be excited at that critical value (Noyes et al. 1984, Soon et al. 1993). For stars with $Ro < 0.65$, the amplitude of the modulation does not show a correlation with the Rossby number and the existence of an amplitude saturation amongst the fastest rotators ($P_{rot} < 10 - 12$ hrs) is still open to debate (O'Dell et al. 1995).

The amplitude of the optical flux modulation implies that a large fraction of the stellar photosphere is covered by starspots. Filling factors up to $\sim 45 - 50\%$ of the stellar disk have been derived by O'Neal et al. (1996) for II Peg. Such a large spot coverage may result from a large number of small solar-sized spots or from a relatively small number of huge starspots representing giant counterparts of sunspots. Doppler imaging indeed gives evidence for a fine structure of the spot pattern in, e.g., AB Dor where spots with a diameter of $\sim 5° - 10°$ have been detected by Donati & Cameron (1997) in the equatorial region. The eclipse mapping of the active components of RS CVn and AR Lac also shows a variation of the distribution of the spotted area vs. longitude down to the resolution limit allowed by the quality of the data and method capability, suggesting the presence of spots as small as $10° - 20°$.

3.1. Distribution in Longitude

Sunspot groups show a marked tendency to appear within pre-existing active regions forming complexes of activity which can last up to 5 – 6 solar rotations (Harvey & Zwaan 1993). Long-lasting active longitudes have been found only for the location of the energetic flares and seem to rotate at a different rate than the photosphere (Bai 1988, Jetsu et al. 1997).

In active stars the presence of a detectable optical modulation indicates a non-uniform distribution of the spots in longitude. The presence of active longitudes, where new spots preferentially appear, may however be hidden by the redistribution produced by differential rotation (Eaton et al. 1996, Hall 1996). In active stars belonging to close binary systems, it may be envisaged that the differential rotation and the hydromagnetic dynamo may be affected by tidal effects, and the instability leading to the emersion of the flux tubes may be different in a tidally distorted star than in a single object. However, present observations do not support a universal and simple behaviour for the location of starspots in close binaries (see Henry et al. 1995). For the K0 IV component of AR Lac, the analysis of about 25 y of optical photometry has provided evidence for a large and persistent active region located around the substellar point, i.e., the point where the line joining the centers of the two components intersects the surface of the star. Two other active longitudes showing a less persistent spot activity have been found at $\sim 120°$ from the most active one (Lanza et al. 1998a). A somewhat similar behaviour may be present in HK Lac, where two active longitudes separated by $\sim 110°$ appear to have been present for more than 30 y. On the contrary the K2 IV active component of RS CVn shows evidence for a migration of the most active longitude with respect to the orbital time reference frame. Recently, Berdyugina & Tuominen (1998) have suggested that in EI Eri and II Peg (and possibly in σ Gem and HR 7275) the most active longitude may switch by $\sim 180°$ in a period of a few months in a way reminiscent of the so-called flip-flop phenomenon in the single FK Com stars (Jetsu et al. 1993). It is important to note that when only sparse data are available for a given star, it is very difficult to discriminate between a regular migration of the active longitudes and a switching of the activity among different fixed longitudes. Therefore, a great improvement in our observational understanding of active longitudes is expected in the forthcoming years as more extended and nearly continuous data sets will be made available through the extensive use of automated photometric telescopes.

On the theoretical side, mean field dynamo models for single stars assuming a dynamo action extended over the entire convective envelope and allowing for a non-axisymmetric field distribution have been explored by Moss et al. (1991, 1995). When the differential rotation is not large ($\Delta\Omega/\Omega < 0.1$), non-axisymmetric field modes with an azimuthal wavenumber $m = 1$ can be excited at lower dynamo numbers than the modes with $m = 0$. The extension of such an approach to binaries is still very preliminary, but Moss & Tuominen (1997) suggested a preference for the mean field to concentrate around the line joining the centers of the two components. However, other relevant aspects of the starspot longitude distributions are not explained by these models, in particular the possible migration of the active longitudes versus time and the flip-flop phenomenon.

3.2. Distribution in Latitude

An interesting and still somewhat controversial issue is the existence and nature of polar spots, i.e., large spotted caps centered over the stellar poles which are often derived from photometric and Doppler imaging maps (Byrne 1992, 1996). A new approach to the problem was proposed by Hatzes et al. (1996) who investigated the variation of the Ca I $\lambda 6439$ Å line profile versus the inclination of the stellar axis and found that stars with lower inclination tend to have flat-bottomed line cores. Such a behaviour can be easily accounted for by the change of spot visibility when located close to the rotation pole, while different explanations require ad hoc and somewhat unrealistic assumptions. Hatzes (1998) suggested also that the latitudinal distribution of the spot activity may depend on the rotation period with faster rotators showing a tendency toward the formation of a polar spotted cap, whereas slower rotators may show a high latitude belt of activity but not truly polar spots.

The reliability of high latitude and polar structures has been discussed also by Strassmeier & Rice (1998). They showed the presence of a high-latitude spot in all the Doppler images derived from different lines on spectra of the same star, irrespectively of the change in the equivalent width and sensitivity of the line profile to the temperature gradient in the atmosphere. Bruls et al. (1998) studied the effects of the chromospheric temperature increase on the line profiles, including also NLTE effects, and concluded that it is unlikely that polar spots are an artifact due to misinterpretation of the spectral signature of chromospheric activity.

Polar spots do not represent the only kind of photospheric activity in very active stars since spots are indeed distributed over a wide range of latitudes as shown by Doppler imaging and eclipse mapping models. In AB Dor, Donati & Cameron (1997) found evidence of spot activity ranging from the equator to the pole and Strassmeier & Rice (1998) found both nearly equatorial and high latitude spots on the single young G dwarf EK Dra. Eclipse mapping shows low latitude spots on the K2IV active component of RS CVn and the K0IV component of AR Lac (Rodonò et al. 1995, Lanza et al. 1998a). Recently Vogt et al. (1998) presented a long-term Doppler imaging monitoring of the active component of HR 1099, which strongly suggests that starspots form at low or intermediate latitudes, then slowly migrates towards the pole along spiral patterns and on time scales of a few years.

From the theoretical point of view, the existence of polar spots was discussed by Schüssler & Solanki (1992) who pointed out that, in a rapid rotator with a dynamo operating near the base of a deep convection zone, the effect of the Coriolis force is to make flux tubes to emerge nearly parallel to the rotation axis, thus producing high latitude activity even if the dynamo amplifies the field only at low latitudes. Such a result was confirmed by more detailed non-linear models for flux tube instability and eruption by Schüssler (1996), Schüssler et al. (1996) and DeLuca et al. (1997). Such a modelling approach has proven to be successful in the solar case where it appears to be capable of reproducing the sunspot behaviour (Caligari et al. 1995). Its extension to rapidly rotating stars, however, does not seem to be capable of accounting for the observed low latitude activity.

In order to explain low latitude spots with the emersion of flux tubes from the base of the convection zone of rapidly rotating stars, it is necessary to postulate that the initial intensity of the field be $> 10^6$ G. This requirement may pose a serious problem for a stable storage of the field during the amplification phase, even in a deep overshoot layer, and exceeds the energy available from differential rotation for the field amplification, unless it is concentrated into a very small volume, as in the case of extremely thin flux tubes (DeLuca et al. 1993). Convective motions and meridional circulation may not significantly affect the motion of flux tubes with a field intensity significantly higher than the equipartition value and, therefore, their inclusion can not easily reconcile the model with observations. Conversely, a distributed dynamo operating in the bulk of the convection zone seems to be a simpler explanation for the observed wide range of spot latitudes, also because its field intensity is nearly in equipartition with the convective motions and, therefore, convection and meridional circulation may significantly affect the spatial distribution of the emerging flux.

Moreover, it is important to consider that emerged flux tubes may become decoupled from the deeper layers because of field instabilities and diffusion within time scales significantly shorter than their lifetimes. Therefore, the distribution and motions of old spots might be related to the surface flow more than to the deep dynamo properties (Schüssler 1987), thus continuous monitoring of active stars is needed to constraint the distribution and the properties of the young starspots, which are more likely to provide us with information on the dynamo processes.

4. Differential Rotation

Hall (1991a) and Henry et al. (1995) estimated stellar differential rotation using few-spot models and showed that the relative amplitude of the differential rotation $\Delta\Omega/\Omega$ decreases with increasing angular velocity Ω as: $\Delta\Omega/\Omega \sim \Omega^{-0.76\pm0.06}$. There may be also a correlation with the Roche lobe filling factor in binaries in the sense that differential rotation may be reduced for near contact stars.

The use of starspots as a tracer for stellar rotation is based on the solar analogy, but the lack of spatial resolution and the spot growth and decay, as well as their systematic proper motions, may significantly affect the derived values of the differential rotation and may also hide the true relationship between $\Delta\Omega/\Omega$ and Ω, if these effects depend on the stellar angular velocity (Lanza et al. 1992, 1994, Donahue 1993).

Doppler imaging may provide us with resolved maps, which can be cross-correlated to derive information on stellar differential rotation. Hatzes (1998) summarized some of the latest results which confirm that in very active stars the relative amplitude of the latitudinal differential rotation $\Delta\Omega/\Omega$ is actually one or two orders of magnitude smaller than in the Sun. The same conclusion was reached by Donati & Cameron (1997) and Vogt et al. (1998), who studied AB Dor and HR 1099, respectively. Vogt et al. (1998) cast doubts on the reliability of the differential rotation amplitude derived from photometric modelling and suggest also that spots may not be good tracers of the rotation of the unmagnetized plasma, because they seem to display systematic motions both in latitude and longitude. An interesting result obtained from Doppler imaging cross-correlations is that on some stars the latitudinal differential rotation seems to be in the opposite sense with respect to the Sun, i.e., with the pole rotating faster than the equator.

Theoretical models of stellar rotation are still very preliminary. They adopt the mean field approach and a simplified theory of the interaction between rotation and turbulent convection (Kitchatinov & Rüdiger 1993). The very first model computations did not take into account the feedback effect of the meridional circulation induced by the differential rotation itself, thus leading to an increase of $\Delta\Omega/\Omega$ with increasing Ω (Küker et al. 1993). More recent models, which include such an effect, indeed show a trend in qualitative agreement with the observations, although they are not capable of explaining the polar acceleration observed in some stars (Rüdiger et al. 1998).

5. Activity Cycles

The application of the solar analogy to active stars has stimulated a search for long-term variations of the total spot area, which may be analogous to the 11-yr solar cycle. Also the dynamo theory supports the existence of cycles in stars and predicts various types of behaviour, which depend on the model assumptions and stellar parameters (see below).

The simplest and less indirect proxies of activity cycles are the quasi-periodic modulation of the total starspot area, as derived from spot models, and the variation of the mean stellar magnitude. It is important to note that in the case of the Sun the average bolometric luminosity increases when there are more spots, an effect due to the larger area and brightness of the faculae and photospheric magnetic elements (Foukal & Lean 1988). In the case of very active stars, however, the colour indexes indicate that cool spots usually dominate the photometric modulation (Henry et al. 1995, cf. also Foukal 1993, Solanki 1999).

In addition to the long-term variation of the spotted area, Henry et al. (1995) have proposed other indicators of activity cycles, i.e., the modulation of the mean colour indexes and the modulation of the orbital period in close binaries. Spruit (1982) and Spruit & Weiss (1986) convincingly ruled out any detectable change of the color indexes along an activity cycle due to the surface redistribution of the flux blocked by starspots. Moreover, the observed long-term variation of the $B-V$ index is usually very small and may be affected by factors not directly connected with spot activity (cf. Amado & Byrne 1997). Therefore, color index variations should not be regarded as a good proxy indicators of activity cycles.

The modulation of the orbital period in close binaries, having at least one component with an outer convective envelope, is a well observed phenomenon and has been suggested to be associated with magnetic activity cycles (Hall 1989). A possible explanation has been proposed by Applegate (1992) and rediscussed by Lanza et al. (1998b), who considered the variation of the gravitational quadrupole moment of an active star. This variation is proposed to be induced by changes of the angular momentum and magnetic field distribution within its convective envelope which may occur along an activity cycle. The variation of the gravitational quadrupole moment induces a variation of the gravitational acceleration of the companion star, thus leading to the observed change of the

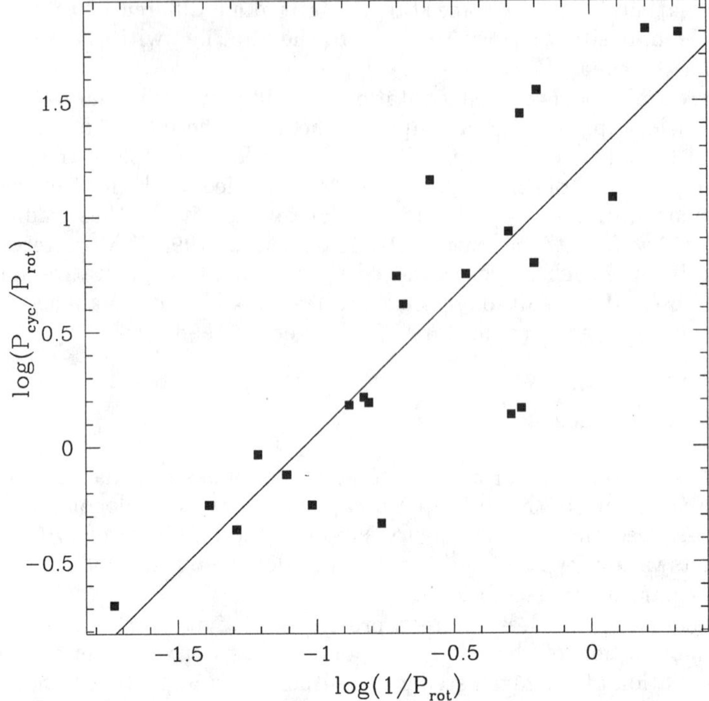

Figure 1. The quantity $\log(P_{cyc}/P_{rot})$ vs. $\log(1/P_{rot})$ for the spotted stars for which data on the activity cycles are available (see text). The line of regression is also plotted.

orbital period. The attractive feature of this mechanism is that it does not require any spin-orbit exchange of angular momentum, an exchange that would imply time scales of the order of thousands of years, not of a few decades as indicated by the available observations.

The relationship between the presumably cyclic orbital period modulation and the activity cycle, as derived from starspot area changes, is not yet observationally clear. There seem to be two different scenarios: a) for RS CVn it is found that the period of the starspot cycle is a half of the orbital period modulation (Rodonò et al. 1995), a behaviour which may be tentatively interpreted by the torsional oscillation model proposed by Lanza et al. (1998b), and there are preliminary indications of a similiar behaviour also for AR Lac (Lanza et al. 1998a) and RT Lac (Keskin et al. 1994); on the contrary, b) for V471 Tau, and possibly CG Cyg, the period of the light modulation equals that of the orbital period modulation (Ibanoglu et al. 1994, Hall 1991b).

It is important to clarify this issue, both observationally and theoretically, in order to use the orbital period modulation as a proxy for the magnetic cycle period also in those active stars whose photometric modulation is very difficult to detect, as the secondaries of Algols and cataclismic variables.

Baliunas et al. (1996) suggested to use the slope of the correlation $\log(P_{cyc}/P_{rot})$ vs. $\log P_{rot}$, where P_{cyc} is the activity cycle period and P_{rot} the rotation period, to discriminate between different kinds of dynamo models, specifically, between dynamos distributed throughout an extended portion of the convective envelope and those working in a thin shell. Following Brandenburg et al. (1994), we assume that the activity cycle period scales with the dynamo number D as $P_{cyc} = 2\pi R^2 D^{-n}/\eta$, where R is the star's radius and η the turbulent magnetic diffusivity. The dynamo number is defined as $D = \alpha\Omega' R^4/\eta^2$, where α is the magnitude of the α effect and Ω' a measure of the differential rotation. The scaling exponent is $n \sim 0.5$ for an extended dynamo, and $n \sim 0.15$ for a dynamo in a shell at the base of the convection zone.

In a rapidly rotating star it is possible to assume that α is at a constant value (saturation of the α effect) whereas $\eta \sim \Omega^{-1}$, according to the turbulence model adopted by Rüdiger & Brandenburg (1995). If we assume that the differential rotation scales as $\Omega' \sim \Omega^{0.25}$, according to the relation given by Henry et al. (1995), then we find that $(P_{cyc}/P_{rot}) \sim (1/P_{rot})^{0.9}$ for a distributed dynamo, and $(P_{cyc}/P_{rot}) \sim (1/P_{rot})^{1.7}$ for a shell dynamo.

In Figure 1 we plot the correlation between $\log(P_{cyc}/P_{rot})$ and $\log(1/P_{rot})$, as derived from observations (Henry et al. 1995, Rodonò et al. 1995, Lanza et al. 1998a, Oláh & Kolláth 1999). The slope of the observed correlation is 1.2 ± 0.15, just in between the slopes predicted for the two models. Therefore, we can not presently draw any firm conclusion about the kind of dynamo at work in very active stars. However, the correlation coefficient of the linear regression is 0.86, thus suggesting that this approach may indeed provide us with information on the dynamo when more precise determinations of the differential rotation and dynamo parameters become available.

6. Conclusion and Future Perspective

In the Sun, a shell dynamo based on the $\alpha - \Omega$ mechanism seems to be capable of explaining the main features of the 11-yr cycle, even if several aspects still need to be clarified (e.g., Rüdiger & Brandenburg 1995, Schlichenmaier & Stix 1995). However, in the most active stars it appears difficult to reconcile the observed latitudinal range of starspots with the prediction of thin shell dynamo models. Moreover, the relative amplitude of the differential rotation is observed to decrease with increasing rotation rate, whereas the α effect should increase and possibly saturate in fast rotators (Brandenburg et al. 1994). Therefore, in addition to a solar-like dynamo, probably working in the overshoot zone, a distributed $\alpha^2\Omega$ dynamo is likely to be present in very active stars (Brandenburg et al. 1990, Moss et al. 1995). An $\alpha^2\Omega$ dynamo may be invoked to interpret, at least qualitatively, the existence of a threshold for a sizeable light modulation and possibly other properties of the spot pattern, such as the existence of active longitudes. Moreover, a distributed dynamo seems to be a good candidate for explaining the observed orbital period modulation in active close binaries (cf. Lanza et al. 1998b).

In the near future, significant progress is expected from new observational approaches introduced in recent years and from long-term studies of selected objects. Systematic photometric and spectroscopic monitoring will provide data on the emergence latitudes, proper motions and rotational properties of young spots, which are likely to be the most relevant to constrain dynamo models, while Zeeman Doppler imaging will provide maps of the surface fields to be compared with the solar behaviour. Reliable information on the surface differential rotation will be obtained through cross-correlation of Doppler maps, thus providing a key parameter for dynamo models. Long-term studies will provide additional information on stellar activity cycles and their connection with the orbital period modulation in close binaries. These data may help us to understand the effects of a strong dynamo on the energy and angular momentum balances in very active stars.

Acknowledgments. Active star research at Catania Astrophysical Observatory and Institute of Astronomy of Catania University is funded by MURST (*Ministero dell'Università e della Ricerca Scientifica e Tecnologica*), CNAA (*Consorzio Nazionale per l'Astronomia e l'Astrofisica*) and the *Regione Siciliana*.

References

Amado, P.J. & Byrne, P.B., 1997, A&A 319, 967

Applegate, J.H., 1992, ApJ 385, 621

Baliunas, S.L. & Jastrow, R., 1990, Nature 348, 520

Baliunas, S.L., Donahue, R.A., Soon, W.H., et al., 1995, ApJ 438, 269

Baliunas, S.L., Nesme-Ribes, E., Sokoloff, D. & Soon, W.H., 1996, ApJ 460, 848

Berdyugina, S.V. & Tuominen, I., 1998, A&A 336, L25

Brandenburg, A., Charbonneau, P., Kitchatinov, L.L. & Rüdiger, G., 1994, ASP Conf. Ser. vol. 64, p. 354

Brandenburg, A., Moss, D., Rüdiger G. & Tuominen I., 1991, Geophys. Astrophys. Fluid Dyn. 61, 179

Bruls, J.H.M.J., Solanki, S.K. & Schüssler, M., 1998, A&A 336, 231

Budding, E., 1996, in Stellar Surface Structure, IAU Symp. 176, K.G. Strassmeier & J.L. Linsky (Eds.), Kluwer Ac. Publ., Dordrecht, p. 95

Byrne, P.B., 1992, in Surface Inhomogeneities in Late-Type Stars, P.B. Byrne & D.J. Mullan (Eds.), LNP 397, Springer-Verlag, Berlin, p. 3

Byrne, P.B., 1996, in Stellar Surface Structure, IAU Symp. 176, K.G. Strassmeier & J.L. Linsky (Eds.), Kluwer Ac. Publ., Dordrecht, p. 299

Byrne, P.B. & Mullan, D.J., 1992, Surface Inhomogeneities in Late-Type Stars, LNP 397, Springer-Verlag, Berlin

Byrne, P.B. & Rodonò, M., 1983, Activity in Red-Dwarf Stars, IAU Coll. 71, D. Reidel Publ. Co., Dordrecht

Byrne, P.B., et al., 1995, A&A 299, 115

Caligari, P., Moreno-Insertis, F. & Schüssler, M., 1995, ApJ 441, 886

Cameron, A.C. & Hilditch, R.W., 1997, MNRAS 287, 567

DeLuca, E.E., Fisher, G.H. & Patten, B.M., 1993, ApJ 411, 383

DeLuca, E.E., Fan, Y. & Saar, S.H., 1997, A&A 481, 369

Donahue, R.A., 1993, PhD Thesis, New Mexico State Univ., Las Cruces, New Mexico

Donahue, R.A., 1996, in Stellar Surface Structure, IAU Symp. 176, K.G. Strassmeier & J.L. Linsky (Eds.), Kluwer Ac. Publ., Dordrecht, p. 261

Donahue, R.A., Saar, S.H. & Baliunas, S. L., 1996, ApJ 466, 384

Donati, J.-F., 1999, these proceedings

Donati, J.-F., Cameron, A. C., 1997, MNRAS 291, 1

Eaton, J.A., Henry, G.W. & Fekel, F.C., 1996, ApJ 462, 888

Foukal, P., 1993, Sol. Phys. 128, 219

Foukal, P. & Lean, J., 1988, ApJ 328, 347

Hall, D.S., 1989, Space Sci.Rev. 50, 219

Hall, D.S., 1991a, in The Sun and Cool Stars: Activity, Magnetism, Dynamo, IAU Coll. 130, I. Tuominen, D. Moss and G. Rüdiger (Eds.), Springer-Verlag, Berlin, p. 353

Hall, D.S., 1991b, ApJ 380, L85

Hall, D.S., 1994, Mem. S. A. It. 65, 73

Hall, D.S., 1996, in Stellar Surface Structure, IAU Symp. 176, K.G. Strassmeier & J.L. Linsky (Eds.), Kluwer Ac. Publ., Dordrecht, p. 217

Harvey, K.L. & Zwaan, C., 1993, Sol. Phys. 148, 85

Hatzes, A.P., 1998, A&A 330, 541

Hatzes, A.P., Vogt, S.S., Ramseyer, T.F. & Misch, A., 1996, ApJ 469, 808

Henry, G.W., Eaton, J.E., Hamer, J. & Hall, D.S., 1995, ApJS 97, 513

Ibanoglu, C., Keskin, V., Akan, M., Evren, S. & Tunca, Z., 1994, A&A 281, 811

Jetsu, L., Pelt, J. & Tuominen, I., 1993, A&A 278, 449

Jetsu, L., Pohjolainen, S., Pelt, J. & Tuominen, I., 1997, A&A 318, 293

Keskin, V., Ibanoglu, C., Akan, M., Evren, S. & Tunca, Z., 1994, A&A 287, 817

Kitchatinov, L.L. & Rüdiger, G., 1993, A&A 276, 96

Kosovichev, A.G., 1996, ApJ 469, L61

Krause, F. & Rädler, K.-H., 1980, Mean-Field Magnetohydrodynamics and Dynamo Theory, Pergamon Press, Oxford

Küker, M., Rüdiger, G. & Kitchatinov, L.L., 1993, A&A 279, L1

Lanza, A.F., Rodonò, M. & Zappalà, R. A., 1992, in Surface Inhomogeneities in Late-Type Stars, P.B. Byrne & D.J. Mullan (Eds.), LNP 397, Springer-Verlag, Berlin, p. 298

Lanza, A.F., Rodonò, M. & Zappalà, R. A., 1994, A&A 290, 861

Lanza, A.F., Catalano, S., Cutispoto, G., Pagano, I. & Rodonò, M., 1998a, A&A 332, 540

Lanza, A.F., Rodonò, M. & Rosner, R., 1998b, MNRAS 296, 803

Lockwood, G.W., Skiff, B.A. & Radick, R. R., 1997, ApJ 485, 789

Moffatt, H.K., 1978, Magnetic Field Generation in Electrically Conducting Fluids, Cambridge Univ. Press, Cambridge

Moss, D., Tuominen, I. & Brandenburg A., 1991, A&A 245, 129

Moss, D., et al., 1995, A&A 294, 155
Moss, D. & Tuominen, I., 1997, A&A 321, 151
Noyes, R.W., Weiss, N.O. & Vaughan, A.H., 1984, ApJ 287, 769
O'Dell, M.A., Panagi, P.M., Hendry, M.A. & Cameron, A.C., 1995, A&A 294, 715
Oláh, K., et al., 1997, A&A 321, 811
Oláh, K. & Kolláth, Z., 1999, these proceedings
O'Neal, D., Saar, S.H. & Neff, J.E., 1996, ApJ 463, 766
Piskunov, N.E., Tuominen, I. & Vilhu, O., 1990, A&A 230, 363
Radick, R.R., Lockwood, G.W., Skiff, B.A. & Baliunas, S.L., 1998, ApJS 118, 239
Rice, J.B., Wehlau, W.H. & Khokhlova, V. L., 1989, A&A 208, 179
Rodonò M., 1992, in Evolutionary Processes in Interacting Binary Stars, IAU Symp. 151, Y. Kondo, R.S. Sistero & J. Polidan (Eds.), Kluwer Ac. Publ., Dordrecht, p. 71.
Rodonò, M., Lanza, A.F. & Catalano, S., 1995, A&A 301, 75
Rosner, R. & Weiss, N.O., 1992, in Harvey K. L. (Ed.), The Solar Cycle, ASP Conf. Ser. vol. 27, p. 511
Rüdiger, G. & Brandenburg, A., 1995, A&A 296, 557
Rüdiger, G., von Rekowski, B., Donahue, R. A. & Baliunas, S. L., 1998, ApJ 494, 691
Schlichenmaier, R. & Stix, M., 1995, A&A 302, 264
Schüssler, M., 1987, in B.R. Durney & S. Sofia (Eds.), The Internal Solar Angular Velocity, D. Reidel Publ. Co., Dordrecht, p. 303
Schüssler, M., 1996, in Stellar Surface Structure, IAU Symp. 176, K.G. Strassmeier & J.L. Linsky (Eds.), Kluwer Ac. Publ., Dordrecht, p. 269
Schüssler, M. & Solanki, S.K., 1992, A&A 264, L13
Schüssler, M., Caligari, P., Ferriz-Mas A., Solanki, S.K. & Stix, M., 1996, A&A 314, 503
Solanki, S.K., 1999, these proceedings
Soon, W.H., Baliunas, S.L. & Zhang, Q., 1993, ApJ 414, L33
Spruit, H.C., 1982, A&A 108, 348
Spruit, H.C. & Weiss, A., 1986, A&A 166, 167
Stix, M., 1976 A&A 47, 243
Strassmeier, K. G. & Linsky, J.L., 1996, Stellar Surface Structure, IAU Symp. 176, Kluwer Ac. Publ., Dordrecht
Strassmeier, K.G., Hall, D.S. & Henry, G.W., 1994, A&A 282, 535
Strassmeier, K.G. & Rice, J.B., 1998, A&A 330, 685
Vincent, A., Piskunov, N.E. & Tuominen, I., 1993, A&A 278, 523
Vogt, S.S., Hatzes, A.P., Misch, A.A. & Kürster, M., 1998, ApJS, in press
Vogt, S.S., Penrod, G.D. & Hatzes, A. P., 1987, ApJ 321, 496

Solar and Stellar Activity: Similarities and Differences
ASP Conference Series, Vol. 158, 1999
C.J. Butler and J.G. Doyle, eds.

New Perspectives on Solar Prominences

B. Schmieder

Observatoire de Paris, 92195 Meudon Cedex Principal, France

Abstract.
New observations of solar prominences obtained during multi-wavelength campaigns with the *SOHO* and *Yohkoh* space missions bring new insight into the physical conditions of prominences and their environment. The new data provides important constraints for theory. In this framework, 3-D models of the magnetic configuration of filaments have been developed recently. They allow us to consider non-potential configurations and electric current distributions. We will present some new developments of 3-D models of a bipolar magnetic field and twisted flux tube which explains the presence of fine structures in prominences. The models also explains how the feet reach the photosphere and how prominences erupt as a part of a coronal mass ejection (CME). The new models bring valuable information on the global solar magnetic field with direct implications for dynamo theories and on the helicity of the heliosphere.

1. Introduction

This is not an exhaustive review on prominences but a subjective summary of papers presented during the IAU 167 meeting held in Aussois in April 1997. The meeting results are supplemented by some papers appearing recently in the literature. I will draw heavily from some reviews by Berger, Démoulin, Martin & McAllister, Seehafer, Vial, Webb. Brendan Byrne was fascinated by solar prominences and tried in different ways to detect similar prominences on stars (see Byrne et al. 1998 and Cameron this issue). We had long conversations about the similarities of these objects, and we discussed what star prominence studies can bring to solar physics. I will review in the first chapter some general properties of solar prominences. In the second chapter I will present the status of magnetic field observations and theories of prominences. In the last chapter I will draw attention to the now well established rules concerning the chirality of prominence signatures and then conclude with a discussion on how these statements can be interpreted.

2. Structure and Dynamics of Filaments

2.1. General properties

Prominences are thin structures consisting of cold plasma embedded in the hot corona (Figure 1). Their general shape has been known for nearly a century (e.g.

d'Azambuja and d'Azambuja 1948). When observed on the disk, prominences are often referred to as filaments. They form in channels with aligned magnetic flux. Mainly two types of prominences can be considered: the quiescent prominences and the active-region prominences. The latter ones are lower in altitude, and smaller in all their dimensions (see Table 1). The magnetic field of a quiescent prominence is typically 5- 40 G and the direction is mainly along the length of the prominence. The field is almost horizontal, but in the horizontal plane it is inclined to the filament axis at an angle of about 15^0.

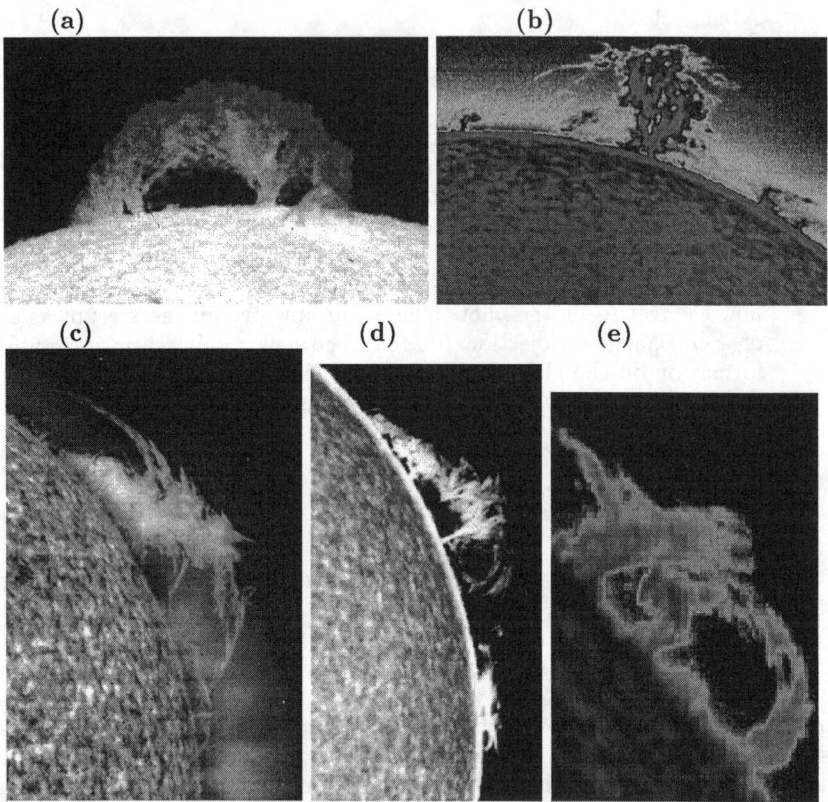

Figure 1. Observations of prominences in Hα (a), Ca II line (b) and (d) with the Meudon spectroheliograph and in UV lines: (c) with EIT He II 304Å and (e) with SUMER C III 977Å.

2.2. Diagnostic of Thermodynamic Parameters

In order to understand the existence of cool material in the corona, it is essential to precisely measure its temperature and the variation of temperature within the structure, especially at the interface regions. Table 1 is a summary of the Hvar reference Atmosphere of Quiescent Prominences (Engvold et al. 1990). Since this publication, new investigations have been completed. I will quote observations in the He 10830 Å line which allow one to see clearly the corridor of the filament (Harvey & Gaizauskas 1998), the observations obtained in He II 304Å with EIT which has provided many movies of eruptive prominences and allows one to better understand the flow inside prominences. The *SOHO*/SUMER spectrograph has tracked filaments and prominences using the Lyman series window. Lyman lines show a central reversal in filaments. NLTE computations show that the reversed part of the profile is due to absorption by the filament overlying the chromosphere. The ratio of the peaks gives an indications of the velocities inside the filament. To fit all the Lyman series we need to introduce a prominence-corona transition region in the isothermal model (Schmieder et al. 1998). Using *SOHO*/CDS spectrometer observations, Kucera et al. (1998a) derived the optical thickness of a prominence from the analysis of the absorption of the coronal emission lines by Hydrogen (H^0 upper bound at $\lambda < 911$ Å) and Helium (He^0 upper bound at $\lambda < 504$ Å) continua.

2.3. Fibrils and Plagettes in the Filament Channel

Filaments are always found above inversion lines of the photospheric magnetic field in regions, called corridors or filament channels, which are nearly free of chromospheric fibrils. They are also characterized on either side by the presence of Hα fibrils, nearly aligned with the inversion line, indicating a high magnetic shear (Foukal 1971, Rompolt 1990). The corridor is nearly free of vertical magnetic field flux except small parasitic polarities (Martin 1990). At the locations of these small polarities we observed *the plagettes* from which fibrils stream parallel to the filament (Figure 2). Two morphological forms of filament channels and filaments have been identified. They are called *dextral and sinistral.* An observer standing on the positive polarity side of a dextral/sinistral filament would see the lateral feet pointing to the right/left along the filament channel. These names correspond to right-bearing/left-bearing
prominences if the prominence feet bear off to the right/left of the main axis. All dextral filaments are right-bearing. Most quiescent filaments in the northern/southern hemisphere are *dextral/sinistral* at mid- and high latitudes, regardless of the cycle (Martin et al. 1994). Observations show that there is a good correlation between the direction of the magnetic field along the filament as inferred from the plagettes and as inferred from the lateral prominence feet.

The filament channel is larger in radio wavelength than in optical lines (Nobeyama observations).

2.4. Coronal Arcades

The X-ray observations made by *Yohkoh*/SXT have told much about the coronal arcades overlying filament channels (McAllister et al. 1998). The skew of the coronal arcades changes according to the hemisphere. Skew was defined as the acute angle between the loops and the polarity inversion or filament axis. If the arcade loops cross over the filament in the sense of threads of a left-hand/right-hand screw they defined the loops as left-skewed/right-skewed. Left skewed arcades overlie dextral filaments (Martin & McAllister et al. 1995).

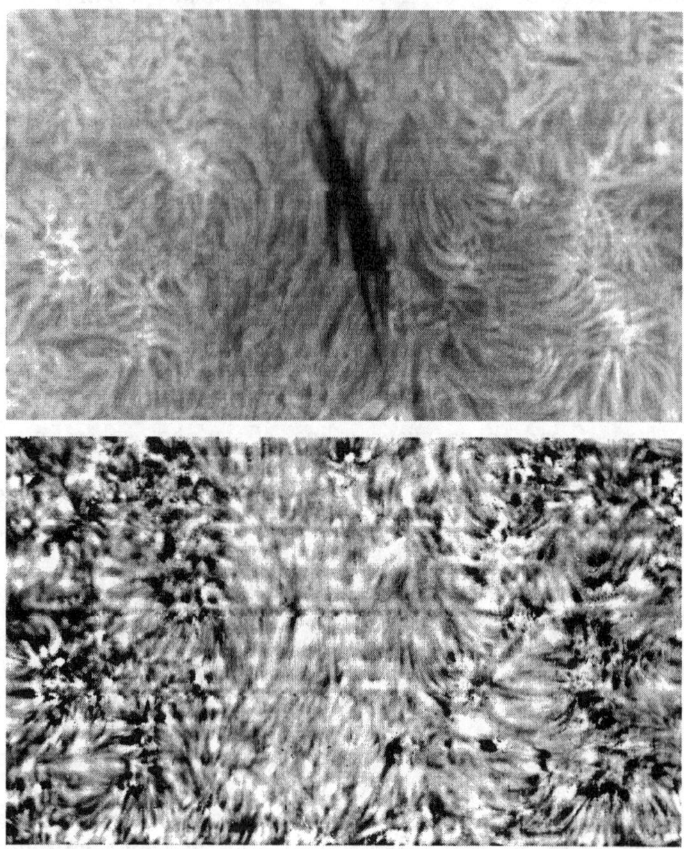

Figure 2. Observation of a filament in the northern hemisphere with the MSDP spectrograph at Pic du Midi: (top panel) – intensity in Hα. Note that it is a dextral filament with right bearing, note also the plagettes in the filaments channel and the fibrils aligned along the main filament axis. (bottom panel) – Doppler-shifts of the filaments and the chromosphere. Note that the velocities are reduced in the filament channel and small cells of higher up and down velocities are located just at the location of the feet. Note aligned velocity structures parallel to the main filament axis in the filament channel (Schmieder et al. 1991).

Table 1. Identity card of prominences

	Quiescent Prominence	Remarks
length	60 - 600 Mm	less for active-region filament
wide	4 - 15 Mm	
distance between feet	30 Mm	size of the supergranules
density	10^{10} - 10^{11} cm^{-3}	less in prominence-corona region (10^8- 10^9 in cavity)
temperature	5000 - 8000 K	higher T (PCTR) observed in UV lines
magnetic field	4 -50 G	100 G in active-region prominence
angle (axis of prom./B)	15 - 25 degrees	Hanle effect
configuration	inverse	normal for active region filament
magnetic flux	2.5×10^{19} Mx	for active-region filament $F = 10^{20}$- 10^{21} Mx
flows	vertical 0.5 -5 $km\ s^{-1}$, horizontal 10-20 $km\ s^{-1}$	60 $km\ s^{-1}$ in active prominences
Diagnostics	Hydrogen lines (Hα - Lyman lines), Ca II radio $\lambda < 1cm$	NLTE radiative transfer physical parameters (Nobeyama Observatory)

2.5. Velocities for Quiescent Prominences

According to Doppler-shift measurements only weak vertical velocities are observed ($< 5\ km\ s^{-1}$), higher horizontal velocities are detected reaching 20 to 30 $km\ s^{-1}$ (Schmieder et al. 1989). In activated filaments even higher velocities are observed, particularly inside the lateral feet (Figure 2b).

2.6. Twisted Ropes in Eruptive prominence

Many observations suggest a helical-like pattern during eruption of prominences (Rompolt 1990, Vršnak et al. 1991, Kucera et al. 1998b). The helical structure of prominences is more obvious at the limb (Figure 3). On the disk Raadu et al. (1988) observed blueshifts and redshifts along disturbed filaments suggesting twisted motion along the axis. Filippov (1994) discovered a pattern typical of a twisted flux rope with fishbone-like structures on both sides of the inversion line in a Hα filament. Twisted configurations are also identified in CMEs with in situ measurements from Ulysses (Bothmer et al. 1996). Finally the emergence of twisted flux ropes is also identified by vector field measurements (Leka et al. 1996). Models of eruptive filaments predict such twist (Figure 4)

3. Magnetic Field

3.1. Observations

Hanle Effect In prominences the Zeeman effect only allows the measurement of the longitudinal component of the magnetic field (Kim 1990). Radio wavelengths provide information on the field strength (Apushkinskii et al. 1990). The Hanle effect gives the three components of the field and the electron density from the polarization components in two spectral lines (Bommier et al. 1994). The main results of Hanle measurements are the following: the prominence field has the opposite direction to that expected from simple extrapolation of the photospheric field. It is what is called "Inverse type" (Leroy et al. 1984), the orthogonal component (B_{per}) is opposite to the field of a simple arcade and also the field component parallel to the prominence ($B_{//}$) is opposite to that of an arcade sheared by differential rotation.

Prominence Magnetic Field It is now well accepted that the magnetic field in prominences is nearly horizontal, homogeneous on the scale of a few arc seconds, and increasing with height (Leroy et al. 1983).

Presence of Magnetic Dip The gravitational scale height of prominence material ($\sim 200 - 500$ km) is much lower that the typical height of prominences ($10^4 - 10^5$ km). This implies that the plasma pressure cannot provide the mechanism of support. The observed velocities (a few $km\ s^{-1}$) are much less than the free fall velocities and usually the velocity pattern does not show such flows except in arch filament systems or post-flare loops, structures that we do not classify commonly as quiescent prominences. The cold material should be supported by the magnetic field. If enough mass can be brought into the top of the sheared magnetic arcade in less than one hour, the gravity force can bend the top of the field lines and form a dip (Dere et al. 1990, Fiedler & Hood 1993). The observations of dips are not evident but they could be suggested in some data (Rompolt 1990).

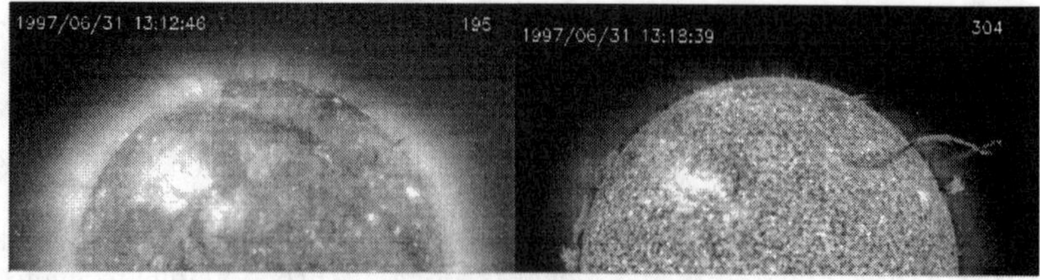

Figure 3. Eruptive prominence observed in Fe IX 195 Å and He II 304 Å by *SOHO*/EIT on May 31, 1997. Note the helical structure of the prominence observed in He II (Courtesy of C.Delannée).

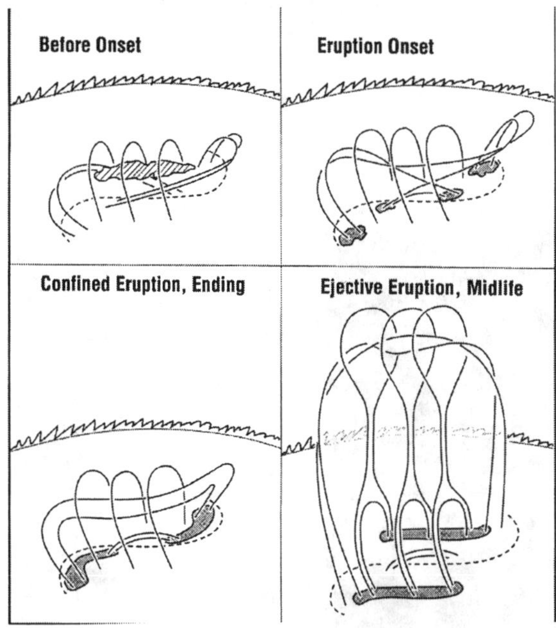

Figure 4. Sketch of an eruptive filament (Courtesy of R.Moore)

3.2. Magnetic Configurations supporting Prominences

Since the well known 2D model of Kippenhahn & Schlüter (1957), several models have been proposed in 2 1/2-D and then in 3-D. Two types of models are considered: arcade-like configurations and twisted flux tubes.

Model in 3-D with Overlying Arcade Arcade models in 2 1/2-D cannot have a dip at the top of the field lines even when the magnetic shear increases (Amari et al. 1991). In 3-D it can occur. The overlying arcade locally compresses the central part of an underlying sheared arcade (Antiochos et al. 1994). The latter model gives an inverse polarity prominence with a magnetic field nearly aligned with the photospheric inversion line.

Quadrupolar configurations These configurations naturally have a dip. Malherbe & Priest (1983) and later Démoulin & Priest (1993), Uchida (1998), developed such models. The presence of a corridor free of significant field is needed to have a prominence extension reaching the chromosphere and converging motions to provide mass supply. These models are perfectly adapted for active region filaments (Tang 1987) or filaments between two active regions (MacKay et al. 1998). The presence of "dual arcades" observed by *Yohkoh*/SXT is a signature of magnetic reconnection in quadrupolar configurations (Uchida 1998).

Models in 3-D with Twisted Flux Tubes This class of model provides the needed dipped magnetic field and seem to be well supported by recent observations. The twisted configuration can be obtained by photospheric twisting motions (Priest

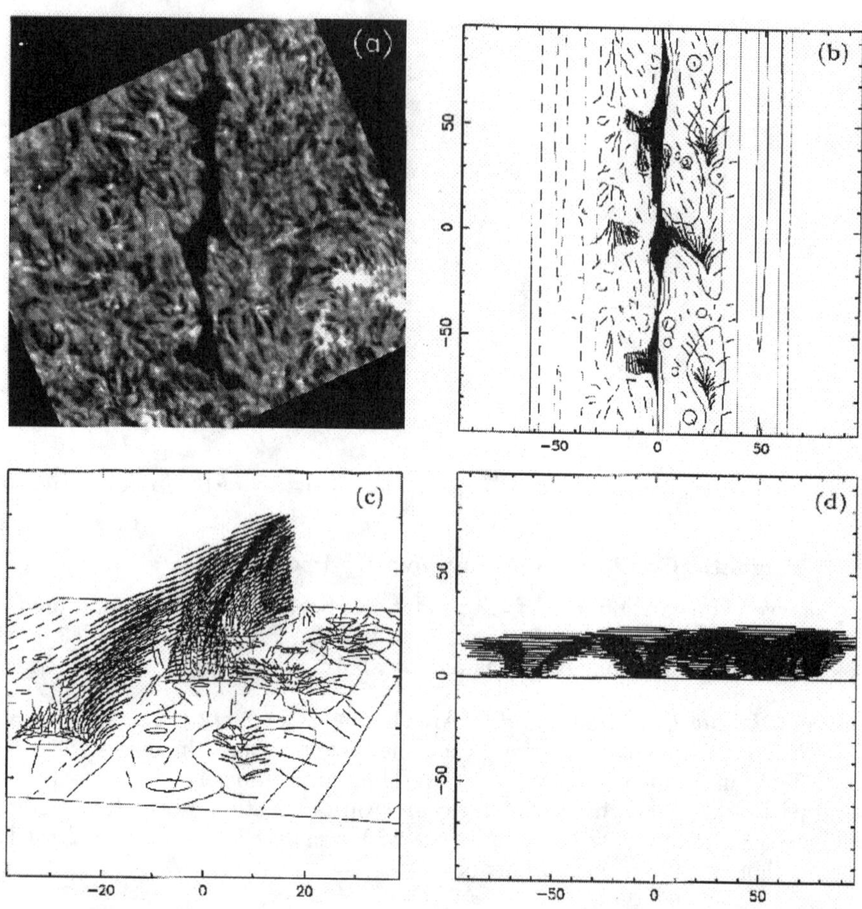

Figure 5. Model of bald patch filament in 3-D: (a) Observations of a Hα filament with the MSDP spectrograph on the VTT at Tenerife, (b) MDI magnetogram and extrapolated magnetic field lines drawn only in the dips, (c) 3-D view, (d) side view (Aulanier et al. 1998 c)

et al. 1989), by converging motions in a sheared arcade with magnetic reconnection at the inversion line (van Ballegooijen & Martens 1989), by resistive instability in a sheared arcade or by relaxation and accumulation of helicity (Démoulin & Priest 1989, Rust & Kumar 1994) or by emergence from the convective zone (Low 1994, 1996). Prominence feet appear as a direct consequence of the parasitic polarities present in the filament channel. Priest et al. (1996) used the emergence of small loops to create feet or "barbs". The feet would be due to the reconnection occurring between the emerging dipole and the main filament field. Aulanier & Démoulin (1998 a,b) showed that the feet appear naturally above secondary photospheric magnetic inversion lines produced by the parasitic polarities which create local quadrupolar configuration able to support the plasma in the feet.

3-D model of Prominence with a bald patch A recent model has been developed by Aulanier et al. (1998c, 1999) which shows the importance of field topology. They consider a bipolar configuration imposing a twisted flux tube using the equations of linear magneto-hydrostatics (lmhs) for the extrapolation (Low 1992). Pressure and gravity are taken into account. They use directly the observed longitudinal fields, taken from magnetograms obtained by *SOHO*/MDI comparing their field lines with the filament observations obtained on the ground (VTT, Tenerife) and in space (SUMER). The dips of the field lines where material can be supported is examined. In Figure 5 they draw only the portion of the field lines dipped enough (2Mm) finding good agreement between the shape and location of the dipped field lines and the shape of the filament body (Figure 5). The lateral feet are explained by the presence of parasitic polarities close to the filament axis. When you consider low field lines with dips tangent to the photosphere, you can define a region containing all these tangent points where the field is purely horizontal. Such a place is called a "bald patch" (location in the photosphere free of vertical magnetic field lines). The overlay shows that the lateral feet correspond to such locations on each side of the inversion line where material can be readily supported. Many Hα dark chromospheric fibrils correspond to such locations. The magnetic field can be defined by magnetic surfaces. Because of the presence of parasitic polarities in the filament channel, separatrices do exist in many places. When the bald patch corresponds to a part of the intersection of the separatrice with the photosphere, a current layer can exist over the bald patch and the plasma along the field lines can be heated by ohmic dissipation. This explains the presence of the heated plasma observed at UV temperatures along the legs of the arcades overlaying filaments observed with SUMER in Si IV (Kucera et al. 1999a).

4. Magnetic Helicity and Filaments

The observations show specific relationships of chirality between filaments, coronal arcades, active regions, coronal mass ejections and magnetic clouds.

4.1. Chirality of Filaments and Arcades

The chirality relationship of filaments and overlying arcades is in the inverse sense, as shown above. It applies to both quiescent and dynamic arcades that evolved after filament eruption/coronal mass ejection events. Many of the dynamic arcades evolved such that the later-forming loops showed apparent rotation relative to the earlier-forming loops. This apparent rotation also has a specific sense. Left-skewed/right-skewed arcades turn in the counterclockwise/clockwise sense decreasing the magnetic shear (Martin 1998).

4.2. Shear Angle of Active Regions

Conspicuous fibrils are observed near sunspots penumbras. Richardson (1941), recognizing a similarity between fibril pattern and the pattern of iron fileings around bar magnets, made an extensive study of the orientation of the superpenumbra fibrils and found that the majority are clockwise for sunspots in the southern hemisphere. This was confirmed later by Rust and Martin (1994). On the other hand Seehafer (1990) estimated the electric current helicity in 16 active regions and concluded that the electric current helicity is predominantly positive/negative in the southern/northern hemisphere. This result was confirmed also from a comparison between Mees vector magnetograms and current-free configurations (Pevtsov et al. 1994).

4.3. Chirality of large scale X-ray Structures

Very large scale X-ray structures have been noticed to exhibit a slender S or backward-S shape as a function of hemisphere (Rust & Kumar 1996, Moore et al. 1997). Reverse S-shaped structures predominate in the northern hemisphere as do S-shaped structures in the southern hemisphere. These S-shape patterns also apply to many filament channels.

4.4. Helical Structure of Magnetic Cloud

Interplanetary clouds are the ejected parts of coronal mass ejections as first concluded by Marubashi (1986). Many such plasma clouds have helical structure detected by magnetometers on board spacecrafts (Burlaga 1991, Webb 1998). Gosling (1990) proposed that helical magnetic fields in interplanetary clouds originate from magnetic reconnection in ascending coronal arcades. Based on the same scenario, Martin & McAllister (1998) anticipate that left-skewed coronal arcades become left-helical CMEs and are detectable later as left-helical interplanetary clouds. Rust (1994) found that the majority of interplanetary clouds with left/right-helical structures come from the northern/southern hemispheres, but he proposed that the helicity of the clouds reflects the helicity of the filaments before they erupt, which in turn reflects the helicity of the solar dynamo generated fields.

4.5. Twisted Configurations

Modelling a filament as an initially untwisted axial magnetic field aligned with the inversion field, Martin & McAllister (1998) claim that dextral filaments develop right-helical twist on the sun and that the corresponding interplanetary clouds have the helicity opposite to the CMEs after they have been detached from the Sun by reconnection. This solution is controversial and another solution of all these laws of chirality is proposed based on the idea that the filament lies in the lower portion of a large-scale helical magnetic field which supports the filament against gravity (Rust & Kumar 1994, Aulanier et al. 1998 c, 1999). However Martin & McAllister did not support the idea of a twisted flux rope because of fine structure observations and vertical velocities in the barbs.

5. Discussion on Helicity

A natural way to explain how both magnetic field components, parallel and normal to the prominence, are inverse is to use a quadrupolar field sheared by the differential rotation (Priest et al. 1996). However, this gives a magnetic helicity dominance in each hemisphere contraty to observations. Berger (1998) reviewed different statements that are well established to sketch a solution. Helicity is conserved in ideal MHD, and it is probably well conserved on the Sun, so any positive helicity generated by plasma motions, for example, must be offset by negative helicty generation. Berger proposed to explain the sign of the helicity, namely, Northern hemisphere negative, Southern Hemisphere positive by taking into account the transfer of the helicity either into or from the solar wind or across the equator, beneath the surface. His solution to helicity in filaments is based on the emergence of twisted flux tubes. The same approach was advocated by Rust (1994). Reconnection below the solar surface could create twisted flux tubes which emerge due to bouyancy. By transfer and redistribution of the helicity and by using faster rotation at the equator one can reproduce at least the observed global helicity pattern.

Démoulin (1998) explained that the right sign of helicity could be obtained with Coriolis forces acting on diverging supergranular flows (Priest et al. 1989). This process is very efficient and can accumulate helicity in twisted tubes but since localized motions give neutralized currents, it is difficult to reproduce the dominance of one polarity as observed in each hemisphere. If a local dynamo cannot give a good explanation, then global dynamo models should be investigated for their possible helicity-producing qualities.

6. Conclusion

The large amount of data concerning helical features, from filaments to interplanetary clouds, have shown a great organizing principle of nature. We need models that can fit into a unified theory. We have shown the importance of twisted flux tube emergence, the presence of parasitic polarities in the inversion line corridor, the aligned fibrils in the filament corridor and all the chirality laws. Many questions are unsolved and we need further observations with multiwavelength instruments to test the existing different proposed approaches.

References

Amari T., Démoulin P., Browning P., Hood A.W. & Priest E.R. 1991, A&A, 241, 604

Antiochos S.K., Dahlburg R.B. & Klimchuck J.A. 1994, ApJ, 420, L41

Anzer U. 1972, Solar Phys., 24, 324

Apushkinskii G.P., Nesterov N.S., Topchilo N.A. & Tsyganov A.N. 1990, SvA, 34(5), 530

Aulanier G., Démoulin P.1998a, PASP, 150, 186

Aulanier G., Démoulin P.1998b, A&A, 329, 1125

Aulanier G., Démoulin P., van Driel-Gesztelyi L., Mein P & deForest C., 1998c, A&A, 335,309

Aulanier G., Démoulin P., Mein N., van Driel-Gesztelyi L., Mein P & Schmieder B. 1999, Solar Phys., in press

Berger M. 1998, PASP, 150,102

Bommier V., Leroy J.L. 1998, PASP, 150, 434

Bommier V., Landi Degl'Innocenti E., Leroy J.L. & Sahal-Bréchot S. 1994, Solar Phys., 154, 231

Bothmer V., Desai M.I., Marsden R.G. et al. 1996, A&A, 316, 493

Burlaga L.F. 1991, in Physics of the Inner Heliosphere II, eds. R.Schwenn, E.Marsch, Springer Verlag, New York,1

Byrne P.B., M.T.Eibe & van den Oord G.H.J. 1998, Publications of the ASP, 150, 227

d'Azambuja L., d'Azambuja M. 1948, Annales de l'Observatoire de Paris, 6,7

Démoulin, P. 1998, Publications of the ASP, 150, 78

Démoulin P. & Priest E.R. 1989, A&A, 214, 360

Démoulin P. & Priest E.R. 1993, Solar Phys., 144, 283

Dere K.P., Schmieder B. & Alissandrakis C. 1990, A&A, 233, 207

Engvold O. 1990, in V. Ruždjak, E.

Fiedler R.A.S. & Hood A.W. 1993, Solar Phys., 146, 297

Filippov B.P. 1994, Astron. Letters, 20, 665

Foukal P. 1971, Solar Phys., 19, 59

Gosling J.T. 1990, in Physics of Magnetic Flux Ropes (eds) C.T.Russell, E.R. Priest, L.C.Lee, Geophys.Mono. Ser. 58, 343

Harvey K. and Gaizauskas V. 1998, Publications of the ASP,

House L.L. & Berger M.A. 1987, ApJ, 323, 406 Corona

Kim I.S. 1990, IAU Colloq. 117, Springer-Verlag, 49

Kippenhahn R. & Schlüter A. 1957, Zs. Ap. 43, 36

Kucera T.A., Andretta V. & Poland A.I. 1998 a, Solar Phys.,in press

Kucera T.A., Poland A.I., Wiik J.E., Schmieder B. & Simnett G. 1998b, PASP, 150, 318

Leka K.D., Canfield R.C., McClymont A.N. & van Driel-Gesztelyi L. 1996, ApJ, 462, 547

Leroy J.L., Bommier V. & Sahal-Bréchot S. 1983, Solar Phys., 83, 135

Leroy J.L., Bommier V. & Sahal-Bréchot S. 1984, A&A, 131, 33

Low B.C. 1992, ApJ, 399, 300

Low B.C. 1994, Plasma Phys., 1, 1684

Low B.C. 1996, Solar Phys., 167, 217

MacKay D.H. & Priest, E.R. 1996, Solar Phys., 167, 281

MacKay D.H., Gaizauskas V., Ricckard G.J. & Priest, E.R. 1997, ApJ, 486, 534

Mc Allister A.H., Hundhausen A.J., MacKay D.H. & Priest, E.R. 1998, PASP, 150, 43

Malherbe J.M. & Priest E.R. 1983, A&A, 123, 80

Marabashi K. 1997 in Coronal Mass ejections Geophys. Mono.Ser., 99, 147

Martin S.F. 1990, IAU Colloq. 117, Springer-Verlag, 1

Martin S.F., Bilimoria R., Tracadas P.W. 1994, in R. Rutten & C. Schrijvers (eds.), Solar Surface Magnetism, Kluwer Ac. Pub., 303

Martin S.F. & Mc Allister A.H. 1995, in Proc. Colloq. IAU 153

Martin S.F. 1998, Publications of the ASP, 150, 419

Moore R.L., Schmieder B., Hathaway D.H., Tarbell T.D. 1997, Solar Phys., 176, 249

Pevtsov A.A., Canfield R.C. , Metcalf T.R. 1995, ApJ, 440, L109

Priest E.R., Hood A.W. & Anzer U. 1989, ApJ, 344, 1010

Priest E.R., Van Ballegooijen A.A. & MacKay D.H. 1996, ApJ, 460, 530

Raadu M.A., Schmieder B., Mein N. & Gesztelyi L. 1988, A&A, 197, 289

Richardson R.S. 1941, ApJ, 41, 24

Rompolt B. 1990, Hvar Obs. Bull. Vol. 14, 1, 37

Rust D.M. & Kumar A. 1994, Solar Phys., 155, 69

Rust D.M. & Martin S.F. 1994, in Solar Active Region Evolution, Publications of the ASP 68, 337

Schmieder B. 1989, in E.R. Priest (ed.), *Dynamics and Structure of Quiescent Solar Prominences*, Kluwer Academic Publishers, Dordrecht, Holland, 15

Schmieder B. 1990, IAU Colloq. 117, Springer-Verlag, 85

Schmieder B., Raadu M.A. & Wiik J.E. 1991, A&A, 252, 353

Seehafer N. 1990, Solar Phys., 125, 219

Seehafer N. 1998, Publications of the ASP, 150, 407

Tang F. 1987, Solar Phys., 107, 233

Uchida Y. 1998, PASP, 150, 163

van Ballegooijen A.A., Martens P.C.H. 1989, ApJ, 343, 971

van Ballegooijen A.A., Martens P.C.H. 1990, ApJ, 361, 283

Vial J.C. 1998, PASP, 150, 175

Vršnak B. 1998, PASP, 150, 302

Vršnak B., Ruždjak V. & Rompolt B. 1991, Solar Phys., 136, 151

Webb D. 1998, PASP, 150, 463

Solar and Stellar Activity: Similarities and Differences
ASP Conference Series, Vol. 158, 1999
C.J. Butler and J.G. Doyle, eds.

Stellar Prominences

Andrew Collier Cameron

School of Physics and Astronomy, University of St Andrews, North Haugh, St Andrews, Fife, KY16 9SS, Scotland

Abstract. I review recent observational and theoretical progress in the study of condensations of cool material trapped in the coronae of rapidly rotating stars. Such condensations were discovered over a decade ago, producing transient moving absorption features in the optical Balmer-series lines. These condensations appear to be in a physical state similar to that of solar prominence material, in the sense that they are sufficiently cool and dense to contain substantial amounts of neutral hydrogen. The strong coronal magnetic fields of the stars in which they occur serve both to confine and to insulate the cool material, even though the ambient corona is typically at a temperature of several million degrees. There, however, the resemblance to solar prominences ends. The coronal condensations (sometimes dubbed "slingshot prominences") form at a variety of distances from the star, but are observed to favour formation sites near the Keplerian co-rotation radius. Simple support models involving purely dipolar fields can explain the existence of stable mechanical equilibria for sheet-like prominences in the equatorial plane. More complex field configurations show a wider variety of stable formation sites, both inside the co-rotation radius and out of the stellar equatorial plane. The variety of behaviour exhibited by otherwise similar stars viewed at different inclination angles thus provides important observational clues to the three-dimensional structure of the coronal magnetic field. The temporal evolution of their optical signatures shows major changes in the coronal topology on timescales as short as a day or two. Recent attempts to use Doppler imaging to track fluid shear in the underlying photosphere show that the surface of the star evolves on a more leisurely timescale of weeks. New efforts to extrapolate the 3-D structure of the coronal field, using prominence maps augmented by Zeeman-Doppler images of the surface field, should allow us to determine the rate at which large-scale coronal structure is de-stabilised by surface differential rotation.

1. Introduction

Over the last 15 years, discoveries of condensations of neutral material in the coronae of late-type stars other than the Sun have led to the recognition that stellar coronae, like the solar corona, are multi-phase media. These stellar coronal condensations have often been dubbed "prominences", on the grounds that they appear to consist of dense, relatively cool material, embedded in the hot, ambient medium of a stellar corona and giving rise to optically thick Balmer emission or (more usually) absorption. The implied presence of strong magnetic fields in the active stars where they are observed, suggests that the condensations are both confined and supported by the field, and insulated by relatively poor cross-field thermal conductivity. There, however, the resemblance to solar prominences probably ends. Indeed, whenever Brendan Byrne spoke on this subject he always emphasized the looseness of the solar analogy when applied to the physics of cool condensations in stellar coronae.

In this review, I shall begin with a brief survey of the observational evidence for stellar phenomena that have qualitative resemblances to both active-region and quiescent prominences on the Sun. Mostly, however, I wish to emphasize the utility of the longer-lived stellar coronal condensations as tracers of stellar coronal structure. The spatial distribution of the condensations themselves provides us with a valuable diagnostic of the complexity of the coronal field topology in active stars. In the rapidly rotating young stars where the "slingshot prominences" occur, they serve as tracers of both the radial extent and longitudinal structure of the closed-field corona. When observed repeatedly, their evolution from one rotation cycle to the next provides us with detailed insights into the rate of evolution of the global topology of the coronal magnetic field.

2. Active Stellar Prominences

There is also considerable indirect evidence for mass motions of cool material associated with stellar flares. Haisch et al (1983) observed a soft X-ray deficit at low X-ray energies following a flare on Proxima Cen, which they interpreted as a prominence eruption accompanying a two-ribbon flare. Doyle et al (1989) saw evidence of surge-like activity in the form of transient absorption features in IUE spectra of the Mg II h & k lines, blue-shifted by 80 km s^{-1}, following a flare in the RS CVn subgiant II Peg. Byrne (1987), on the other hand, interpreted transient red-shifted Hα absorptions in the same star in terms of post-flare loops with a 50 km s^{-1} downflow. These dynamic phenomena can provide useful insights into flare energy budgets. By modelling Doppler shifts and line asymmetries in a set of multi-wavelength flare observations of AD Leo in terms of an erupting filament followed by a slow coronal mass ejection, Houdebine et al (1993) concluded that in this case the kinetic energy of the mass motions dominated the flare energy budget.

3. Quiescent Stellar Prominences

Several different types of observation have yielded exidence for much larger and perhaps longer-lived cool structures in stellar coronae. Many of the most compelling early observations came from pencil-beam studies of eclipsing binary systems where a hot, compact companion is viewed through the corona of a larger, late-type companion for part of the orbit. Schröder (1983) attributed post-eclipse dips in the UV continuum of the eclipsing K supergiant + main-sequence B-star system 32 Cygni, to Rayleigh scattering in the H I ground state, and inferred a neutral H column density of 10^{24} cm^{-2} for a cool prominence-like structure located well above the stellar limb.

The eclipsing K0V + white dwarf binary V471 Tau is particularly well suited to such pencil-beam studies, having an orbital period close to half a day. Jensen et al (1986) reported the presence of soft X-ray dips repeating on successive orbits, at phases where the white dwarf was seen through the K star's corona, and inferred a column density $N_H \sim 10^{20}$ cm^{-2}. At about the same time, Guinan et al (1987) found transient line absorption in UV lines including Si II, Si III, Si IV, C II, C III and C IV, suggesting a clumpy distribution of material at a variety of temperatures along the lines of sight traced out by the white dwarf as it passed behind the K star's corona.

There is also strong evidence of intra-system material in many RS CVn systems. In particular, anomalous Hα absorption was seen in eight of ten eclipsing systems surveyed by Hall & Ramsey (1992, 1994). Similar condensations, but at a temperatures closer to 2×10^4 K, have been invoked by Steeghs et al (1996), as an explanation of peculiar low-velocity emission lines located near the inner Lagrangian points of the dwarf novae IP Peg and SS Cyg in their outburst states.

4. Coronal Condensations in Single Stars

The other major line of investigation into stellar coronal condensations has been the study of Hα absorption transients in ultra-rapidly rotating young main-sequence stars with rotation periods in the range from 0.4 to 1.0 d. Cameron & Robinson (1989a,b; CR89a,b) used time-resolved spectroscopy to infer the presence of clouds of Hα-absorbing material in enforced co-rotation with the marginally pre-main sequence K0 star AB Doradûs.

The spatial distribution of Hα absorbing clouds around AB Dor and similar stars is easily inferred from time-series spectra. A typical time-series Hα observation is shown in Figure 1 together with a simultaneous time-series showing the average photospheric line profile. In both time-series, transient features are seen to drift across the line profile, initially appearing blueshifted by approximately $-v \sin i$ from the stellar centre-of-mass velocity, and drifting toward the red until they disappear at $+v \sin i$. The principal disturbances seen in the photospheric line profiles are the bright "bumps" produced by dark spots on the stellar surface. The rates at which these features drift across the profile are therefore typical of features located on the stellar surface. The dark transient features seen in the Hα time-series, on the other hand, cross from $-v \sin i$ to $+v \sin i$ in little more than an hour in many cases. Since the features are seen to recur at intervals of one stellar rotation, the drift rate when the feature is in the centre of the profile is directly proportional to the distance of the feature from the stellar rotation axis:

$$\dot{v} = \Omega^2 \varpi \sin i$$

where ϖ is the cylindrical radial coordinate of the cloud, Ω is the star's rotational angular velocity, and i is the inclination of the stellar rotation axis to the line of sight.

By measuring the drift rates of individual transients, the distances of clouds from the rotation axis can be expressed in units of the stellar equatorial radius

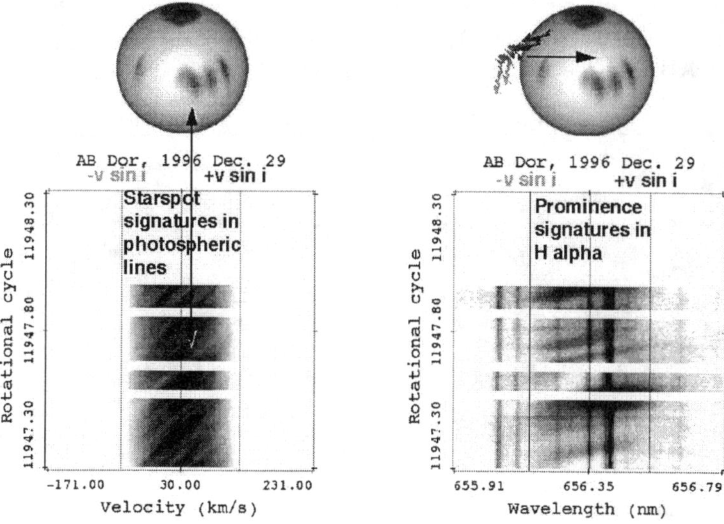

Figure 1. Spectral signatures of (a) photospheric starspots in combined profile of 1500 photospheric absorption lines, and (b) Hα absorption transients due to corotating circumstellar clouds, in time-series spectra secured at the AAT on 1996 December 29 by Donati et al (1998). Note the rapidity with which the Hα transients cross the disc, relative to the stellar surface features.

in terms of the directly measurable quantities $\dot{v}$, $v \sin i$ and Ω:

$$\frac{\varpi}{R_*} = \frac{\dot{v}}{\Omega v \sin i}.$$

In their first studies of this phenomenon, CR89a,b found the clouds to lie at a range of distances from the stellar rotation axis, with a strong concentration near the co-rotation surface which in AB Dor lies some 2.7 stellar radii from the stellar rotation axis. Since then, the circumstellar cloud system of AB Dor has been observed almost annually, and similar phenomena have been observed in nearly a dozen stars by various authors whose efforts are summarized in Table 1.

The physics of the structure and formation of these condensations is also of interest in its own right. While the fine dynamical and spatial structure of solar prominences is directly resolvable on the Sun, only indirect measures are available on stars. By observing transients simultaneously in Mg II h & k, Ca II H & K and Hα, Collier Cameron et at (1990) were able to determine that a typical prominence transient arises in a cloud with a column density $N_H \sim 10^{20}$ cm^{-2}, at a temperature $T \sim 8000 - 9000$ K, and with a projected cross-sectional area $A \sim 3 \times 10^{21}$ cm^2. This implies a total mass $M \sim 2 - 6 \times 10^{17}$ g, two to three orders of magnitude greater than a typical solar quiescent prominence. There are typically 6 to 8 such condensations in the observable slice of the corona at any given time, at cylindrical radii ranging from 2 to 8 R_*.

Table 1. Rapidly rotating (pre-) main-sequence stars surveyed to date, for AB Dor-like Hα absorption transients.

Star	Sp. type	P_{rot}	$v \sin i$	$R \sin i$	Proms?	References
HE 373					Yes	a
HE 520					Yes	a,b
HE 622					Yes	a
HE 699					Yes	a,b
PZ Tel	K0V	0.95	70	1.31	Yes	c
AB Dor	K0V	0.515	91	0.93	Yes	d,e,f,g,h
Speedy Mic	K0V	0.380	140	1.05	Yes	i
BD+22 4409	K3V	0.424	69	0.58	No	j,k
HK Aqr	M1V	0.431	69	0.59	Yes	l
RE J1816+541	M1V	0.459	61	0.55	Yes	m

[a]Collier Cameron & Woods 1992
[b]Barnes et al 1998
[c]Barnes et al 1999, in preparation
[d]Collier Cameron & Robinson 1989a
[e]Collier Cameron & Robinson 1989b
[f]Collier Cameron et al 1990
[g]Donati & Cameron 1997
[h]Donati et al 1998
[i]Jeffries 1993
[j]Jeffries et al 1994
[k]Eibe et al 1998
[l]Byrne, Eibe & Rolleston 1996
[m]Eibe 1998

In a recent theoretical study of the thermal stability of stellar coronal loops, Ferreira & Mendoza-Briceno (1997) found that the formation of fine, sheet-like condensations near the loop summit, with temperatures close to those observed, should be a common feature of such large loops. These occur for a wide variety of surface conditions; their precise locations relative to the loop summit were found to depend on whether or not the loop heating rate was symmetric about the summit.

Collier Cameron et at (1990) also found indirect evidence of time-varying non-thermal velocity fields, whose origin may be akin to the oscillatory disturbancess often observed in filamentary substructures of solar quiescent prominences. The evidence for such filamentation was strengthened by recent HST observations (F. Walter et al, in preparation) which showed transient absorption features in the UV resonance lines of C IV coinciding with Hα transients observed simultaneously from the ground. This observation is strongly reminiscent of the results obtained by Guinan et al in their pencil-beam study of V471 Tau, and it is tempting to conclude that both groups were observing similar coronal structures. Since the pressure scale height within such a prominence is expected to be of order hundreds of kilometres – three orders of magnitude less than the typical linear dimensions of the stellar condensations – a picture is beginning to emerge in which cool material at 8000 to 9000K co-exists with material at both

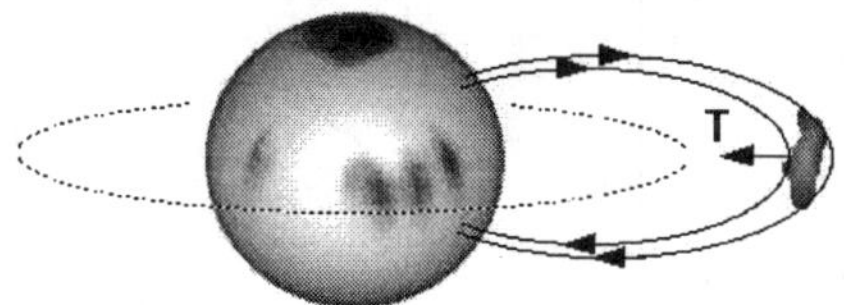

Figure 2. Schematic illustration of magnetic confinement of a coronal condensation outside the co-rotation radius, by a simple dipole-like magnetic field.

transition-region and coronal temperatures. This picture is once again broadly consistent with what we see in solar quiescent prominences.

5. Magnetic Confinement

As described above, the radial accelerations of the clouds' absorption signatures show that most of the prominences lie at cylindrical radii ϖ near (but some inside and some substantially outside) the equatorial co-rotation radius, defined simply as the radius $a = (GM_*/\Omega^2)^{1/3}$ at which the orbital period of a test particle in a circular orbit is equal to the star's axial rotation period. If a cloud is located significantly outside the co-rotation radius, the inward gravitational force on the plasma is insufficient to keep its material in a synchronous orbit. An additional inward force is needed to keep the cloud in co-rotation with the star. CR89a proposed that this force was supplied by the magnetic tension force due to the curvature of the magnetic field lines near the summit of a closed magnetic loop, effectively anchoring the cloud to the stellar surface (Figure 2).

The dipole-like structure of the magnetic loop shown schematically in Figure 2 is undoubtedly overly simplistic, in that it suggests that the centrifugal support mechanism can only work for clouds located in the stellar equatorial plane and outside the stellar co-rotation radius. The presence of clouds at radii substantially *inside* the co-rotation radius has been reported by Byrne, Eibe & Rolleston (1996) in the single M1V rapid rotator HK Aqr, and by Eibe (1998) in the single M1V rapid rotator RE J1816+541.

Ferreira (1998) has studied the conditions for stable centrifugal support of sheet-like prominences in rapidly rotating stellar coronae with multipolar configurations. He pointed out that for support to be possible, the component of effective gravity (incorporating centrifugal forces) *along the direction of the local field* must be in stable equilibrium. He found that stable equilibria exist inside corotation even for a dipole (Figure 3, left panel) or dipole-sextupole (Figure 3, right panel) configuration. Ferreira's work also shows that stable prominence locations can also be found substantially above or below the stellar equatorial plane, if the field has a strong quadrupolar component. This is particularly important for the case of AB Dor, whose axial inclination of 60° precludes the possibility that clouds in the equatorial plane near the co-rotation radius could transit the stellar disc at all. The transients that we see in AB Dor must therefore

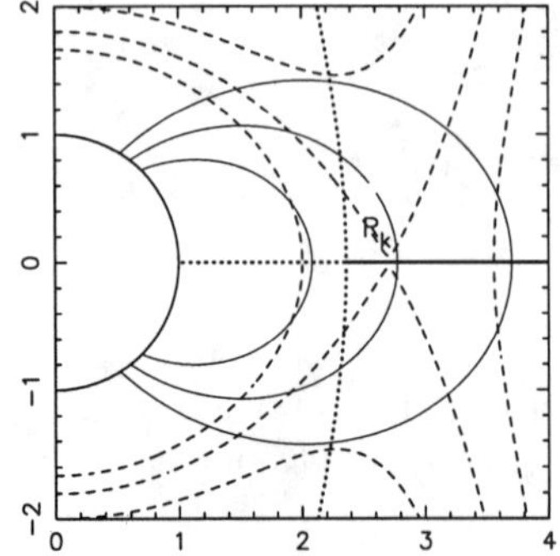

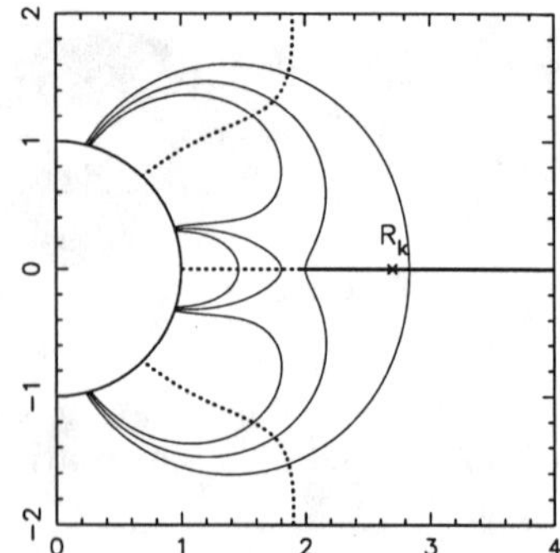

Figure 3. Stable locations for prominence support in (a) a dipole field, and (b) a quadrupole-sextupole field as computed by Ferreira (1998). Note that increased field complexity allows prominence support at cylindrical radii substantially less than the equatorial co-rotaton radius R_K.

arise in clouds that are supported a stellar radius or more above the equatorial plane. Ferreira's work shows that the coronal magnetic field must exhibit a complex multipolar topology at distances two or more stellar radii above the photosphere, if clouds are to be stably supported as far inside corotation or as far above the equatorial plane as we observe them.

There do, however, appear to be limits to the distance above the equatorial plane at which prominence support is possible. The main evidence for this comes from the lack of observed prominence activity in the rapidly rotating mid-K dwarf BD+22°4409, whose axial inclination is, at 45° to 50° (Jeffries et al 1994), demonstrably the lowest among the ultra-fast rotators in the solar neighbourhood. In a detailed study of Hα profiles secured over two nights in 1993 August, Eibe et al (1998) found no evidence for transient AB Dor-like absorption features. They found, however, that the Hα emission profile is anomalously narrow, with a full width at half maximum intensity substantially less than the stellar $v \sin i$. They also remarked that the profile was asymmetric to the blue at all rotation phases, suggesting that the chromospheric emission profile is partially masked by persistent red-shifted absorption arising in a curtain of down-flowing material containing significant amounts of neutral hydrogen. Similar red-shifted absorption, suggesting sporadic low-level mass infall, has been noted in several weak-line T Tauri stars (Wolk & Walter 1996), and in II Peg (Byrne 1987). This led Walter & Byrne (1998) to suggest that cool material may be continually condensing in the corona. This idea makes the implicit assumption that local density enhancements high in the corona suffer increased radiative losses which can trigger a local thermal collapse. The growing cool condensation is subsequently fed by hot coronal material blowing up the legs of the now under-pressured loop. The prominence mass is thus supplied from the stellar surface via upflows, the Coriolis forces on which may even destabilize the loop unless the field is sufficiently strong (Ferreira & Jardine 1996).

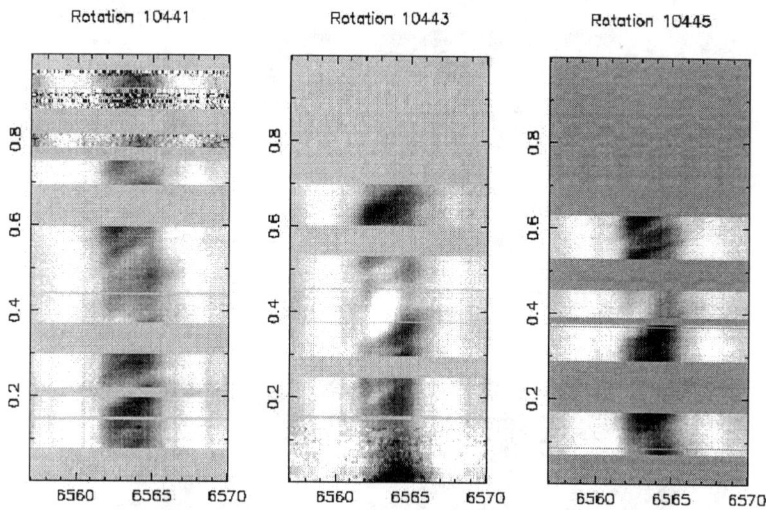

Figure 4. Three time-series of Hα spectra of AB Dor, secured on the 4-m telescope at CTIO on 1994 November 15, 16 and 17. The distribution of transients is shown at intervals of 2 stellar rotations.

Material condensing in a region of stable mechanical equilibrium will collect as a prominence-like cloud. Elsewhere, however, the dense unsupported material will rain back down on to the stellar surface, giving red-shifted absorptions at velocities approaching the free-fall value.

6. Prominence Formation and Evolution

While a few large cloud complexes on AB Dor have been observed to persist for a week or more, individual substructures within them seldom last for more than a few stellar rotations. New sub-structures form and old ones disappear at the rate of about one per day. The fact that there are usually 6 to 8 substructures in the observable slice of the corona at any given time supports the idea that any individual substructure has a typical lifetime of only a few days. This rapid evolution of the prominence system is clearly seen in Figure 4, which shows the pattern of Hα transients observed on three consecutive nights at CTIO in 1994 November. Significant growth is apparent in the complex of transients spanning phases 0.15 and 0.35, between nights 2 and 3. A diffuse but significant transient, seen at phase 0.45 on the first night, weakens and disappears over the ensuing 48 hours. There is an abrupt increase in the drift rate of the absorption transient at rotation phase 0.55, on the third night of observation. While this cloud had been sitting near the co-rotation radius previously, in the course of two or at most three stellar rotations it increased its distance from the rotation axis to 6 stellar radii. Similar behaviour has been observed on two other occasions, in

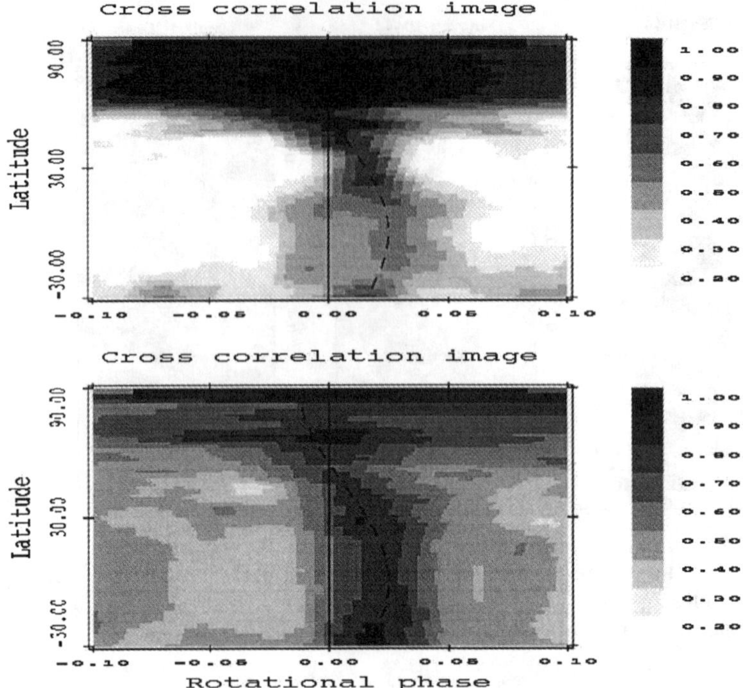

Figure 5. The surface differential rotation pattern on AB Doradus, as determined from cross-correlation functions for the starspot distribution (top) and magnetic flux distribution (bottom) in constant-latitude strips taken from surface images of AB Dor secured on 1996 December 23 to 29 (from Donati et al 1998).

1986 November (CR89b) and in 1996 December (Donati et al 1998), and usually signals an abrupt disappearance of the transient.

All this activity is consistent with new clouds forming and old ones disappearing on timescales of order 24 hours. The disappearance of an old cloud may occur either violently, with the cloud moving rapidly outward, or passively as a gradual weakening *in situ.* If indeed material is continually condensing out of the large-scale corona, and settling temporarily in whatever potential mimima the shifting geometry of the corona can provide, the evolution of the condensations provides a direct observational diagnostic of the rate at which the large-scale structure of the corona is evolving.

7. Relationship to Surface Structures

This immediately raises the question of what drives this rapid reorganization of the large-scale field. One possible candidate is shearing of the footpoints of the large-scale loop structures by surface differential rotation. The rate of surface shear due to differential rotation on AB Dor has been determined recently by Donati & Collier Cameron (1997) to be very similar to that of the Sun (see Figure 5), so that significant shearing of footpoints only occurs on timescales of weeks to months. The growth and decay of the smallest spots resolvable by Doppler imaging on other stars similar to AB Dor also appears to take place on timescales of order 3 to 4 weeks (Barnes et al 1998). Evidently the coronal geometry evolves at a rate far faster than either surface differential rotation or the growth and decay of small-scale magnetic-field concentrations at the stellar surface. Shearing of loop footpoints by differential rotation alone could in theory suffice to cause a loss of mechanical and/or thermal equilibrium (Jardine & Collier Cameron 1991, Unruh & Jardine 1997). The loop would, however, have to be close to a loss of equilibrium for the relatively slow shear due to differential rotation to destabilise it on the observed timescales.

Further evidence for the relationship between coronal and surface structures emerged recently, again from the 1995 and 1996 studies of AB Dor. Donati & Collier Cameron (1997) used tomographic back-projection methods to determine the stellar longitudes of a number of condensations over several nights in both observing seasons. When compared with the surface differential rotation pattern, the rotation rate of the observable part of the prominence system was found to be significantly slower than that of the stellar surface at the equator, but within the range of values found at higher latitudes. In both seasons, the prominence rotation rate was found to be comparable to that of the stellar surface at latitudes of order 60° to 70°. Interestingly, this range of latitudes coincides with the strongest band of regions of strong radial field in the Zeeman-Doppler images, but avoids the strongest regions of azimuthal field. The radial field at this latitude also is highly complex, with a strong east-west pattern of alternating magnetic polarities.

The solar analogy, in which quiescent prominences tend to form along magnetic polarity inversion lines separating adjacent unipolar regions on the solar surface, does not hold in an obvious fashion for these large condensations near the co-rotation radius. Preliminary studies by Jardine and co-workers at St Andrews (M. Jardine 1998, personal communication), using potential-field extrapolation from the Zeeman-Doppler images to explore the complexity of the large-scale field, suggest that the stable prominence support sites near the co-rotation radius bear little obvious relation to the locations of magnetic polarity inversion lines at the stellar surface. The loops that connect adjacent regions of opposing magnetic polarity tend to be too short and low-lying to support prominences at any significant distance above the surface. This is a consequence of the complexity of alternating sectoral polarity structure seen at the latitudes where the loops seem to be anchored. Those loops with summit heights approaching or exceeding the co-rotation radius tend to have widely separated footpoints, which may span several polarity reversals at the level of the photosphere.

8. Summary and Future Prospects

Coronal condensations show great promise as a tool for probing the radial extent and geometric complexity of the closed-field coronae surrounding rapidly rotating late-type stars. The fact that prominences are observed both within

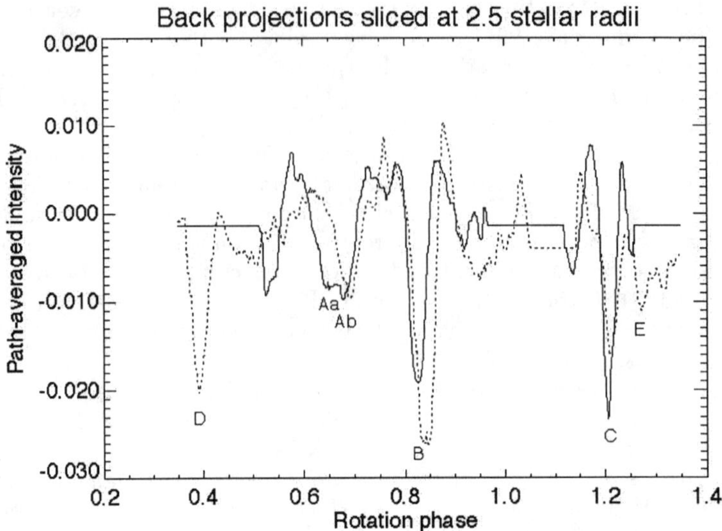

Figure 6. Phase drift of prominences on AB Dor, observed at the AAT from 1995 Dec 7 (solid) to 11 (dotted). The mean rotation rate of the prominence system lags the equator, and matches the surface rotation at latitude 60° to 70° (from Donati & Cameron 1997).

the co-rotation radius, and in regions located significantly above and below the equatorial plane, requires complex field topologies to provide mechanical support support. The red-shifted Balmer-line absorption seen in BD+22°4409 and many RS CVn systems suggest that coronal condensations form throughout the extended stellar corona in these systems too, giving continuous downflows at high latitudes and in other regions where the field cannot give mechanical support. The global structure of the prominence system appears to evolve faster than surface structure. Whether this indicates that the coronal field is continually destabilised by surface shear, or whether other mechanisms are needed to drive the evolution of the corona, remains an open question. Nonetheless, the combination of global field models derived from Zeeman-Doppler imaging with prominence evolution studies is proving to be a powerful method for studying the evolving three-dimensional topology of active stellar coronae.

References

Barnes J.R., Collier Cameron A., Unruh Y.C., Donati J.-F. & Hussain G.A.J., 1998, MNRAS, 299, 904

Byrne P., Eibe M. & Rolleston W., 1996, A&A, 311, 651

Byrne P. B., 1987, in Linsky J.L. & Stencel R.E., eds, Lecture Notes in Physics, Vol. 291: Cool Stars, Stellar Systems and the Sun. Springer-Verlag, Berlin, p. 491

Collier Cameron A. & Robinson R.D., 1989a, MNRAS, 236, 57

Collier Cameron A. & Robinson R.D., 1989b, MNRAS, 238, 657

Collier Cameron A. & Woods J.A., 1992, MNRAS, 258, 360

Collier Cameron A., Duncan D.K., Ehrenfreund P., Foing B.H., Kuntz K.D., Penston M.V., Robinson R.D. & Soderblom D.R., 1990, MNRAS, 247, 415

Donati J.-F. & Collier Cameron A., 1997, MNRAS, 291, 1

Donati J.-F., Collier Cameron A., Hussain G. A.J. & Semel M., 1998, MNRAS, In press

Doyle J.G., Byrne P.B. & van den Oord G.H.J., 1989, A&A, 223, 219

Eibe M.T., Byrne P.B., Jeffries R.D. & Gunn A.G., 1998, A&A, In press

Eibe M.T., 1998, A&A, 337, 757

Ferreira J.M. & Jardine M., 1996, A&A, 305, 265

Ferreira J.M. & Mendoza-Briceno C., 1997, A&A, 327, 252

Ferreira J.M., 1998, in New Perspectives on Solar Prominences (IAU Colloquium 167). ASP Conference Series, San Francisco, p. 239

Guinan E.F., Wacker S.W., Baliunas S.L., Loeser J.G. & Raymond J.C., 1987, in New Insights in Astrophysics: Eight Years of UV Astronomy with IUE. ESA SP-263, p. 197

Haisch B.M., Linsky J.L., Bornmann P.L., Stencel R.E., Antiochos S.A., Golub L. & Vaiana G.S., 1983, ApJ, 267, 280

Hall J.C. & Ramsey L., 1992, AJ, 104, 1942

Hall J.C. & Ramsey L., 1994, AJ, 107, 1149

Houdebine E.R., Foing B.H., Doyle J.G. & Rodono M., 1993, A&A, 278, 109

Jardine M.M. & Collier Cameron A., 1991, Solar Phys., 131, 269

Jeffries R.D., Byrne P.B., Doyle J.G., Anders G.J., James D.J. & Lanzafame A. C., 1994, MNRAS, 270, 153

Jeffries R.D., 1993, MNRAS, 262, 369

Jensen K.A., Swank J.H., Petre R., Guinan E.F., Sion E.M. & Shipman H.L., 1986, ApJ, 309, L27

Schröder K.P., 1983, A&A, 124, L16

Steeghs D., Horne K., Marsh T.R. & Donati J.F., 1996, MNRAS, 281, 626

Unruh Y. & Jardine M., 1997, A&A, 321, 177

Walter F.M. & Byrne P.B., 1998, in Tenth Cambridge Workshop on Cool Stars, Stellar Systems, and the Sun. ASP Conference Series, San Francisco, p. 1458

Wolk S.J. & Walter F.M., 1996, AJ, 111, 2066

Solar and Stellar Activity: Similarities and Differences
ASP Conference Series, Vol. 158, 1999
C.J. Butler and J.G. Doyle, eds.

Do Polar Spots Exist?

V. Aarum

Institute of Theoretical Astrophysics, University of Oslo, Norway

Abstract. Large, stable polar spots have been reported for 9 of the 13 RS CVn systems to which the Doppler imaging technique has been applied. Additional flux contributions from companion stars in multiple systems constitute a substantial error source, and we present our own work as well as examine previous work by others in this context.

1. Introduction

Polar spots are controversial due to the sensitivity of the Doppler imaging technique to stationary perturbations in the spectral line cores. The spectral distortion caused by a polar spot will appear in the centre of the line profiles at all phases. Such a distortion or 'filling in' may arise from other effects. As the reference line profiles of the unspotted star, line profiles from a plan parallel LTE model atmosphere or spun up line profiles of a slowly rotating (assumed unspotted star of the same spectral type) are usually used. Byrne (1996) has pointed out that there are unknown differences between the atmospheres of rapidly rotating, active stars and those of slowly rotating, inactive stars. One cannot rule out that the active star's atmosphere itself can cause a 'filling in' of the line profiles. Another effect, which has been given little attention in the literature, is the additional flux contribution from companion stars. Not only can the spectral lines from the secondary blend those from the spotted primary, the added continuum flux also makes the lines from the primary appear shallower. The latter effect is hard to distinguish from that of a polar spot, and it cannot be neglected even in cases where the composite spectra show no lines from the secondary.

2. Current Study of UX Ari

UX Ari consists of 3 stellar components. The primary (K0 IV) is the spotted, cooler component of the RS CVn system constituted by the primary and the secondary (G5 V). Little is known about the tertiary.

Spectral observations of UX Ari were inverted using the inversion routine described by Berdyugina (1998). The resulting stellar surface maps (Aarum 1997) do not show a polar spot, as previously reported by Vogt & Hatzes (1991).

The radial velocity of each component was measured in the reference frame of the system by determining the shifts of the spectral lines in the wavelength region 5800 Å to 7500 Å. From our measured velocity amplitudes v_1 and v_2 we find the mass ratio $\frac{M_1}{M_2} = \frac{v_2}{v_1} = 1.16$. If the primary has a normal mass for a K0 subgiant (Allen 1973), the derived mass ratio suggests that the secondary is a subgiant rather than a main sequence star (Carlos & Popper 1971).

We have developed a separation technique for spectra of multiple systems based on comparing the spectral line depths of the component to be separated to the spectral line depths of a single, inactive star of the same spectral type. Its application results in the relative continuum flux contribution from each component as functions of both wavelength and orbital phase.

If the binary components are at the same distance from the Earth, their continuum flux ratio is roughly given by their Planck functions and radii:

$$\eta \equiv \frac{\mathcal{F}_{2,c}}{\mathcal{F}_{1,c}} = \frac{B_{\lambda,2} R_2^2}{B_{\lambda,1} R_1^2}.$$

$\mathcal{F}_{i,c}$, $B_{\lambda,i}$ and R_i are the continuum flux, the Planck function and the radius, respectively, of component i. The temperatures of the primary and the secondary of UX Ari taken from Vogt & Hatzes (1991) and their radii taken from Elias et al (1995) yield a continuum flux ratio of $\eta = 0.05$ at $\lambda = 6400$ Å. Applying our separation technique yields $\eta = 0.5$. This result is in agreement with the measured mass ratio and corresponds to a radius of the secondary that lies between that of a G5 V star and that of a G5 IV star (Allen 1973), if the radius of the primary and the temperatures of the two stars are kept constant.

3. Results of Previous Work

Table 1 lists the results of Doppler imaging of 13 RS CVn systems including 4 observations in addition to those provided by Strassmeier (1996b). All the stars listed in Table 1 are RS CVn-type stars, which means that they are all binaries. The first 7 stars are single-lined, and none of the corresponding papers state whether or not a possible continuum flux contribution from the secondary has been considered. In the likelihood that it has not been accounted for, it is possible that the polar spots reported in these cases are artifacts originating from excess flux from a secondary component.

The most striking feature of Table 1 is that the excess flux from companion stars is almost never reported. Only 4 of the papers cited state that this flux was removed and how it was done. It is therefore uncertain at this moment whether or not the derived polar spots can be largely ascribed to this effect.

As can be seen from Table 1, the secondary of an RS CVn system can contribute a considerable flux to the composite spectrum (see e.g. HD 155555 or SV Cam). This supports the importance of considering the effect of companion stars in preparing the spectral observations for Doppler imaging.

The limit of the continuum flux ratio below which the excess flux from a secondary may be neglected, i.e. its 'critical' value, can be estimated using a polar spot like the one reported by Vogt & Hatzes (1996). The line distortion caused by this spot corresponds to $\eta = 0.13$. This spot is small and has a low contrast compared to other reported polar spots, so its effect is not strong. Most of the double-lined systems in Table 1 have $\eta > 0.13$, so the effect could resemble that of a polar spot if not considered and corrected for in the analysis. If excess flux from companion stars is so important, the question of whether or

Table 1. The 13 RS CVn systems to which the Doppler imaging technique has been applied. The column marked 'C' denotes whether or not flux contributions from companion stars were considered, and if so, how they were removed: 'n' = not given, 'o' = removed using observed spectra, 'c' = removed using synthetic spectra. Masses and radii are taken from la Dous & Gimenéz (1994); temperatures are taken from Allen (1973). Reference codes: A97: Aarum (1997), H98: Hatzes (1998), He97: Hempelmann et al (1997), S96a: Strassmeier (1996a), S96b: Strassmeier (1996b), VH91: Vogt & Hatzes (1991), WS98: Weber & Strassmeier (1998)

Star	Spectral type	η	$\frac{M_1}{M_2}$	C	Polar spot?	Ref
σ Gem	K1 III/?	?	?	n	no	S96b[c]
HU Vir	K0 IV/?	?	?	n	yes	S96b[c]
HU Vir	K0 IV/?	?	?	n	yes	H98
DM UMa	K0–1 III–IV/?	?	?	n	yes	S96b[c]
II Peg	K2 IV–V/?	?	?	n	yes	S96b[c]
IN Com	G5 III–IV/hwd(/?)	?	?	n	yes	S96b[c]
UZ Lib	K0 III/?	?	?	n	yes	S96a
EI Eri	G5 IV/?	?	?	n	yes	S96b[c]
EI Eri	G5 IV/?	?	?	n	yes	S96b[c]
V711 Tau	K1 IV/G5 IV	0.23	1.3	n	yes	S96b[c]
V711 Tau	K1 IV/G5 IV	0.23	1.3	n	no	S96b[c]
V711 Tau	K1 IV/G5 IV	0.23	1.3	n	yes	S96b[c]
V711 Tau	K1 IV/G5 IV	0.23	1.3	o	yes	S96b[c]
V711 Tau	K1 IV/G5 IV	0.23	1.3	o	yes	S96b[c]
V711 Tau	K1 IV/G5 IV	0.23	1.3	n	yes	S96b[c]
CF Tuc	K4 IV/G0 V	0.52	1.14	n	no	S96b[c]
HD 155555	K0 IV–V/G5 IV	3.4	0.9	n	yes	S96b[c]
SV Cam	K4 V/G2–3 V	4.6	0.72	n	no	He97
IL Hya	K0 III–IV/?	?	?	n	yes	WS98
UX Ari	K0 IV/G5 V/?	0.05[a]	1.12	c	yes	VH91
UX Ari	K0 IV/G5 V/?	0.5[b]	1.16[b]	o	no	A97

[a]Radii taken from Elias et al (1995), temperatures taken from VH91.

[b]Masses, radii and temperatures taken from Aarum (1997).

[c]The reference to the original paper is given by Strassmeier (1996b).

not single stars show polar spots seems appropriate. Strassmeier (1996b) lists observations of 4 single stars. There are a total of 12 observations of these stars, and 5 of them report a polar spot. Single stars do not show polar spots to the same degree as RS CVn systems (5 of 12 vs 15 of 20), suggesting that excess flux from companion stars does indeed play a role in influencing the results.

4. Conclusions

We have seen that excess flux from companion stars in binary systems constitutes an error source in Doppler imaging that has been given little attention in the literature. The effect of excess flux from unspotted components resembles that of a polar spot. The treatment of excess flux from companion stars should be given more attention in the future. Polar spots have been reported in 15 of the 20 cases where the Doppler imaging technique has been applied to RS CVn systems. For single stars, polar spots have been reported in 5 of 12 cases. Although the statistics is small, this gives support to the idea that excess flux from companion stars affects the results of Doppler imaging.

Acknowledgments. The author wishes to thank Oddbjørn Engvold for valuable comments on the manuscript. This research has made use of the Simbad database, operated at CDS, Strasbourg, France. The project is supported by the Norwegian Research Council under project number 122520/431.

References

Aarum, V. 1997, Study of starspots on UX Arietis using Doppler imaging, candidatus scientiarum thesis, University of Oslo

Allen, C.W. 1973, Astrophysical quantities, London: Athlone Press

Berdyugina, S.V. 1998, A&A, 338, 97

Byrne, P.B. 1996, in IAU Symp 176, Stellar Surface Structure, K.G. Strassmeier & J.L. Linsky, Kluwer Academic Publishers, 299

Carlos, R.C., & Popper, D.M. 1971, PASP, 83, 504

la Dous, C., & Gimenéz, A. 1994, IUE – ULDA Access Guide No 5, Chromospherically Active Binary Stars, ESA SP–1181

Elias II, N.M., Quirrenbach, A., Witzel, A., Naundorf, C.E., Wegner, R., Guinan, E.F., & McCook, G.P. 1995, ApJ, 439, 983

Hatzes, A.P. 1998, A&A, 330, 541

Hempelmann, A., Hatzes, A.P., Kürster, M., Patkós, L. 1997, A&A, 317, 125

Strassmeier, K.G. 1996a, A&A, 314, 558

Strassmeier, K.G., 1996b, in IAU Symp 176, Stellar Surface Structure, K.G. Strassmeier & J.L. Linsky, Kluwer Academic Publishers, 289

Vogt, S.S., & Hatzes, A.P. 1991, in IAU Coll 130, The Sun and Cool Stars: Activity, Magnetism, Dynamos, I. Tuominen, D. Moss & G. Rüdiger, 297

Vogt, S.S., & Hatzes, A.P. 1996, in IAU Symp 176, Stellar Surface Structure, K.G. Strassmeier & J.L. Linsky, Kluwer Academic Publishers, 245

Weber, M., & Strassmeier, K.G. 1998, A&A, 330, 1029

Solar and Stellar Activity: Similarities and Differences
ASP Conference Series, Vol. 158, 1999
C.J. Butler and J.G. Doyle, eds.

Observations of Hydrogen and Helium Continua in Solar Prominences

Vincenzo Andretta[1], Therese A. Kucera[2] & Arthur I. Poland

NASA/Goddard Space Flight Center - Greenbelt, Maryland - USA

Abstract. Observations in the Extreme Ultraviolet of solar prominences provide new and potentially very interesting diagnostics for the density and ionization state of prominences. In this paper we show some examples and preliminary results.

1. Introduction

Solar prominences show a remarkable variety of appearances when observed in the Extreme Ultraviolet (EUV). In particular, many coronal lines often reveal dark features associated with prominences. While such features have already been observed during the *Skylab* mission (Orral & Schmahl, 1976), it is the *Solar Heliospheric Observatory* (SOHO; Domingo, Fleck & Poland, 1995), with its superior wavelength coverage and spatial resolution of its instruments that has permitted a detailed study of this phenomenon.

An example of a prominence observed in the EUV is given in Figure 1. Those images show peak intensities of selected lines in a raster taken with the *Normal Incidence Spectrometer* (NIS) of CDS (*Coronal Diagnostic Spectrometer*; Harrison et al., 1995) aboard SOHO. These data will be further described and discussed in the following sections (especially in §3.). The leftmost panel shows the prominence in emission as it appears when observed in relatively "cool" lines. The rightmost two panels show intensity maps in more typically coronal lines (formation temperatures around 10^6 K or higher). These latter images clearly show the prominence as a silhouette against the bright coronal background.

A possible interpretation of such dark features involves bound-free absorption by relatively cool prominence material. The main species that can absorb photons at EUV wavelengths are hydrogen atoms, helium atoms and ions. In order to give a rough estimate of the importance of this effect, Table 1 lists values of the optical depths that could be considered representative for a large quiescent prominence, at the wavelengths of the three coronal bands of SOHO/EIT (Delaboudinière et al., 1995). In deriving these values, we have assumed a line-of-sight size of the prominence $L \approx 5000$ Km, hydrogen number density $n(\mathrm{H}) \approx 10^{10}$ cm^{-3}, and helium abundance $n(\mathrm{He})/n(\mathrm{H}) \approx 0.1$. From the knowledge of the cross section, $\sigma(\lambda)$, the optical depth is then $\tau(\lambda) \approx \sigma(\lambda) \times n \times L$.

[1]National Research Council Resident Research Associate

[2]Space Applications Corporation.

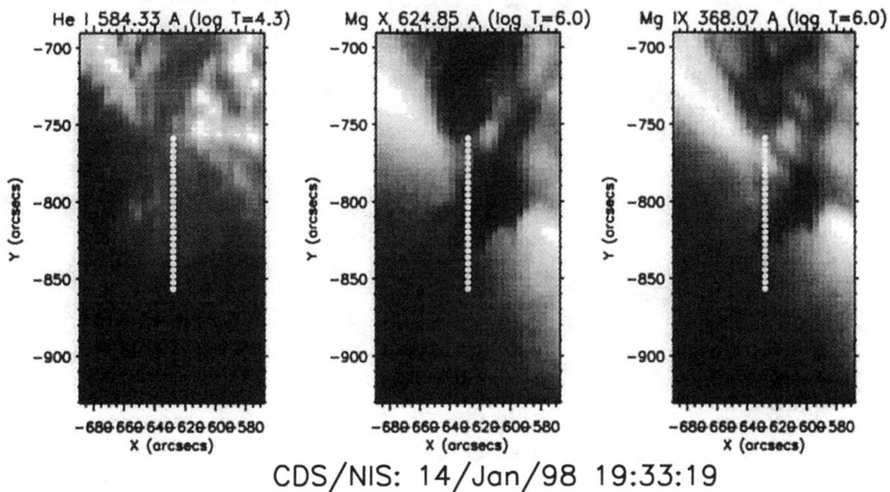

Figure 1. SOHO/CDS observations of a quiescent prominence. Dots indicate the positions of the GIS slit.

Table 1. Representative values of absorption optical depths at the wavelengths of the three coronal filters of EIT:

Optical depths				[Ionization state]
λ	=171 Å	195 Å	284 Å	
$\tau(\mathrm{H}^0)$ =	**0.26**	**0.38**	**1.18**	$[n(\mathrm{H}^0)/n(\mathrm{H}) = 1]$
$\tau(\mathrm{He}^0)$=	**0.46**	**0.61**	**1.30**	$[n(\mathrm{He}^0)/n(\mathrm{He}) = 1]$
$\tau(\mathrm{He}^+)$=	**0.36**	**0.52**	-	$[n(\mathrm{He}^+)/n(\mathrm{He}) = 1]$

At wavelengths observed by CDS/NIS ($\lambda > 310$ Å), the optical depths will be even larger (cross sections scale as $\sim \lambda^3$ for H^0 and He^+, and as $\sim \lambda^2$ for He^0).

2. A basic quantitative model

In Kucera, Andretta & Poland (1998) we have introduced a simple model to derive quantitative information from such observations of quiescent prominences. One of the basic assumptions of the method is that there is no significant emission from the prominence in lines forming at $\log T \gtrsim 5.9$. This assumption makes the interpretation of the absorption features much easier. In fact, the observed intensity, I_{obs}, in each given coronal line is then:

$$I_{\mathrm{obs}} = I_{\mathrm{b}}\,[f\,\mathrm{e}^{-\tau} + (1-f)] + I_{\mathrm{f}}$$

where I_{b} and I_{f} are, respectively, the background and foreground intensity (see Figure 2), and f is the fraction of the detector pixel covering the prominence.

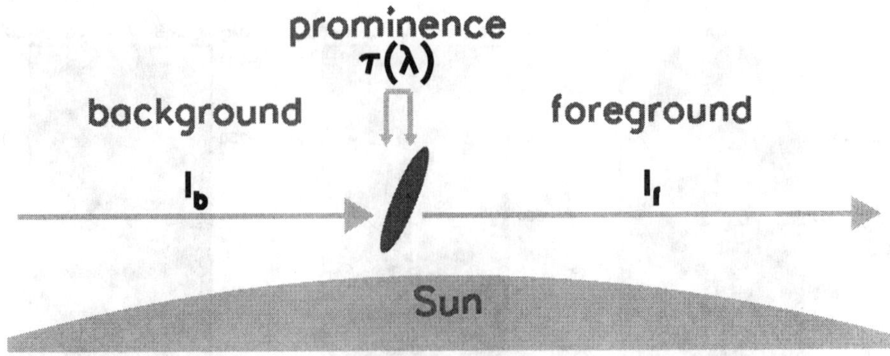

Figure 2. Schematic representation of the geometry assumed for the derivation of basic absorption model.

The total optical depth of the prominence material is $\tau(\lambda) = \sigma(\mathrm{H}^0)\xi(\mathrm{H}^0) + \sigma(\mathrm{He}^0)\xi(\mathrm{He}^0) + \sigma(\mathrm{He}^+)\xi(\mathrm{He}^+)$, where ξ is the column depth (in cm^{-2}) of the absorbers. In terms of observed absorption depth, $d \equiv 1 - I_{\mathrm{obs}}/(I_{\mathrm{b}} + I_{\mathrm{f}})$, we have:

$$d(\lambda) = G\left[1 - \mathrm{e}^{-\tau(\lambda)}\right] ,$$

where we have introduced a "geometrical factor", $G \equiv f\, I_{\mathrm{b}}(I_{\mathrm{b}} + I_{\mathrm{f}})$, that encompasses the essential geometry of the problem.

If we make the further assumption that the "geometrical factor" G is approximately the same for all the lines, a measure of the absorption depth as function of wavelength allows, in principle, the determination of the column densities of the absorbing species.

3. Extending the method to shorter wavelengths

Despite its simplicity, the technique illustrated in the previous section seems to be able to reproduce EUV observations in the CDS/NIS wavelength range, at least in some cases, as discussed in Kucera, Andretta & Poland (1998). However, CDS/NIS observations are limited to wavelengths greater than 310 Å, and therefore do not permit an estimate of the column depth of He^+ (the photoionization threshold of that ion is at 228 Å). Such an estimate is possible, on the other hand, by using the other CDS spectrometer, whose four spectral bands extend from ≈ 150 Å to ≈ 780 Å.

We therefore planned new SOHO observations of quiescent prominences with both the NIS and GIS instruments, with the additional support of EIT. The observations (coordinated in the framework of SOHO *Joint Observing Program* #63) were carried out on January 12–16, January 21, February 20 and April 27/28 1998. In all cases, each NIS raster was followed by a sequence of GIS spectra at several positions along a line, mimicking a "slit spectrum" (an example is shown in Figure 1).

Work is still in progress, but our preliminary analysis of the GIS spectra seems to indicate that the model can be applied to those data as well. An

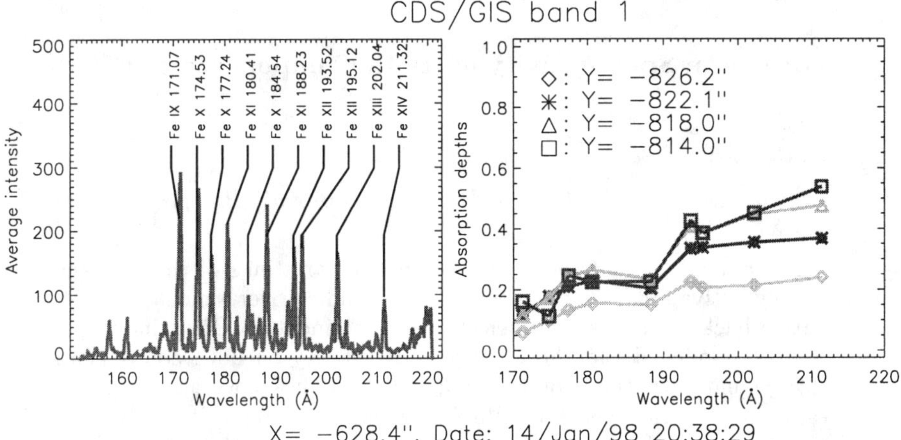

Figure 3. Absorption depths determined from CDS/GIS data (wavelength band 1) at several positions in the sequence of spectra (see also Fig. 1). The left-hand panel shows the average spectrum, with identification of the main component of the most intense features.

example is shown in Figure 3. From that figure the increase of absorption with wavelength, as expected from the continuum absorption model, is clearly visible.

4. Remarks on a work in progress

In conclusion, we have shown that the two CDS detectors can allow the determination, at least in principle, of the ionization state of helium, as well as the column density of neutral hydrogen. If an estimate of the electron density is available from other diagnostics, this method promises to be a useful tool that can be used as a base for determining several fundamental properties of prominences (ionization state, helium elemental abundance, filling factor).

Acknowledgments. This work was carried out while one of the authors (V. A.) held a National Research Council–NASA Research Associateship at Goddard Space Flight Center. SOHO is a project of international cooperation between NASA and ESA.

References

Delaboudinière, J.-P., et al. 1995, Solar Phys., 162, 291

Domingo, V., Fleck, B., & Poland, A. I. 1995, Solar Phys., 162, 1

Harrison, R. A., et al. 1995, Solar Phys., 162, 233

Kucera, T. A., Andretta, V., & Poland, A. I. 1998, Solar Phys., in press

Orrall, F. Q., & Schmahl, E. J. 1976, Solar Phys., 50, 365

Solar and Stellar Activity: Similarities and Differences
ASP Conference Series, Vol. 158, 1999
C.J. Butler and J.G. Doyle, eds.

Long-term Spot Activity on Both Components of BY Dra

Zs. Kővári

Konkoly Observatory, 1525 Budapest, P.O. Box 67, Hungary

Abstract. In this paper the photometric variations of the well-known doubly active binary system BY Dra are studied over almost thirty years dating back to the first photoelectric measurements of Chugainov in 1965.

We modelled the photometric spot activity on both components of the system with the spot temperatures and spot coverage ratios being estimated on the basis of suitable two-colour observations. Spot temperatures were found to be at around 3600K on both stars. The large amplitude long-term variability in the mean system brightness suggests a considerable change in the overall spottedness of the stars. Though the cooler component is less luminous, it seems to be the more spotted. Spot coverage ratios of 2-15% were obtained for the primary and 14-35% for the secondary component, relating to the visible parts of the stellar surfaces.

The larger coverage together with a low inclination of 30° requires spotted regions at high ($\geq 70°$) stellar latitudes, thus providing indirect evidence of polar spots on the stars.

1. Introduction

As has been noted several times before (e.g., in Pettersen et al. 1992, Rodonò & Cutispoto 1992, Kővári, Oláh & Guinan 1995, Kővári & Oláh 1996, hereafter Paper I, Oláh & Kővári 1997) spottedness should have been considered on both components of such doubly active binaries ('DABs', Oláh & Kővári 1997) as BY Dra, where the system is formed by two similar active late-type components. Moreover, IUE measurements demonstrated the existence of chromospheric activity on *both* components of BY Dra (Oláh & Kővári 1997).

In this paper the results of modelling long-term photometric spot activity on both components are given, using a technique introduced in Paper I. Fitting the rotational light and colour variations with our model, spot temperatures and spot coverage ratios were determined for the two stars using suitable B and V observations from almost three decades dating back to the first photoelectric measurements of Chugainov (1973) in 1965. In § 3, the results will be displayed and discussed.

2. Observations and the Model

For the present investigation, simultaneous photoelectric observations between 1965 and 1993 in B and V colours were used. Data sources are: Chugainov (1973, 1976), Melkonian et al. (1981), Rodonò et al. (1986), Pettersen et al. (1992), Kővári & Oláh (1996), Oláh (1998). In Paper I the method of modelling double activity has already been presented and, as an illustration, applied for three suitable datasets of BY Dra from 1965, 1988 and 1991. In this work those results are considerably widened and refined.

For spot temperature determination the $B-V$ colour index curves were used. Previously, as unspotted Δ magnitudes, the record values from 1993 were taken. For the present investigation, the maximum amplitude versus minimum brightness were plotted for each available light curve; then the trend of the diagram was extrapolated back until the amplitude became zero (for the method see Oláh et al. 1997). As a result, somewhat brighter values were taken for Δm_{unsp}, than those used in Paper I (see below).

3. Results and Discussion

In the panels of Figure 1, the model fits for B, V and $B-V$ are displayed. The changes in spot temperatures and spot coverage ratios from one observational season to another are displayed in Figure 2. It can be seen that spot temperatures fall mostly around 3600K, independent of the effective temperatures of the stars. This result reinforces the idea, that spot temperatures in late-type dwarfs do not differ much from 3500K (cf., Vogt 1981, Poe & Eaton 1985). Comparing the three results for the years 1965, 1988 and 1991 with the results in Paper I, one can notice a considerable decrease in the spot temperatures, relative to the former values. Nevertheless, the *only* change between the modelling processes carried out in Paper I and here was the use of different reference (unspotted) magnitudes. In this paper brighter unspotted magnitudes were used (by 0.05 mag. and 0.045 mag. in B and V respectively) that resulted in about 700-900K cooler spots on the primary and 0-300K cooler spots on the secondary. This result proves again, that photometric spot modelling and spot temperature determination is sensitive on the unspotted brightness (cf. Kővári & Bartus 1997).

Spot coverages fall in the range 2-15% and 14-35%, relative to the visible stellar surfaces of the primary and the secondary star, respectively, suggesting generally a higher level of spottedness on the cooler component. In 1971-74 and 1993 the system brightness was close to the maximum level reached in 1993-94 (see Figure 1 in Paper I). For those periods, the results showed the lowest spot coverage on the primary, whilst the secondary was covered by huge dark areas. During those years the secondary component was the more active star. This was suggested previously by Pettersen et al. (1992), who cited Vogt's (1980) observation concerning the lack of H_α emission coming from the primary in 1974 whilst the emission from the secondary was at its normal (i.e., active) level. The quantitative results of this paper fully confirm that picture.

As indirect evidence for the existence of spotted regions near to or covering the stellar poles, the large amplitude long-term change in the mean system

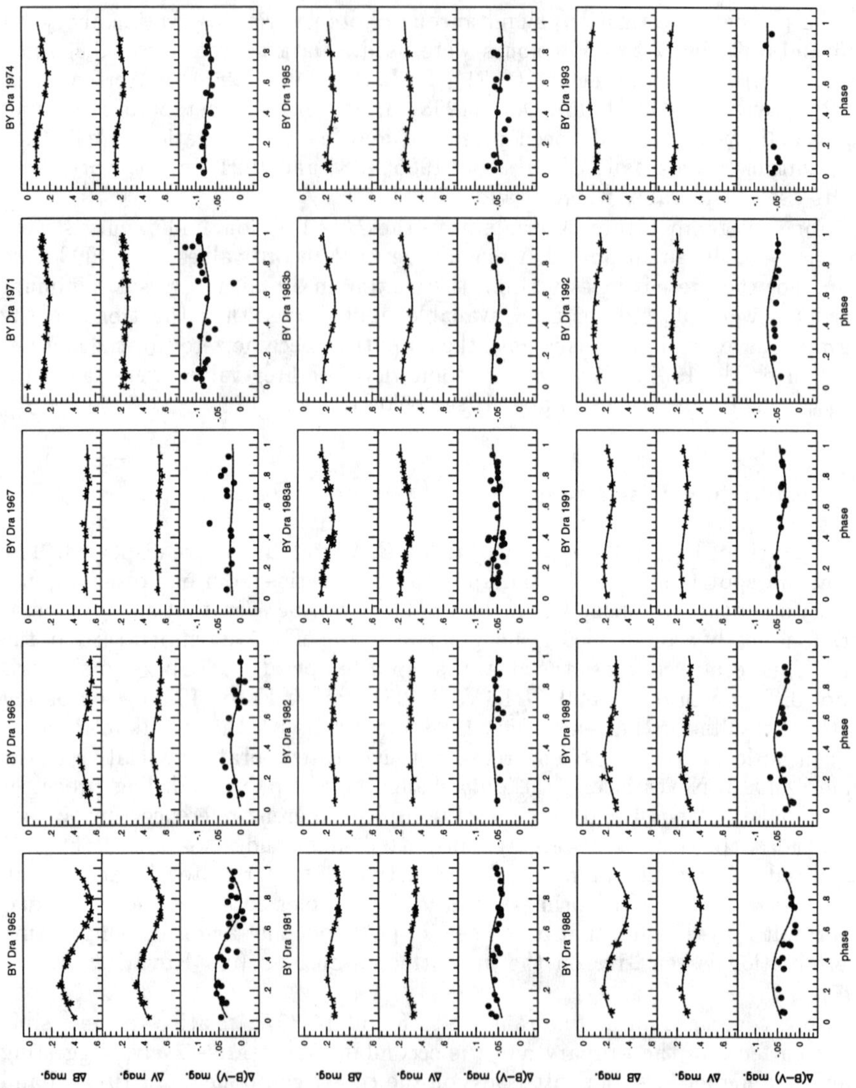

Figure 1. Modelled light and colour curves of BY Dra in the years 1965-1993. The phase values for the phase-averaged light and colour curves are calculated using an arbitrary epoch at $E_0 = 2439000.0$.

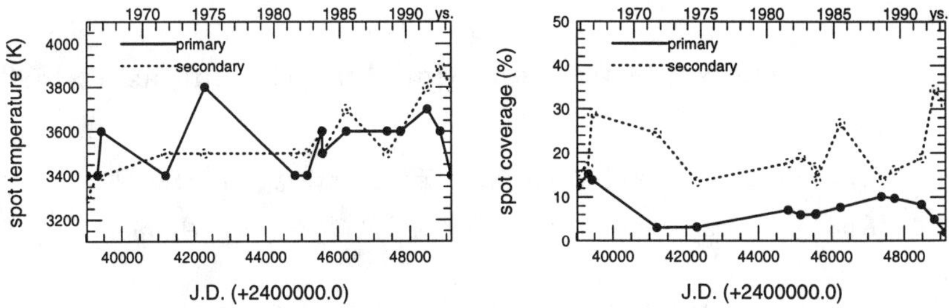

Figure 2. Long-term change in spot temperatures and spot coverage for the two components of BY Dra. The spot coverage values on the secondary component generally exceed the values obtained for the primary.

brightness is mentioned (see Figure 1 in Paper I) which suggests a considerable change in the overall spottedness on the stars. This, beside a low inclination value of 30°, requires spottedness at high ($\geq 70°$) stellar latitudes.

In this study the photometric variations of BY Dra were modelled successfully as a DAB, over almost thirty years. We conclude that spot activity on *both* components of BY Dra seems to be the most likely interpretation for the long-term photometric behaviour.

Acknowledgments. The author wish to thank Dr. Katalin Oláh for her kind help. Financial support from the Hungarian Government through OTKA T-019640 and F-019642 is also acknowledged.

References

Chugainov, P.F. 1973, Izv. Krymsk. Astron. Obs., 48, 3

Chugainov, P.F. 1976, ibid. Tom. LIV, p. 85

Kővári, Zs., Oláh, K., Guinan, E.F. 1995 in:*Stellar Surface Structure, Poster Proceedings*, IAU Symp. No. 176, ed. K.G. Strassmeier, p. 159

Kővári, Zs. & Oláh, K. 1996, A&A, 305, 811

Kővári, Zs. & Bartus, J. 1997, A&A, 323, 801

Melkonian, A.S., et al. 1981, Astrofizika, 17, 215

Oláh, K., Kővári, Zs. 1997, in:*The Earth and the Universe*, Aristotle Univ. of Thessaloniki, Thessaloniki: Ziti Editions, eds. G. Asteriadis et al., p. 151

Oláh, K. 1998, priv. comm.

Pettersen, B.R., Oláh, K. & Sandmann, W.H. 1992, A&AS, 96, 497

Poe, C.H. & Eaton, J.A. 1985, ApJ, 289, 644

Rodonò, M., Cutispoto, G. 1992, A&AS, 95, 55

Rodonò, M., et al. 1986, A&A, 165, 135

Vogt, S.S. 1980, ApJ, 240, 567

Vogt, S.S. 1981, ApJ, 250, 327

Solar and Stellar Activity: Similarities and Differences
ASP Conference Series, Vol. 158, 1999
C.J. Butler and J.G. Doyle, eds.

What can we Learn from Modelling the Sun as a Star?

K. Oláh, L. van Driel-Gesztelyi[1], Zs. Kővári & J. Bartus

Konkoly Observatory, 1525 Budapest, P.O. Box 67, Hungary

Abstract. We present models of active areas on the Sun using software originally written for starspot modelling. As data we used the one-dimensional measures of the solar radio (10.7cm, DRAO, Canada) and soft X-ray (*GOES* satellites) radiation. We compare the results with direct images from Nobeyama and *Yohkoh.*

Radio and X-ray data were used because their response to magnetic activity is orders of magnitude stronger, and the disturbed/emitting areas are considerably larger than that of the optical continuum radiation (i.e sunspots). Thus, the size of these regions and the amplitudes of the measured variability are similar to those we attribute to starspots.

We find that the contrast between the bright active regions and the undisturbed areas on the Sun is systematically decreasing in time following roughly a hyperbola. A similar change in the contrast term might be present when modelling dark starspots, in the opposite sense. We present, as a possible stellar analogue, the change of the flux ratio on V833 Tau in 1993-1994.

1. Introduction

To compare the result of starspot modelling with the direct image of an active star is an old dream. In photometric starspot modelling, according to the limited information content of the one dimensional data, simplified assumptions are used: usually two, sometimes three dark areas are supposed to exist to describe the light variation. Close to the activity minimum in 1996, the Sun had one main big (plus 1-2 small) active region (AR). Daily direct images in several wavelengths are available from this time, as well as radio (DRAO, Canada) and X-ray (*GOES* satellites) one dimensional data. The amplitude of the stellar rotational modulation is comparable to the solar radio or X-ray data. The simple nature of the solar activity in 1996 and the available data gave us the possibility to analyse the solar one-dimensional observations using spot modelling programs written for stars, and then to compare the results with the direct images.

---

[1]Observatoire de Paris, DASOP, F-92195 Meudon Cedex, France

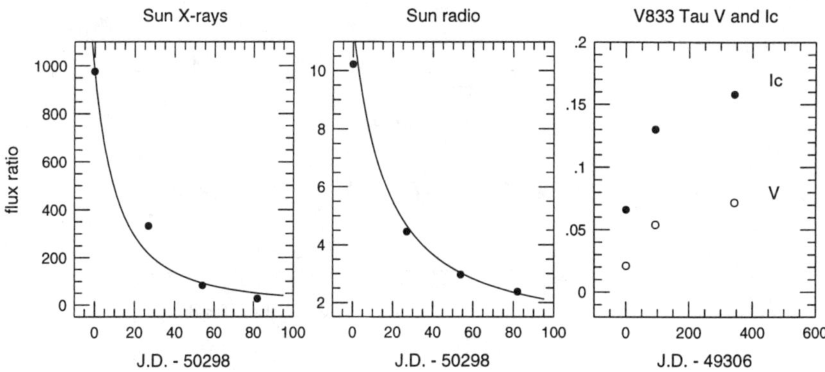

Figure 1. Flux ratio changes on the Sun in X-rays and radio wavelengths and on V833 Tau in the optical and near infrared wavelengths between 1993 Nov.–1994 Nov.

2. Method

For modelling the solar radio and X-ray data we used a software package TISMO (*TI*me-series *S*pot *MO*delling, Bartus, 1996) without major modification. The GOES9 X-ray data were averaged forming 3 daily values, that resulted in a similar number of datapoints for the two datasets in the given time-interval. We removed the micro- and macro-flares from the original data. In the course of our work we found that it was misleading to apply a constant contrast term (flux ratio between the AR and the background) for all the data. As this quantity continuously decreased during the four rotations modelled, we modified the modelling program to take this variability into account. The asymmetric shape of the radio and X-ray curves showed that, beside a major active area on the Sun the presence of an additional small AR was necessary to get an improved fit for the data. Indeed, a small AR appeared on the Sun in August 7, 1996. Thus, we used a two-spot model fit the solar data.

3. Results and Discussion

We studied NOAA 7978, a single major active region on the Sun near sunspot minimum between July-October, 1996.

First we used constant flux ratios: 3.0 for radio, and 236 for X-rays. These were derived from average intensity ratios between the AR and the undisturbed surface that we found from the data, and supposing an $\approx$5% spot area, typical for the large ARs. The resulting spot sizes were decreasing as the amplitude of the variation decreased. Comparing the results with the direct images we found that the sizes of the ARs were not decreasing, but increasing. The only possible reason for this discrepancy was the change in the contrast term between the spot and the background that affects strongly the spot sizes. With the help of the *Yohkoh* direct images and the GOES9 and DRAO X-ray and radio fluxes we

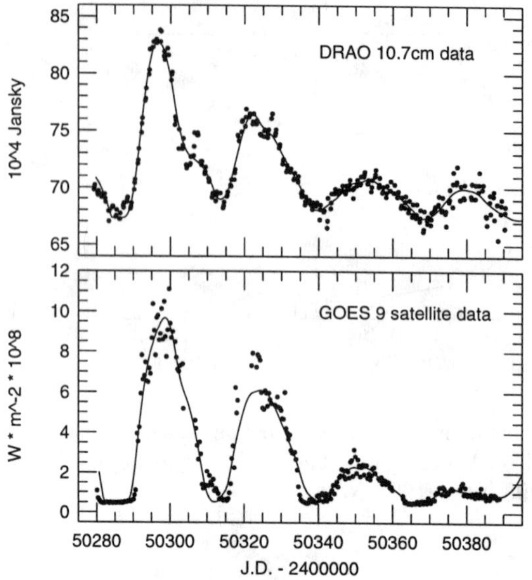

Figure 2. Modelled radio and X-ray variability of the Sun.

found flux ratios for the time of all meridian passages of the big active region. We approximated the flux ratio (FR) change as a function of time as:

$$\mathrm{FR(t)} = \mathrm{c_1} + \frac{\mathrm{c_2}}{(\mathrm{t_0} - \mathrm{t})^{\mathrm{c_3}}} \tag{1}$$

where c_1 is the value that FR approaches in time, t_0 is the starting time of the event, c_3 describes the speed of the decrease, and c_2 is a constant determined by least-squares fits. This result is displayed in Figure 1, where we plotted the flux ratio change in radio and X-ray wavelengths on the Sun (left and middle panel).

It is interesting to compare this result with the flux ratio (starspot temperature) changes on V833 Tau between 1993 Nov. – 1994 Nov (Figure 1, third panel). Very possibly a new spot was formed on the star in 1993 Nov. that was also accompanied by a huge flare. During about a year the changes of the flux ratio in V and I_C colours run in a similar way to those on the Sun in radio and X-rays, but of course, in the opposite sense, and on a longer time-scale. The flux ratio in V833 Tau tends to a value possibly characteristic for a 'quiet' state of the spotted star, observed also in 1990.

Figure 2 shows the fitted radio and X-ray data, and Figure 3 displays the resulted spot latitudes and radii, using fixed and variable FRs. When we used FRs derived with the above equation, we got more reliable spot sizes as is seen in Figure 3.

The main goal of our work was to access the reliability of our modelling which we check by comparing the results with direct images of the Sun. The Nobeyama radio images (at 1.8 cm) of the Sun show very similar spot sizes as our 10.7cm modelling results. Similarly, ARs from *Yohkoh* images are in good

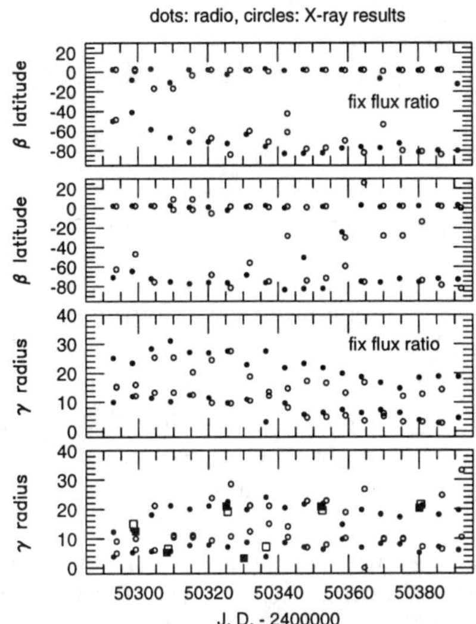

Figure 3. Computed latitudes and sizes of the spots on the Sun. Large filled and empty squares represent determinations of spot sizes from direct radio and X-ray images, respectively, at or close to the meridian passage of the spots.

agreement with the modelling results using the GOES9 data. We found the biggest discrepancy was between the modelled and imaged latitude values as expected, since this is the most unstable spot modelling parameter, especially in the case of high inclination (see Kővári & Bartus, 1997). The very low negative latitude of the big spot which resulted from the modelling is explained by the fact that, when none of the two suspected ARs are in view, small, short-lived ARs appear on the visible hemisphere. Therefore the flux does not decline to the unspotted level (around J.D. 50314 = Aug. 18). To account for this, the big spot was placed near the south pole, so part of it could be visible from the other hemisphere of the Sun. We think this is an important finding when concerning polar spots on stars. We conclude, that if we have good value for the contrast between the spot and the undisturbed surface, and knowing well the flux of the undisturbed surface, we shall get a good result for the spot size.

Acknowledgments. Thanks are due to D. Marik for his kind help in data handling. Financial support from the Hungarian Government through OTKA T-019640, OTKA T-026165, OTKA F-019642 and AKP 97-58 2,2 is acknowledged.

References

Bartus J. 1996, Occ. Techn. Notes at Konkoly Obs. No. 1/96

Kővári Zs. & Bartus J. 1997, A&A, 323, 801

Oláh K., Strassmeier K. G., Kővári Zs. & Guinan E. F., 1999, in preparation

Solar and Stellar Activity: Similarities and Differences
ASP Conference Series, Vol. 158, 1999
C.J. Butler and J.G. Doyle, eds.

Rotation and Cycle Lengths of Active Stars

K. Oláh & Zs. Kolláth

Konkoly Observatory, 1525 Budapest, P.O. Box 67, Hungary

Abstract. We demonstrate that low frequencies in the Fourier spectra of the light variability of active stars originate only from the area changes of the active regions. The peaks corresponding to long-term trends and to activity cycles appear together and it is a complicated procedure to distinguish between them. We stress that a cycle is probably present in the observed light variability only if its length is definitely less than half of the time base.

On this basis we discuss the long-term behaviour of nine well-observed, fast rotating active stars. The short-period systems all have long-term variability with 1-3 years characteristic cycle lengths that is not strictly periodic. Stars with longer rotational periods have definitely longer cycle lengths, but the time base of the available photoelectric observations is still insufficient to find well-defined quasi-periods. The decades-long cycle lengths based on photographic patrol plates for some fast rotating stars could be similar to the several hundred years' cyclic variability of the Sun.

1. Introduction

The search for solar-type activity cycles on stars begun about 30 years ago. From the beginning it was evident, that apart from the rotational modulation, longer time-scale variations can be observed in the light variability of most active stars. It turned out, however, that the derived cycle lengths were close to the solar value, i.e. varied between ≈6-7 to ≈15 years, regardless of the mass, spectral type and rotational rate of the objects. This result was curious, and the question arose whether a derived cycle length was real or only an artifact of the time-base of the observations? The publication of an extensive photometric database of active stars by Strassmeier et al. (1997) enabled us to investigate this problem for nine objects (see Table 1).

2. Method

We used Fourier analysis to study the possible presence of low frequencies (long periods) in the datasets. Our main goal was to find the shortest possible quasi-periods that are still much longer than the rotation period. First the data were pre-whitened with the long-term trends. The length of these trends in most cases were commensurable with the length of the datasets themselves. After pre-whitening, the repeated analysis still showed low frequency peaks. We rigorously

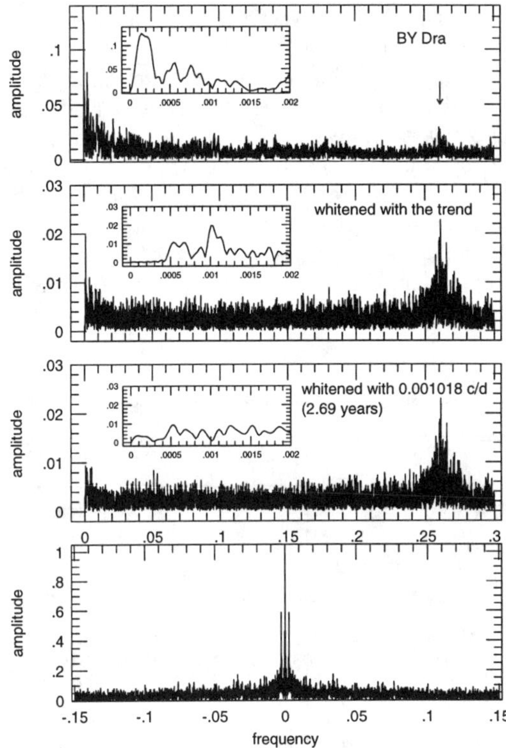

Figure 1. Example of low frequencies present in the observations of BY Dra. The arrow mark the rotational frequency.

studied their reality, making the pre-whitening in different ways. A cycle was considered real when it repeated more than twice in the observed time interval. Figure 1 shows an example series of power spectra of BY Dra.

We have tested the possible origin of the observed long-term variabilities. The long-period RS CVn system HK Lac has a nearly 30 years long photoelectric dataset, which was recently modelled and published by Oláh et al. (1997). Using the times of those observations we generated synthetic lightcurves using the spot modelling results to study if the changes of the spots's locations and/or areas cause the long-term cycles. We generated test data with fixed spot coordinates and/or sizes. The period analysis of these synthetic data showed that with fixed coordinates (longitude and/or latitude) the low frequency peaks of the power spectra were very similar to that obtained for the original dataset. But if the sizes of the spots were fixed the low frequency peaks disappeared. Also, the power spectrum of the mean J.D. of the lightcurves and the corresponding spot radius showed a similar low-frequency pattern to the original dataset. This test thus directly proves that the long-term change originates from the variability of the spot coverage and *not* from spot migrations.

3. Results and Discussion

The results of our period analysis are given in Table 1 and displayed in Figure 2. The last column of the table gives how much longer the time-base is than the resulting cycle length. If this number is larger than 2, then we may consider the cycle length real. It is seen, that the short period systems have cycle lengths of about 1.5-3 years. These cycles are noisy and not easy to detect. For the longer period systems (P_{rot} >10 days) we could not find similar length cycles. These stars seem to have at least 2-3 times longer cycles than the short period ones, but these cycle lengths in most cases approach the time-base of the data. A general tendency, however, is seen from Figure 2: longer period active stars have longer cycle lengths, that, we stress, are the shortest possible cycles. We cannot rule out longer quasiperiodic variability superimposed on these cycles.

Table 1. Basic stellar data and results

Star	spectral class	$P_{(orb)}$	$P_{(rot)}$	$\tau_c(NE)$	Rossby no.	cycle length(c)	time base(t)	t/c
		(days)	(days)	(days)		(years)	(years)	
V833 Tau	K5V	1.7880	1.7936	28.31	0.063	2.62	9.4	3.6
EI Eri	G5IV	1.9472	1.9527	19.32	0.099	2.68	16.4	6.1
						1.67		9.8
BY Dra	K4V+K7.5V	5.7951	3.8285	32.29	0.119	2.69	27.8	10.3
LQ Hya	K2V	single	1.6009	23.34	0.069	10	13.6	1.4
						2.8		4.9
HU Vir	K0IV	10.3876	10.42	79.43	0.131	5.8	14.1	2.4
IL Hya	K1III	12.908	12.791	107.90	0.119	9.7	17.5	1.8
VY Ari	K3-4 IV-V	13.198	16.199	67.76	0.239	15	22.2	1.5
HK Lac	K0III	24.4284	24.378	90.57	0.269	13.6	28.4	2.8
						8.7		3.3
IM Peg	K2 II-III	24.6488	24.494	43.75	0.560	$\leq$30	25.7	0.9

The long-term changes of the stars investigated seem to have multiple timescales. On EI Eri, a cycle length of about 13 years is very probable, apart from shorter (1-2 years) cycles. On V833 Tau, the observations show a quasi-period of about 2.5 years, and a longer trend is seen from the 10 years long photoelectric observations, but no period can be given; however, from photographic observations a cyclic variability in the order of 50-70 years was found (Bondar 1995). The long-term behaviour of HK Lac has multiple timescales as well: the shortest possible cycle length is about 8 years, but longer period variations on more than a 30 years timescale may also be present. On IM Peg we did not find cycles shorter than 30 years, that is comparable with the time-base of the observations.

This phenomenon could be compared with the multi-periodic behaviour of the Sun: beside the ≈11-years long cycle (which varies between 9-13 years), the Gleissberg cycle (≈90 years), and a rough cycle of about a few hundred years (Wolf, Spörer, Maunder minima) can also be found in the solar activity. With an $\alpha - \omega$ boundary layer dynamo, assuming that the α effect operates also in the convective envelope, Zangrilli & Bianchini's (1997) calculations implies long periods with cycle lengths (60-70 years and 200-300 years) resembling the observed ones, that modulated the 11-years solar cycle.

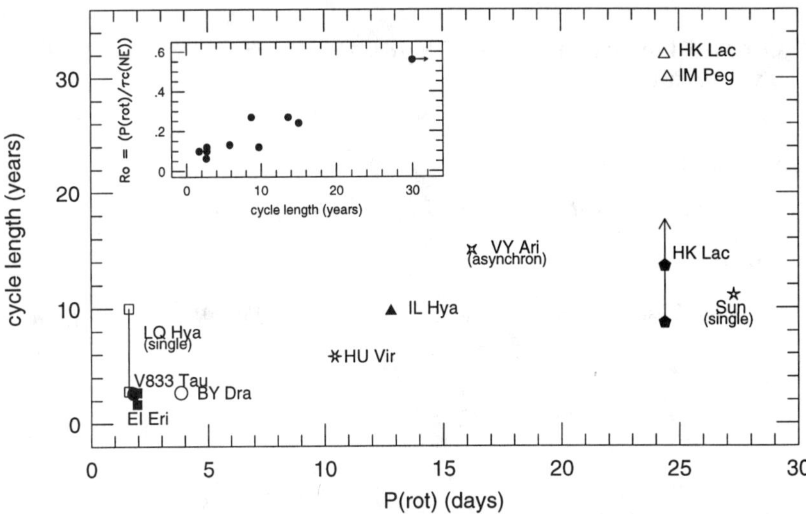

Figure 2. Cycle lengths of nine well-observed active stars. The Sun is plotted for comparison. The inset figure shows the cycle lengths vs. Rossby numbers for the binaries.

In Figure 2 (inset) we plot the Rossby number $\mathrm{Ro} = \mathrm{P_{rot}}/\tau_\mathrm{c}(\mathrm{NE})$, where τ_c is the convective turnover time of the binary star as a function of the derived cycle length. The non-empirical (NE) Rossby numbers were taken from Gunn et al. (1998): five of our program stars were included in their calculations, and for the rest we interpolated $\tau_c(NE)$ through $\tau_c(E)$ which was calculated using Noyes et al.'s (1984) $(B-V)-\tau_c$ relationship. The Rossby number increase with increasing cycle length. A similar phenomenon was found by Ossendrijver (1997) but for slowly rotating single stars. That result cannot be applied directly to our stars because of the physical differences between the program stars, but the gross similarity of these results is encouraging.

Acknowledgments. We thank Zs. Kővári for helpful discussions. Financial support from the Hungarian Government through OTKA T-019640 is acknowledged.

References

Bondar, N.I. 1995, A&AS, 111, 259

Gunn, A.G., Mitrou, C.K. & Doyle, J.G. 1998, MNRAS, 296, 150

Noyes, R.W. et al. 1984, ApJ, 279, 763

Oláh, K. et al. 1997, A&A, 321, 811

Ossendrijver, A.J.H. 1997, A&A, 323, 151

Strassmeier, K.G. et al. 1997, A&AS, 125, 11

Zangrilli, L. & Bianchini, A. 1997, Mem. S.A.It. Vol. 68, 479

Solar and Stellar Activity: Similarities and Differences
ASP Conference Series, Vol. 158, 1999
C.J. Butler and J.G. Doyle, eds.

On the Determination of Spot Parameters in Active Late Type Stars

M. Zboril

Astronomical Institute, Tatranská Lomnica, 059 60, Slovakia

P.B. Byrne

Armagh Observatory, College Hill, Armagh, BT61 9DG, N. Ireland

P.J. Amado

Osservatorio Astrofisico di Catania, Cittá Universitaria, I 95125, Italy

A.G. Gunn

NRAL, Jodrell Bank, Cheshire, SK11 9DL, UK

Abstract. Photometric and spectroscopic variability of active late-type stars are frequently interpreted as evidence for magnetic activity analogous to solar activity. In particular cool, dark spots with enhanced photospheric magnetic fields are invoked to explain both photometric and spectroscopic variability. In one of the diagnostic methods, the features such as TiO and VO in spectra were suggested to detect spots in stellar atmosphere of cool stars. In this contribution we discuss the TiO band technique and its limits using our own spectroscopic material.

1. Introduction

Photometric and spectroscopic variability of late type stars is frequently interpreted as evidence for magnetic activity analogous to solar activity. Techniques to detect spots are basically divided into three: photometry, Doppler imaging and TiO band spectroscopy. Spectral features such as Titanium Oxide (TiO) or Vanadium Oxide (VO) molecular bands (Ramsey & Nations 1980) were suggested as proxies to detect cool spots on the atmospheres of active stars since these molecules can only exist on those parts of the stellar atmosphere cooler than 3500 K. This spectroscopic method was explored by several authors and applied to a sample of chromospherically active late type stars. In the recent publications (Neff et al. 1995, O'Neal et al. 1997) the authors used TiO spectra at the 7055Å and 8860Å spectral regions to examine both spot temperature and areas on a number of chromospherically active stars. We have obtained high signal-to-noise, high resolution spectra of a sample of field K and M dwarfs to study TiO bands, in particular the TiO 7055Å and 8860Å bands. In this contribution we examine the calibration of TiO 7055Å using our own spectroscopic material and take a very preliminary look at the active star EQ Vir as well.

2. Observations

During two nights in June 1994 we observed 9 dwarf stars with spectral types between K0 and M5 using the 4.2m WHT in the Canary Islands. The instrument was equipped with the Utrecht Echelle Spectrograph (UES) providing the resolving power ~30000 and signal-to-noise ~200 in the continuum near 7055Å region. Table 1 gives some basic characteristics of the programme stars, i.e. spectral

Table 1. Programme stars-WHT94.

Gliese No.	Sp. Type	B-V	R-I	TiO band depth
411	M2V	+1.51	+0.92	15.0
517	K5V	+1.20	+0.54	5.5
569	M0V	+1.48	+0.80	22.0
673	K7V	+1.36	+0.60	4.0
687AB	M3.5V	+1.50	+1.10	28.0
699	M5V	+1.73	+1.25	38.0
702A	K0V	+0.86	+0.26	1.0
820A	K5V	+1.17	+0.47	2.4
896A	M4V	+1.71	+1.28	36.0

type, photometric indices from Gliese & Jehreiss (1991) and our measured TiO band-depth index. The spectra were extracted from the echellograms and routinely reduced. The metallicity for these stars was studied in Zboril & Byrne (1998) and was found not to vary within the individual stars too much, from -0.20 up to -0.30 dex. Another set of observations of relatively high spectral resolution (0.08Å) is available from AAT (observations made in January 1996) and from WHT+UES (June 1996) (resolution 0.08Å). The INT (September 1997) data using MUSICOS spectrograph, on the other hand, are associated with stars known to have very different metallicities and surface gravities but are still being processed and will become available in winter 1998.

3. TiO λ7055 Å index

To define the empirical band-depth index given in Table 1, we have followed Neff et al's procedure, i.e. we measured the difference between pseudo-continuum levels on either side of the bandhead, expressed as a percent of the continuum level to the blue of the bandhead. The exact spectral regions used to determine this ratio were 7043-7048 Å and 7060-7065 Å. To inter-compare our results directly we plot in Figure 1 the dependence of band-depth index with the $(R–I)$ colour index rather than with the effective temperature.

EQ Vir is a young, single, rapidly rotating, flare and variable star. The photometric period is 3.96 days (Bopp & Fekel 1977). It also possesses strong chromospheric and transition emission lines. Saar et al. (1986) derived a value for the mean magnetic field of 2.5 kG, covering 80% (!) of the stellar surface. Figure 2 shows differences in TiO band-depth index and the hydrogen alpha profile within 2 consecutive nights, i.e. phase difference 0.25.

Table 2. Programme stars-AAT96 & WHT96.

HD/Gl No.	Sp. Type	B-V	R-I	TiO depth	phase
HD45184	G2V	0.62	—	1.1	
HD26965	K1V	0.82	0.45	1.1	
HD22049	K2V	0.88	0.47	1.4	
Gl204	K5V	1.11	0.43	2.3	
Gl205	M1.5V	1.49	0.85	19.5	
HD33833	G7III	0.96	—	1.2	
HD29085	G6III	0.98	0.54	1.0	
HD34649	K2.5III	1.28	0.63	1.0	
HD38099	K4III	1.48	—	8.3	
HD147379	M0Vvar	1.41	0.68	9.0	
HD151288	K7.5Ve	1.37	0.57	2.5	
Gl517	K5Ve	1.20	0.54	6.5	0.00[1]
Gl517	K5Ve	1.20	0.54	6.5,7.0	0.25

[1] Note: JD0 adopted equals to JD of the night.

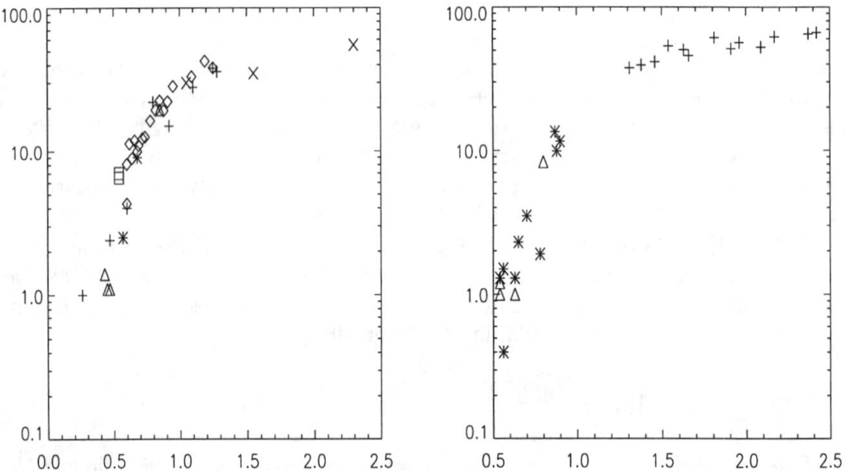

Figure 1. Left figure: TiO 7055 Å band depth index vs. (R–I) for dwarf stars. WHT94 data (+), WHT96 data (∗), AAT96 data (△), Neff & O'Neal 1995 data (◇), Phoenix data (x) and EQ Vir data (□). Right figure: the relationship for giant stars. O'Neal & Neff 1997 data (+), Neff & O'Neal 1995 data (∗), AAT96 data (△).

4. Results

We have evaluated numerically the dependence of the band index upon the ($R-I$) colour index by making a polynomial fit to the data for both dwarfs and giants.

$$D_{7055}^{dw,giant} = \sum a_i \cdot (\log (R - I))^i \tag{1}$$

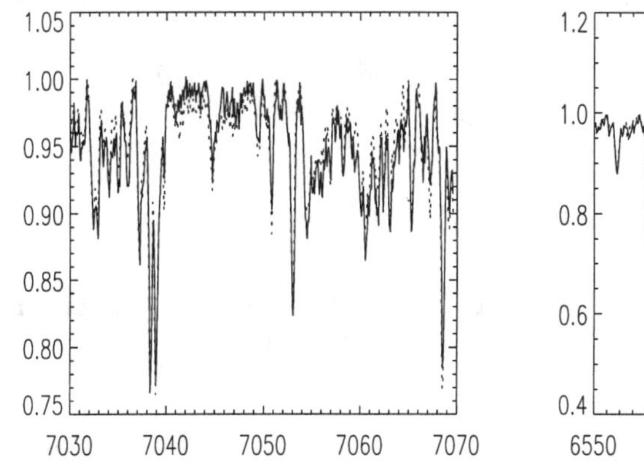

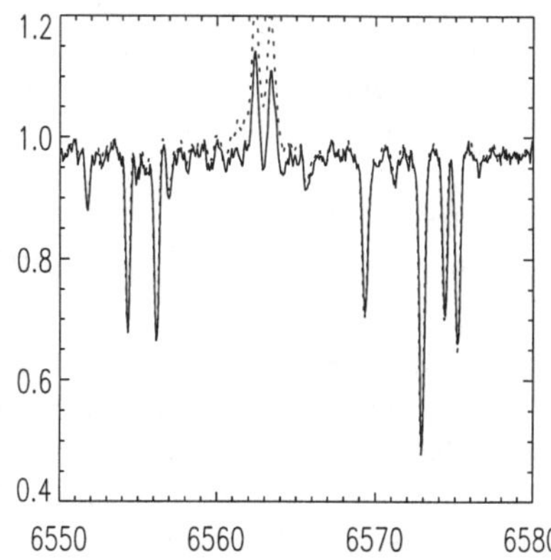

Figure 2. Left figure: TiO 7055 Å index progress for EQ Vir in two consecutive nights. Right figure: hydrogen alpha profile progress.

The relationships saturate for very cool stars which means that the error bar on the measured band-depth index will respond in a large error in the temperature estimate. The hot stars in the calibration, on the other hand, display a scatter making the calibration less precise as well. Our spectroscopic observations seem to be integrated very well into the previous calibrations. The band-depth indexes for EQ Vir suggest later spectral type while the H α emission changed dramatically in two consecutive nights covering 0.25 of the photometric phase. A more detailed account of this work will appear in forthcoming paper.

Acknowledgments. We acknowledge gratefully the supervising by Dr. P. B. Byrne. MZB and AGG were supported by UK PPARC grants while PJA by an Armagh Observatory studentship. Drs. M. T. Eibe and W.R.J.R. Rolleston are cordially thanked for the observations of EQ Vir and WHT95 datasets.

References

Bopp, B. W., Fekel, F. 1977, AJ, 82, 490

Gliese, W., Jahreiss, H. 1991, *3rd Catalogue of Nearby Stars*

Neff, J.E., O'Neal, D., Saar, S.H. 1995, ApJ, 452, 879

O'Neal, D., Saar, S.H., Neff, J.E. 1997, AJ, 113, 1129

Ramsey, L. W., Nations, H. L. 1980, ApJ, 239, L121

Saar, S. H., Linsky, J. F., Beckers, J. M. 1986, ApJ, 302, 777

Zboril, M. & Byrne, P. B. 1998, MNRAS, 299, 753

Solar and Stellar Activity: Similarities and Differences
ASP Conference Series, Vol. 158, 1999
C.J. Butler and J.G. Doyle, eds.

Can Chromospheric Activity Mimic a Polar Spot?

J.H.M.J. Bruls & M. Schüssler

Kiepenheuer-Institut für Sonnenphysik, Schöneckstr. 6, D-79104 Freiburg, Germany

S.K. Solanki

Institut für Astronomie, ETH-Zentrum, CH-8092 Zürich, Switzerland

Abstract. Doppler imaging studies indicate that most rapidly rotating cool stars seem to have high-latitude (polar) spots. It has been proposed, however, that the flat-bottomed cores of spectral lines that are interpreted as due to polar spots, might also be caused by chromospheric activity. A thorough NLTE radiative transfer analysis of 14 of the most-used Doppler-imaging lines shows that chromospheric activity can produce filling in of the profiles of the strongest lines only, and that flat-bottomed lines occur only if the activity is concentrated towards the poles. In the observations, however, also the weaker lines have flat-bottomed cores, so that it is unlikely that polar spots are an artifact due to misinterpretation of the spectral signature of chromospheric activity. On the other hand, we cannot exclude that chromospheric activity provides part of the filling in of the cores of some stronger lines.

1. Introduction

Many of the stars accessible to Doppler imaging, necessarily rapidly rotating and thus with high activity levels, have line profiles with a flat-bottomed core. This filling in of the line profile is nearly time-independent and it is generally attributed to large, stable and long-lived starspots at high latitudes: polar spots. In addition to the imaging results, the presence of spectral lines that are formed at lower than photospheric temperatures indicates the presence of extended cool spots on such stars. A theoretical explanation why those spots can be encountered at high latitudes has also been given (Schüssler & Solanki 1992, Schüssler et al. 1996). Nevertheless, polar spots have been a source of controversy, and in particular it has been speculated (e.g., Byrne 1992) that the filling in of the spectral line cores — the most direct evidence for polar spots — could also be due to chromospheric activity. Since the high rotation rate may influence the star's atmosphere, it is conceivable that lines that are considered photospheric in non-active cool stars show cores filled-in by chromospheric emission in more active stars (Basri et al. 1989).

From a radiative transfer point of view, one of the most compelling arguments against chromospheric activity being the cause of the flat-bottomed line profiles lies in the fact that they occur almost independently of the line strength.

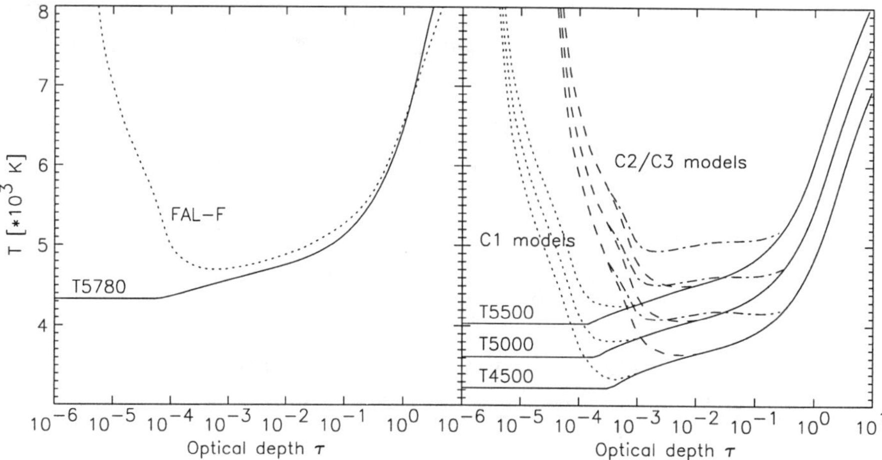

Figure 1. Temperature, T, as a function of optical depth τ at 5000 Å. Left: the solar models used to derive the temperature correction, ΔT, to represent chromospheric activity. Right: stellar models. Solid curves: inactive stellar atmosphere models with effective temperatures of 4500, 5000 and 5500 K; dotted curves: models with standard chromospheric activity (C1); dashed curves: models with increased chromospheric activity (C2); dot-dashed curves: C2 models with enhanced photospheric temperatures (C3).

This means that whatever is causing the filling-in of the line cores, it must be something that works over a significant height range in the atmosphere, not just in the chromosphere. We investigate the sensitivity of the profiles of 14 widely-used Ca I and Fe I Doppler imaging lines to chromospheric activity; the spread of line strengths is such that the formation heights of these lines cover a sufficiently large range in the atmosphere. A detailed account of this analysis has been given in Bruls et al. (1998); here we present the main results.

2. Chromospheric Activity Models

We use three atmospheric models of inactive stars with $\log(g) = 3.8$ and effective temperatures of 4500, 5000 and 5500 K, and the quiet Sun model T5780 (Figure 1), with a line blanketed radiative equilibrium model similar to those described by Edvardsson et al. (1993). An outward extension (isothermal layer in hydrostatic equilibrium) was appended to these models to allow some of the stronger lines to become optically thin.

Stellar activity is modelled by adding a schematic temperature rise $\Delta T(\tau)$ to an inactive stellar model and recomputing the hydrostatic equilibrium. Apart from a small offset to match the mid- and upper photospheric temperatures, the basic chromospheric temperature enhancement, ΔT^{C1}, is obtained by subtract-

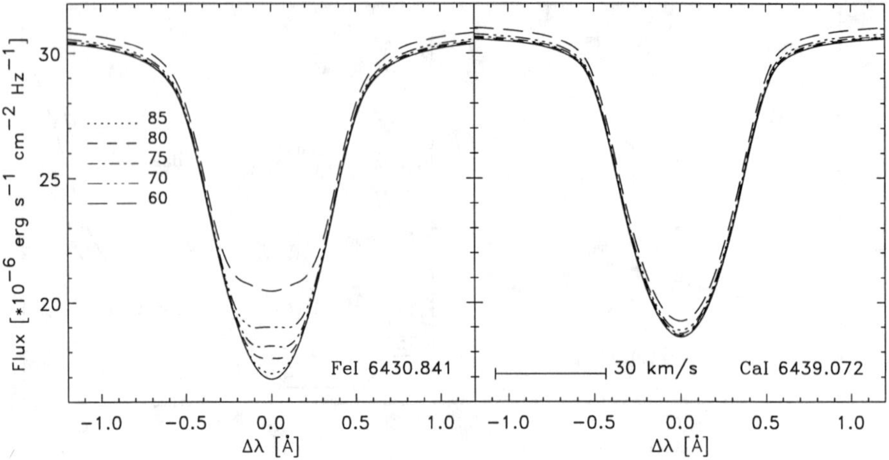

Figure 2. Rotationally-broadened line profiles of Fe I 6430 Å (left) and Ca I 6439 Å (right) for a star with $T_{\rm eff} = 4500$ K and a rotation rate of 30 km/s with T4500/C3 active regions of varying extent placed at the poles. The various curves are labelled with the latitude (in degrees) above which the T4500/C3 activity model is applied and below which the inactive T4500 model is present. The star's rotation axis is inclined by 45° w.r.t. the line of sight.

ing the temperature of the T5780 model from that of the FAL-F network model of Fontenla et al. (1991). As expected, the C1 models are similar to the very low ('basal') activity dwarf M star models of Mauas et al. (1997). The C2 models are obtained by shifting $\Delta T^{\rm C1}$ downward by one decade in optical depth, simulating a very early chromospheric temperature rise. The C3 models, finally, also have a photospheric temperature enhancement. $\Delta T^{\rm C3}$ was chosen such that $T(\tau)$ is almost constant throughout the photosphere; it is stronger than the photospheric temperature enhancement exhibited by FAL-F relative to T5780.

3. Computed Line Profiles

The line profiles emerging from the above model atmospheres (without rotation) can be summarized as follows. Only very few lines are sensitive to a chromospheric temperature rise. The two most prominent Doppler imaging lines, Fe I 6430 Å and Ca I 6439 Å, can actually be taken as representative of two groups of lines that sense the chromosphere to some extent. Fe I 6430 Å already notices a C1 type chromosphere, and its influence strongly increases with decreasing effective temperature; both the disk-center and the limb profiles are affected. Ca I 6439 Å, on the other hand, is only weakly sensitive. It requires the strong C2 chromospheric activity in combination with the T4500 model to induce noticeable line profile changes, and even then they occur mainly near the limb.

In the present case, we assume the chromospheric activity to be homogeneously distributed over the stellar surface, it requires an intensity enhancement near disk center (where the Doppler shift is negligible) and preferably in the line core to produce a rotationally-broadened disk-averaged profile with an enhanced core intensity. Only the Fe I 6430 Å line satisfies this requirement, but its line core does not have a flat bottom, because the strong limb emission also enhances the wings of the disk-averaged profile. The only way around this is to concentrate the chromospheric activity near the stellar poles, so that the rotation does not shift the line parts with largest emission into the wings of the disk-averaged profile. Due to projection effects, the influence on the line profile is more clearly visible if the rotation axis of the star is not perpendicular to the line of sight. The left panel of Figure 2 shows the line profiles of Fe I 6430 Å from the T4500 model with T4500/C3 activity within a certain area around the stellar poles. Indeed, the line core is filled-in and the remainder of the line profile is unmodified with respect to the inactive star's profile, but more importantly, for a certain size range of the polar active region, the core is virtually flat. The same results for activity type C2. The right panel of Figure 2 displays the corresponding profiles for the Ca I 6439 Å line. This line does not have a flat-bottomed core.

4. Conclusions

Figure 2 represents sort of a worst-case scenario — low effective temperature, extreme amounts of chromospheric activity concentrated towards the poles and a favorable inclination angle — and even then we find that we can produce flat-bottomed line cores only in a few lines.

Our analysis suggests that chromospheric activity cannot be the main cause of the flat-bottomed line cores that are observed on many active stars. We cannot exclude, however, that polar active regions, a magnetic phenomenon (as are starspots), contribute to the line core filling in stronger lines.

References

Basri G., Wilcots E. & Stout N., 1989, PASP, 101, 528

Bruls J. H. M. J., Solanki S. K. & Schüssler M., 1998, A&A, 336, 231

Byrne P. B., 1992, in P. B. Byrne & D. J. Mullan (eds.), Surface Inhomogeneities on Late-Type Stars, Springer-Verlag, p. 3

Edvardsson B., Andersen J. & Gustafsson B. et al., 1993, A&A, 275, 101

Fontenla J. M., Avrett E. H.& Loeser R., 1991, ApJ, 377, 712

Mauas P. J. D., Falchi A., Pasquini L.& Pallavicini R., 1997, A&A, 236, 249

Schüssler M. & Solanki S. K., 1992, A&A, 264, L13

Schüssler M., Caligari P., Ferriz-Mas A., Solanki S. K. & Stix M., 1996, A&A, 314, 503

Part 4

PHYSICAL MANIFESTATIONS OF ACTIVITY: FLARES

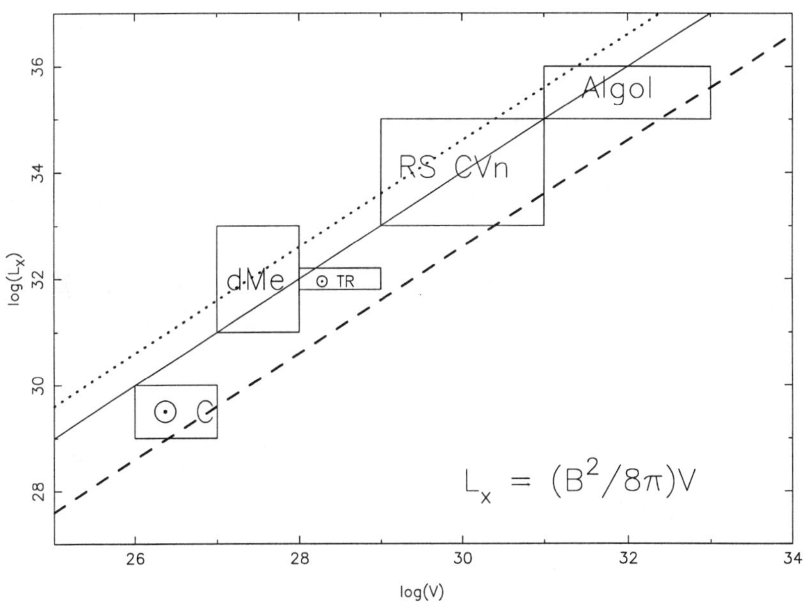

Typical ranges of the X-ray losses and the flare volumes for dMe stars, RS CVn-systems and Algols as determined for *EXOSAT* observations. Also shown are the ranges for solar compact (C) and solar two-ribbon (TR) flares as observed during the solar maximum year in 1980. Older *EXOSAT* data are used to have a sample observed with a single instrument. The volumes have been obtained from the decay phase of stellar flares and have been measured for solar flares. The lines indicate the relation $L_X = (B^2/8\pi)V$ for values of the magnetic field at the flare site of 100 G (dashed), 500 G (solid) and 1000 G (dotted). The typical field strength is indicative at which height in the corona a flare occurs.

Solar and Stellar Activity: Similarities and Differences
ASP Conference Series, Vol. 158, 1999
C.J. Butler and J.G. Doyle, eds.

Stellar Flares: Theory and Observations

G.H.J. van den Oord

Sterrekundig Instituut Utrecht, P.O. Box 80.000, 3508 TA Utrecht, The Netherlands

Abstract. In this review various aspects of stellar flares are discussed.

1. Introduction

Over the past years missions like the *EUVE*, *ROSAT*, *ASCA*, *GINGA*, *SAX* and *HST* have resulted in observations of a number of stellar flares. Despite the new wavelength domains opened up by some of these missions, and the improved spectral and temporal resolution, the picture of stellar flares which emerged during the *EXOSAT* era has not changed dramatically. Therefore I restrict myself in this short review to general considerations about (stellar) flares. Papers describing recent observations of flares can nowadays easily be found with ADS on the WWW. For older, but still up-to-date, reviews of flares I refer to IAU Coll. 104 on Solar and Stellar Flares; Haisch, Strong and Rodonò (1991), Kuijpers (1992), van den Oord (1993) and Haisch & Schmitt (1996).

2. Stellar and solar flares: general considerations

It is often claimed in observing proposals that stellar flares provide unique opportunities to study the process of magnetic energy release on different scales than provided by laboratory experiments and solar flares. This claim can however be questioned. Even with the impressive spatial resolution available for solar flare observations, the spatial scales on which energy release takes place have not been resolved yet. In the case of solar flares the general consensus is that during a flare magnetic energy, stored in the pre-flare configuration, is released on a relatively short time scale. It is the specific details of the magnetic topology and the local thermodynamical conditions which determine the magnitude, the appearance and the time evolution of a flare. Because both factors vary strongly, both spatially and in time, each flare has a different appearance. This has led to an enormous variety of specific models for solar flares although basically two classes can be identified. Firstly there are the models in which a global instability occurs leading to a large scale eruption of a magnetic structure. A second class of models are those in which emerging flux (Heyvaerts & Priest 1977) or photospheric motions force two, topologically different, magnetic structures to interact (see Sakai & de Jager 1996).

Over the past decade it also has become clear that it is often difficult to distinguish between flares and coronal heating. Both phenomenon are related to magnetic energy release and it seems that energy release occurs on all scales ranging from micro- and nano-flares up to large (two-ribbon) flares and coronal mass ejections. The number of events versus the energy is distributed according to a power law (for a discussion see Haisch & Schmitt 1996). To understand all these different types of magnetic energy release it is of paramount importance to study the magnetic topology, a subject which is receiving an increasing amount of attention. The magnetic field can be described in terms of flux systems which are groups of field lines which show an identical topological behaviour. Each flux system is bounded by a separator surface. Because at a separator surface two flux systems with a different topological behaviour meet, the behaviour of the field does not have to be continuous when crossing the separator surface. This implies that currents run over the separator surface leading to diffusion of the field in such a way that the discontinuities of the field at the separator surface disappear. Under normal coronal conditions this process takes a long time but whenever the resistivity is locally enhanced the process can occur faster and result in (observable) heating.

A special case exists when two separator surfaces cross, resulting in the presence of a neutral line (in two-dimensions the neutral line corresponds to an X-type neutral point). Such a line is intrinsically unstable and will collapse into a current sheet provided that the sources of the field are able to move (Syrovatskii 1981). In the case of the emerging flux model or loop coalescence models it is the photospheric motions which provide this freedom of movement while in the case of, e.g., two-ribbon flares it is the unstable current carrying filament which provides this motion. Directly related to this motion is the rate at which magnetic field can be transported into the thus created current sheet in order to provide the energy for a flare. In the case of loop coalescence this rate is basically determined by the velocity of the photospheric foot points of the loops involved, which is relatively slow. In the case of a filament eruption it is the eruption of the current carrying filament which automatically results in transport of field into the current sheet. The large velocity of the filament results in a relatively fast transport of field into the current sheet which explains why this type of flare generates more power than flares driven by photospheric motions. In other words, energetic events require a fast transport of field into the current sheet and this can best be achieved if the sources of the field (the currents) are able to move rapidly. This requires preferentially coronal sources of the field, so coronal current systems.

It is nowadays increasingly recognized that the three dimensional (3-D) aspect of the field plays an important role. From a topological point of view this has lead to new insights. In 2-D the basic ingredients of the field are flux systems, separators and neutral points (X- and O-type). In 3-D there exists a much larger class of magnetic structures including various types of magnetic nulls (Lau & Finn 1990), quasi-separatrix surfaces and singular field lines. Unfortunately, simulations of 3-D reconnection are still difficult to perform although some analytical results are available (Priest & Démoulin 1995). Another development is the fundamental insight that the whole concept of describing a magnetic structure by depicting the field lines may not always work. The concept of a field line is useful when the magnetic resistivity is low and, especially, the geometrical scales are large. Under coronal conditions the resistivity is comparable to copper at room temperatures. So an important requirement is that the scales are large (absence of thin current sheets). This whole idea is however based on the use of Ohm's law. If one applies the generalized Ohm's law there are various terms which can contribute to a generalized resistivity, not necessarily related to Ohmic or anomalous dissipation. Integrating along a field line into a volume with a non-zero generalized resistivity shows that a field line can loose its identity everywhere in the corona and not necessarily at special places like current sheets (Hesse & Schindler 1988). In fact, if the concept of generalized resistivity applies, it is possible to have topological changes in the coronal magnetic field which are not accompanied by Ohmic heating.

Another important concept which has appeared in the (solar) physics literature is that of magnetic helicity. Under ideal MHD conditions helicity is a conserved quantity. This has some important consequences. Without conservation of helicity the lowest energetic state of a field is that of a non-linear force-free field. When the magnetic energy is minimized with the constraint that the helicity is conserved, the lowest accessible energetic state is that of a linear force-free field. Also it can be shown that due to flares the helicity does not change much because the magnetic energy decays faster than the helicity (Berger 1988,1991). The concept of helicity is however extremely difficult to apply because helicity is a non-local quantity which is not gauge-invariant (one can use a helicity density but the interesting MHD-constraint only applies to the volume-integrated helicity density). To apply the concept of helicity to the coronal field requires that special attention is paid to the photospheric boundary (Heyvaerts & Priest 1984). The problem of the gauge-invariance can be solved by working with a relative helicity. Still, there remains the problem that a star like, e.g., the Sun has an extremely dynamic surface. MDI observations (*SOHO*) have clearly demonstrated that even the quiet Sun is extremely dynamic, with emerging ephemeral regions and flux cancellations in the super-granular lanes (Schrijver et al. 1998). All this activity is accompanied with a continuous change of the helicity which makes it difficult to apply this concept in real situations.

Helicity is only one of a number of ideal MHD constraints. Another constraint which has, relatively recently, received attention is the magnetic crossing number (the number of projected crossings of field lines in a structure when viewed from different directions). Clearly, the crossing number is related to the complexity of the field and therefore to the energy contents of the field (Berger 1994).

In order to determine the energy in the field, and the possible occurrence of reconnections, one really has to know the exact structure of the field which is, in fact, impossible. Above I stated that topological concepts like flux systems are important for describing the field. A flux system can however only be defined when one knows the connectivity of the field lines (e.g., which patches of opposite polarity in the photosphere are connected by one or more field lines). There are two ways of approaching this problem (Aly 1984), often referred to as boundary value problems one and two (BVP1 and BVP2). BVP1 describes the situation that, given the instantaneous components of the magnetic field in the photosphere, one calculates the structure of the overlying coronal field. BVP1 is the problem one solves when extrapolating the field using magnetogram data. Depending on the type of magnetogram (normal component or vector) the resulting field is potential or (non-)linear force-free. One problem with this method is that one misses the presence of singular lines which may be present in the corona. In other words: the solutions are always relatively smooth. The biggest problem is however that BVP1 is not the correct problem to solve. The correct description of the (evolution of) the coronal field requires solving BVP2. In that type of boundary value problem the magnetic connectivity of field lines is taken into account. For example, the photospheric foot points of a flux system can be displaced by photospheric flows. In BVP2 this effect of the history of the photospheric boundary conditions is taken into account while in BVP1 it is not. One is therefore faced with the undesirable situation that the correct problem to solve cannot be solved because, next to the horrendous mathematical problems, it requires continuous information about the photospheric changes.

It will be clear that large flares require the presence of large current sheets in which magnetic reconnection takes place. There has been, for a long time, a debate whether flares are thermal or non-thermal in character. This is the question of energy partitioning: does reconnection mainly result in bulk heating of the plasma or in accelerated particles? The obvious answer is, of course, that this varies for different flares (Benka & Holman 1994). The reconnection region can be considered as a MHD shock which can be described, like an ordinary shock, without knowing in detail the internal structure. The global field topology determines the current which must be present in the current sheet. Once the drift speed of the current-carrying particles becomes too large, a variety of plasma instabilities can be excited resulting in anomalous resistivity and, thus, strong local heating. Because the current has to be maintained through a region of enhanced resistivity a strong (inductive) electric field is present. At least a fraction of the particles (in the tail of the distribution) will experience the effect of run-away acceleration by this electric field. It is interesting to speculate that this results in a decrease of the electric field because the required current is then carried by a smaller number of high-speed electrons instead of a slowly drifting distribution. After a short time the cycle will start again, explaining the bursty character of flares during their impulsive phase. In this model the energy release results basically in heating and non-thermal particles. This was strongly debated after the *Solar Maximum Mission* at the beginning of the eighties. The *Yohkoh* mission, has resulted in models which include the jets of plasma which emanate from the reconnection region (Masuda et al. 1994). These jets impinge on the surrounding fields, create shocks and substantial heating, and are used to explain bright loop top sources. It is remarkable that the plasma jets are only applied in flare models since the *Yohkoh* mission while theoretical reconnection models have included these jets for decades.

There are cases in which the reconnection process primarily results in particle acceleration. A good example is the case of inertial resistivity. Loosely speaking this concept can be applied in curved current sheets. The accompanying curvature of the field results in drifts of the accelerated particles in such a way that they leave the current sheet before crossing a mean free path. So this is an example of collisionless reconnection. Because the particle leave the sheet with a higher momentum than they possessed when they entered the sheet, work has been done and therefore energy is dissipated. This explains the name inertial resistivity. In these models electrons and protons leave the sheet with equal momentum so that the bulk of the energy resides in the proton population (Martens & Young 1990).

The number of accelerated electrons is large and can amount to 10^{36} electrons s^{-1}. In the eighties there was a strong debate how such beams, which present huge currents, can propagate. It is however easy to show that the electrons in the ambient plasma will be accelerated in the opposite direction to the beam direction (for proton beams in equal direction) so that the net current remains low. There is a maximum beam current which can propagate through a loop which depends on the ratio of the densities of the beam particles and the ambient electrons. Once the beam becomes so strong that the drift speed of the ambient electrons exceeds a threshold, anomalous resistivity will develop. The electric field, which forces the ambient electrons through the resistive volume, will, at the same time, rapidly decelerate the beam electrons. In this case a beam can result in a thermal flare in the corona (van den Oord 1990).

From a thermodynamical point of view, all solar flares follow more or less the same scenario in their evolution. An impulsive phase can be identified, characterized by non-thermal emission, during which thermal conduction and, especially, non-thermal particles are present. This marks the phase of violent energy release. The non-thermal particles enter the lower atmospheric layers (chromosphere and sometimes the photosphere) and emit non-thermal bremsstrahlung when they are collisionally braked in these dense layers. The non-thermal emission represents, however, only a small fraction of the energy of a particle so that most of its energy is transferred to the ambient medium which is heated. The heating corresponds to an effective increase of the scale height so that the heated plasma expands upward. The accompanying density increase in the chromosphere and corona increases the radiative losses from these layers. Because of the relative high density of the plasma, it is quickly thermalized so that the emission has a thermal character. This radiative cooling of the plasma is the gradual or thermal phase of the flare. Because of the cooling, the effective scale height starts to decrease and the plasma drains from the corona. A problem with observing the thermal phase is that one often has to integrate for some time to get a good signal-to-noise ratio for a spectrum. In the mean time the cooling plasma goes through various ionization stages so that a single or two-temperature fit to a spectrum may give erroneous results.

A point to keep in mind is that, for the flare volume as a whole, thermal conduction merely acts as a energy redistribution mechanism and not as a sink. The only true losses are radiative and in the form of the kinetic energy of ejecta. It is often claimed that most of the flare energy resides initially in non-thermal particles. Because these deposit the energy which is later radiated during the gradual phase, one expects a correlation between the total non-thermal and thermal emission of a flare (the Neupert effect; Neupert 1968). This correlation has been found for solar flares and supports the above described evaporation/condensation scenario. It does not provide information about the total energy released in a flare because the work done by Lorentz forces to accelerate, e.g., a filament is not accounted for.

Let us now turn to stellar flares. Observations of these phenomena are characterized by a lack of spatial resolution and moderate time and spectral resolution. This implies that an observer receives only information about the gross characteristics of a flare. If different layers have different temperatures and densities the interpretation of the observed spectrum depends strongly on the observing wavelength and method of analysis. Missions with an improved spectral resolution like the *Extreme Ultra-Violet Explorer* (*EUVE*), and in the near future *AXAF* and *XMM*, provide slightly more information in the form of a differential emission measure distribution (DEM). This function describes the relative contribution by plasma at different temperatures to the observed spectrum. There are however a number of problems related to the DEM. The (inversion) process of obtaining a DEM from an observed spectrum is ill-conditioned so that the specific shape depends on the method of analysis (this is, of course, also true for one- and two-temperature fits). Another problem is that flare temperatures can be quite high, often exceeding 50 MK. At such temperatures the spectra contain almost no lines and consist of a (flat) featureless continuum. This makes it difficult to determine a temperature and emission measure because there is a trade-off between these quantities. Sometimes flare temperatures of 100 MK are cited in the literature. This number is however the maximum temperature included in the present-day spectral codes and should not be taken to literally.

Keeping in mind the limited information available for stellar flares, when compared to solar flares, it turns out that stellar flares are basically scaled-up versions of solar flares. With this I mean that no dramatic new concepts have to be introduced to understand these flares. An exception to this rule are flares on accretion disks near compact objects. These flares occur in radiation dominated environments so that Compton drag in current sheets can provide the necessary resistivity (van Oss et al. 1993). Other possible exceptions are flares in binary systems, for which duplicity may play a role, and flares near T Tauri stars for which disk-star interactions, mediated by the magnetic field, may play a role. In this review I restrict myself, however, to flares on single dwarf and main-sequence stars and flares in Algol and RS Cvn-type binaries.

Because stellar flares often involve much more energy than solar flares it is interesting to consider the underlying reasons. The energy available in a flaring volume V equals $(B^2/8\pi)V$ with B some average field strength in the volume. To explain energetic flares B has to be large, V has to be large or both. For most objects listed above, the typical photospheric field strengths are always of the order of one to a few kiloGauss, roughly speaking the equipartition field strength at the top of the convection zone. Let ℓ be the typical size of the active regions on a star. This size is at maximum about the depth of the convection zone and scales, therefore, roughly with the stellar radius. It is also the height above the photosphere above which the field obtains a dipole character and no stable equilibria are possible (van Tend & Kuperus 1978). For flares at typical heights comparable to ℓ above the photosphere we then find that the flare energy scales approximately like $B^2_{\rm surf}V \approx B^2_{\rm surf}\ell^3 \approx 10^{37}(R_*/R_{\rm sun})^2$ erg. This is sufficient energy to power most flares which have been observed. This relation does not imply, however, that the typical flare energy scales with the size of the star. Only the maximum available energy does. In binary systems a special situation exists between both stars. The interstellar volume between the stars is constant but the magnetic field can reach much higher values than if the binary companion is absent. The field cannot expand freely because of the presence of the companion so that a compression of the field occurs and the magnetic energy density goes up (van den Oord 1988).

A problem with considerations of the above type is that, given all uncertainties, the result is that there is always enough magnetic energy available to explain the energetics of observed flares. As such it is not a critical test. The actual flare energy also depends on the typical height of the current sheet(s). For flares close to the surface the local magnetic energy density is higher resulting in a higher power during the primary energy release. In Figure 1 I show the typical X-ray luminosities L_X for flares on various objects as a function of the volume V involved. The lines depict the relation $L_X = (B^2/8\pi)V$ for field strengths of 100 G (dashed), 500 G (solid) and 1000 G (dotted). The figure shows that flares on dMe stars are as energetic as solar two-ribbon flares. Because of the smaller volume involved the typical field strength for these flares is higher indicating that they occur lower in the atmosphere where the field strength is closer to its surface value. High surface filling factors of the field are in favour of this picture in the sense that the presence of a large number of active regions (dipoles) implies that higher multipole components dominate the behaviour of the field so that the spatial scales involved are smaller than in the case of the presence of one bipolar structure.

The actual estimates of how much energy is involved in observed flares is very inaccurate because (1) different observations have a different wavelength coverage (so that one can miss important emissions), (2) new wavelength domains, like the EUV, have been opened up, and (3) there is a considerable variation in the way different authors speculate about unobserved losses. The result of this is that each year a new 'largest' flare for a class of objects is published. The only thing one can conclude is that multi-wavelength campaigns are extremely important to determine the energetics of flares.

Since stellar flares are so energetic, when compared to solar flares, it is commonly assumed that the flare topology is comparable to that of the largest solar flares: the two-ribbon flares. This makes sense because in such a field topology, considerable amounts of pre-flare energy can be stored in a (current) filament and there is the possibility of a global instability. Basically two models appear in the stellar flare literature, that of Kopp & Poletto (1984) and that by Kaastra (1985). The Kopp & Poletto model is a sequence of MHD equilibria with decreasing magnetic energy. It is assumed that if this sequence occurs, the change in magnetic energy corresponds to the dissipated energy. The Kaastra model corresponds to a more realistic topology with a filament, a current sheet, a possibility for an instability, a pre-flare energy storage, etc.. It has also been studied numerically (Forbes & Isenberg 1991) and as an equivalent electric circuit (Martens & Kuin 1989). An attractive aspect of the model is that it explains in a natural way how magnetic energy is transported into the current sheet. Both models permit us to calculate the dissipated power once a velocity has been introduced and both models predict the formation of an arcade of loops. In the Kopp & Poletto model a time dependent velocity law can be introduced while in the Kaastra model the velocity has to be constant because of the absence of accelerations (force-free evolution). Both models contain so many free parameters that they can be easily adjusted to an observed flare. Because they involve large structures, they are useful for explaining long-lasting flares.

Short and intense flares can be simply modeled by considering an equivalent circuit of a loop. Such a circuit contains an inductance $\mathcal{L}$, two resistances (one

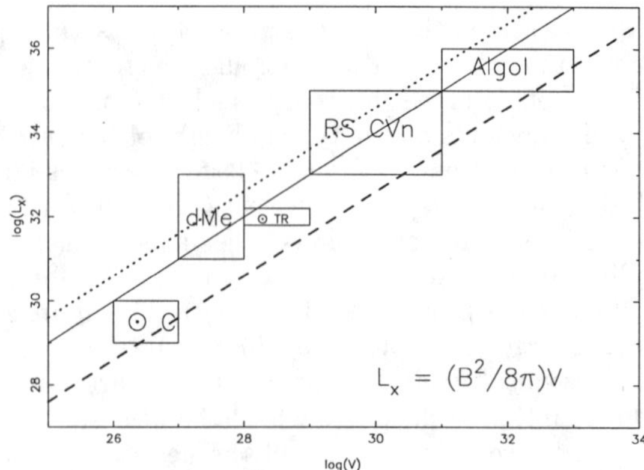

Figure 1. Typical ranges of the X-ray losses and the flare volumes for dMe stars, RS CVn-systems and Algols as determined for *EXOSAT* observations. Also shown are the ranges for solar compact (C) and solar two-ribbon (TR) flares as observed during the solar maximum year in 1980. Older *EXOSAT* data are used to have a sample observed with a single instrument. The volumes have been obtained from the decay phase of stellar flares and have been measured for solar flares. The lines indicate the relation $L_X = (B^2/8\pi)V$ for values of the magnetic field at the flare site of 100 G (dashed), 500 G (solid) and 1000 G (dotted). The typical field strength is indicative at which height in the corona a flare occurs.

coronal R_{cor} and one photospheric R_{phot}) and a photospheric load (van den Oord et al. 1988). In quiescent coronal conditions the coronal resistance is low and the characteristic $\mathcal{L}/R_{phot}$ time scale corresponds to the time scale for energy storage. During a flare, anomalous resistivity develops and the coronal resistance starts to dominate in the circuit. The $\mathcal{L}/R_{cor}$ time scale is then the time scale for dissipation during the impulsive phase. The free parameter in this model is the inductance $\mathcal{L} \approx \ell/c^2$ with ℓ the length of the circuit. The actual processes which occur in a real magnetic loop when anomalous resistivity develops have been described by Melrose (1992). He demonstrates that Alfvénic fronts are created (rotational discontinuities) which 'bounce' back and forward in the loop. The Poynting flux is directed towards the dissipative volume and provides the energy.

Flares manifest themselves clearly at X-ray wavelengths. The temperatures and emission measures do not permit the derivation of thermodynamical properties of flares unless the volume is known. The general method which is used to estimate the volume considers the decay phase of flares and equates the decay time of the observed flux (and possibly of the temperature and emission measure) to a typical loss time or model. There are three schools of thought with little interaction. Van den Oord & Mewe (1989) consider the case that cooling loops behave as quasi-stationary loops with some possibility for additional heating. Hawley & Fisher (1990, 1994) use various parts of the energy equation to describe the various phases of a flare. The Palermo group (Reale et al. 1997, 1998) derive scaling laws from numerical simulations of the decay phase. It would be useful to analyze a specific flare with all three methods to check whether the results agree. On the other hand one can argue that if the radiative losses are the dominant loss mechanism, the combination of the observed decay time, temperature and emission measure yields a good estimate of the volume. The situation changes when thermal conduction plays a role since the conductive cooling time depends on the temperature gradient length scale which is difficult to estimate.

An intriguing recent result is that in the coronae of Algol and RS CVn-type stars there is evidence for metal under-abundances with respect to the photospheric values. A possible explanation for this observation is that this is purely a result of hydrostatic settling (van den Oord & Mewe, 1998). The scale heights of individual ion species depend on their charge and mass. This implies that the abundances one derives from observations depend on the actual ions 'observed'. Gravitational settling take however a long time and is only important in long, low-density, loops as found on Algols and RS CVns. In these systems chromospheric evaporation during flares destroys the stratification and the abundances should obtain their photospheric values. This has indeed been observed (Mewe et al. 1997, Güdel et al. 1998).

A field of research which has developed rather independently is that of radio observations (apart from some multi-wavelength campaigns). Radio flares are signatures of non-thermal particle distributions. The emission can be incoherent, mainly synchrotron emission, or coherent (see van den Oord 1996). In the latter case the emission is thought to correspond to the fundamental or first harmonic of the gyro-frequency or plasma frequency. Coherent emission reveals itself through high brightness temperatures, high degrees of polarization and strong time variability. It is always an indication of the presence of particle distributions with a considerable amount of free energy (so non-Maxwellian distributions). This free energy can be converted directly to electromagnetic radiation (e.g., the electron-cyclotron maser) or result in the excitation of longitudinal plasma waves. Interactions of these waves then result in (transverse) electromagnetic radiation. In the case of incoherent emission, mainly synchrotron radiation, it is difficult to interpret the observations because of radiative transfer effects and the unknown source structure (density and field topology). Despite these difficulties with the interpretation of the observations it is found that many objects are surrounded by halos of energetic particles (Mutel et al. 1984, 1985, Lestrade et al. 1984ab, Massi et al. 1988, Trigilio et al. 1993, Benz et al. 1998).

A really surprising result is the general relation between the soft X-ray luminosity L_{X} and the 5–10 GHz radio luminosity L_{R} in the form $L_{\mathrm{X}}/L_{\mathrm{R}} \approx 10^{15.5}$. This relation holds over many orders of magnitude and for many classes of objects (Güdel & Benz 1993, Benz & Güdel 1994). The X-ray emission is optically thin and scales with the emission measure. If the radio emission is optically thin gyro-emission it is proportional to the total number of fast particles. These particles have to be energized during flares or form an energetic tail created during the process of coronal heating. Particle acceleration is never a smooth, continuous process in nature. This implies that the number of energetic particles can vary strongly in time which would destroy the tight relation. It is therefore more likely that the radio emission originates from a large reservoir of energetic particles surrounding the star. The total number of particles does not vary a lot due to new supplies of particles and the particles slowly lose their energy due to collisional and synchrotron losses. An alternative interpretation is that the radio emission is optically thick (see van den Oord 1996). The advantage of assuming optically thick emission from the non-thermal particle population is that an optically thick source is less sensitive to variations in the number of non-thermal particles. In any case, there seems a trend that if the average heating in a corona increases, the X-ray luminosity goes up and at the same time the number of non-thermal particles. A similar trend is found between the time-averaged flare energy in the U-band and the quiescent X-ray emission on flare stars (Doyle & Butler 1985, Whitehouse 1985). All these relations have been found for groups of stars. Observations of individual stars indicate that there is also a time correlation between the X-ray variability and variability of the Balmer lines on time scales of several minutes (e.g., Butler et al. 1986).

In the case of stars, the line and continuum emission at optical wavelengths are the best tracers of non-thermal particles next to the radio emission. There exists an enormous number of observations based on broadband photometry. The difficulty with these type of observations is that these are difficult to interpret since line and continuum emission cannot be separated. Black-body fits result in temperatures of 5000 – $\sim$20000 K and surface areas involved of less than a percent to a few percent of the visible surface. By assuming a black-body the flare area is minimized since a (optically thick) black-body is the most efficient radiator per unit surface. In order to compensate black-body losses much higher beam fluxes are required than in the case that part of the emission originates from lines (van den Oord et al. 1996). It is interesting to observe that, if during the impulsive phase the beam energy is primarily used to ionize the deeper atmospheric layers, while at the same time recombination occurs, the resulting temperatures are always comparable to those found for solar white light flares. The ionization process acts as a thermostat. The problem with these types of global considerations is that they do not explain the details in optical flare spectra. For a discussion of the spectroscopic signatures of optical flares I refer to Houdebine (1992). Most models for optical flares which have appeared in the literature are highly idealized and treat the spectrum emanating from an idealized slab of plasma in which a spatially varying heating occurs because of stopped beam particles, thermal conduction and (X-ray) backwarming (e.g., Hawley & Fisher 1994, Mauas & Falchi 1996). A more promising way to go ahead has been presented during the conference by Abbett & Hawley (1999) who has coupled an hydrodynamical and a radiative transfer code.

A problem which always will be present is that the distribution of the particles which excite an optical flare are not known. Hard X-ray spectra are not available and even if they were be observed, the inversion process necessary to derive distribution functions from these spectra suffers from the same problems as DEM analyses. There has been a debate whether the particle beams consist of protons instead of electrons. Proton beams can explain the lack of correlation between radio and optical flares because the radiation efficiency of protons at radio wavelengths is negligibly small. On the other hand, for various reasons, mainly related to radiative transfer effects, electron beams do not necessarily have to result in observable radio emission (van den Oord 1996). At the moment the evidence for proton beams is weak although Brosius et al. (1995)

claim that proton beams can explain red-shifted Lyman-α radiation in stellar flares. Montes & Ramsey (1998) observed a Li I enhancement during a long-duration stellar flare which they ascribe to spallation reactions. This suggests the presence of accelerated ions.

3. Conclusions

The study of stellar flares is a difficult branch of astronomy. Flares cannot be predicted and are one-time events, each time with a different signature. The overall picture is that scaled-up models of solar flares seem capable of accounting for the global characteristics. The problem with applying these models is that there are (too) many free parameters while the data are not as detailed as for solar flares. This precludes a critical test. On the observational side much can be gained by obtaining multi-wavelength observations in concerted efforts while a unified approach should be used for the analyses of flares.

Acknowledgments. The author acknowledges financial support from Armagh Observatory which made it possible to attend the meeting.

References

Aly, J.J. 1984, ApJ, 283, 349

Abbett, W.P. & Hawley, S.L., 1999, these proceedings

Benka, S.G. & Holman G.B. 1994, 435, 469

Benz, A.O., Conway, J. & Güdel, M. 1998, A&A, 331, 596

Berger, M.A. 1988, A&A, 201, 355

Berger, M.A. 1991, in *Mechanisms of Chromospheric and Coronal Heating*, Eds. Ulmschneider, P., Priest, E.R. & Rosner, R., Springer-Verlag, p. 570

Berger, M.A. 1994, Space Science Rev., 68, 3

Brosius, J.W., Robinson, R.D. & Maran, S.P. 1995, ApJ, 441, 385

Benz, A.O., Güdel, M. 1994, A&A, 285, 621

Butler, C. J., Rodonò, M., Foing, B.H. & Haisch, B. M. 1986, Nature, 321, 679

Doyle, J.G. & Butler, C. J. 1985, Nature, 313, 378

Fisher, G.H. & Hawley, S.L. 1990, ApJ, 357, 243

Forbes, T.G., Isenberg, P.A. 1991, ApJ, 373, 294

Güdel, M. & Benz, A.O. 1993, ApJ, 405, L63

Güdel, M., Linsky, J.L., Brown, A. & Nagase, F. 1998, ApJ, in press

Haisch, B., Strong, K.T. & Rodonò, M. 1991, ARA&A, 29, 275

Haisch, B. & Schmitt, J.H.M.M. 1996, PASP, 108, 113

Hawley, S.L. & Fisher, G.H. 1994, ApJ, 426, 387

Hawley, S.L., et al. 1995, ApJ, 453, 464

Hesse, M. & Schindler, K. 1988, J. Geoph. Res., 93, 5559

Heyvaerts, J. & Priest, E.R. 1977, ApJ, 216, 123

Heyvaerts, J. & Priest, E.R. 1984, A&A, 137, 63

Houdebine, E.R. 1992, Irish Astron. J., 20, 213
Kaastra, J.S. 1985, Solar flares: an electrodynamical model, Thesis, University of Utrecht
Kopp, R. & Poletto, G. 1984, Sol. Phys., 93, 351
Kuijpers, J. 1992, in The Sun, A Laboratory for Astrophysics, J.T. Schmeltz & J.C. Brown, Kluwer, 1992, 535
IAU Coll. 104 on Solar and Stellar Flares, 1989, B.M. Haisch, M. Rodonò, Kluwer
Lau, Y-T. & Finn, J.M. 1990, ApJ, 350, 672
Lestrade, J.-F., et al. 1984a, ApJ, 282, L23
Lestrade, J.-F., et al. 1984b, ApJ, 279, 184
Martens, P.C.H. & Kuin, N.P.M. 1989, Solar Phys., 122, 263
Martens, P.C.H., Young, A. 1990, ApJS, 73, 333
Massi, M., et al. 1988, A&A, 197, 200
Masuda, S., et al. 1994, Nature, 371, 495
Mauas, P.J.D., Falchi, A. 1996, A&A, 310, 245
Melrose, D.B. 1992, ApJ, 387, 403
Mewe, R., et al. 1997, A&A, 320, 147
Montes, D. & Ramsey, L.W. 1998, A&A, 340, L5
Mutel, R.L., et al. 1984, ApJ, 278, 220
Mutel, R.L., et al. 1985, ApJ, 289, 262
Neupert, W.M. 1968, ApJ, 153, L59
Priest, E.R., Démoulin, P. 1995, JGR, 100, A12, 23443
Reale, F., Betta, R., Peres, G., Serio, S. & McTiernan, J. 1997, A&A, 325, 782
Reale, F. & Micela, F. 1998, A&A, 334, 1028
Sakai, J.-I. & de Jager, C. 1996, Space Science Rev., 77, 1
Schrijver, C.J., et al. 1998, Nature, 394,152
Syrovatskii, S.I. 1981, ARA&A, 19, 163
Trigilio, C., Umana, G. & Migenes, V. 1993, MNRAS, 260, 903
van den Oord, G.H.J. 1988, A&A, 205, 167
van den Oord, G.H.J. 1990, A&A, 234, 496
van den Oord, G.H.J. 1993, Adv. Space Res., 13(9), 143
van den Oord, G.H.J. 1996, in Radio emission from the Sun and the Stars, A.R. Taylor, J.M. Paredes, PASP Conf. Ser., 93, 263
van den Oord, G.H.J., Mewe, R. & Brinkman, A.C. 1988, A&A, 205, 181
van den Oord, G.H.J. & Mewe, R. 1989, A&A, 213, 245
van den Oord, G.H.J., et al. 1996, A&A, 310, 908
van den Oord, G.H.J. & Mewe, R., 1999, A&A, submitted
van Oss, R.F., van den Oord, G.H.J. & Kuperus, M. 1993, A&A, 270, 275
van Tend, W. & Kuperus, M. 1978, Solar Phys., 59, 115
Whitehouse, D.R. 1985, A&A, 145, 449

Solar and Stellar Activity: Similarities and Differences
ASP Conference Series, Vol. 158, 1999
C.J. Butler and J.G. Doyle, eds.

The Nature of Flares and Flare-like Phenomena on the Sun

Kenneth J. H. Phillips

Space Science Dept., Rutherford Appleton Laboratory, Chilton, Didcot, Oxon. OX11 0QX, U.K.

Abstract. Although solar activity has been at a low level for the past few years (1994–1997), much work has been done that increases our understanding of one of the major mysteries of solar physics, i.e. solar flares. Such events involve the sudden release of energies of up to 10^{32} ergs in a few tens of minutes. Many examples of flare-like phenomena are now recognized in which much smaller (100 million times or so) amounts of energy are released. We discuss the nature of these events and their relation to the quiet-Sun corona.

1. Introduction

Solar flares are among the most spectacular of time-dependent phenomena that can be studied from the Earth, and the literature describing the observational and theoretical aspects of them is now enormous. We can describe a solar flare as a sudden release of up to 10^{32} ergs, with energy emitted as electromagnetic radiation extending from gamma-rays to very low-frequency radio waves, as well as waves, shock waves, mass motions and energetic particles. A lower limit to the amount of energy released is not well defined. Great advances in our knowledge of flares are made each time data from a new spacecraft are returned, particularly in X-rays, for it is in that region of the spectrum that the major energy release takes place. But optical observations, particularly when combined with spacecraft observations, continue to give us a lot of information which we must take account of in formulating theories that describe what is going on. It is also evident that what we call flares are really the tip of a large ice-berg which encompasses a whole range of flare-like phenomena where the energy releases are much smaller. The recent solar minimum (1994–1997) has allowed us to observe very small scale phenomena that are flare-like in their time evolution and other characteristics: only their released energies (as small as 10^{24} ergs) and location (small bipolar regions rather than fully fledged active regions) distinguish them from flares proper.

To review such a vast literature, even that produced over the past few years, is a daunting task, so I have deliberately limited myself to a few items that seem to me to have novel interest. This contribution, then, should be read alongside other reviews on related topics to get a rounded view of the subject. The subjects I shall discuss are the following: (1) Observation of high-velocity X-ray plasmoids; (2) New flare models from the *Yohkoh* X-ray data; (3) Flare loop-top sources; (4) Flare-like phenomena observed in the quiet Sun.

2. Observations of High-Velocity X-ray Plasmoids

The Soft X-ray Telescope (SXT) on board the Japanese *Yohkoh* spacecraft has obtained many thousands of solar images since the spacecraft was launched in 1991, and the data amassed represent one of the most important solar archives we have. Solar images are either whole Sun (Full Frame Images, FFI) or a partial frame image (PFI) centred on a particular region of interest. When a flare occurs, SXT is commanded to form a series of images in particular wavelength bands (defined by filters in a rotatable wheel assembly) that cover the region of interest only. The cadence of the images in a particular band can be as small as a few seconds, and with a spatial resolution of 2.45″ we are able to follow in detail the evolution of a soft X-ray flare. A particularly useful view is offered by limb flares, for we can then see radial motions of plasma. Recent papers by Ohyama & Shibata (1997, 1998) discuss the pre-flare heating and mass motions in two well-observed limb flares which occurred in 1992/93, when the Sun was near the peak of its activity cycle. It is clear from these observations that there is much activity *before* the main flare event itself. For both flares, a large blob of hot plasma (or "plasmoid") is seen to be ejected from the top of the main flare region which is already rising before the flare's impulsive stage, defined by hard X-ray emission seen with the Hard X-ray Telescope (HXT), also on *Yohkoh*. In one case the rising velocity is only 10 km/s, but in the other 250 km/s. In both cases, there is a sudden acceleration at the impulsive stage.

Rough values of temperature may be obtained from the ratio of photon fluxes in two neighbouring wavelength band-passes. Following this procedure, it is deduced that the temperature of the rising blob is approximately 10 MK, even before the main rise. Again, rough densities N_e may be derived from the emission measure N_e^2V (V is the volume) and assuming a unit filling factor and simple geometry; these are of order 10^{10} cm^{-3}. From these values it is possible to get the masses of the accelerated plasmoids, which are $10^{13} - 10^{14}$ g. Below the ejected plasmoid, one sees a much more slowly rising loop structure forming the main part of the soft X-ray flare. The kinetic energy of the plasmoid is significantly less than the thermal energy of the main X-ray loop ($3N_ekT_eV$).

According to Ohyama & Shibata (1997), the plasmoid's motion seems to be almost a by-product of the flare activity itself, given the relatively small kinetic energy, which is in contrast to some previous flare theories (e.g. Hirayama 1991) in which the plasmoid deforms overlying magnetic fields so that reconnection occurs to fuel the main flare activity. Rather, it would appear that there is a large-scale disturbance in the field that gives rise to both heating of the X-ray flare loop and the ejection of the plasmoid.

3. New Flare Models

This naturally leads on to empirical models of such flares involving reconnecting field lines. These have been widely discussed for many years, but the era of X-ray observations from the *Solar Maximum Mission*, a NASA spacecraft which flew in the 1980's, focused much attention on the possibility that instabilities like the Petschek mechanism operate in flares. Two extremely well observed flares in the *Yohkoh* data in particular lead one to believe that reconnection of field lines are the primary source of energy driving the flare. These were a long-duration flare on 1992 February 21, in which a slowly rising loop shows a cusp-like structure at the loop apex, and a much more impulsive flare on 1992 January 13 (Masuda et al. 1994) seen on the limb with the source of hard X-rays as imaged by the HXT instrument is clearly observed to be above the soft X-ray loop. The original theory of the Petschek mechanism (Petschek 1964) called for reconnection to take place in a very confined volume, with a pair of slow shocks to convert the magnetic energy in the volume ($B^2/8\pi$) to heat and produce the kinetic energy of the moving plasma. It should be noted that the ideas behind the model have undergone much modification in recent years (e.g. Cargill & Priest 1983).

A number of papers have appeared which give a plausible magnetic configuration of the observed limb flares (e.g. Tsuneta 1997; Tsuneta & Naito 1998). One of the observations that these models are at pains to reproduce is the high-temperature ridges that are deduced from SXT images in two X-ray band-passes. These have temperatures of up to 20 MK, and are located well above the bright soft X-ray loop that is generally observed and which forms the bulk of the soft X-ray emission. In a region immediately above the soft X-ray loop apex is the hard X-ray source. Like the bright soft X-ray loop, this is generally slowly rising, with a velocity 10 $km\ s^{-1}$. As indicated in the previous section, a plasmoid that is accelerated to very high velocities (100 $km\ s^{-1}$ or larger) is located high above the main part of the flare.

Figure 1 shows the model explaining the main observational features according to Tsuneta (1997). A Petschek-type of magnetic field reconnection takes place at the X-point indicated in the figure, with fast-moving plasma ejected from the reconnection site between the two slow shocks. The plasmoid on this picture is actually a rising loop seen edge-on, at least these are the circumstances held to occur for a limb flare on 1991 December 2. The plasma of the hot ridges (where the temperature is up to 20 MK) is heated by the slow shocks. The shocks also extend upwards and so there should be hot ridges in regions above the X-point, a feature not generally seen by the *Yohkoh* SXT instrument. This could be because of a low density and therefore small emission measure. Note that plasma is "squeezed" out from the reconnection volume at immense speeds, according to Figure 1 (from Tsuneta 1997), at $\sim$ 1000 km/s, in both upward and downward directions. Tsuneta (1997) finds that Hirayama's (1991) picture, in which the plasmoid as it is ejected creates an X-point, is consistent with the flare he studies. This fact is in contrast to the finding of Ohyama & Shibata (1998) who as mentioned above argue that the plasmoid's kinetic energy is small compared with the thermal energy of the main X-ray loop and so it is unlikely that the plasmoid creates circumstances leading to reconnection.

The high-temperature ridges mentioned earlier are worthy of discussion. There is a possibility of a spatial shift between the taking of images in successive SXT filters, which has consequences on temperature maps that are commonly produced from SXT data (Siakowski et al. 1996). The *Yohkoh* spacecraft pointing is subject to a jitter with amplitude of $1.6''$ when the SXT filter wheel is moved from one filter to another in a typical sequence, as these authors show. This jitter may produce spurious temperature features if a temperature map based on filter ratios is then formed. This important matter clearly deserves further attention.

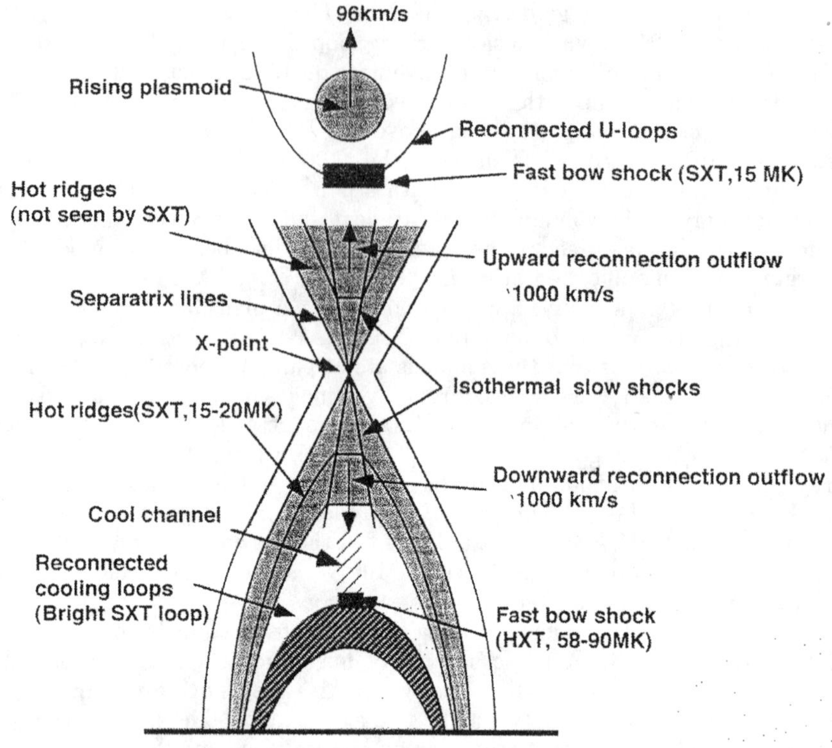

Figure 1. The flare model of Tsuneta (1997). Outflows occur between the slow shocks, giving downward and upward flows, which when they collide with the rising plasmoid and the bright X-ray loop below cause a soft X-ray source and hard X-ray source respectively. Plasma is also heated by the same shocks to give the hot ridges. (Figure courtesy S. Tsuneta.)

If the plasma were heated to X-ray temperatures, say $T_e \sim 20$ MK, any X-ray line spectra would show large Doppler displacements, substantially larger than those commonly seen with the Bragg Crystal Spectrometer (BCS) on *Yohkoh*. The displacements would be both to the "blue" and "red" (i.e. to shorter or longer wavelengths) – either might be visible, but because the resonance line of e.g. Fe XXV has several satellite lines to the red but none to the blue, the blue-shifted component (produced by the upward-moving plasma) would be more readily detectable. Since in general such large shifts are not detected, the plasma cannot be heated to X-ray temperatures by the reconnection process.

As the solar cycle proceeds towards a maximum in a couple of years from now, there will be further opportunities to observe X-ray flares on the limb from *Yohkoh*. We might then have complementary observations from *SOHO* which could help decide the accuracy of the models described above. The LASCO coronagraph, for example, could help in studying the nature of ejected plasma (spherical or loop geometry) if the plasmoid is still identifiable by the time it reaches the field of view of LASCO.

4. Soft X-ray Flare Loop-Top Sources

A feature of soft X-ray flares as noted in *Yohkoh* SXT images is the presence of bright emission kernels or "loop-top sources" (Acton et al. 1992, Feldman 1994, 1995). For impulsive flares, the source diameters are $\sim$ 2000 km (approximately one SXT pixel), while for long-lasting flares, the source is perhaps 10 000 km across. They are detectable in images from the *Solar Maximum Mission* and the S082A instrument on the *Skylab* mission which flew in the 1970's. Some of these features could well be explained by geometrical effects (Alexander & Katsev 1996) since the emission is optically thin in X-rays, but for many flares the geometry is sufficiently simple that one can be reasonably sure that the loop-top source is a real phenomenon.

The plasma parameters can be derived from SXT observations of these features, using intensity ratios when two or more SXT filters are used in rapid succession when observing the flare. However, it is more satisfactory to use line spectra as observed by the BCS instrument on *Yohkoh*, the intensity of the satellite lines to the resonance line for the ions the BCS observes giving the electron temperature directly. Unfortunately, the BCS instrument has virtually no spatial resolution, but observations of limb flares, in particular a series of such flares over a period of hours as an active region was rotating over the west limb, has given important clues (Khan et al. 1995). In particular, it is established from these observations that the bulk of the BCS emission (at least during the decay phase) was from the loop-top, not the foot-points of the loop, and that temperatures between 10 and 20 MK were observed. Additionally, there is significant non-thermal broadening of the X-ray lines, indicating turbulence, with velocities of up to $\sim 200\ km\ s^{-1}$.

At first sight, it is puzzling why such sources manage to persist over any length of time, particularly for the long-duration events, where the sources appear to retain their identities for tens of minutes at least. If we assume a regular topology for the magnetic flux tube that flare loops appear to outline, then if the sources are bright because they are denser than their surroundings, a lifetime of no more than $L/\sqrt{(}2kT_e/m_{\rm ion})$ would be expected (L is the loop semi-length, $m_{\rm ion}$ the proton mass) since the dense plasma would expand at approximately the ion sound speed. For $L \sim 20\,000$ km and $T_e \sim 20$ MK, this time is about half a minute. If on the other hand the sources are bright because they are hot, their lifetime would be governed by conduction to the cooler plasma in the loop legs. For Spitzer conductivity, the time would be roughly $3N_e k l^2/(\kappa_0 T_e^{5/2})$ (l being the loop-top source radius, and Spitzer conductivity $\kappa = \kappa_0 T_e^{5/2}$, with $\kappa_0 = 10^6$ c.g.s. units. With $l \sim 1\,000$ km and $N_e \sim 10^{12}$ cm^{-3}, this time is only 2 or 3 s. Thus loop-top sources need to be maintained in some way over the flare decay.

Recent work by Jakimiec et al. (1998) has examined a number of flares observed by the *Yohkoh* instruments. One item of interest is that, whereas BCS observations give flare temperatures much higher than, and therefore seemingly unrelated to, SXT temperatures, Jakimiec et al. found a strong correlation between the emission measures N_e^2V from the two instruments for a large number of flares at their maximum. This is an important clue, like the one mentioned above from the work of Khan et al. (1995), that the loop-top source is multi-thermal, containing plasma with temperatures of only a few MK co-existing with much hotter plasma, with $T_e \sim 20$ MK.

A few loop-top sources are sufficiently extended that we may see the temperature distribution across them using SXT filter intensity ratios. It is found that the temperature is very uniform indeed. Of course, we are unable to do the corresponding process for BCS as it has no spatial resolution.

A model that explains this distribution is one in which the magnetic field geometry is highly complex inside the loop-top source (Figure 2). This is the one advanced by Jakimiec et al. (1998), based on previous work along the same lines. It is hypothesized that the field lines are so tangled inside the loop-top source that small current sheets are formed (see inset in figure). These current sheets are responsible for heating and moving the plasma to create random velocities, or as seen with the BCS "turbulent" velocities. As the heated plasma cools to, e.g. ~ 5 MK, newly created hot (~ 20 MK) is formed, so that viewed from *Yohkoh* hot and cooler plasma co-exist. The tangled magnetic field is randomly attached to the field of the loop legs which has a regular topology. At the edges one might expect enhanced heating because of this.

A simple analysis shows that a uniform distribution of temperature would result if there were indeed a heating function that is enhanced at the edges of the source, with an energy loss term that is given, not by conduction as commonly assumed, but instead by turbulence, with the energy flux given by $F = -\kappa_t dT_e/dr$ (r is the radial variable measured from the source centre and $\kappa_t = \rho c_p \bar{v} l_m/2$, c_p being the specific heat at constant pressure, ρ the density, $\bar{v}$ the mean velocity of turbulent elements, and l_m the mixing length).

5. Flare-like Phenomena on the Quiet Sun

The past few years (1994–1997) have seen a low level of solar activity enabling detailed investigations of the quiet Sun with instruments on both the *SOHO* and *Yohkoh* spacecraft. As has long been suspected e.g., from optical observations, the "quiet" Sun is in fact highly dynamic, with the activity often in the form of small-scale energy releases. Ultraviolet observations with *Solar Maximum Mission* (Porter et al. 1987), for example, showed the existence of tiny brightenings, some lasting only 2 minutes, located along the neutral lines of small magnetic bipoles in quiet-Sun regions. Extensive work has been done using *Yohkoh* soft X-ray data on active-region transient brightenings (e.g. Shimizu & Tsuneta 1997). We shall not consider these here, except to note that the energy of such phenomena, including larger flares, falls short of the energy requirements of a typical active region by an order of magnitude or so. Interest in these small energy releases in the quiet Sun and active regions centres on whether they are able to supply energy to heat the general corona and active region coronal loops respectively along the lines of the "nano-flare" theory of Parker (1988). We will discuss this below.

For soft X-rays, there are several periods over the past solar minimum when activity was very low (at the *GOES* A level, or 100 times smaller than the level

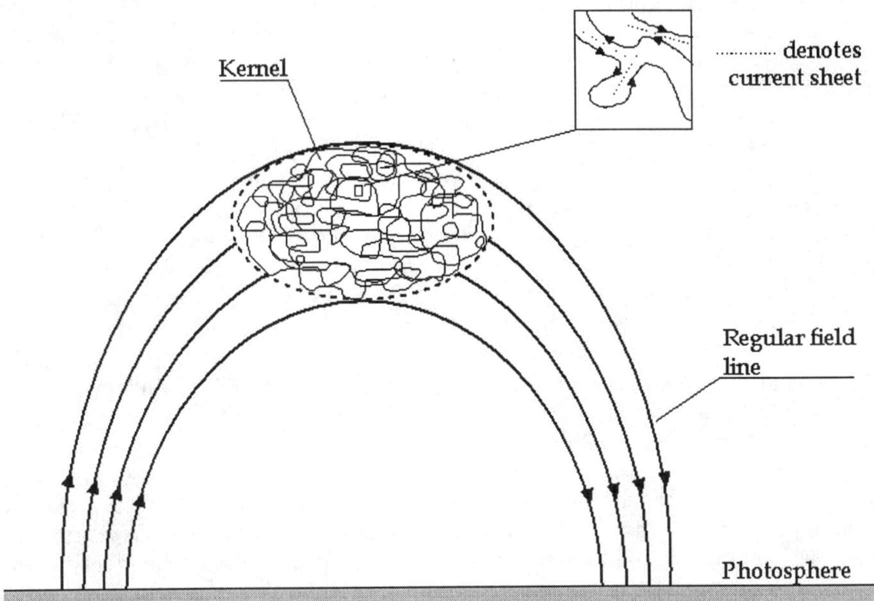

Figure 2. Model to explain the loop-top sources in soft X-ray flares (Jakimiec et al. 1998). The magnetic field lines of the loop legs have a regular topology, but in the loop-top region they are tangled and form current sheets (see inset) which give rise to plasma heating to 20MK (this plasma cools to much lower temperatures, so that high and low temperatures appear to co-exist in the same plasma). (Figure courtesy J. Jakimiec)

of a small C-level flare) and *Yohkoh* SXT was making observations. Krucker et al. (1997) report on observations of the quiet corona made in February 1995 with both *Yohkoh* and the Very Large Array (VLA) which was tuned to three centimetre wavelengths. They saw four small "network" flares with emission in both X-rays and cm-radio waves in small X-ray bright points. The X-ray emission in the AlMg and Al0.1 filters of SXT, for instance, are only a few tens of DN s^{-1} (a large active region might have a flux of several hundred DN s^{-1}). The VLA emission shows impulsive bursts at the onset of each event, like the impulsive X-ray bursts in fully fledged flares. The thermal energies of the flares, measured from the X-ray temperatures and emission measures, are as small as a few 10^{25} ergs, or 10^{-7} of a large, X-class (or optical 3B) flare. The deduced temperatures are extremely low, 1.1 – 1.7 MK, compared with active-region flares. More extensive observations are reported by Krucker & Benz (1998) using the EUV imaging experiment EIT on *SOHO*, which obtains images of the solar corona at a wavelength of 195 Å every approximately 20 minutes. The thermal energies go down to only 8 10^{24} ergs in this analysis. When the flare frequency is plotted as a function of thermal energy, a power-law distribution results, with power equal to about 2.5.

Tiny coronal "flashes" have been noted by Koutchmy et al. (1997) in polar coronal holes, with short durations (down to 1.5 minutes) and fluxes of only 10^{24} ergs, so only 10 times the energy of a nano-flare. These are the smallest quiet-Sun transient brightenings reported so far. There is a possibility that there is an association with X-ray jets, miniature versions of those noted by Shibata et al. (1994). Other work by Preś & Phillips (1998) (see also Preś & Phillips, these Proceedings) indicates that network flares occur in coronal bright points, as defined by EIT Fe XII emission, which are correlated with magnetic bipoles in which the components are either converging or diverging. This work also shows an intimate relation of the magnetic flux found from analysis of data from the Michelson Doppler Imager (MDI) on *SOHO* and the EIT Fe XII emission.

Meanwhile, there is much work on the incidence of small transient brightenings in the quiet-Sun transition region and chromosphere using *SOHO* data. Harrison (1997) reported on small brightenings which he calls "blinkers", visible in data from the Coronal Diagnostic Spectrometer (CDS). Work is underway (Harra-Murnion et al. 1998) to establish the frequencies of such brightenings. It is clear from preliminary results (Fig. 3) that EUV brightenings are much more common in transition-region lines such as O V (formed at a temperature of about 300 000 K) than the chromospherically formed He I line. Over the two-hour period that Figure 3 covers, many tens of brightenings can be identified by eye. The observations in this case were made with a $4'' \times 4'$ slit located across a quiet-Sun region in a N-S direction, while the Sun slowly rotates across the observed region. Features drift across at the slow rate of $9''$ per hour. As network features are generally several arcseconds across, only the gradual rise and fall of emission in the intensity image are due to these features drifting across; the more rapidly varying features are due to real brightenings. The equivalent velocity map shows a very loose correlation of downward velocity and intensity.

Analysers of spacecraft data are apt to forget the fact that coronal emission can be readily observed by ground-based instruments. Work by Airapetian & Smartt (1995) on the Fe XIV green line and Fe X red line images show the occurrence of loop interactions which take the form of transient brightenings both within and outside of active regions. Although these authors emphasize the large differences of the physical conditions of these transient brightenings with active-region flares, there is otherwise little difference between them and what is observed for network flares seen with the *Yohkoh* SXT instrument.

Are these events larger manifestations of the nano-flares mentioned by Parker (1983) and others over the years in the context of coronal heating? Several of the authors above (e.g. Krucker & Benz 1998, Airapetian & Smartt 1995) discuss the energy distribution of events they observe. There was some pessimism, notably expressed by Hudson (1991), that small events could ever be pertinent to the coronal heating problem because of the steepness of their energy spectrum. In general, the energy distribution is observed to be power-law, i.e. $dN/dW \propto W^{-k}$ ergs s^{-1} ($N(W)dW$ is the number density of flares with energies between W and $W + dW$). The energy of the total of all flares, if available to heat the corona, will be $P = \int W(dN/dW)dW$ integrating from $W_{\rm min}$ to $W_{\rm max}$. If $k < 2$ very small events will not contribute much to the total energy, but if $k > 2$ they will. Unfortunately, it seems as if most flare energy distributions have $k \sim 1.8$, tantalisingly close to 2 but still a little short of it.

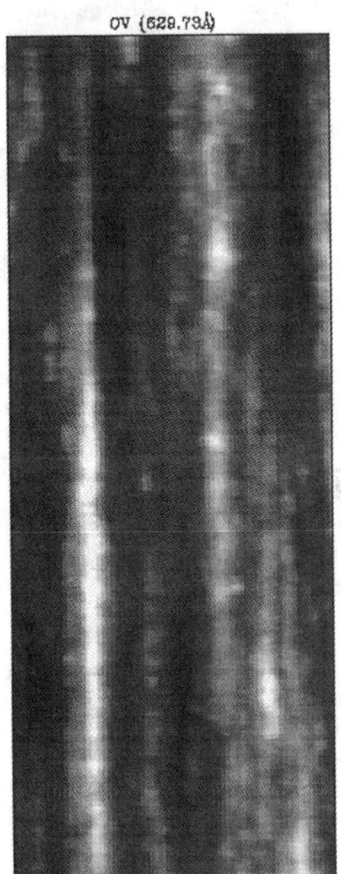

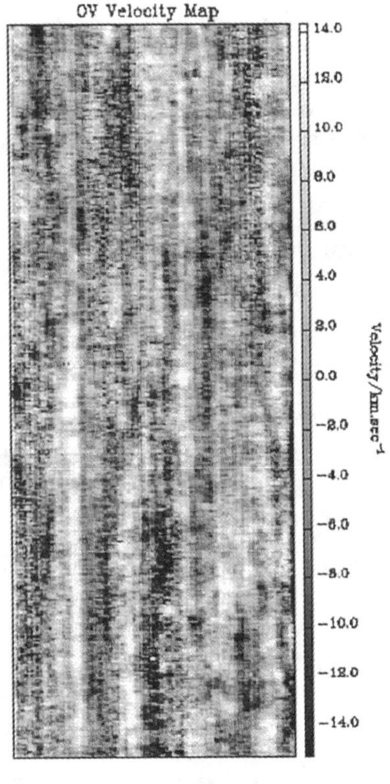

Figure 3. Intensity and velocity scans of a quiet Sun region in the O v line at 629.73 Å. The intensity scan on the left is formed by the *SOHO* CDS $4'' \times 4'$ slit placed N.-S. over a 2-hour period. Solar features slowly rotate past the slit (at a rate of $9''$ per hour). Time forms the vertical axis, with the slit dimension along the horizontal axis. The vertical bright strips are network features drifting past the slit, with short flare-like brightenings appearing as bright points. The righthand plot shows the corresponding velocity scan (bright areas = downward velocities and vice versa) for this same area. (*Figure courtesy of P. T. Gallagher*)

The recent results of Krucker & Benz (1998) puts k at 2.3 to 2.5, so this is a little more reassuring that small events could possibly be responsible for coronal heating.

Nevertheless, as Hugh Hudson and I have noted in private discussions, we should note that the energy of small events either escapes radiatively directly or is transferred conductively along field lines to the chromosphere where it is presumably radiatively lost in optical wavelengths, so that nano-flares, microflares etc. are really coronal "cooling" agents! Also, Hugh Hudson points out that small events in active regions, for instance, occur on different field lines from the main active region loops, so a direct connection does not seem to occur. There are thus some problems to be solved before we conclude that the nano-flare hypothesis is the answer to the coronal heating problem.

6. Summary

Flares and flare-like phenomena are still as topical as ever, judging by the current literature on these subjects. Here I have reviewed some aspects which to me are important, not only for the physics of flares but for the more general and equally vexing problem of coronal heating. There is no doubt that the X-ray observations from *Yohkoh* are helping us to make big strides in our understanding of flares. There are plenty of very good examples of limb flares, e.g., in the SXT data set which allows us a glimpse of the possible reconnection processes. An example of the latter is the observation of X-ray plasmoids from above the main emitting region, though whether the rising plasmoid creates a reconnection X-point or the reconnection causes the plasmoid to rise is moot at present. Some recent models use the Petschek mechanism in more or less its basic form, though MHD theory has now advanced considerably since Petschek's original paper, so some consideration should be given to this. The observation of hot ridges in the *Yohkoh* SXT data is an important ingredient of the models being put forward, but it is also important that possible spacecraft jitter be allowed for in the data analysis for determining the reality of the ridges. Some further work on this would appear to be advisable. The observation of loop-top sources can be explained by a turbulent magnetic configuration at the loop top in which current sheets are continually formed, giving rise to plasma heating and motions. It is possible that this region of turbulence would fit into the models discussed in Section 3. Finally, there is a large family of flare-like brightenings, observed in X-rays, extreme ultraviolet and visible-wavelength radiation. They seem to be distinct in some respects, but future investigations will probably show that some types are related or even different aspects of the same phenomenon. It is uncertain whether these events are important for coronal heating, but recent evidence about their energy spectrum would suggest that they could be.

Acknowledgments. I thank Dr Paweł Preś for help with the figures and for helpful discussions. I also thank Prof. J. Jakimiec, Prof. S. Tsuneta, and Mr P. T. Gallagher for allowing me to use figures from their work.

References

Acton, L. W., Feldman, U. & Bruner, M. E. 1992, PASJ, 44, L71

Airapetian, V. S. & Smartt, R. N. 1995, ApJ, 445, 489

Alexander, D. & Katsev, S. 1996, Sol. Phys., 167, 153

Cargill, P. J. & Priest, E. R. 1983, ApJ, 266, 383

Feldman, U., Hiei, E., Phillips, K. J. H., Brown, C. M. & Lang, J. 1994, ApJ, 421, 843

Feldman, U., Seely, J. F. & Doschek, G. A. 1995, ApJ, 446, 860

Harra-Murnion, L. K., et al. 1998, in preparation

Harrison, R. A. 1997, Sol. Phys., 175, 467

Hirayama, T. 1991, in Flare Physics in Solar Activity Maximum 22 ed. Y Uchida et al. (Berlin: Springer), 197

Hudson, H. S. 1991, Sol. Phys., 133, 357

Jakimiec, J., Tomczak, M., Falewicz, R., Phillips, K. J. H. & Fludra, A. 1998, A&A, 334, 1112

Khan, J. I., Harra-Murnion, L. K., Hudson, H. S., Lemen, J. R. & Sterling, A. C. 1995, ApJ, 452, L153

Koutchmy, S., Hara, H., Suematsu, Y. & Reardon, K. 1997, A&A, 320, L33

Krucker, S., Benz, A. O., Bastian, T. S. & Acton, L. W. 1997, ApJ, 488, 499

Krucker, S. & Benz, A. O. 1998, ApJ, 501, L213

Masuda, S., Kosugi, T., Hara, H., Tsuneta, S. & Ogawara, Y. 1994, Nature, 371, 495

Ohyama, M. & Shibata, K. 1998, ApJ, 499,934

Ohyama, M. & Shibata, K. 1997, PASJ, 49, 249

Parker, E. N. 1988, ApJ, 330, 474

Petschek, H. E. 1964, AAS-NASA Symp. on Solar Flares, ed. W. N. Hess (NASAS SP-50), 425

Porter, J. G., Moore, R. L., Reichmann, E. J., Engvold, O. & Harvey, K. L. 1987, ApJ, 323, 380

Preś, P. & Phillips, K. J. H. 1998, ApJ, in press

Shibata, K. et al. 1992, PASJ, 44, L173

Shimizu, T. & Tsuneta, S. 1997, ApJ, 486, 1045

Siakowski, M., Sylwester, J., Jakimiec, J. & Tomczak, M. 1996, Acta Astron. 46, 15

Tsuneta, S. 1997, ApJ, 483, 507

Tsuneta, S. & Naito, T. 1998, ApJ, 495, L67

Solar and Stellar Activity: Similarities and Differences
ASP Conference Series, Vol. 158, 1999
C.J. Butler and J.G. Doyle, eds.

Dynamic Solar Flare Model Atmospheres

William P. Abbett[1] & Suzanne L. Hawley[1]

Michigan State University, Dept. of Physics and Astronomy, East Lansing, MI 48824-1116, USA

Abstract. A dynamic theoretical model of a flare loop from its footpoints in the photosphere to its apex in the corona is presented, and the effects of non-thermal heating of the lower atmosphere by accelerated electrons and soft X-ray irradiation from the flare heated transition region and corona are investigated. Important chromospheric transitions are treated in non-LTE.

1. Introduction

A solar flare is believed to be the result of a sudden release of free magnetic energy at a localized point in the solar corona. Exactly how this energy is liberated into the solar atmosphere is not well understood, but its effects include the rapid acceleration of charged particles along lines of magnetic field. The peak energy deposition of a strong flux of non-thermal electrons will occur in the upper chromosphere, triggering downward moving condensations along with explosive evaporation of chromospheric material to coronal temperatures (Fisher et al. (1985), Zarro et al. (1988), Fisher (1987)). White light continuum emission is observed to emanate from localized areas around the footpoints of flare loops, and this emission tends to be both spatially and temporally correlated with the hard X-ray radiation (Hudson et al. (1992)). This suggests that the emission occurs at the point where the accelerated electrons impact the pre-flare chromosphere.

It is the response of the chromosphere to the deposition of flare energy that ultimately determines the characteristics of the excess optical line and continuum radiation. Since the radiation field is often decoupled from local conditions in the chromosphere, and since mass motions and strong velocity gradients develop as material is rapidly accelerated through atmospheric plasma, the formalism of non-LTE (non-local thermodynamic equilibrium) radiation hydrodynamics is required to properly model the structure, energetics, and optical emission of the atmosphere.

We model a magnetically confined flare loop from its coronal apex to its footpoints in the photosphere. We introduce a flux of energetic, accelerated electrons at the top of the loop, and compute the dynamic, energetic, and radiative response of the atmosphere to this non-thermal heating. Soft X-ray irradia-

---

[1]Current address: Space Sciences Laboratory, University of California, Berkeley CA 94720-7450

tion of the lower atmosphere due to the flare heated corona is self-consistently included in the models.

2. Methodology

A modified version of the non-LTE radiative hydrodynamic code of Carlsson & Stein (1997) is used to obtain the simultaneous solution of the plane-parallel equations of mass, momentum, and internal energy conservation along with the level population equations, and the equation of radiative transfer. The equations of radiation hydrodynamics, together with the charge conservation equation, are solved implicitly on an adaptive grid via Newton-Raphson iteration. Advected quantities are treated using the second-order upwind technique of van Leer (1977). The adaptive mesh follows the prescription of Dorfi & Drury (1987), and is of critical importance when modeling the dynamics of the lower atmosphere during flares. Features in the chromosphere, such as strong shocks and compression waves, develop quickly and require dense spatial distributions of grid points in order to be properly resolved.

We use the method of Hawley & Fisher (1994) to calculate the energy deposition rate at each height in the atmosphere due to thick target heating from a flux of energetic, non-thermal electrons introduced at the loop apex. To model the soft X-ray flux, we use the optically thin, thermal emission spectra of Raymond & Smith (1977) along with the temperature and electron density stratification of the atmosphere to calculate the emitted power per wavelength band at each atmospheric layer. The soft X-ray flux incident on any other portion of the atmosphere due to the aggregate contribution of the emitting regions is then calculated using a plane-parallel approximation if an emitting region is nearby, and the point source approximation of Gan & Fang (1990) if the emitting region is distant. Once the flux is determined, the heating rate is calculated as in Hawley & Fisher (1992).

The total flare heating at each level of the atmosphere is a sum of the thermal contribution from soft X-ray irradiation and the non-thermal contribution of the accelerated electrons. The flare heating is included in the equation of internal energy conservation as an external source, ensuring detailed energy balance at all depths in the atmosphere and at all times during the flare.

An important part of the model atmosphere calculation is the non-LTE treatment of the radiation field. Atoms important to the chromospheric energy balance are treated in non-LTE: a six-level hydrogen atom, a six level singly ionized calcium atom, a nine level helium atom and a four level singly ionized magnesium atom. Complete redistribution is assumed for all lines. For the Lyman transitions, the effects of partial redistribution are mimicked by truncating the profiles at ten Doppler widths (see Milkey & Mihalas (1973)). Other atomic species are included in the calculation as background continua in LTE using the Uppsala opacity package of Gustafsson (1973). Optically thin radiative cooling due to bremsstrahlung and coronal abundances of carbon, oxygen, neon, and iron is included in the calculation of the total radiative loss rate.

Two basic methods can be used to model the initial, pre-flare state of the solar chromosphere. They have been dubbed by Ricchiazzi & Canfield (1983) as the "semi-empirical" approach and the "synthetic" approach. Semi-empirical models ignore energy transport in the atmosphere in favor of a fixed temperature and density stratification which reproduces observed optical emission from static solutions to the equations of statistical equilibrium and radiative transport. Synthetic models differ in that the atmospheric structure is produced as a result of a self-consistent solution of the basic equations of hydrodynamics and radiation transport. One advantage inherent in using a synthetic pre-flare model rather than a pre-existing semi-empirical state is that dynamic effects not directly related to the influx of flare energy are eliminated from the flare simulation. For this reason, we generated a synthetic pre-flare state to use as the initial starting atmosphere for the dynamic run.

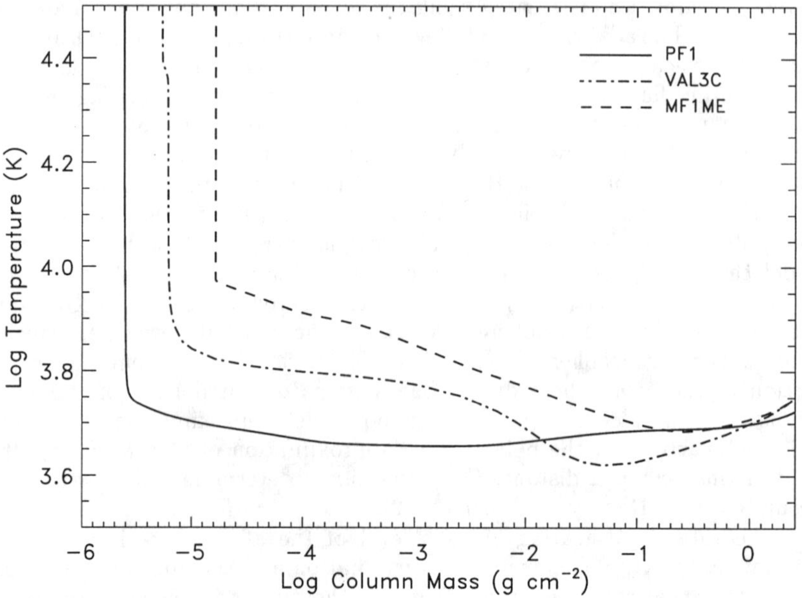

Figure 1. The temperature stratification as a function of log column mass (g cm $^{-2}$) for the three model atmospheres PF1, VAL3C, and MF1ME.

To generate the pre-flare state, a transition region and corona were added to the initial state of Carlsson & Stein (1997) and this starting approximation was allowed to relax to a state of hydrostatic and energetic equilibrium. During this procedure, the upper boundary (the apex of a symmetric coronal loop) was held at 10^6 K, and an assumed amount of non-radiative quiescent heating per unit mass was applied to several grid zones adjacent to the lower boundary in order to fix the gradient of the temperature at the base of the photosphere. No other external driving mechanism or source of heating was provided. The resultant temperature and electron density stratification of the relaxed state (hereafter referred to as the PF1 atmosphere) is shown in Figure 1, and is compared to the standard semi-empirical VAL3C chromospheric model of Vernazza et al. (1981)

and to the semi-empirical active atmosphere of Metcalf (1990) (MF1ME) that was used as the pre-flare atmosphere of Hawley & Fisher (1994). We see that PF1 is not as active as the pre-flare state MF1ME, and is somewhat quieter than the VAL3C quiescent state.

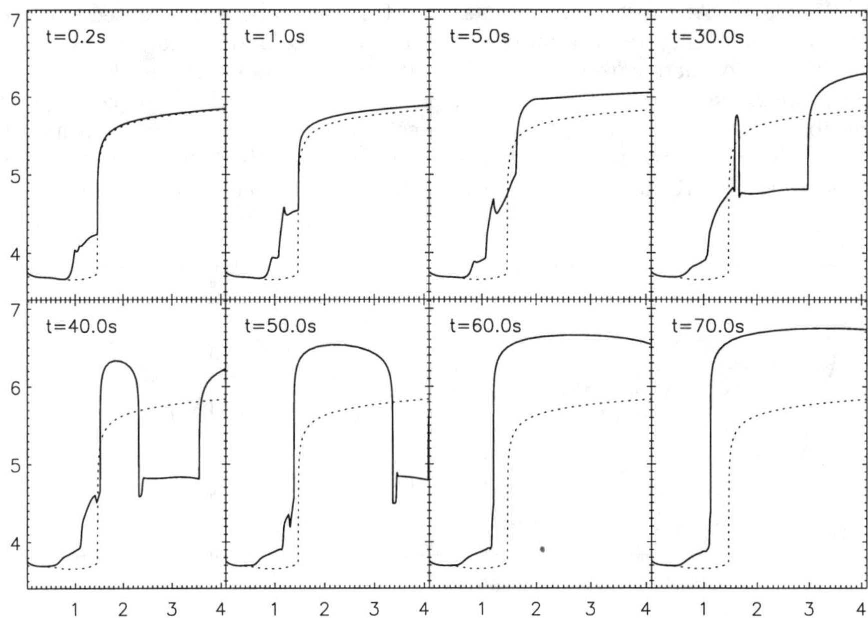

Figure 2. The F10 atmosphere. The *solid* line represents the log of the temperature, T (K), as a function of height above $\tau_{5000} = 1$, z (megameters), at time t. The *dotted* line denotes the pre-flare temperature structure (PF1).

3. Impulsive Phase Dynamics

To investigate the properties of the lower atmosphere during the impulsive phase of a moderately strong solar flare, we consider a constant electron energy flux of 10^{10} ergs cm^{-2} s^{-1} (hereafter called run F10) applied to the PF1 atmosphere for seventy seconds. The F10 run can be described as having two distinct dynamic phases, each with a definite observational signature: a *gentle* phase, characterized by stages of near-equilibrium, and an *explosive* phase characterized by large material flows, and strong hydrodynamic waves and shocks. These features are illustrated in Figures 2 & 3 which show the temperature and electron density evolution of the atmosphere. Figure 3 also shows the beam heating profile. The gentle phase is represented in the first panels of each figure, while the explosive evaporation is shown in panels four through six. The last two panels depict the new flare-heated atmosphere with a deeper transition region and hotter corona than the pre-flare state.

To understand this behavior, we follow the progress of the beam heating process in the lower atmosphere. The accelerated electrons initially deposit energy in the upper chromosphere, and rapidly heat a broad region to $\approx 10^4$ K (see the first frame of Figure 2 at 0.2 seconds, and note the region between ≈ 1.1 and 1.5 Mm). At this temperature, atomic transitions between excited states of hydrogen efficiently radiate away a majority of the non-thermal energy input, and the atmosphere enters a short-lived state of near-equilibrium. However, in the localized region where non-thermal energy deposition is highest (corresponding to the penetration depth of 20 keV electrons), the temperature continues to climb. Figure 3 shows that after 0.2 seconds of flare heating, this point is located approximately 1.1 megameters above the height where $\tau_{5000} = 1$.

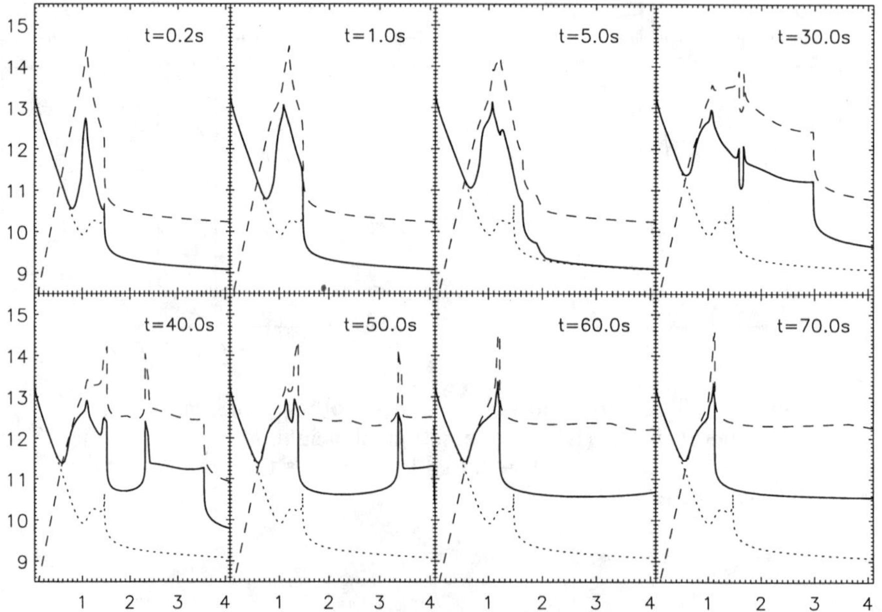

Figure 3. The F10 atmosphere. The *solid* line represents the log of the electron density as a function of height, z (Mm), at time t ($\log n_e$, $[n_e]$=cm $^{-3}$). The *dotted* line denotes the pre-flare electron density, and the *dashed* line is the log of the non-thermal electron heating, scaled so that it will fit on the plot ($\log Q_{e}-+11.5$, $[Q_{e}-]$=ergs cm $^{-3}$ s^{-1}).

Although the local temperature minimum close to the lower boundary of the chromospheric plateau (seen in the first frame of Figure 2) looks like a chromospheric condensation, it is not the same phenomenon. During the gentle phase, this temperature minimum develops without a substantial increase in the local gas density, and without the downward motion and steep velocity gradients usually associated with a fully developed condensation. Thus, the optical line emission formed in this region is not significantly Doppler shifted, and the profiles remain essentially symmetric about their nominal central wavelengths.

The increasing number of collisional ionizations throughout the temperature plateau elevates the density of protons in the region. This effects a change in the thermalization depth of energetic electrons, since it is considerably more likely for a high energy electron to undergo a Coulomb interaction and thermalize in regions where proton densities are high. Thus, even though there is little material flow in this region, in time the non-thermal heating profile broadens slightly, and the point of maximum energy deposition slowly moves upward toward the corona.

By 0.3 seconds, a significant fraction of hydrogen in the upper chromosphere is ionized, and the bulk of the flare energy can no longer be effectively radiated away. The plasma in the upper chromosphere rapidly heats to $\approx 10^5$ K where collisionally excited resonance transitions of abundant metal ions dominate the radiative cooling (eg. C IV), and the radiative loss function attains its maximum value. The atmosphere then enters a second, longer-lasting state of near-equilibrium (see panel 2 of Figure 2), and once again, the energy input of the flare electrons is balanced by radiative losses. The development of two distinct temperature plateaus (rather than a single plateau at $\approx 10^5$ K) is characteristic of the gentle phase, and is a direct result of the upward drift of the non-thermal heating profile in response to variations in the hydrogen ionization fraction.

The gentle phase persists until the atmosphere is no longer able to effectively radiate away the non-thermal energy input. In the F10 run, this occurs after approximately 27 seconds of impulsive heating, at which time the atmosphere proceeds to the explosive phase. During this phase, the bulk of the energy from the accelerated electrons can no longer be radiated away, and instead is able to do work in the atmosphere. Material is forced away from the point of maximum flare heating, forming a shock front that moves rapidly upward into the corona, and a narrow compression wave that propagates downward into the chromosphere (see panel 5 of Figure 2 at 1.4 and 2.3 Mm). Within the downward moving wave, high gas densities lead to increased radiative cooling, and a local temperature minimum appears, forming a chromospheric condensation. Note that although the non-thermal electron energy flux in run F10 is a factor of three below the explosive limit of Fisher et al. (1985), the atmosphere eventually undergoes explosive evaporation.

Much of the excess optical emission during the explosive phase originates in the high density condensation (and fast upward-moving evaporation). The emergent intensity reflects the material motion and steep velocity gradients of these regions; the observed line profiles are highly Doppler shifted and are asymmetric about their nominal central wavelengths.

4. Continuum Emission

Figure 4 shows the flare continuum emission at the early and late stages of the F10 run, and Figure 5 shows the light curve of the continuum at 5000 Å and in the near wings of Hα. Notable features include an initial *reduction* in the continuum intensity (eg. the top panel of Figure 5 during the first second of flare heating) which is not mirrored in the Hα line wings. Later in the flare, the 5000 Å continuum undergoes a one percent brightening over its pre-flare value. This effect is much more pronounced when the electron energy flux is an order of magnitude higher than in run F10.

This behavior can be qualitatively understood as follows. The influx of flare energy from the non-thermal electrons increases collisional rates in the upper chromosphere, thus elevating the population densities of high-energy bound states of hydrogen. The number of photoionizations from the upper states then increases, and photons from the photosphere that normally would escape and be seen as continuum radiation get absorbed high in the chromosphere. This results in a decrease in the observed continuum intensity as shown in the first panel of Figure 4.

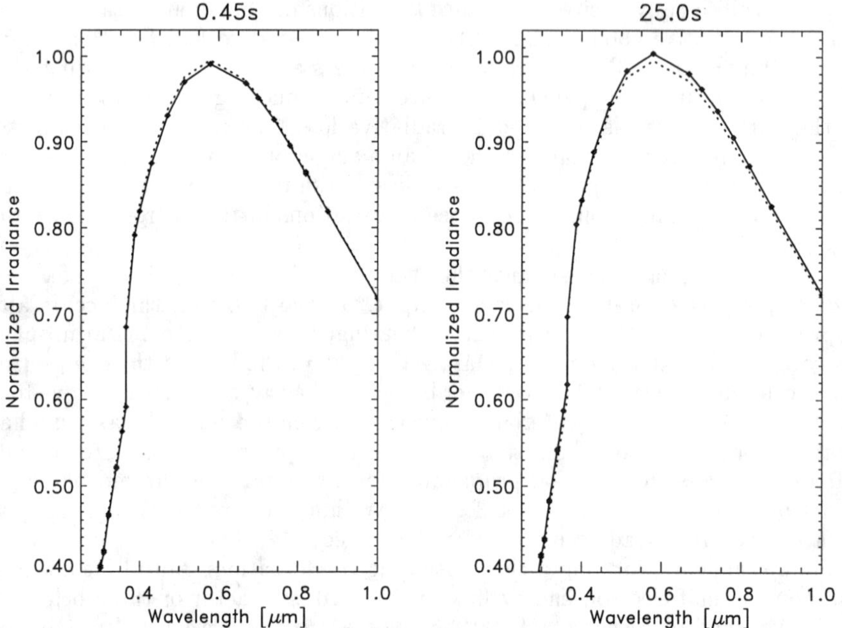

Figure 4. *Solid* lines show the Balmer and Paschen continua (normalized to the gradual phase maximum) for the F10 run. The *dotted* line is the continuum emission of the pre-flare atmosphere.

However, this population excess is short lived. The high number of photoionizations liberates electrons into the continuum, quickly depopulating bound states of the hydrogen atom. The number of photoionizations is thus reduced, just as increases in the electron density cause a sharp rise in the number of recombinations in the same region. The net result is a switch over between the dominance of photoionization absorption and recombination emission in the region of the low temperature ($\approx 10^4$ K) plateau.

The excess emission in the near wings of Hα originates in the same region, but begins almost immediately after the onset of non-thermal heating (the radiative balance in this case being controlled by the relative strength of the non-local radiation field, and the local value of the Planck function). Thus, the time lag that occurs between Hα wing emission and the Paschen brightening is almost entirely controlled by the amount of time it takes for recombination radiation to dominate the chromospheric plateau. We note that a time lag similar to this has been observed in the strong white light flare of March 7, 1989 observed by Neidig et al. (1993).

By late in the explosive phase (the dynamics of which are shown in the lower panels of Figures 2 & 3), the hydrogen ionization fraction in the upper chromosphere is approximately two orders of magnitude greater than during the early gentle phase. The increase in electron density throughout the upper chromosphere and the elevated gas density within the chromospheric condensation both contribute to high levels of hydrogen recombination radiation emanating from the upper chromosphere. We discuss in detail the formation of the continuum and line emission during the explosive phase in Abbett & Hawley (1998).

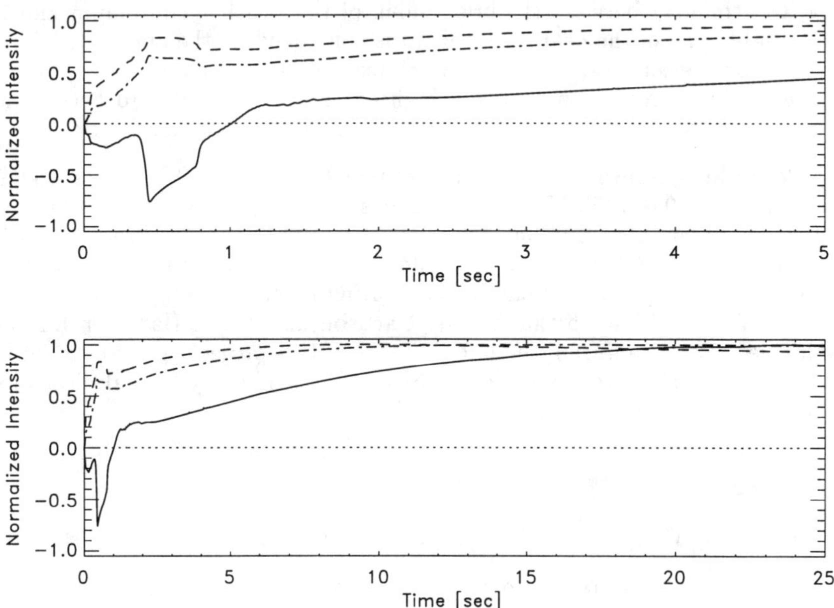

Figure 5. Continuum intensity (normalized between the preflare value and the maximum value at 25.0s) at 5000 Å (*solid line*) for the F10 run. The normalized Hα line wing intensities are shown at +2.5 Å (*dashed line*) and -2.5 Å (*dotted-dashed line*). For this case, the lag between the 5000 Å continuum brightening and the Hα wing brightening is $\approx$ 1.0s.

5. Conclusions

The numerical simulations support the hypothesis that a sudden flux of non-thermal electrons is a major factor in the appearance of optical continuum emission at the footpoints of magnetic loops during strong flares, and that this emission originates throughout a broad temperature plateau formed locally in the upper chromosphere immediately after flare onset. Specifically, the primary conclusions of this analysis are the following:

- An impulsive event can be described as having two phases, a *gentle* phase characterized by a state of near equilibrium, and an *explosive* phase characterized by large material flows, and strong hydrodynamic waves and shocks.

- Hydrogen recombination radiation from the chromospheric plateau (gentle phase) and the chromospheric condensation (explosive phase) is the primary cause of the white light continuum brightening observed during strong flares.

- The time lag between the brightening of the Paschen continuum and the brightening of the near wings of Hα is controlled by the amount of time it takes for electron densities in the plateau to become high enough, and the densities of hydrogen atoms in high energy bound states to become low enough, to allow recombination radiation to dominate the region.

Acknowledgments. This work was funded in part by NSF grants AST 96-16886 and AST 94-57455. The computations described here were partially supported by the National Computational Science Alliance and utilized the NCSA SGI/CRAY Power Challenge Array. Additional computations were carried out on the SUN Enterprise-6000 parallel computing facility at Michigan State University. We would like to thank Mats Carlsson and Viggo Hansteen for their invaluable assistance during the initial stages of this project, and we would like to acknowledge Bob Stein and George Fisher for their many helpful discussions during the course of this work.

References

Abbett, W. P., & Hawley, S. L., 1999, in prep.

Carlsson, M., & Stein, R. F., 1997, ApJ, 481, 500.

Dorfi, E. A., & Drury, L. O., 1987, JcP, 69, 175.

Fisher, G. H., Canfield, R. C., & McClymont, A. N., 1985, ApJ, 289, 425.

Fisher, G. H., 1987, 317, 502.

Gan, W. Q., & Fang, C. 1990, ApJ, 358, 328.

Gustafsson, B., 1973, A Fortran Program for Calculating "Continuous" Absorption Coefficients of Stellar Atmospheres, Uppsala Astron. Obs., Ann. 5, No. 6, Uppsala: Landstingets Vergstader.

Hawley, S. L., & Fisher, G. H., 1992, ApJ, 78, 565.

Hawley, S. L., & Fisher, G. H., 1994, ApJ, 426, 387.

Hudson, H. S., Acton, L. W., Hirayama, T., & Uchida, Y., 1992, Publ. Astron. Soc. Japan, 44, L77.

Metcalf, T. R., 1990, PhD Thesis, University of California, San Diego.

Milkey, R. W., & Mihalas, D., 1973, ApJ, 185, 709.

Neidig, D. F., Kiplinger, A. L., Cohl, H. S., & Wiborg, P. H., 1993, ApJ, 406, 306.

Raymond, J. C., & Smith, B. W., 1977, ApJS, 35, 419.

Ricchiazzi, P. J., & Canfield R. C., 1983, ApJ, 272, 739.

van Leer, B., 1977, JcP, 23, 276.

Vernazza, J. E., Avrett, E. H., & Loeser, R., 1981, ApJS, 45, 635.

Zarro, D. M., Canfield, R. C., Metcalf, T. R., & Strong, K. T., 1988, ApJ, 324, 582.

Solar and Stellar Activity: Similarities and Differences
ASP Conference Series, Vol. 158, 1999
C.J. Butler and J.G. Doyle, eds.

Observations and Modelling of a Flare on AD Leo

D. Jevremović[1], C.J. Butler & J.G. Doyle

Armagh Observatory, College Hill, Armagh BT61 9DG, N. Ireland

Abstract. We present high temporal (~ 3sec), medium spectral resolution ($\Delta\lambda/\lambda \sim 5000 - 7000$) spectrophotometric observations of a flare on AD Leo. A two magnitude flare in U was recorded at 00:45 UT on March 14 1998 using the William Herschel Telescope (WHT) and both arms of the ISIS spectrograph. The high time resolution was achieved using the CCD readout in the Low Smear drift mode. The rising phase of the flare lasted fifteen seconds and the decay phase around 20 minutes. We estimate the total energy of the flare to be 10^{32} erg. We present gas-dynamic calculations of the response of the stellar atmosphere to heating by an electron beam.

1. Introduction

Flares on late type stars are widely believed to arise from magnetic processes similar to those that drive the better known solar flares. This, canonical, view is based on the similar characteristics of their optical, ultraviolet and X-ray emission. As observational data becomes available with a time resolution compatible with the time-scale of flare events it should eventually be possible to explore the physical parameters of the flares in greater detail and to make a more meaningful comparison between solar and stellar events.

2. Observations

Spectrophotometric observations were carried out during the nights March 11/12 - 13/14 1998 using WHT and both arms of the ISIS spectrograph. The spectral coverage (using the R600B and R600R gratings) was 3600-4400 Å in the blue and 5850-6620 Å in the red. This setup covered all the Balmer lines (except Hβ), the sodium doublet and the Ca II H & K lines. The wavelength resolution was ~ 0.8 Å/pixel on the TEK1024 CCD detectors(pixel size 0.24μ). High temporal resolution (~ 3s = 2s exposure +1s readout) was achieved using the low smear drift mode described by Rutten et al. (1997). We have used windows of 1k×20 pixels in both spectral regions.

Spectra were reduced in the standard manner (bias and flat-field correction, extraction, wavelength and flux calibration. The slit width was set to 2 arcsec

---

[1] Belgrade Observatory, Volgina 7, 11000 Belgrade, Yugoslavia

allowing us to achieve spectrophotometric calibration with an estimated error of ±5%

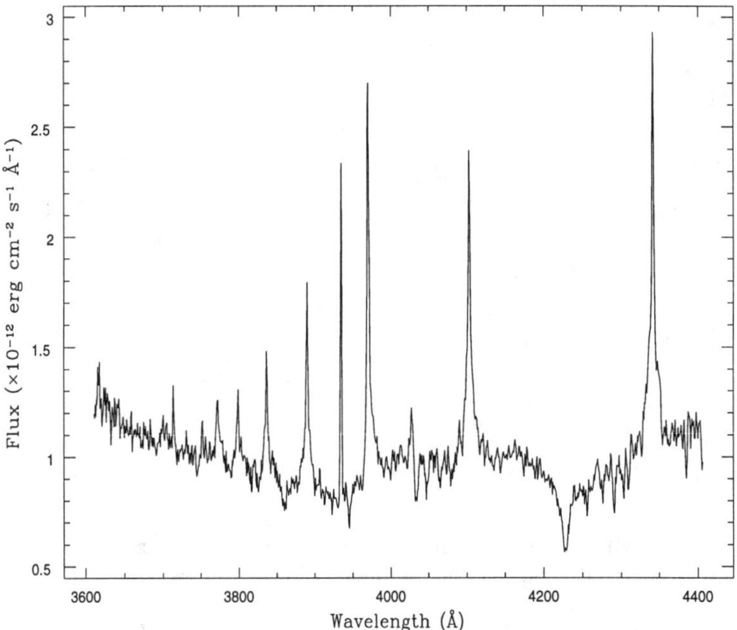

Figure 1. The AD Leo spectrum as recorded around flare maximum

A two magnitude flare (in U) was detected at UT 00:45 on March 14 1998. The pre-flare rise lasted around two minutes and the impulsive phase around 15 seconds followed by a decay of about 15 minutes. The time profile of the flux in Hδ is shown in Figure 2.

From the integrated fluxes in time and assuming that one third of energy is radiated in the blue/UV we conclude that total energy radiated during this flare was approximately 10^{32} erg.

3. Modelling

The set of equations which govern the behaviour of a simple fluid, two temperature plasma are the following (a) the mass conservation equation, (b) momentum conservation and (c) the energy conservation equation. These are solved with column depth as a Lagrangian coordinate using an adaptive mesh. Our code is coupled with the radiative transfer code MULTI (Carlson, 1986) which provides regular updates of the radiative losses for H, Ca and Mg. The code uses a grid of 150 points and re-meshing is accomplished using changes in temperature as the control function. The initial atmosphere (dM3.5e) is based on the photospheric structure of Allard & Hauschildt (1996), with an attached chromosphere and

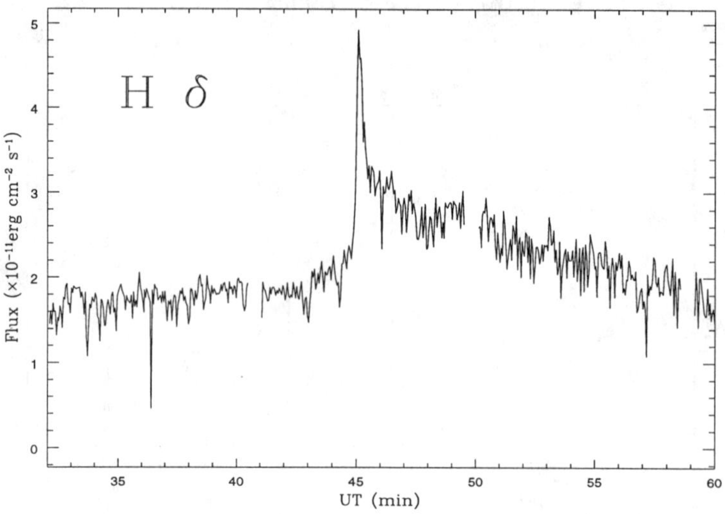

Figure 2. The time profile of H δ

transition region with a temperature minimum at log m_0=1 and transition region at logm_{tr}=-3.8 (T=8500K). In the initial chromosphere we kept $dT/dlogm$ constant.

We assume that the stellar atmosphere is heated by the beam of energetic electrons with a spectral index $\delta = 5$ and a cutoff energy of 20 keV. We calculated the energy deposited in the atmosphere according to the standard approach (eg. Emslie 1978, Emslie & Brown (1981), Hawley & Fisher 1994):

$$P_e(N) = \frac{\gamma(N)(\delta-2)}{2\mu_0} B_{x_c}\left(\frac{\delta}{2}, \frac{1}{3}\right) \frac{F_{20}}{E_c^2} \left[\frac{N^*(N)}{N_c^*}\right]^{-\delta/2} \tag{1}$$

We consider the following case: a beam flux of 5 10^{11} erg cm^{-3}s^{-1}, a triangular shape for the heating function and a duration of the heating of 15 seconds. The intensities relative to the pre-flare level of Hα and Hγ computed by this model, compared to the observational data are shown in Figure 3.

4. Conclusion

Our observations are believed to be the first optical observations of a stellar flare with a time resolution better than 5s. Observations of solar flares in the optical region with this time resolution are still very rare. We have developed a new code for the analysis of the atmospheric response to heating during stellar flares. The introduction of full radiative transfer allows us to follow the spectral features during the course of the flare. The first results show that it is possible to simulate the main features occurring during the stellar flare using our simplified model. In future work will explore a wider set of input parameters, and, in particular, different heating functions, as Figure 3 indicates continual heating.

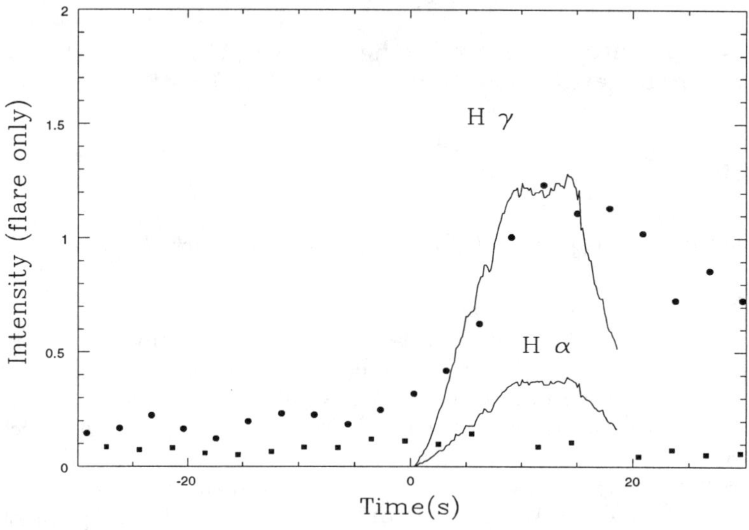

Figure 3. Evolution of the relative intensities in Hα and Hγ lines: dots (observational data), solid line (model calculation).

References

Allard, F. & Hauschildt, P.H., 1995, ApJ, 445, 433

Carlson, M., 1986, Uppsala Spec. Rept. No. 33

Hawley, S.L. & Fisher, G.H, 1994, ApJ, 426, 387

Emslie, A.G., 1978, ApJ, 224, 241

Rutten, R., Gribbin, F., Ives, D., Bennet, T. & Dhillon, V., 1997, Drift-Mode CCD Readout Users' Manual, ING

Solar and Stellar Activity: Similarities and Differences
ASP Conference Series, Vol. 158, 1999
C.J. Butler and J.G. Doyle, eds.

A Long-Duration Flare in the X-ray/EUV selected Chromospherically Active Binary 2RE J0743+224

D. Montes[1,2] & L.W. Ramsey[2]

The Pennsylvania State University, Department of Astronomy and Astrophysics, 525 Davey Laboratory, University Park, PA 16802, USA

Abstract.
E J0743+224 (BD +23 1799) is a chromospherically active star, selected by X-ray and EUV emission, detected in the Einstein Slew Survey and *ROSAT* Wide Field Camara (WFC) all sky survey, and classified as a single-lined spectroscopic binary. However, the high resolution echelle spectroscopic observations we present here reveal it as a double-lined spectroscopic binary. A dramatic increase in the chromospheric emissions (Hα and Ca II IRT lines) is detected during the observations. We interpret this behaviour as an unusual long-duration (> 8 days) flare based on (a) the temporal evolution of the event, (b) the broad component observed in the Hα line profile, (c) the detection of the He I D_3 line in emission and (d) a filled-in He I λ6678 Å line. We detect a Li I λ6708 Å line enhancement which is clearly related with the temporal evolution of the flare. The maximum Li I enhancement occurs just after the maximum chromospheric emission observed in the flare. We suggest that this Li I is produced by spallation reactions in the flare.

1. Introduction

2RE J0743+224 (BD +23 1799, SAO 79647) is a EUV source detected during the *ROSAT* Wide Field Camera (WFC) all sky survey (Pound et al. 1993; Pye et al. 1995). The WFC optical identification program (Mason et al. 1995) identified it as a chromospherically active star with EW(Ca II H) = 4.1 and EW(Hα) = 1.8. The optical spectroscopic observations presented by Jeffries et al. (1995) confirmed it as chromospherically active with strong Hα emission (EW(Hα) = 1.4). These authors found it to be a single-lined spectroscopic binary with spectral type K0 and a rotational velocity, $v\sin i$, of 13 km s^{-1}. No orbital solution is reported and only a lower limit of 3 days to the period is given. This star was also detected in X-rays by the Einstein Slew Survey (Elvis et al. 1992) receiving the name 1ES 0740+22.8 and confirmed as chromospherically

---

[1]Departamento de Astrofísica, Facultad de Físicas, Universidad Complutense de Madrid, E-28040 Madrid, Spain

[2]Guest observer at McDonald Observatory

active by Schachter et al. (1996) who give an estimated spectral type K0III/IV and a measured $v\sin i$ of 17.0 km s^{-1}.

2. Observations

We present here high resolution (0.16 Å) echelle spectroscopic observations of this chromospherically active binary (hereafter CAB), obtained during a 10 night run 12-21 January 1998 using the 2.1m telescope at McDonald Observatory and the Sandiford Cassegrain Echelle Spectrograph. These observations reveal it to be a double-lined spectroscopic binary (SB2) with a orbital period of 10 days. We have determined the chromospheric contribution in the Hα and Ca II IRT lines using the spectral subtraction technique (Huenemoerder & Ramsey 1987; Montes et al. 1997). The synthesized spectrum was constructed using the program STARMOD developed at Penn State (Barden 1985). The best fits to the 2RE J0743+224 spectra were obtained using a K1 III star for the primary and a K5 V for the secondary, with a relative contribution to the continuum of 0.85 /0.15. In Figure 1 we plot the observed spectra together with the K1 III reference star in the left panel and the subtracted spectra in the right panel.

3. Results

A dramatic increase in the chromospheric emission (Hα and Ca II IRT lines) is detected during the observations (see Figure 1) The increase of the emission starts on the 3rd night (1998 January 15) and reached maximum on the 5th night. At the end of the observations (1998 January 22), the chromospheric lines had not yet recovered their quiescent value. Thus the total duration of the event was larger than 8 days. Several arguments favor the interpretation of this behavior as an unusually long-duration flare:

(1) The temporal evolution of the event is similar to that observed in other solar and stellar flares, with an initial impulsive phase characterized by a strong increase in the chromospheric lines. The Hα emission EW increases by a factor of $\approx$5 in only one day (from the 2nd to the 3rd night) and by a factor of 7 at the maximum. After this the emission decreases slowly (gradual phase) until the end of the observations. The time evolution of the EW(Hα) during the flare is displayed in Figure 2.

(2) A broad component in the Hα line profile is observed just at the beginning of the event. A two Gaussian components fit to the subtracted spectra is displayed in the right panel of Figure 1.

(3) The detection of the He I D_3 in emission and a filled-in He I λ6678 Å as have been observed in other solar and stellar flares (Zirin 1988; Huenemoerder & Ramsey 1987; Montes et al. 1996; 1997; 1998)

This is an unusual long-duration flare and very different from the largest flares observed in the Sun ($\approx$ hours). However, long-lasting (2 to 9 days) flares have also been observed in other CAB, e.g. II Peg (Berdyugina et al. 1998a), AR Lac (Ottmann & Schmitt 1994), HK Lac (Catalano & Frasca 1994), HR 5110 (Graffagnino et al. 1995), CF Tuc (Kürster & Schmitt 1996), and HU Vir (Endl et al. 1997); and in FK Com-type stars such as YY Men (Cutispoto et al. 1992).

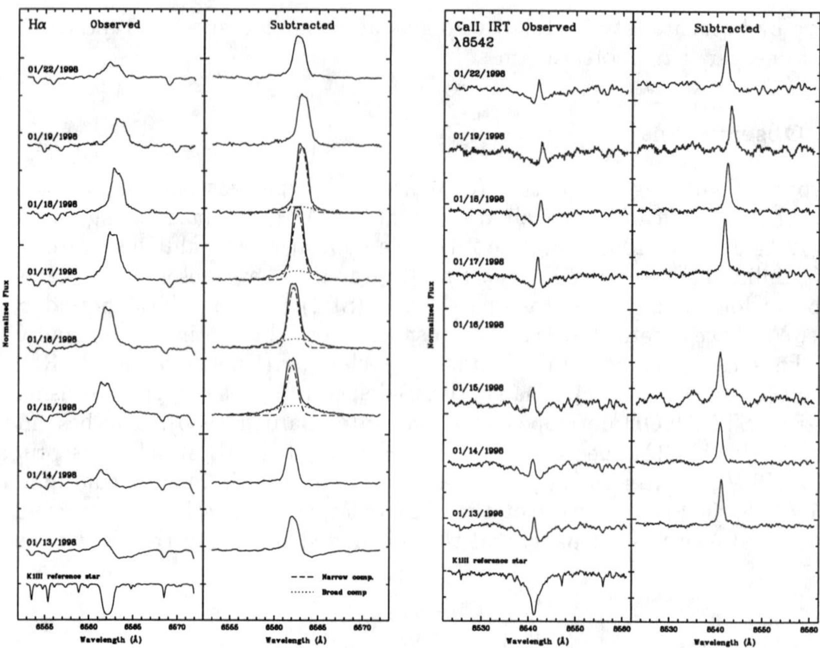

Figure 1. Hα and Ca II IRT observed spectra (left panel), and after the spectral subtraction (right panel)

The analysis of the TiO 7055Å (O'Neal et al. 1996) band indicates that a fraction of the stellar surface is covered by starspots. Strong changes are observed in the depth of this band from night to night in the same way that the chromospheric lines change. This suggest that the chromospheric region responsible for the flare is spatially coincident with photospheric cool spot(s) and could be produced, as have been observed in the largest solar flares, by the interaction of new emerging magnetic flux with old magnetic structures associated with the spot groups.

We detect a Li I λ6708 Å line enhancement which is clearly related with the temporal evolution of the flare. The maximum Li I enhancement (40% in EW) occurs just after the maximum chromospheric emission observed in the flare. A significant increase of the ^{6}Li/^{7}Li isotopic ratio is also detected. No significant simultaneous variations are detected in other photospheric lines. Neither line blends nor starspots seem to be the primary cause of the observed Li I line variation. From all this we suggest that the Li I enhancement is produced by spallation reactions during the flare.

Acknowledgments. This work was supported by the Universidad Complutense de Madrid and the Spanish Dirección General de Investigación Científica y Técnica (DGICYT) under grant PB94-0263, and by National Science Foundation (NSF) grant AST 92-18008. We thank the Staff of McDonald Observatory for their allocation of observing time and their assistance with our observations.

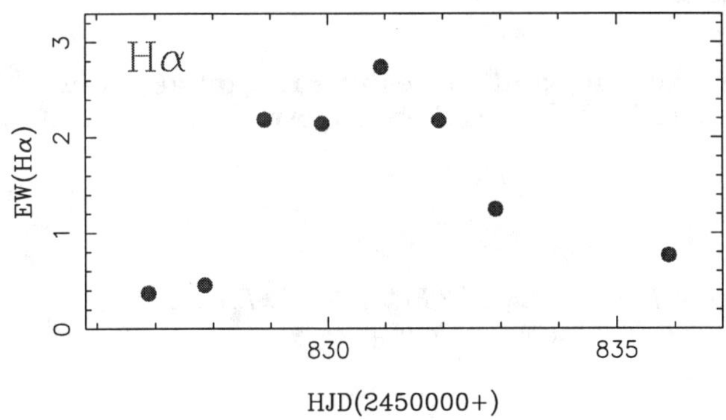

Figure 2. Temporal evolution of the Hα EW during the flare

References

Barden S.C., 1985, ApJ, 295, 162

Berdyugina S.V., Ilyin I. & Tuominen I., 1998a, in: Cool Stars, Stellar Systems, and the Sun, Tenth Cambridge Workshop, J.A. Bookbinder & R.A. Donahue, eds., ASP Conf. Ser. 155, San Francisco: ASP, CD-1477

Catalano S. & Frasca A., 1994, A&A 287, 575

Cutispoto G., Pagano I. & Rodono M., 1992, A&A 263, L3

Elvis M., Plummer D., Schachter J. & Fabbiano G., 1992, ApJS 80, 257

Endl M., Strassmeier K.G. & Kürster M., 1997, A&A 328, 565

Graffagnino V.G., Wonnacott D. & Schaeidt S.i, 1995, MNRAS 275, 129

Huenemoerder D.P. & Ramsey L.W., 1987, ApJ 319, 392

Jeffries R.D., Bertram D. & Spurgeon B.R., 1995, MNRAS 276, 397

Kürster M. & Schmitt J.H.M.M. 1996, A&A, 311, 211

Mason K.O., et al., 1995, MNRAS 274, 1194

Montes D., Sanz-Forcada J., Fernández-Figueroa M.J. & Lorente R., 1996, A&A 310, L29

Montes D., Fernández-Figueroa M.J., De Castro E. & Sanz-Forcada J., 1997, A&AS 125, 263

Montes D., Saar S.H., Collier Cameron A. & Unruh Y.C., 1998, MNRAS (in press)

O'Neal D., Saar S.H. & Neff J.E., 1996, ApJ 463, 766

Ottmann, R. & Schmitt, J.H.M.M. 1994, A&A, 283, 871

Pounds K.A., et al. 1993, MNRAS 260, 77a

Pye J.P., et al. 1995, MNRAS 274, 1165

Schachter J., 1996, ApJ 463, 747

Zirin H., 1988, in Astrophysics of the Sun, (Cambridge University Press)

Solar and Stellar Activity: Similarities and Differences
ASP Conference Series, Vol. 158, 1999
C.J. Butler and J.G. Doyle, eds.

Circumstellar Activity and Flares in FK Comae: New Results from the ESA-MUSICOS Spectrograph on the INT

J.M. Oliveira[1] & B.H. Foing[2]

ESA Space Science Department, ESTEC/SCI-SO, P.O. Box 299, NL-2200 AG Noordwijk, The Netherlands

Abstract.
We present results on the variability and phase behaviour of the Balmer lines in FK Comae. We confirm that these lines are highly variable, with excess emission that originates from extended structures and exhibit clear signs of rotational modulation. We have described the profiles of these lines with a multi-component analysis.

A large flare event lasting several days was detected in both Balmer lines and in the HeI D3 line. The energy released during this flare in Hα was $\sim 10^{37}$ erg, while the total energy in the optical region was estimated to be 10^{39} *erg*, the second largest flare energy ever observed on a cool star.

1. Introduction

FK Comae (HD 117555) is a rapidly rotating and apparently single G5 II giant. This star has an extreme rotational velocity of $v \sin i = 162.5 \pm 3.5$ km s^{-1} (Huenemoerder et al. 1993). The photometric and rotational period are 2.400 day (Jetsu et al. 1993). The Hα emission line in FK Comae is extremely broad and asymmetric and has variations on different time scales. The same characteristics are present in higher Balmer lines (like Hβ). The phase modulation in these lines indicates that the emitting material is co-rotating with the star.

2. Multi-component Analysis

In May 1997, 16 FK Comae spectra of Hα, Hβ and HeI D3 were obtained at the INT, with the ESA-MUSICOS echelle spectrograph. Part of this data set is shown in Figure 1. Additional spectra were obtained in June 1997 on the OHP.

We used a multi-component approach to model this dataset. It consists of fitting a variable number of discrete Gaussian components to each of the spectral lines, representing the emitting structures (Oliveira & Foing 1998). Figure 2

---

[1]Centro de Astrofísica da Universidade do Porto, Rua das Estrelas s/n, PT-4150 Porto, Portugal

[2]on leave from Institut d'Astrophysique Spatiale, CNRS/Univ. Paris XI , Bât. 121, Campus d'Orsay, F-91405 Orsay Cedex, France

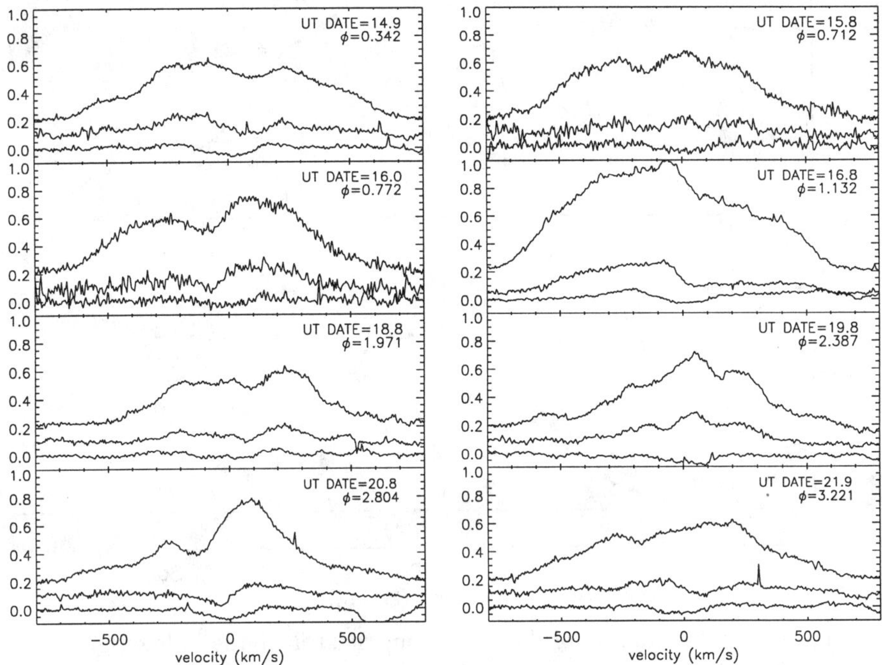

Figure 1. The Hα, Hβ and HeI D3 average spectra from May 97 (from top to bottom in each graph), after removing the rotationally broadened template. The spectra were displaced vertically for clarity. In each graph, the phase and UT date are also indicated. The ephemeris used is from Chugainov (1976): 2 442 192.345 + 2.400E. The integer rotation number ordinate describes the continuity of the data set, with the first spectrum of 14 May as reference. Velocity components are noticeable in the spectra and their variations in velocity and intensity were followed (Oliveira & Foing 1998).

shows an example of this multi-component fit, emphasizing the same components in both Balmer lines.

We also analysed the phase behaviour of these structures. The velocity curves that describe the phase behaviour of the emitting structures, as they co-rotate with the star, have equations of the type:

$$\frac{R}{R_*} \times v_{eq} \sin i \cos\theta \sin(\phi + \phi_0) = v_c \times \sin(\phi + \phi_0)$$

where v_{eq} is the equatorial rotational velocity, θ is the stellar latitude, ϕ_0 the stellar longitude and v_c is the velocity excursion in km s^{-1}.

In Figure 3, we show several plots that characterize the Gaussian components fitted to Hα and Hβ: from bottom to top, the peak intensity vs velocity,

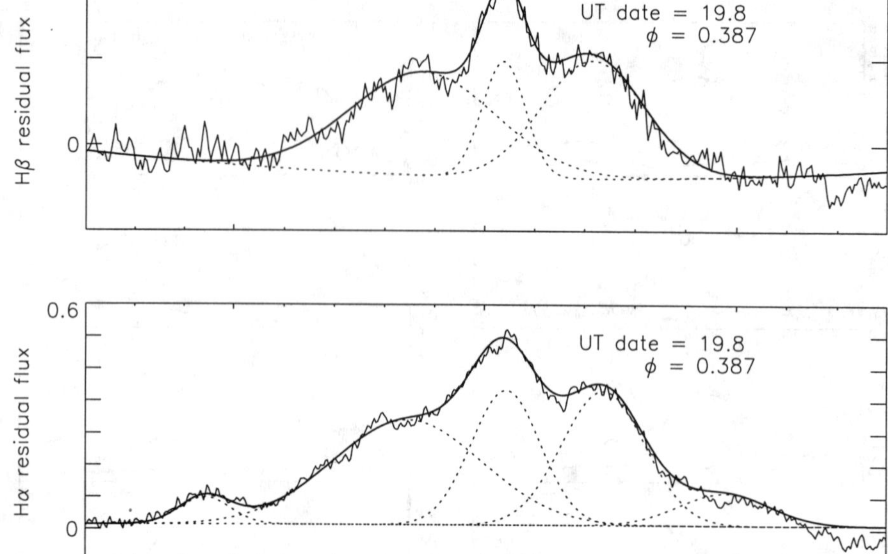

Figure 2. An example of the multi-component fit to the Balmer lines (from top to bottom Hβ and Hα). It is clear that the same components are present in both spectral lines, except for the higher velocity ones, that are masked in the noise in Hβ (Oliveira & Foing 1998).

and velocity vs phase for both lines. As can be seen in this figure, three structures could be followed in phase, corresponding to $v = 245 \times \sin(\phi + 0.62)$, $v = 285 \times \sin(\phi)$ and $v = 487 \times \sin(\phi)$, both in H$\alpha$ and Hβ. The same components and phase behaviour are observed for these two spectral lines.

3. Large Long Duration Flare detected on 16 May 97

A large flare event was detected in the Hα and Hβ lines and also in He I D3 (Oliveira & Foing 1998). The velocity curve that describes the phase evolution of the flaring structure in Hα and Hβ (Figure 3), corresponds to $v_c \sim 245$ kms^{-1} (R cos $\theta = 1.50$ R$_*$) and $\phi_0 = 0.62$. At five other phase positions the same structure is identified. This structure was first detected on May 15 and it was still present on May 21.

This flare event caused an increase in the equivalent width of 7 ± 1 Å in Hα and 1.2 ±0.2 Å in Hβ. Assuming a distance of 215 pc (Huenemoerder et al. 1993), the energy released at flare maximum is 8 ± 1 10^{31} $erg\ s^{-1}$ for Hα and 1.1 ± 0.1 10^{31} $erg\ s^{-1}$ for Hβ. The total energy released in Hα during the flare is 11 ± 3 10^{36} erg. Assuming a flare ratio UBVRI/Hα of 120 (e.g. Avrett et al. 1986), the continuum UBVRI counterpart of the FK Comae flare would amount to losses of 1.3 10^{39} erg. In comparison with the observed giant flares

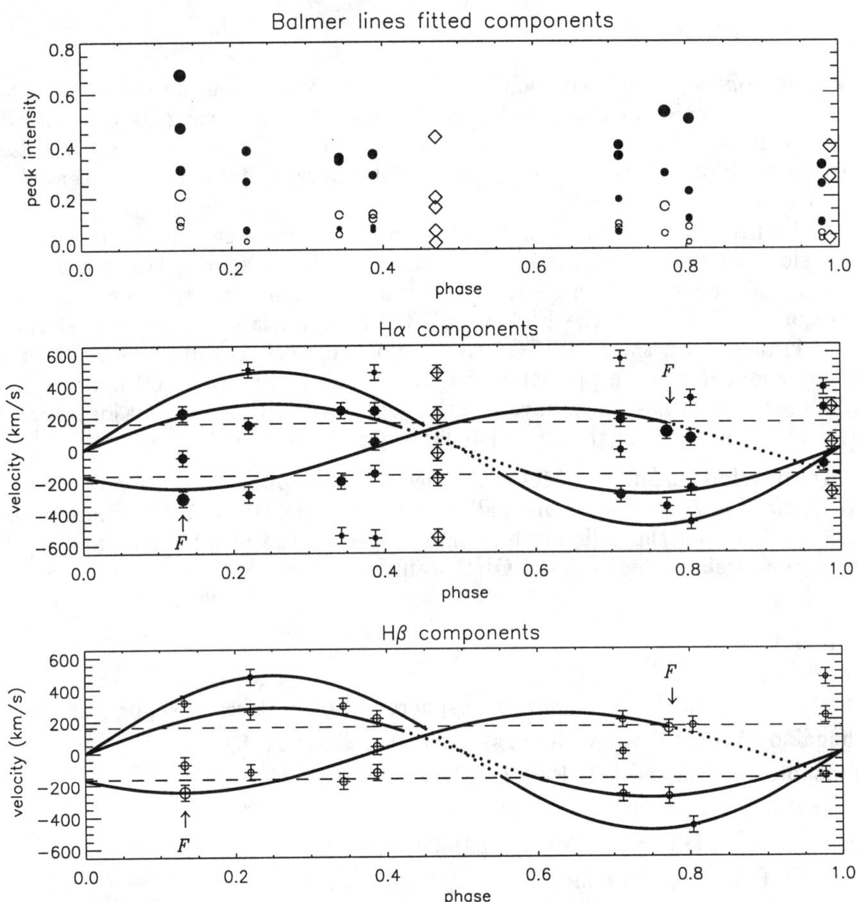

Figure 3. Phase behaviour of Balmer line components (Oliveira & Foing 1998). From top to bottom: peak intensity plotted against phase with the size of the symbols representing the corresponding intensity, • represents Hα and ∘ Hβ from INT May 97 and ⋄ Hα from the OHP June 97; velocity evolution of the Hα components with phase; velocity evolution of the Hβ components with phase. The estimated maximum error in the velocity determinations is about 50 km s^{-1}. In the bottom two graphs, three velocity curves are over-plotted, corresponding to different distances from the rotational axis and longitudes. The dotted part of these curves represents the occultation of the structures by the star. The points indicated with F are flare spectra. The dashed lines represent $\pm\ v \sin\ i$ from the rotational axis.

UBVRI radiative losses of 1.8 10^{39} *erg* and 1.2 10^{38} *erg* found respectively for YY Men (Cutispoto et al. 1992) and HR 1099 (Foing et al. 1994), this would make this flare the second largest flare energy observed.

4. Conclusions

We have applied a multi-component analysis to identify and describe the phase behaviour of emission components in Hα and Hβ. A strong flare event was detected in these two Balmer lines and in HeI D3. The total energy released during the flare in Hα is $\sim 10^{37}$ erg, one of the largest flare energy observed on a cool star.

The need is clear for a multi-site continuous data set to disentangle the intrinsic variations from rotational modulation. We have organized a multi-site campaign in February/March 1998, involving the following telescopes and observatories: INT (with the ESA-MUSICOS spectrograph), Telescope Bernard Lyot, France (with the MUSICOS spectropolarimeter), David Dunlop Observatory (Canada), Russian Special Astrophysical Observatory (SAO) and McDonald Observatory. The phase coverage thus obtained represents a major improvement in the analysis of the circumstellar emission on FK Comae.

Acknowledgments. JMO acknowledges the support of the *Fundação para a Ciência e a Tecnologia* (Portugal) under the grant BD9577/96. The authors wish to thank all the collaborators that helped in the observations and in the analysis as well as the INT and OHP staff.

References

Avrett E.H. et al, 1986,"Lower Atmosphere in Solar Flares", Neidig D.F. (ed.)
Chugainov P.F., 1976, Izv. Krymsk. Astrofiz. Obs. 54, 89
Cutispoto G., Pagano I. & Rodono M., 1992, A&A 263, L3
Foing B.H., et al., 1994, A&A 292, 543
Huenemoerder D.P. et al., 1993, ApJ 404, 316
Jetsu L., Pelt J. & Tuominen I., 1993, A&A 278, 449
Oliveira J.M. & Foing B.H., 1998, A&A accepted

Solar and Stellar Activity: Similarities and Differences
ASP Conference Series, Vol. 158, 1999
C.J. Butler and J.G. Doyle, eds.

Benchmarking the MEKAL Spectral Synthesis Code with High Resolution Solar X-ray Spectra

K.J.H. Phillips

Space Science Department, CLRC Rutherford Appleton Laboratory, Chilton, Didcot, Oxon. OX11 0QX, U.K.

L.K. Harra-Murnion

Mullard Space Science Laboratory, Holmbury St Mary, Dorking, Surrey RH5 6NT, U.K.

R. Mewe & J. Kaastra

SRON Laboratory for Space Research, Sorbonnelaan 2, 3584 CA Utrecht, The Netherlands

Abstract. Since very high resolution soft X-ray spectra of non-solar astronomical sources will soon be available, it is desirable to examine the accuracy of spectral codes. In this paper, a benchmark study of the MEKAL code, extensively used in the past for spectra from *ASCA* and other spacecraft, is presented using high resolution solar flare spectra.

1. Introduction

The next generation of space missions (NASA's Advanced X-ray Astrophysics Facility (*AXAF*) and ESA's *X-ray Multi-Mirror Mission* (*XMM*)) will provide the opportunity to measure with a resolution of $\Delta\lambda \leq 0.05$ Å in the range $\approx$ 1-140 Å.

As much of the line emission data, in particular the line wavelengths, are derived from atomic physics calculations, an observational *benchmark*, in which high-resolution spectra are compared with MEKAL synthesized spectra, is clearly desirable.

Very high-resolution soft X-ray spectra have been obtained in the past using crystal spectrometers dedicated to solar active regions and flares. Among these are spectra from the Flat Crystal Spectrometer (FCS), part of the X-ray Polychromator which was on board the *Solar Maximum Mission* spacecraft (*SMM*), which operated between 1980 and 1989. This paper is concerned with a benchmark study of the MEKAL code using solar flare spectra from the FCS instrument. Adjustments to the atomic code will be discussed in later sections.

Table 1. Examples of the corrections to the wavelengths

Ion	λ_{orig} Å	λ_{corr} Å
Fe XVIII	11.761	11.741
Fe XVII	12.134	12.124
Fe XVII	12.274	12.264
Fe XVII	13.795	13.825
Fe XVII	15.272	15.265
Fe XVII	16.796	16.780
Fe XVII	17.071	17.055
Fe XVII	17.119	17.100

2. FCS Solar Flare Spectra

The *SMM* spacecraft operated fully for nine months in 1980, from the time of launch to the time when an attitude control unit on the spacecraft failed, and from 1984 when Space Shuttle astronauts repaired the attitude control unit to 1989 when the spacecraft re-entered the Earth's atmosphere. The wavelength coverage of the Flat Crystal Spectrometer (FCS) was 1.5–20 Å.

Despite operational problems, spectra taken on two occasions are suitable for analysis and comparison with MEKAL theoretical spectra.

Flare 1 : Flare 1 was during the decay of an **M3 flare** on August 25 1980. A single long spectral scan covering the full range of the FCS was performed, starting at 13:10 U.T. and lasting for approximately 20 minutes. There is significant line emission in the range 5–20 Å which is covered by four of the seven FCS channels. A large number of the lines in the 10.5–17 Å range are due to Fe ions, with the majority due to 2–3 transitions in ions in the range Fe XVII–Fe XIX.

Flare 2 : Several short spectral scans were made during a M4.5 flare on July 2 1985 between 21:19 U.T. and 21:41 U.T., which included the peak at about 21:25 U.T. Subsequently, much higher-excitation lines are apparent in these spectra, notably in the 7.8–10 Å range which includes intense 2–4 lines in various Fe ions (predominantly from Fe XIX to Fe XXIV).

The accuracy of relative wavelengths given by (Phillips et al, 1982) is less than 0.002 Å. Absolute wavelengths were obtained using a reference line in each of the seven channels, generally the resonance line ($1s^2\,{}^1S_0 - 1s2p\,{}^1P_0$) of He-like ions, the theoretical wavelengths of which were known to 0.001 Å or less.

3. The MEKAL package

Analysis of the soft X-ray and EUV spectra from non-solar astronomical sources in the past has generally made use of the several spectral synthesis codes that are available in the literature. The MEKAL code, used via the SPEX spectral software package (Mewe et al, 1995, Kaastra et al, 1996), has been one of

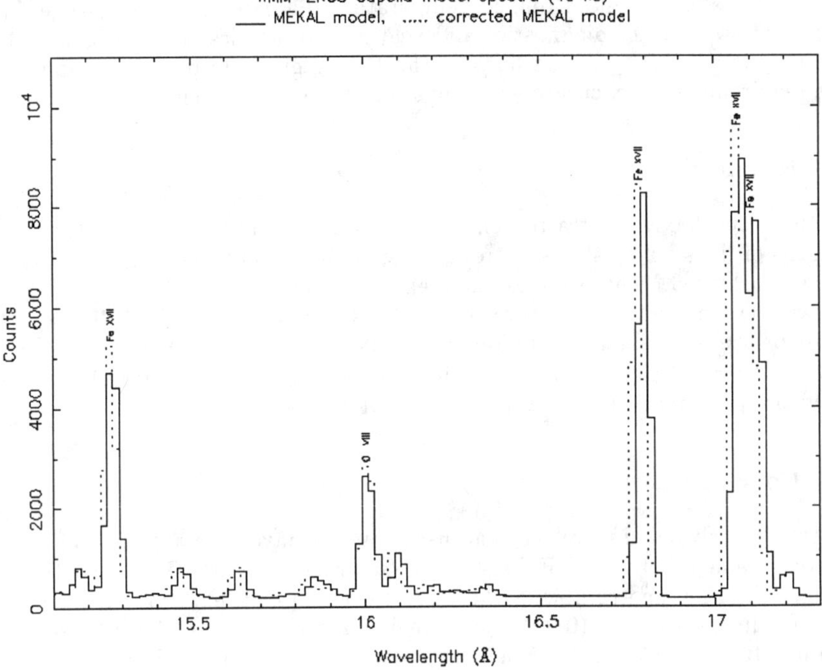

Figure 1. A simulated RGS spectrum of Capella (40 ks) calculated using the previous MEKAL model (solid) and showing the new MEKAL with the wavelength changes added (dashed). This emphasizes the impact which this work will have on future RGS analyses

the most frequently used of these, and contains a considerable amount of data relating to atomic transitions that give rise to both line and continuous spectra.

The original MEKA code (Kaastra & Mewe, 1993) was modified to allow for the fact that comparison of theoretical spectra with those observed by the *ASCA* spacecraft of the central regions of galaxy clusters (Fabian et al, 1994) showed a discrepancy with the intensity ratio of the two groups of spectral lines due to 3–2 and 4–2 transitions in various ions of Fe (specifically Fe XVII–Fe XXIV). The addition of more than 2000 lines in the 7–19 Å range using data from the HULLAC (Hebrew University/Lawrence Livermore Atomic Code) code (Klapisch et al, 1977) has enabled this discrepancy to be largely removed.

4. *XMM* RGS

XMM is designed specifically to investigate in detail the X-ray emission characteristics of cosmic sources. The wavelength band of the Reflection Grating Spectrometer (RGS) is 5-35 Å (0.35-2.5 keV). It has a resolving power of $\geq$

1000 which enables it to resolve the Fe L-shell Fe XVII–Fe XXIV line complex (10-18 Å), allowing more refined temperature diagnostics. The He-like O VII triplet allows density diagnostics and velocity diagnostics from various Fe lines are permitted. Many of the lines in the Fe-L region are resolved emphasising the requirement for accurate wavelengths in the synthetic spectra.

5. Results

We used the *SMM* spectra to benchmark the MEKAL wavelengths. The wavelengths of lines which showed strong mis-matching were changed until the best χ^2 was achieved. The most extreme examples are listed in Table 1. Figure 1 shows how important these changes are for the analysis of *XMM* RGS spectra. Most of the lines are fitted more accurately (e.g. the Mg XI forbidden line at 9.31 Å, which is blended with satellite lines j and k). This naturally produces a substantial change to the calculated DEM.

6. Conclusions

With relatively small wavelength adjustments, we have been able to achieve very good agreement between MEKAL and the solar flare spectra. This is particularly true of the Fe L-line complex, where several Fe ions have $n = 2 - 3$ transitions that fall in the 10–14 Å (0.9–1.2 keV) region and Fe $n = 2 - 4$ lines (observed in the hot July 2, 1985 flare). When the comparison is completed, we shall have a very complete line list in the X-ray spectral region 1.9 – 20 Å and the MEKAL code should give a very accurate representation of the X-ray spectra of sources in which excitation conditions (electron collisional excitation, radiative decay) are like those in solar flares.

Acknowledgments. LKH-M and KJHP thank the British Council for funding to carry out this collaboration. LKH-M acknowledges PPARC for postdoctoral funding.

References

Fabian, A. C., Arnaud, K. A., Bautz, M. W. & Tawara, Y., 1994. ApJ, 436, L63

Kaastra, J. S. & Mewe, R., 1993. Legacy, 3, 16

Kaastra, J. S., Mewe, R. & Nieuwenhuijzen, H., 1996.

Klapisch, M., Schwab, J. L., Fraenkel, J. S. & Oreg, J., 1977. J. Opt. Soc. Am., 61, 148

Mewe, R., Kaastra, J. S. & Liedahl, D., 1995. Legacy, 6, 16

Phillips, K. J. H., Leibacher, J. W., Wolfson, C. J., Parkinson, J. H., Fawcett, B. C., Kent, B. J., Mason, H. E., Acton, L. W., Culhane, J. L. & Gabriel, A. H., 1982. ApJ, 256, 774

Solar and Stellar Activity: Similarities and Differences
ASP Conference Series, Vol. 158, 1999
C.J. Butler and J.G. Doyle, eds.

On the Use of the Rise Phase in Estimating Flare Dimensions

Paweł Preś

Space Science Department, Rutherford Appleton Laboratory, Chilton, Didcot, Oxfordshire OX11 0QX, U.K.
Astronomical Institute, Wrocław University, Kopernika 11, 53-622 Wrocław, Poland

Abstract. Widely applied methods of estimating flare sizes are based on the analysis of the decay phase. Hydrodynamic models of flaring loops also show that the equivalent loop size can be estimated from the analysis of the rate of thermal energy increase during the flare rise phase. Unfortunately, the limited sensitivity of X-ray detectors allows the application of this approach only to solar flares and the most intense stellar events. This disadvantage can be avoided by converting the time evolution of energy to that of X-ray flux.

1. Introduction

It is necessary to know flare dimensions to analyse the physical conditions of flaring plasmas. Unlike solar flares, the size of which can be observed directly by imaging X-ray telescopes, the dimensions of stellar events can only be estimated with the use of flare models. Widely applied methods of estimating stellar flare dimensions are based on the analysis of the evolutional time-scale of the decay phase. It is known that for the case of solar flares the decay phase is often affected by the continued plasma heating that prolongs its evolution (cf. Sylwester et al. 1993a). If this is neglected most of the models lead to an over estimate of the loop length. Recently, Reale et al. (1997) developed a method of estimating how strongly the decay phase evolution was affected by sustained heating. This has been applied by Reale & Micela (1998) to two flares observed by *ROSAT* on AD Leo and CN Leo.

2. Use of the Rise Phase

Sylwester et al. (1993b) proposed an alternative approach to the problem of flare size determination. They showed that during the rise phase the rate of increase of the plasma thermal energy depends on the loop length. Assuming that the flare volume, V, does not change during the flare's evolution the plasma thermal energy can be analysed in terms of the 'thermodynamic measure', $\eta = T\sqrt{\mathcal{E}} = E_{th}/3k\sqrt{V}$. Here the plasma temperature, T, and emission measure, $\mathcal{E}$, are calculated from a single-temperature model fit to the observed X-ray features. They showed that for flares occurring in a loop with constant cross-section and with a constant heating rate that acts in a fraction ρ of the loop, the evolution of η during the rise phase can be well described by the simple equation

$$\eta(t) = \eta_{ss}[1 - \exp(-t\sqrt{\rho}/\tau)].$$

The characteristic energy rise time τ is very close to the thermodynamic timescale introduced by Serio et al. (1991), i.e. it is scalable with the flaring loop semi-length L, while η_{ss} scales with the loop cross-sectional area, A,

$$\begin{aligned} L/\sqrt{\rho} &= \frac{\tau/\sqrt{\rho}\, T_7^{1/2}}{120} 10^9 \text{ cm} \\ A\sqrt{\rho} &= \left(\frac{\eta_{ss}}{1.7\, 10^6 T^3}\right)^2 L/\sqrt{\rho} \text{ cm}^2. \end{aligned} \tag{1}$$

Here T_7 is the plasma temperature expressed in the units of 10^7 K. In the case stellar flares, the value of ρ remains unknown and the above equations allow us to estimate the upper limit of loop length and the lower limit of loop cross-section only. The plasma volume, $V = 2AL$, is however independent of the factor ρ.

This method allows the rise phase to be used as an independent source of estimating the plasma volume. It can describe flares under the following conditions:

1. flares occurring in a single loop or a number of loops with similar lengths and constant cross-section, and
2. flares exhibiting small variations of plasma heating during the rise phase, resulting in small variations of plasma temperature.

Considerable analysis of thermodynamic measure evolution requires frequent and accurate estimates of T and $\mathcal{E}$. Unfortunately, even recent X-ray detectors have sensitivities which are too low for this method to be effectively applied to most stellar events. This disadvantage can however be offset.

From the very start of a flare, the plasma temperature is determined by an equilibrium between the heating rate and thermal conduction cooling, because the conduction along the flare loop is very efficient (Jakimiec et al. 1992). Under these circumstances the above condition (2) involves a constant plasma temperature, $T \approx const$, so that the X-ray flux observed in any spectral feature depends only on the changes of emission measure, $F_X \propto \mathcal{E} \propto \eta^2$. In this case the increase of measured X-ray flux during the rise phase can be expressed by

$$F_X = F_{ss}\,[1 - \exp(-t\sqrt{\rho}/\tau)]^2. \tag{2}$$

This equation allows $\tau/\sqrt{\rho}$ to be estimated from analysis of the light curve. If the plasma temperature during the rise phase is known then the emission measure corresponding to the saturated level can be calculated as $\mathcal{E}_{ss} = F_{ss}/\Phi(T)$, where $\Phi(T)$ is temperature sensitivity of the analysed X-ray feature. These values allow the saturation level of thermodynamic measure, $\eta_{ss} = T\sqrt{\mathcal{E}_{ss}}$ to be calculated.

The above modification allows us to use a simple X-ray light curve instead of a more complicated spectral analysis. An important assumption of constant plasma heating can be checked by performing spectral analyses just a few times during the rise phase.

3. Application

Here we illustrate the method for the case of two previously analysed stellar events observed by the ME instrument on board the *EXOSAT* satellite. The flare on Algol on 19 August 1983 was described by Oord & Mewe (1989) and the flare on σ^2 CrB on 29 September 1983 was analysed by Oord, Mewe & Brinkman (1988). Figure 1 shows fits of the Eq. (2) to the observed ME light curves.

Both flares showed small variations of plasma temperature during the rise phase making them suitable for this analysis. For the Algol flare the plasma temperature varied in the narrow range 70 to 80 MK. Fitting Eq. (2) to the rise phase light curve gives $\tau/\sqrt{\rho} = 20 \pm 1.4$ minutes and $F_{ss} = 11.1 \pm 0.5$ c/s. Assuming the plasma temperature $T = 70$ MK these parameters give the upper limit on the loop length $L/\sqrt{\rho} = 2.6\,10^{10}$ cm. During the rise phase of the σ^2 CrB flare, the plasma temperature varied slightly in the range 60–64 MK. The characteristic energy rise time obtained for this flare is only $\tau/\sqrt{\rho} = 2.8 \pm 0.8$ min, resulting in a loop semi-length $L/\sqrt{\rho}$ of only $4.2\,10^9$ cm.

The upper limits of loop lengths obtained here are less than estimates obtained with the application of previous methods (see Table 1). The flaring loops lengths of Algol and σ^2 CrB event are estimated here to be at least 3 and 5 times shorter respectively. The flare volumes however do not differ significantly. As a result, estimations of plasma density and pressure are similar to those obtained previously.

The discrepancy between the loop length estimates can be explained by the distinct influence of prolonged heating during the decay phase. As Oord & Mewe (1989) showed the Algol flare decayed through the quasi-stationary process which occurs when the plasma heating decays with the time-scale longer than the thermodynamic time-scale. In this case the flare's evolution can be prolonged many times compared to undisturbed evolution. Addition of measurement uncertainties can make flare decay analysis unsuitable for an estimation of its size (Reale et al. 1997).

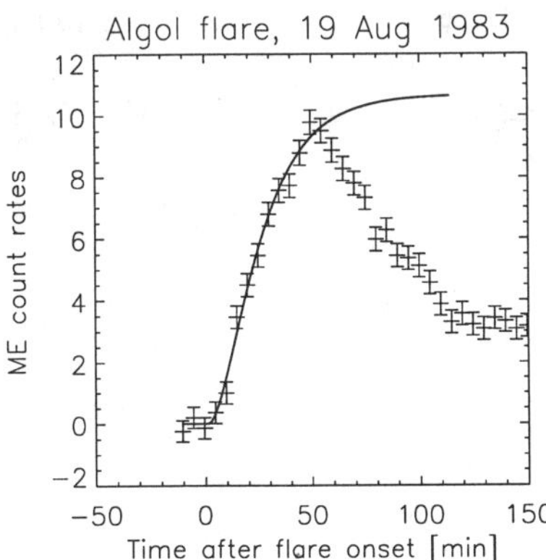

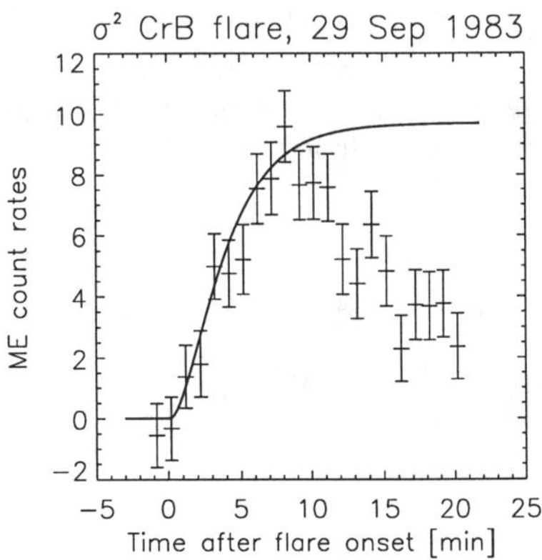

Figure 1. Fits of the Eq. (2) to the rise phase light curves of two flares observed by the ME instrument aboard the EXOSAT satellite

Table 1. Comparison of flare dimensions with previous estimates and list of physical parameters

Star	Date	L [cm]	V [cm^3]	$L/\sqrt{\rho}$ [cm]	V [cm^3]
		Previous results		This paper	
Algol	19/08/83	$8.0\,10^{10}$	$1.4\,10^{31}$	$2.6\,10^{10}$	$1.9\,10^{31}$
σ^2 CrB	29/09/83	$2.3\,10^{10}$	$6.9\,10^{29}$	$4.2\,10^{9}$	$4.9\,10^{29}$

4. Summary

We have proposed a modification of the Sylwester et al. (1993b) method, which extends the range of methods used to estimate the stellar flare sizes. This allows both the rise and decay phases to be used independent of flare size determination. The method is especially valuable when a flare decays through the quasi-stationary cooling and the loop size estimated on the basis of the decay phase analysis is problematic.

References

Jakimiec, J., Sylwester, B., Sylwester, J., Serio, S., Peres, G. & Reale F. 1992, A&A, 253, 269

Oord, G.H.J, R. Mewe, R. & Brinkman, A.C. 1988, A&A, 205, 181

Oord, G.H.J & Mewe, R. 1989 A&A, 213, 245

Reale, F., Betta, R., Peres, G., Serio, S. & McTiernan, J. 1997, A&A, 325, 782

Reale, F. & Micela, G. 1998 A&A, 324, 1028

Serio, S., Reale, F., Jakimiec, J., Sylwester, B. & Sylwester, J. 1991 A&A, 241, 202

Sylwester, B., Sylwester, J., Serio, S., Reale, F., Bentley, R.D. & Fludra, A. 1993a, A&A, 267, 586

Sylwester, J., Sylwester, B., Jakimiec, J., Garcia, H., Serio, S. & Reale, F. 1993b, Adv. Space Res., Vol. 13, No. 9, 307

Solar and Stellar Activity: Similarities and Differences
ASP Conference Series, Vol. 158, 1999
C.J. Butler and J.G. Doyle, eds.

Light Curve Variation and Flare Eruption in the Sun and Stars

L. Teriaca

Armagh Observatory, College Hill, Armagh, BT61 9DG, N. Ireland

S. Catalano

Osservatorio Astrofisico di Catania, Viale A. Doria 6, 95125 Catania, Italy

Abstract. The relation between light curve (sunspot number) changes and flare eruption in some of the most active RS CVn binary systems (and the Sun) is analysed. We have studied the last 20 years of light curves of some of the most active RS CVn binary stars, together with extensive bibliographic research of all the flares observed on these systems. In order to obtain an estimate of the variation of the spot area and/or spot configuration, associated with flare eruptions, we perform spot modelling analysis of the photometric data. We show that a correlation between the flare importance (energy released) and the percentage variation of spot area (ΔA/A) is definitely present for II Peg, and seems to be present for other active binary stars. We also analyse the occurrence of the strongest solar flares against the sunspot number and the C II flux measured by the *UARS SOLSTICE* instrument for the period 31 July 1991 - 9 October 1995.

1. Introduction

In the last few years, the idea that solar and stellar magnetic activity are different aspects of the same phenomenon was widely demonstrated (see papers in Byrne & Rodonó, 1983) and today the *Solar Stellar Connection* is one of the most important tools in the study of solar and stellar atmospheres. In Figure 1, we present a comparison of solar and stellar magnetic diagnostics. It is interesting to observe the high correlation between sunspot number (spot coverage) and C II flux (chromospheric emission) that is also observed in active binary stars as an anti-correlation between Hα emission and V photometry (Catalano et al. 1996). In fact, in these stars, the modulation in the V-band is due to dark spots of magnetic origin on the surface of the active star in analogy with the solar case (Eaton & Hall, 1977; Rodonó et al., 1987). Here we show the correlation between flares and light curve changes in active binary stars (RS CVn stars) and analyse the situation on the Sun, where the Sunspot number and the C II flux are used as a proxy of the active region coverage.

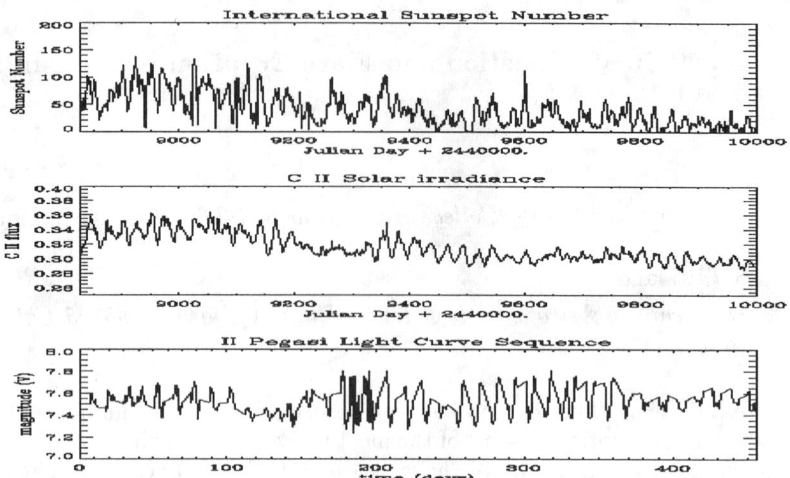

Figure 1. Comparison of solar and stellar activity diagnostics: the Wolf sunspot number (upper panel), the solar C II flux (middle) and (lower) II Peg light curve evolution (after removing time gaps), vs time. Notice the similarity of the light curve amplitude variation.

2. Active Binary Stars

In this section we show some examples of relationships between light curve changes and flares in some active binary stars. In Figure 2 we show the V light curve (Strassmeier 1989) of IM Pegasi during May – July 1985. In this period an UV flare was detected (Buzasi et al., 1989). It is possible to observe a deeper minimum $\sim$ 20 days (0.82 rotations) before the flare eruption. Afterwards, the light curve returns substantially to the previous level and remains unchanged for, at least, 3 rotations. The observed flare seems to be associated with a growing spot and emergence of new magnetic flux, that it is evolving in 2 or 3 rotations. Using a spot modelling technique, we calculate the spot configuration for rotations (a) and (b) of Figure 2.

The same technique was applied to other active binary stars for which flares observations and simultaneous photometry were available. We have examined six cases for II Peg and extended the analysis to UX Ari (four cases), HK Lac and HR 1099 (one case). Particularly relevant is the case of the large flare on HK Lac (1.28 10^{37} ergs in Hα) in September 1989 (Catalano & Frasca, 1994). The flare was preceded by the appearance of a secondary minimum in the V light curve, i.e. the birth of a new spot group.

3. The Sun

We report the strongest X-ray flares ($E_X > 0.2$ ergs cm^{-2} s^{-1}) obtained from Mees Solar Observatory, together with the sunspot number for the period from October 1st 1991 to August 1st 1993. No particular correlation between flares and variation of the sunspot number, from one rotation to another, is observed.

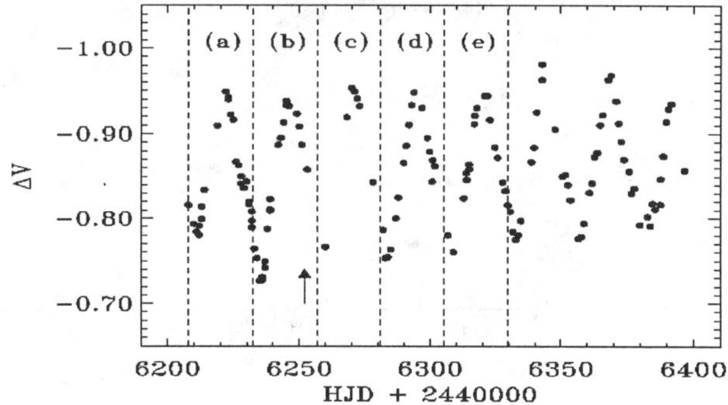

Figure 2. IM Peg light curve from 22 May 1985 to 1 December 1985. Vertical dashed lines marks successive rotations, adopting a photometric period of 24.35 days. The arrow indicate the time of flare eruption.

4. Discussion

Solar flares originate when a new emerging flux tube interact with the pre-existing field (Heyvaerts et al., 1977) and/or when photospheric motions induce an excess of energy (with respect to the current-free configuration) that can be released by magnetic reconnection (Klimchuk et al., 1988). Flare eruption takes place shortly after ($1-2$ days) the emergence of new flux tubes within an *Activity Complex*. The last structures are refreshed by injection of new magnetic flux in the form of bipolar active regions (Parker, 1987).

Magnetic spots are a signature of the presence of a magnetic field, and their evolution is the evidence of the evolution of the magnetic field configuration. To estimate the importance of the variation in the configuration of the magnetic field before and after flare eruption, we calculate, from spot modelling, the quantity:

$$\Delta = \frac{A_{post} - A_{pre}}{A_{pre}} \tag{1}$$

where A_{pre} and A_{post} is the area *pre*– and *post*–flare of the spot that shows the strongest variation. We also estimate the total energy involved in each stellar flare studied. In Figure 3 the total energy E_{Tot} versus Δ for II Pegasi is reported. For other stars we do not have an equally significant number of events, but there is evidence that low energy flares are associated with small (or undetectable) variation of the spot configuration and high energy flares with a strong variation of the spot area. In the solar case we do not observe any correlation. This is in agreement with the low level of activity of the Sun respect to RS CVn binary stars.

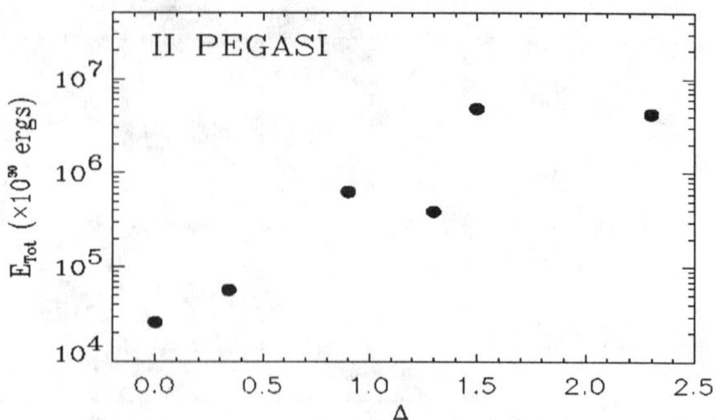

Figure 3. Total flare energy versus percentage variation of the starspot area for II Pegasi.

5. Conclusion

A clear correlation between flare energy and percentage variation of sunspot area is found for II Pegasi and is confirmed by the analysis of other RS CVn stars. On the Sun, we don't observe any clear correlation.

Acknowledgments. Research at Armagh Observatory is grant-aided by the Dept. of Education for N. Ireland while partial support for software and hardware is provided by the STARLINK Project which is funded by the UK PPARC. This work has been supported by the Italian *Ministero dell' Università e della Riscerca Scientifica e Tecnologica,* the *Gruppo Nazionale di Astronomia* of the CNR and by the *Regione Sicilia* who are gratefully acknowledged. We would like to thank the Mees Solar Observatory, University of Hawaii which is supported by NASA grant NAG 5-4941 and NASA contract NAS8-40801.

References

Buzasi D. L., Ramsey L. W. & Huenemoerder D. P., 1987, ApJ 322, 353

Byrne P. B. & Rodonó M., (Eds.) 1983, Activity in Red-Dwarf Stars, IAU Colloquium 71, Reidel, Dordrecht

Catalano S. & Frasca A., 1994, A&A 287, 575

Catalano S., Rodonó M., Frasca A. & Cutispoto G. 1996, in *Stellar surface structure* - IAU Symp.No. 176, K. G. Strassmeier & J. L. Linsky (eds), Kluwer Academic Publisher, Dordrecht, p. 403

Eaton J. A. & Hall D. S., 1977, ApJ 227, 907

Heyvaerts J. et al., 1977, ApJ 216, 123

Klimchuk J. A. et al., 1988, ApJ 335, 456

Parker E. N., 1987, ApJ 312, 868

Rodonó M. et al., 1987, A&A 176, 267

Strassmeier K. G., Hall D. S., Boyd L. J. & Genet R. M., 1989, ApJs 69, 141

Part 5

CHROMOSPHERIC DYNAMICS

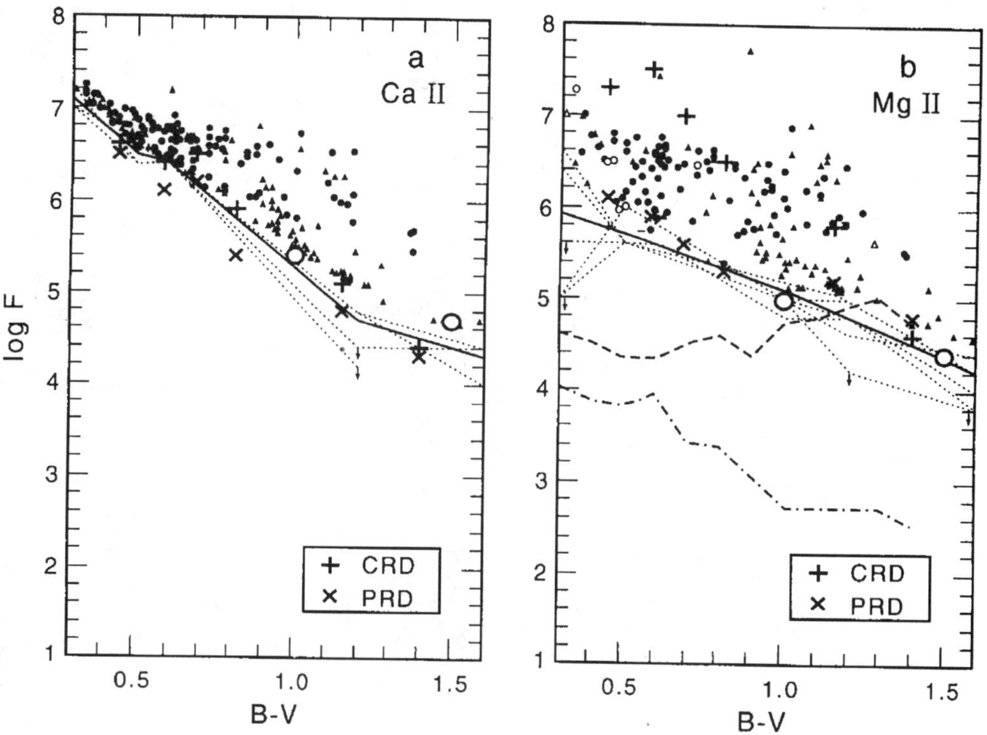

Theoretical and empirical chromospheric emission fluxes in the Ca II and Mg II lines versus colour for late-type stars.

Solar and Stellar Activity: Similarities and Differences
ASP Conference Series, Vol. 158, 1999
C.J. Butler and J.G. Doyle, eds.

Dynamics of the Quiet Solar Chromosphere

Robert J. Rutten

Sterrekundig Instituut, Utrecht, The Netherlands

Bruce W. Lites

High Altitude Observatory, NCAR, Boulder, USA

Thomas E. Berger & Richard A. Shine

Lockheed-Martin Solar and Astrophysics Lab, Palo Alto, USA

Abstract. The solar chromosphere has never been static although it was often modeled so. Even the quiet-sun internetwork chromosphere has become thoroughly dynamic with the acoustic shock interpretation of the Ca II K_{2V} grains. We concentrate on the latter in this brief review. Recent analysis of Advanced Stokes Polarimeter (ASP) data confirms that their excitation is more likely set acoustically than magnetically. *TRACE* imagery permits seeing-free studies of their occurrence patterns.

1. Introduction: Network and Internetwork

Figure 1 is a cartoon paradigm taken from Judge & Peter (1998). It depicts a vertical slice through the quiet solar photosphere and low chromosphere across a single super-granular cell. Clusters of strong-field magnetic elements ("flux-tubes") sit at the cell borders and constitute the "magnetic network" seen on photospheric magnetograms. Due to the relative increase of magnetic pressure over gas pressure, their magnetic fields expand higher up and become space-filling in the chromosphere, combining into a "magnetic canopy" overlying the cell interior. Chromospheric heating (dark shading) occurs presumably preferentially along the cluster axes since the latter show up as bright 1–3 arcsec grainy patches on images taken in Ca II H&K, Ly α, and other ultraviolet lines up to a considerable ionization temperature. These grainy patches constitute the "chromospheric network" which coincides with the magnetic network and outlines, although incompletely, the super-granulation cell boundaries.

The remaining quiet-Sun area beside the network grains is called the internetwork. It intermittently contains bright features of its own, in particular the so-called Ca II K_{2V} grains. They have been given this name because their intensity peaks dramatically just blueward of the Ca II H & K line centers. However, they are also present, at lower contrast and a slight phase shift, on wider-band Ca II K filtergrams such as the ones in Figure 4 and on wide-band near-UV images from *TRACE* such as the ones in Figure 5 (cf. Rutten et al. 1999). We therefore call them "internetwork grains".

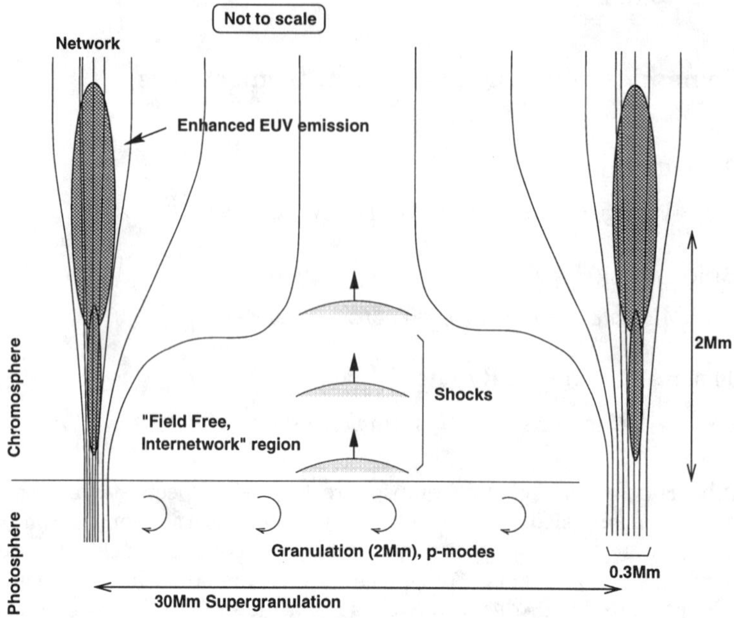

Figure 1. Schematic vertical section through a super-granulation cell with the magnetic network at its borders. From Judge & Peter (1998).

2. Fourier and Grain Signatures

Numerous studies (*e.g.,* Jensen & Orrall 1963, Liu & Sheeley 1971, Lites et al. 1993) have shown that the chromospheric network is characterized Fourier-wise by power at five minutes and longer periodicities, whereas the internetwork modulation tends to peak between five and two minutes. This fact is currently being rediscovered by SUMER spectroscopists. It is illustrated particularly well by spatial Fourier charts derived from *TRACE* image sequences (Figure 2). The network stands out bright in single images and in low-frequency modulation maps, but the contrast with the internetwork modulation amplitudes reverses at higher frequencies.

It is likely that the slow network modulation reflects footpoint buffeting of the magnetic elements by granular dynamics (*e.g.,* Kneer & von Uexküll 1986, Wellstein et al. 1998) rather than an intrinsically oscillatory phenomenon, although long-period waves may play a role at the level of the chromosphere (Lites et al. 1993). In the photosphere, the buffeting is vividly demonstrated by the bright point dynamics in the high-resolution G-band image sequence of Berger & Title (1996) and indeed produces characteristic splitting and merging times per magnetic element of five to seven minutes (Berger et al. 1998).

For the internetwork modulation, commonly called the "chromospheric three-minute" oscillation, the intrinsic oscillatory nature is much more obvious. The best demonstration is obtained by downloading a *TRACE* near-UV image sequence (for example the data taken May 12, 1998 UT 14:31–16:00, of which a

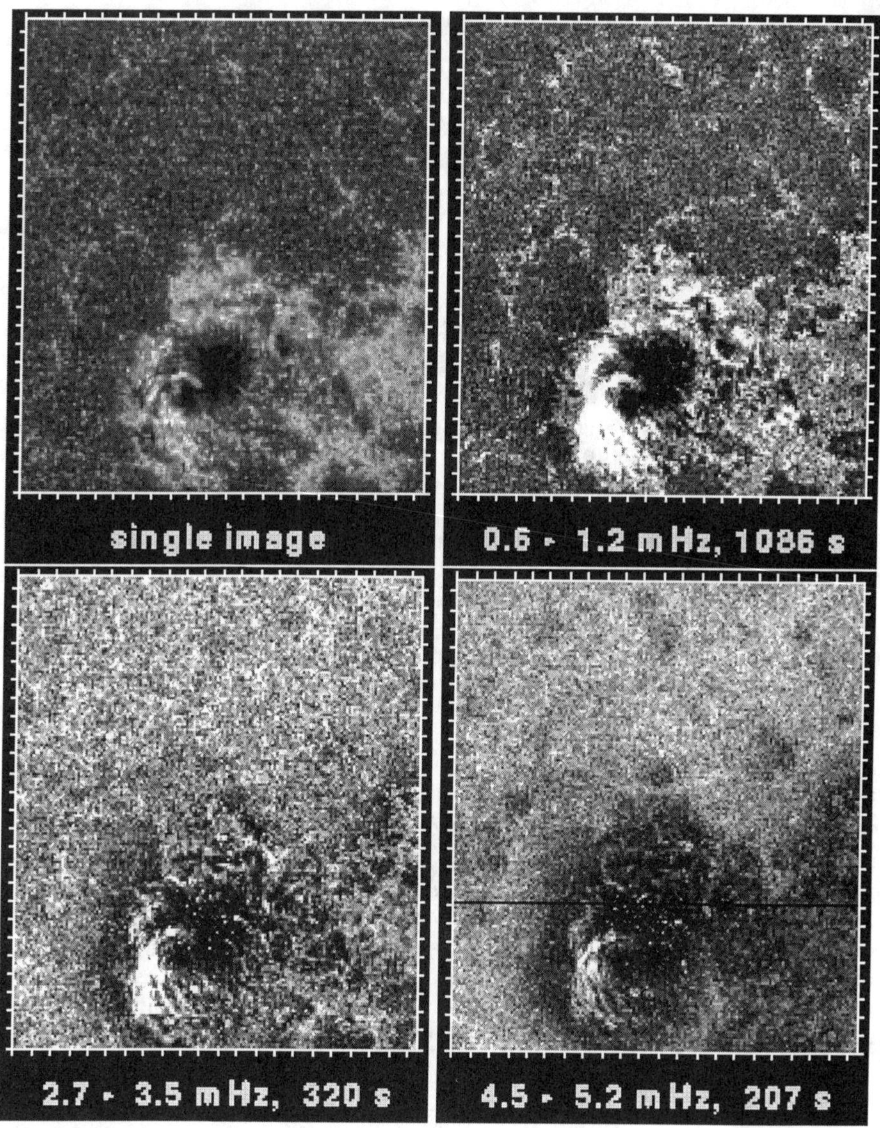

Figure 2. Fourier modulation maps constructed from a *TRACE* 1600 Å image sequence of AR8210 obtained during UT 08:30–11:58 on May 1, 1998. The tick marks are 10 arcsec apart. The first panel contains a sample image. The other panels display relative intensity power per pixel in the indicated frequency bin. A surge occurred during the sequence, producing relatively high power at all frequencies. Penumbral waves cause enhanced power in the second panel. Plage and network have larger power than present in the internetwork at low frequency, but the power contrast flips at higher frequencies where the internetwork modulation gains dominance.

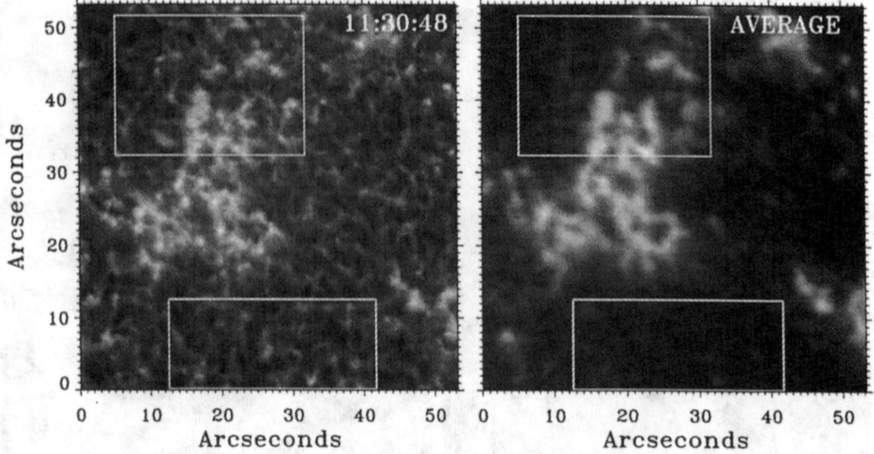

Figure 3. Left: high-resolution Ca II K filtergram taken with the Swedish Vacuum Solar Telescope at La Palma (Loefdahl et al. 1998). Right: 70-min temporal average of the same area, demonstrating the rapid variation of the internetwork brightness pattern (lower box) by showing considerable suppression of its time-averaged contrast relative to the network. From Lites et al. (1999).

small part is shown in Figure 5) and to watch the rapid internetwork morphology changes with time. If such a movie is played fast, the network grains appear as islands of stability in a sea of motion made up of rapidly evolving, spidery patterns that often display apparent supersonic motion and that are very suggestive of wave interference.

Figures 3–5 illustrate the fast internetwork morphology changes and the appearance of internetwork grains as localized pattern enhancements that flash on and off a few times at 2–4 min intervals, frequently in pairs (Figure 4). In addition to these ubiquitous enhancements, "persistent flashers" may occasionally migrate through an internetwork area. They probably represent the chromospheric signature of a newly-emerged strong-field ephemeral region on its way to become part of the network (Brandt *et al.* 1992, 1994).

3. Internetwork Grain Formation and Excitation

The acoustic shock simulations of Carlsson & Stein (1997) have crowned a long sequence of efforts to model chromospheric oscillations, in particular the spectral formation of K_{2V} grains (Athay 1970, Cram 1972, Liu & Skumanich 1974, Mein et al. 1987, Leibacher et al. 1982; Rammacher & Ulmschneider 1992; Fleck & Schmitz 1991b, 1993; Kalkofen et al. 1994; Sutmann & Ulmschneider 1995a, 1995b). The extensive literature on K_{2V} grains has been reviewed by Rutten & Uitenbroek (1991) and Rutten (1994, 1995, 1996). Carlsson and Stein's detailed reproduction of observed spectrally-resolved H_{2V} grain development has ascertained beyond doubt that the grains are caused by weak acoustic shocks

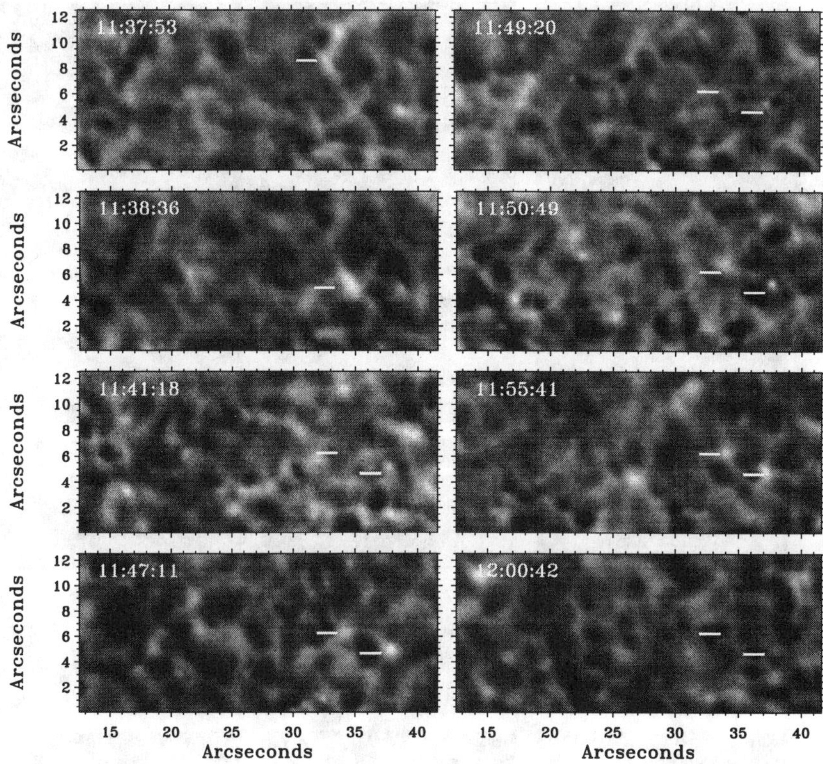

Figure 4. Selected frames from the La Palma Ca II K filtergram sequence of Loefdahl et al. (1998). The area corresponds to the internetwork region outlined by the lower box in Figure 3. The markers in the first two frames point at a brightening which travels along a pattern ridge with apparent supersonic motion (70 $km\ s^{-1}$). The two markers in the other panels (at fixed locations) identify two concurrent recurrent K_{2V} grains which brightened, in phase, a few times at 2–3 min intervals. From Lites et al. (1999).

which propagate upwards through the low chromosphere. The marked H & K profile asymmetry (absence of concurrent H_{2R} peaks) results from substantial line-center opacity shifts caused by higher-located matter that falls back after having been kicked up by earlier shock passages.

Questions yet open are whether internetwork grains appear preferentially at specific places betraying some piston mechanism, to what extent the shocks may be diagnosed in higher layers, and whether the shocks play some significant role, for example in chromospheric heating or in the FIP element segregation observed in the fast solar wind. The issue of acoustic internetwork heating is addressed by Ulmschneider and by Cuntz elsewhere in these proceedings. For the FIP flip issue see Rutten (1998) and other reviews in Fröhlich et al. (1998).

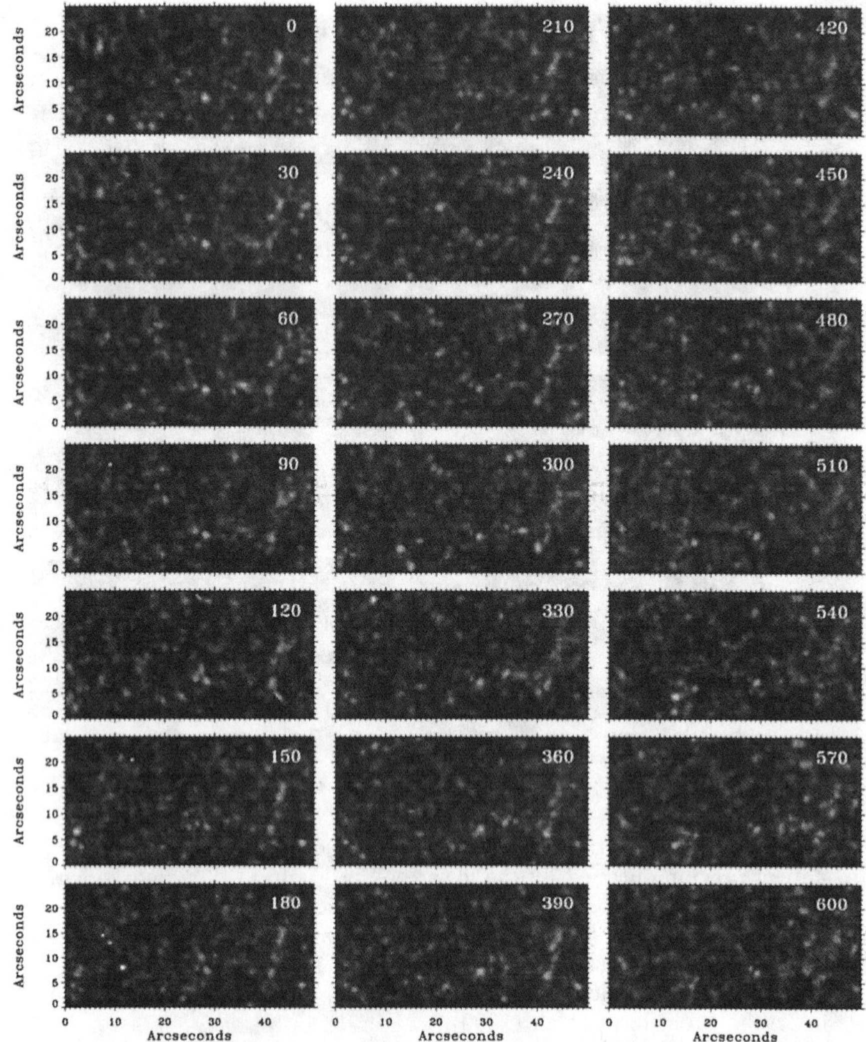

Figure 5. Cutouts from a *TRACE* 1550 Å image sequence (May 12, 1998 UT 15:00:26–15:10:28) illustrating quiet-sun internetwork behavior. The passband contains C IV lines that brighten at magnetic activity and C I lines that provide internetwork grain signatures. The numbers specify elapsed time in seconds.

The question as to what height the shocks penetrate recognizably is a matter of debate in the current literature, mostly on the basis of SUMER spectrometry (*e.g.,* Steffens et al. 1997, Judge et al. 1997, Carlsson et al. 1997, Curdt & Heinzel 1998). Much of the latter is hampered by insufficient solar area, temporal extent, spectral signal and line formation variety, but it is nevertheless clear that the internetwork scene is less clear in higher-formed UV lines than it is in H & K.

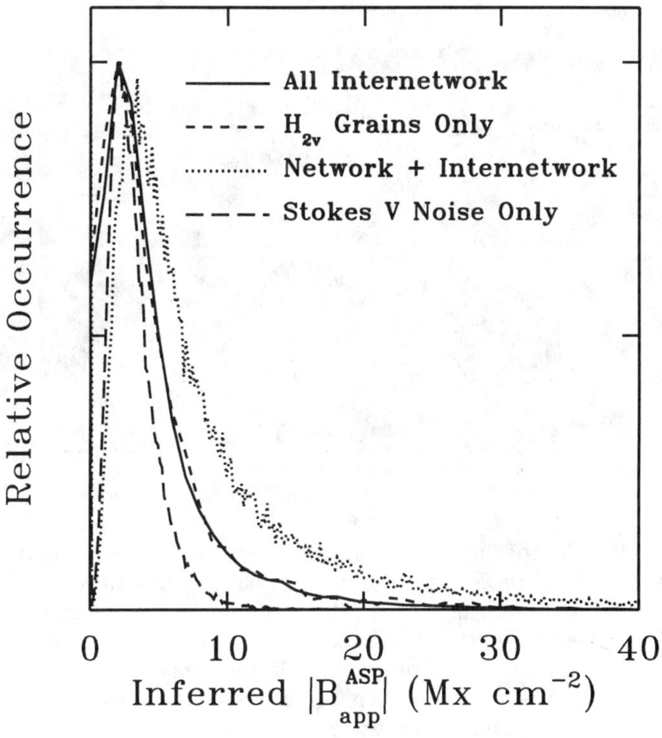

Figure 6. Histograms of the occurrence of weak internetwork fields measured with the Advanced Stokes Polarimeter at the NSO/Sacramento Peak Dunn Telescope. The quantity $|B_{\rm app}^{\rm ASP}|$ represents the apparent flux density derived from profile-integrated Stokes V measurements of the two Fe I 6302 Å lines. The solid curve specifies the flux density distribution for all of an internetwork area, whereas the similar dashed distribution holds for only those space-time locations where a Ca II H_{2V} grain appeared, as measured on simultaneous co-spatial Ca II H spectrograms. The dotted curve includes strong-field network elements (they do not produce kilogauss apparent flux density because the magnetic elements are not resolved; the difference between the measured apparent flux density and the actual intrinsic field strength $B \approx 1400$ Gauss is largely a measure of spatial filling factor within the seeing-smeared and instrument-dependent resolution element). The narrowest distribution is a noise estimate derived from the continuum which has no intrinsic Stokes V signal. Its peak defines the noise threshold of these data. From Lites et al. (1999).

This was already established from rocket data (Hoekzema et al. 1997). Part of the confusion may lie along the line of sight. The Ca II H_{2V} and K_{2V} grain emission senses the upcoming shocks just barely, when they are still quite weak and relatively regular, because the H & K line source functions are dominated by

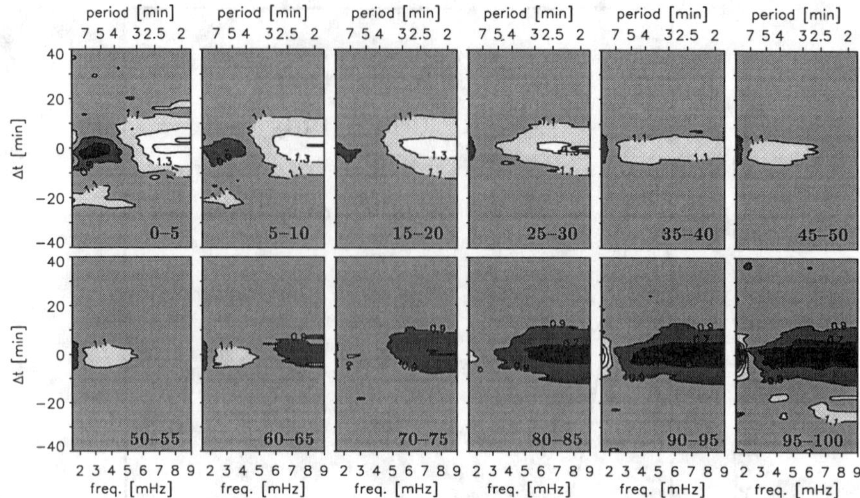

Figure 7. Contour maps measuring the amount of spatial correlation between the brightness of the granulation pattern and the occurrence of large oscillation amplitude, as functions of Fourier frequency (lower abscissae; corresponding periods along the top) and time delay between brightness sampling and Fourier amplitude measurement (ordinate). The time delay is positive when brightness is sampled after Fourier amplitude. The panels divide the granulation brightness range into 5% bins, the top-left one being for image pixels in the darkest intergranular lanes, the bottom-right one for the brightest 5% of all pixels. Contour values exceeding unity (bright) imply that pixels with Fourier amplitude over twice the average, at the given frequency, are preferentially co-aligned with pixels of the given brightness class. The contour value unity (grey) implies absence of systematic co-alignment. Values below unity (dark) imply negative alignment (spatial avoidance). The bright blob at 2–3 min period in the upper-left panel implies that high-frequency waves are preferentially found at the very darkest features. From Hoekzema et al. (1998).

scattering in the higher layers where vertical shock interference produces more irregularity. It is also likely that high-frequency shock sequences interfere with the three-minute ones to a considerably larger extent than simulated by Carlsson & Stein (Ulmschneider, these proceedings). In addition, shock signatures are very much different between optically thick and optically thin line formation. Finally, it is likely that the shock trajectories are not radial as is assumed in 1D modelling or in pixel-by-pixel time sequence and temporal Fourier analyses. The frequent occurrence of K_{2V} grains in synchronized pairs suggests the presence of horizontally spreading disturbances.

The issue of whether internetwork grains betray specific pistons is yet open. A long-standing debate concerns the claim by Sivaraman & Livingston (1982) that K_{2V} grains correlate one-to-one with the presence of enhanced magnetic field (*e.g.*, Rutten & Uitenbroek 1991, Kneer & Von Uexkull 1993, Von Uexkuell & Kneer 1995, Steffens et al. 1996, Hofmann et al. 1996, Remling et al. 1996, Nindos & Zirin 1998, Wellstein et al. 1998). Recently, this claim has been tested directly by Lites et al. (1999) by repeating the measurements of Sivaraman & Livingston (1982) at higher precision and sensitivity. A key result is shown in Figure 6. It indicates that, down to the measurement sensitivity of 3 Mx cm^{-2} in apparent flux density, there is no correlation between the presence of H_{2V} grains and enhanced Stokes V amplitude. A similar absence of significant correlation was found between H_{2V} grains and horizontal internetwork fields diagnosed from linear polarization (Lites et al. 1996). Thus, the occurrence of internetwork grains does not seem to depend on magnetism, apart from the rare persistent flashers.

It seems most likely that turbulent convection supplies the pistons that excite shock sequences producing bright internetwork grains. In particular, the grains may mark locations where "acoustic events" and "intergranular holes" follow on granular collapse (cf. Restaino et al. 1993; Rimmele et al. 1995; Rast 1995; Roudier et al. 1997; Hoekzema & Rutten 1998; Hoekzema et al. 1998; Goode et al. 1998). Figure 7 from Hoekzema et al. (1998) demonstrates in statistical fashion that high Fourier amplitudes in the 2–3 min period range are found preferentially above the darkest features in the granulation. These correspondence charts have been derived from white light images and therefore contain photospheric signal only. Combination of cospatial *TRACE* white light and near-UV image sequences will enable occurrence correlation between chromospheric internetwork grains and photospheric morphology and flow patterns down to granular scales, with sufficiently large data sets to assess the significance of any relation. The internetwork may so become the first domain in which chromospheric and photospheric dynamics are tied together in detail.

References

Athay R. G., 1970, Solar Phys. 11, 347

Berger T. E., Loefdahl M. G., Shine R. S. & Title A. M., 1998, ApJ 495, 973

Berger T. E. & Title A. M., 1996, ApJ 463, 365

Brandt P. N., Rutten R. J., Shine R. A. & Trujillo Bueno J., 1992, in M. S. Giampapa & J. A. Bookbinder (eds.), Cool Stars, Stellar Systems, and the Sun, Proc. Seventh Cambridge Workshop, Astron. Soc. Pac. Conf. Series 26, p. 161

Brandt P. N., Rutten R. J., Shine R. A. & Trujillo Bueno J., 1994, in R. J. Rutten, C. J. Schrijver (eds.), Solar Surface Magnetism, NATO ASI Series C 433, Kluwer, Dordrecht, p. 251

Carlsson M., Judge P. G. & Wilhelm K., 1997, ApJ 486, L63

Carlsson M. & Stein R. F., 1997, ApJ 481, 500

Cram L. E., 1972, Solar Phys. 22, 375

Curdt W. & Heinzel P., 1998, ApJ 503, L95

Fleck B. & Schmitz F., 1991, A&A 250, 235

Fleck B. & Schmitz F., 1993, A&A 273, 671

Fröhlich C., Huber M. C. E., Solanki S. & von Steiger R. (eds.), 1998, Solar Composition and its Evolution – from Core to Corona, Procs. ISSI Workshop, Space Sci. Rev., in press

Goode P. R., Strous L. H., Rimmele T. R. & Stebbins R. T., 1998, ApJ 495, L27

Hoekzema N. M., Brandt P. N. & Rutten R. J., 1998, A&A 333, 322

Hoekzema N. M. & Rutten R. J., 1998, A&A 329, 725

Hoekzema N. M., Rutten R. J. & Cook J. W., 1997, ApJ 474, 518

Hofmann J., Steffens S. & Deubner F. L., 1996, A&A 308, 192

Jensen E. & Orrall F. Q., 1963, PASP 75, 162

Judge P., Carlsson M. & Wilhelm K., 1997, ApJ 490, L195

Judge P. G. & Peter H., 1998, in C. Fröhlich, M. C. E. Huber, S. Solanki & R. von Steiger (eds.), Solar Composition and its Evolution – from Core to Corona, Procs. ISSI Workshop, Space Sci. Rev., in press

Kalkofen W., Rossi P., Bodo G. & Massaglia S., 1994, A&A 284, 976

Kneer F. & von Uexküll M., 1986, A&A 155, 178

Kneer F. & Von Uexkull M., 1993, A&A 274, 584

Leibacher J., Gouttebroze P. & Stein R. F., 1982, ApJ 258, 393

Lites B. W., Leka K. D., Skumanich A., Martinez Pillet V. & Shimizu T., 1996, ApJ 460, 1019

Lites B. W., Rutten R. J. & Berger T. E., 1999, ApJ submitted

Lites B. W., Rutten R. J. & Kalkofen W., 1993, ApJ 414, 345

Liu S. Y. & Sheeley N. R., 1971, Solar Phys. 20, 282

Liu S.-Y. & Skumanich A., 1974, Solar Phys. 38, 105

Loefdahl M. G., Berger T. E., Shine R. S. & Title A. M., 1998, ApJ 495, 965

Mein P., Mein N., Malherbe J. M. & Damé L., 1987, A&A 177, 283

Nindos A. & Zirin H., 1998, Solar Phys. 179, 253

Rammacher W. & Ulmschneider P., 1992, A&A 253, 586

Rast M. P., 1995, ApJ 443, 863

Remling B., Deubner F. L. & Steffens S., 1996, A&A 316, 196

Restaino S. R., Stebbins R. T. & Goode P. R., 1993, ApJ 408, L57

Rimmele T. R., Goode P. R., Harold E. & Stebbins R. T., 1995, ApJ 444, L119

Roudier T., Malherbe J. M., November L., Vigneau J., Coupinot G., Lafon M. & Muller R., 1997, A&A 320, 605

Rutten R. J., 1994, in M. Carlsson (ed.), Chromospheric Dynamics, Proc. Mini-workshop, Inst. Theor. Astrophys., Oslo, p. 25

Rutten R. J., 1995, in J. T. Hoeksema, V. Domingo, B. Fleck & B. Battrick (eds.), Helioseismology, Proc. Fourth SOHO Workshop, ESA SP–376 Vol. 1, ESA Publ. Div., ESTEC, Noordwijk, p. 151

Rutten R. J., 1996, in K. G. Strassmeier & J. L. Linsky (eds.), Stellar Surface Structure, Procs. Symp. 176 IAU, Kluwer, Dordrecht, p. 385

Rutten R. J., 1998, in C. Fröhlich, M. C. E. Huber, S. Solanki & R. von Steiger (eds.), Solar Composition and its Evolution – from Core to Corona, Procs. ISSI Workshop, Space Sci. Rev., in press

Rutten R. J., de Pontieu B. & Lites B. W., 1999, in T. Rimmele, K. Balasubramaniam & R. Radick (eds.), High Resolution Solar Physics: Theory, Observations, and Techniques, Procs. 19th NSO/Sacramento Peak Summer Workshop, Astron. Soc. Pac. Conf. Series, in press

Rutten R. J. & Uitenbroek H., 1991, Solar Phys. 134, 15

Sivaraman K. R. & Livingston W. C., 1982, Solar Phys. 80, 227

Steffens S., Deubner F.-L., Fleck B., Wilhelm K., Schühle U., Curdt W., Harrison R., Gurman J., Thompson B. J., Brekke P., Delaboudinière J.-P., Lemaire P., Hessel B. & Rutten R. J., 1997, in B. Schmieder, J. C. T. del Iniesta & M. Vázquez (eds.), Advances in the physics of sunspots, Procs. First Adv. in Solar Physics Euroconf., Astron. Soc. Pac. Conf. Series 118, p. 284

Steffens S., Hofmann J. & Deubner F. L., 1996, A&A 307, 288

Sutmann G. & Ulmschneider P., 1995a, A&A 294, 232

Sutmann G. & Ulmschneider P., 1995b, A&A 294, 241

Von Uexkuell M. & Kneer F., 1995, A&A 294, 252

Wellstein S., Kneer F. & Von Uexkuell M., 1998, A&A 335, 323

Solar and Stellar Activity: Similarities and Differences
ASP Conference Series, Vol. 158, 1999
C.J. Butler and J.G. Doyle, eds.

Theoretical Chromosphere Models

P. Ulmschneider

Institut f. Theoret. Astrophysik, Univ. Heidelberg, Tiergartenstr. 15, 69121 Heidelberg, Germany

Abstract. Theoretical chromospheric models depend on four fundamental parameters: effective temperature, gravity, metallicity and rotation period. The physical logic is outlined by which these four parameters determine the observed chromospheres. On the basis of the heating mechanism, mechanical energy fluxes are computed for a wide range of late-type stars and the propagation and dissipation of this energy into the outer stellar layers is evaluated. For stars with a low rotation rate, theoretical chromosphere models based on pure acoustic heating for the Sun and other main sequence stars are discussed.

1. Introduction

Despite the fact that the solar chromosphere and corona show large variations in emission across the solar surface as well as during the sunspot cycle, it is well known that the average behaviour of these hot outer stellar layers can only depend on four fundamental parameters, which uniquely define any non-accreting, nondegenerate star. For a single star there are only 3 fundamental parameters, mass, chemical composition and age, while the fourth parameter, rotation, is determined from the angular momentum loss rate and changes in the moment of inertia which are both functions of age. However, because more than 70% of all stars are members of binary or multiple systems, where unknown amounts of orbital angular momentum have been transferred to spin angular momentum, one generally needs to specify the rotation rate independently from the age. Naturally if one considers T-Tauri stars, additional parameters must be specified as e.g. the mass accretion rate. These four fundamental parameters that uniquely specify a star can conveniently be taken to be effective temperature, T_{eff}, surface gravity, g, metallicity, Z_m and rotation period P_{Rot}. The great challenge to the theoretician is to elucidate the physical connection between these four parameters and the observed chromospheres and coronae. It is fortunate that recent advances make it likely that this theoretical aim can be realized in the near future and that reliable theoretical chromosphere models are now in grasping distance.

While in the field of stellar atmospheres the agreement between empirical and theoretical photospheric models based on T_{eff}, g and Z_m is usually very good, we are far from a similar reliability of theoretical chromosphere and corona models. There are three properties which particularly distinguish chromospheres and coronae from photospheres: the *magnetic fields*, the stochastic and secular, spatial and temporal *variability* and the *mechanical heating.* We know that for late-type stars the magnetic fields are not fossil ones, left over from star-formation, but that they are generated by the star itself. The theory which describes how magnetic fields are generated, the dynamo theory, needs two important ingredients: convection and rotation (Choudhuri 1998, Schüssler, Schmidt & Ferriz-Mas 1997, Proctor & Gilbert 1994). Both can be computed or specified on the basis of four fundamental parameters. As a rising convective bubble expands and as horizontal motion on a rotating star leads to Coriolis forces, the generated convective flows are helical. They amplify the magnetic fields and thus create magnetic flux. There is reasonable hope that with massive magnetohydrodynamic simulations the dynamo theory will become highly accurate and predictive in the near future.

The chromospheric and coronal variability derives from the action of the turbulent convection and from magnetic flux emergence. The variability can therefore be explained by modeling the turbulent convection and the time-dependent magnetic flux emergence using the dynamo theory. The same is valid for mechanical heating. It has been known for some time that for the existence of chromospheres and coronae, mechanical heating is absolutely essential. Mechanical energy is produced by wave generation, by foot-point motion of magnetic flux tubes which lead to entangled fields and by magnetic flux emergence. Thus all the peculiar properties of chromospheres should be describable by convection zone models and by the dynamo theory, that is, by theoretical models based on the four fundamental parameters.

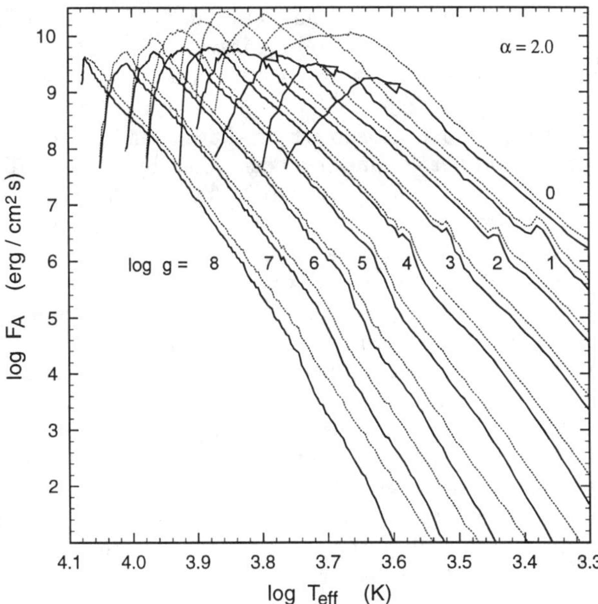

Figure 1. Acoustic fluxes for stars versus T_{eff} for given $\log g$, solar abundance $Z_m = [0]$ and $\alpha = 2.0$.

2. Heating Mechanisms

A survey of the proposed heating mechanisms shows that they can be subdivided into two main types, the *hydrodynamic-* and the *magnetic mechanisms* (see e.g. Narain & Ulmschneider 1996). Each of these major types can be further subdivided into fast and slow mechanisms. Fast hydrodynamic mechanisms (where the periods P are less than the acoustic cut-off period P_A) are the acoustic waves while the slow mechanisms (usually $P > P_A$) are the pulsational waves. The fast magnetic mechanisms, usually called AC (alternating current) mechanisms, are magneto-acoustic-, Alfvén- and surface waves. Slow magnetic mechanisms, also called DC (direct current) mechanisms, are current sheets and micro-flares. Based on successful terrestrial laboratory experiments a large zoo of different magnetic heating mechanisms has been proposed, particularly of the wave type.

The great difficulty in the nearly sixty years since 1941, when Edlén and Grotrian first recognized the solar corona as an extremely hot outer layer, was and is, to identify which heating mechanism is important in the stellar environment and in a given magnetic field configuration on the star. The problem is that it became clear that the heating by mechanical energy typically occurs in regions of very small scales, say in the meter-range on the Sun. This refers to the shock structure region in a shock wave and the current carrying region in a current sheet. Unfortunately, despite of the great recent technological advances to improve the resolution of solar observations it is presently still hopeless to resolve scales in the meter-range. Because of the lacking resolution on the Sun, for instance, it cannot easily be decided whether a non-magnetic region is heated by acoustic shock waves, or by some magnetic mechanism which operates in a weak or unresolved field. Here stellar observations come to the rescue.

It seems utterly surprising that from stellar observations, where entire stars are reduced to point sources, information can be obtained about heating properties in the meter-range. The reason why stellar observations are so crucial in unraveling the important heating mechanisms is that by selecting different stars, the observer can vary T_{eff} by one, g by five, Z_m by three and P_{Rot} by two orders of magnitude. In addition, stellar observations naturally integrate over the surface of the star and thus make easier a comparison with a theory which predicts the average behaviour (plus its range of variability). By computing mechanical energy generation rates and by producing theoretical chromosphere models for different stars it is possible to compare the simulated chromospheric emission fluxes with the observed ones. Any heating mechanism which survives such a rugged comparison over many orders of magnitude in variation of the entire set of fundamental stellar parameters must without doubt be considered identified. This procedure allowed us to finally identify acoustic waves as the basic chromospheric heating mechanism for slowly rotating (large P_{Rot}) non-magnetic stars. Presently the same identification method is also applied to magnetic mechanisms (see the contribution of M. Cuntz, this volume) to unravel the heating mechanism of magnetic chromospheres.

2.1. Acoustic Wave Energy Generation

The theory of sound generation from turbulence was developed in the early 1950's by Lighthill and Proudman and was further extended by Stein to stellar

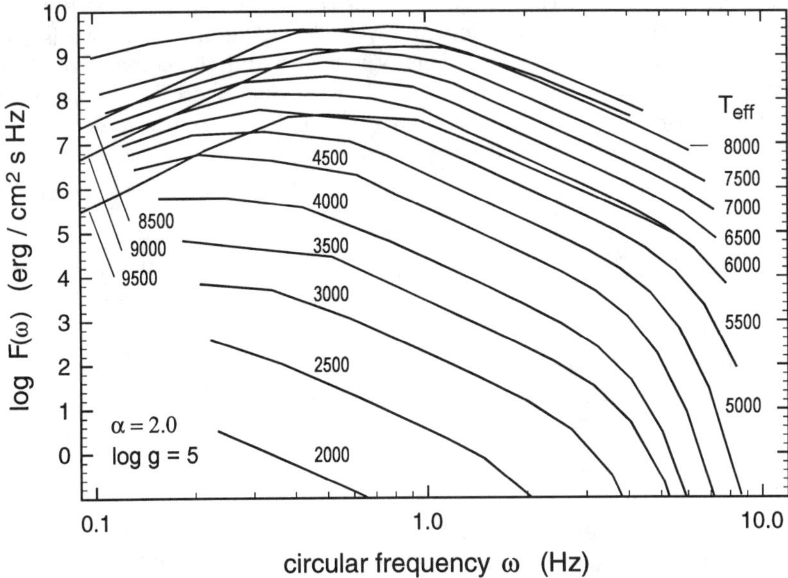

Figure 2. Acoustic spectra for stars versus circular frequency for given T_{eff}, $Z_m = [0]$ and $\alpha = 2.0$, with $\log g = 5$.

convection zones. The Lighthill-Stein theory has been re-discussed in detail by Musielak et al. (1994), who found that the stellar turbulence is best represented by a Kolmogorov-type energy spectrum. By specifying T_{eff}, g, Z_m and the mixing-length parameter α, convection zone models were computed and the emergent acoustic flux as well as the acoustic frequency spectrum evaluated on basis of the Lighthill-Stein theory. For α, the ratio of the mixing-length to the scale height, one typically has $\alpha \approx 1.5$ to 2.0. Figures 1 and 2 show acoustic fluxes as well as acoustic energy spectra computed this way for a large number of late-type stars in the HR-diagram (after Ulmschneider, Theurer & Musielak 1996). While in terrestrial situations, monopole, dipole and quadrupole sound generation is a sequence of progressively less important ways to produce acoustic waves, quadrupole sound generation is the dominant contribution in stellar situations.

As the acoustic energy generation depends mainly on the convective velocity and because convection zone models do not depend on stellar rotation it is clear that the acoustic energy flux does not depend on rotation. However, there is a strong dependence on metallicity. Figure 3 shows the dependence of the acoustic flux on Z_m for main sequence stars (with $\log g = 4.44$) after Ulmschneider et al. (1998). $Z_m = [0]$ means solar metallicity, $Z_m = [-1]$, 1/10 solar metal content etc. It is seen that for stars with $T_{eff} > 5000\ K$, where hydrogen is the main opacity, the acoustic flux does not depend much on Z_m, but that for cooler stars, where the opacity is dominated by metals, the acoustic flux decreases by an order of magnitude for every reduction of the metal content by a factor of ten. This is explained by the fact that a reduction of the surface opacity leads to a shift of the top of the convection zone, the level where the optical depth is roughly unity, to deeper layers where the density is larger and the convective velocities correspondingly smaller. Figures 1 & 3 show that there is a nine orders of magnitude variation of the acoustic flux with the fundamental parameters T_{eff}, g and Z_m, but no variation with P_{Rot}.

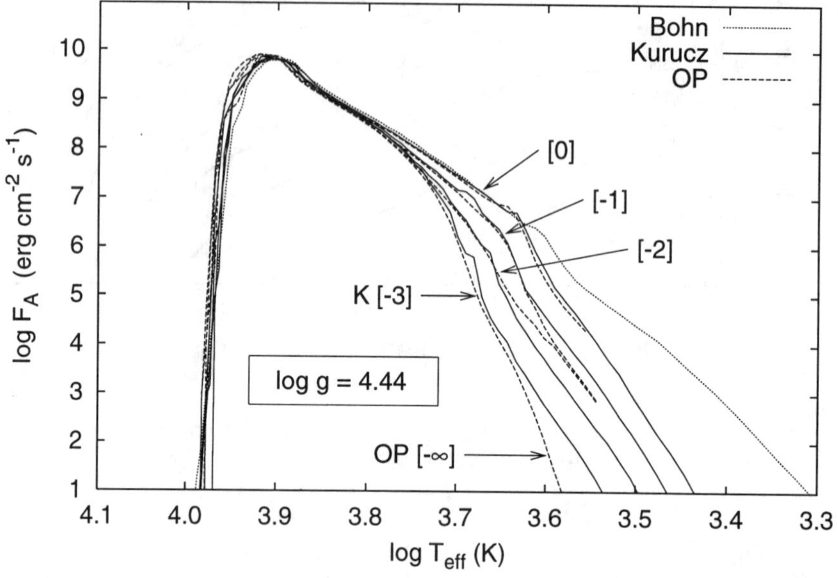

Figure 3. Acoustic fluxes for main sequence stars of $\log g = 4.44$ versus T_{eff} for different metallicity Z_m (indicated in brackets) and $\alpha = 2.0$.

2.2. Magnetic Wave Energy Generation

The magnetic field at the solar photosphere outside of sunspots and pores appears in the shape of flux tubes which are concentrated in the convection zone by the granular and super-granular flows and the convective collapse. The turbulent motions in the non-magnetic convection zone outside strangles and displaces the magnetic flux tubes, which leads to the generation of longitudinal and transverse MHD waves, respectively. Simulating the turbulent motions on the basis of convection zone models, pressure and velocity fluctuations can be applied to the flux tubes and the generated longitudinal MHD tube wave and transverse Alfvén wave fluxes can be computed. Such computations after Huang, Musielak & Ulmschneider (1995) and Ulmschneider & Musielak (1998) are shown in Figure 4. Due to the stochastic nature of the turbulence occasionally large velocity spikes occur which approach 3 km/s, in good agreement with observations. This leads to a very stochastic behaviour of the generated instantaneous wave flux seen in Figure 4.

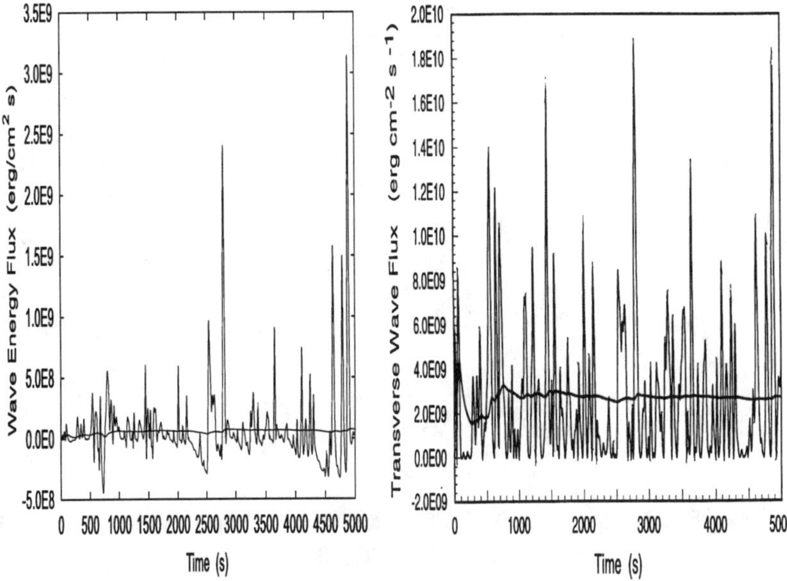

Figure 4. Instant and time-averaged longitudinal and transverse tube wave energy fluxes versus time.

Applying these calculations to stars other than the Sun the magnetic field strength B_0 of the flux tubes at the stellar surface must be known. Here in analogy to the Sun various fractions of the equipartition field strength $B_{eq} = \sqrt{8\pi p_0}$ have been assumed, where p_0 is the gas pressure at the stellar surface. In the future, time-dependent simulations of the convective collapse will provide a definite ratio B_0/B_{eq}. Figure 5 shows longitudinal tube wave fluxes for late-type stars in the HR-diagram computed by Ulmschneider, Musielak & Fawzy (1998). It is seen that these fluxes have a distinctively different behaviour on T_{eff} and g compared to the acoustic fluxes.

3. Wave Propagation

From Figure 2 it is clear that a computation of the acoustic wave propagation must start from the acoustic flux and the acoustic frequency spectrum generated in the stellar convection zone. Calculations using acoustic spectra have been undertaken by e.g. Carlsson & Stein (1994, 1995) and Theuer, Ulmschneider & Kalkofen (1997). In monochromatic wave computations (Buchholz, Ulmschneider & Cuntz 1998), the atmosphere becomes permeated by a series of sawtooth shaped shock waves of constant (limiting) strength which depends on the wave period P. A calculation using an acoustic wave spectrum shows a quite different behaviour as shown in Figure 6 from Theurer (1998). Here the merging of numerous small shocks generates typically a single large, long-period shock, plus a few smaller shocks similar to those in a monochromatic calculation.

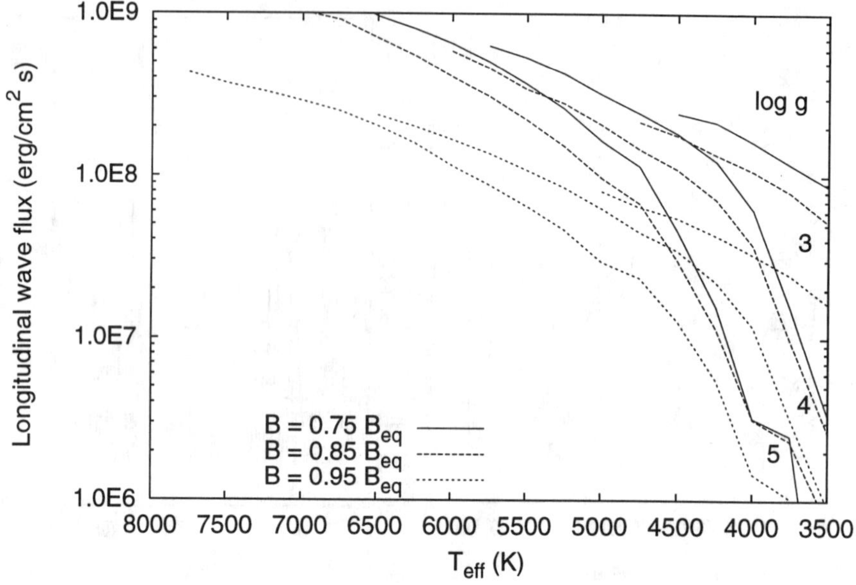

Figure 5. Longitudinal tube wave fluxes for stars versus T_{eff} for given $\log g$, with solar metallicity $Z_m = [0]$, $\alpha = 2$ and different ratios of magnetic field strength B_0/B_{eq}.

Aside of the fact that all computations so far are plane wave calculations which neglect important acoustic refraction and standing wave effects, these most recent calculations still suffer from some unfortunate approximations. The calculations of Theuer et al. (1997) neglect the fully time-dependent ionization of hydrogen, as well as its ionization energy in the energy equation. Both effects lead to the fact that the single large shock in Figure 6 is much too strong and consequently the simulated chromospheric emission is too large. Yet the smaller shocks lower in the atmosphere are adequately described. Stein & Carlsson (1994, 1995), who use a code which includes the fully time-dependent hydrogen ionization and the hydrogen ionization energy, obtained results which suffer from an unfortunate choice of the acoustic input spectrum. Figure 7 (from Theurer et al. 1997) shows two acoustic spectra at height $z = 250\ km$ in the Sun computed on basis of a full Kolmogorov-type acoustic spectrum generated in the convection zone at height $z = -160\ km$. The spectrum on the right hand side would be seen in terrestrial observations because it includes the effect of the modulation transfer function, while the left hand spectrum is actually supposed to be present at $z = 250\ km$ height. The modulation transfer function tells what fraction of a given velocity oscillation can be seen as Doppler shift fluctuation in a spectral line. A low lying Fe-line is formed over a height interval of about 300 km, which for a sound speed of 7 km/s corresponds to the wavelength of a wave with a period of about $P = 40\ s$. For periods $P < 40\ s$, the velocity fluctuations no longer show themselves as Doppler shifts but as line-broadening, which results in

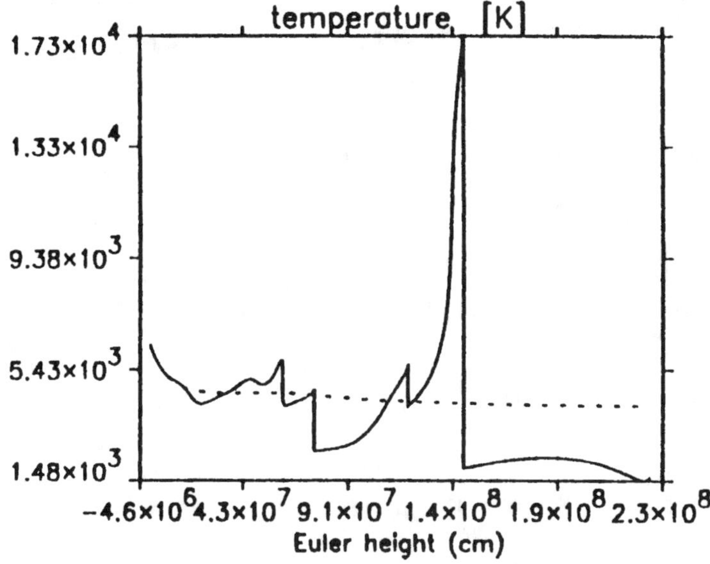

Figure 6. Instantaneous temperature as function of height in the solar atmosphere for an acoustic wave calculation starting from an acoustic spectrum in the convection zone, at time 3680 s. The dotted line is the initial radiative equilibrium temperature distribution.

the decrease of the modulation transfer function to zero. Observing solar velocity fluctuations in a low-lying Fe-line one thus detects only a small part (the long-period part) of the entire acoustic spectrum. Because of this effect the Carlson & Stein calculations, using such observed fluctuations as input, found only the single large long-period shock and no smaller shocks. Although it is clear that these temporary defects will shortly be overcome and wave calculations using the appropriate acoustic spectra will shortly be available, it must be pointed out that monochromatic calculations with adequately chosen wave period, by avoiding the above pitfalls, can provide reasonably reliable results.

4. Line Emission, Comparison of Theory and Observation

After running a radiation-hydrodynamic wave code for a sufficiently long time, the atmosphere reaches a dynamical steady state in which the averaged quantities become time-independent. A powerful diagnostic tool is then to simulate the chromospheric Ca II H&K and Mg II h&k lines and compare their emission flux with observations. Figure 8 (after Theurer 1998) shows the emerging Ca II K line profiles using partial redistribution (PRD) for solar wave calculations using an acoustic spectrum (labeled bKmG) and two monochromatic wave cases with $P = 20$ and 40 s (labeled mP20, mP40), all with the same acoustic energy flux. Due to the single very strong shock both the total emission and

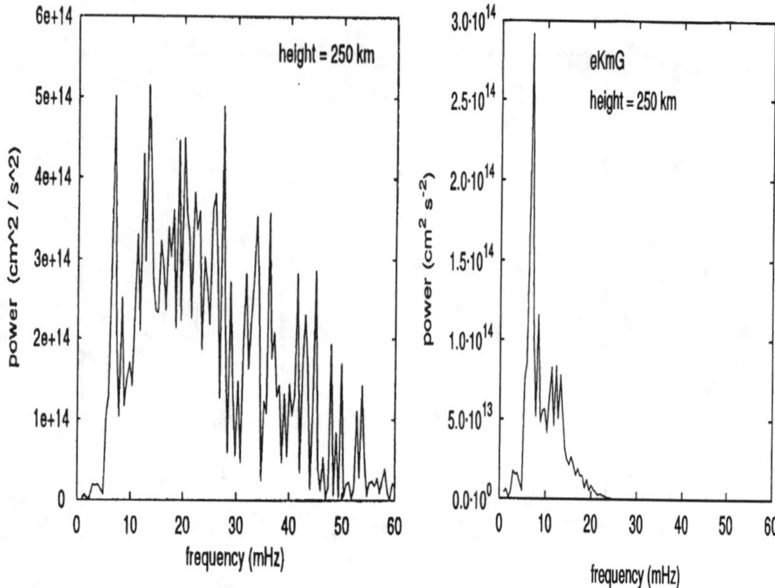

Figure 7. Theoretical acoustic spectra in the solar atmosphere at height $z = 250$ km. The left panel shows the actual spectrum, while the right panel the 'observed' spectrum with the modulation transfer function applied.

the line-asymmetry are exaggerated in the acoustic spectrum case. Aside of this effect, however, the two time-averaged bKmG and mP40 line-profiles appear to be fairly similar, which shows that a monochromatic wave calculation, with the right choice of the wave period (near the maximum of the acoustic spectrum), can give reasonable results.

Figure 9 shows a comparison of the theoretical and observed chromospheric emission fluxes in the Ca II and the Mg II lines of late-type stars versus the color index B-V after Buchholz et al. (1998). Observations indicate that main-sequence (dots) and giant stars (triangles) occupy the same empirical minimum emission line (solid) in the diagram, called *basal flux line*, which is interpreted as the result of pure acoustically heated chromospheres. Dots and triangles above this line represent stars which have additional magnetic heating, correlated with rotation. The theoretical simulations (x's for main-sequence and circles for giant stars) are obtained from the generated acoustic wave fluxes using monochromatic waves. It is seen that the basal flux line is well reproduced. A similar calculation of theoretical Ca II and Mg II line emissions of giants with metallicities $Z_m = [-1]$ and $Z_m = [-2]$ (Cuntz, Rammacher & Ulmschneider 1994) shows that these stars also fall on the basal flux line in agreement with observations. This nice agreement over a wide range of fundamental parameters T_{eff}, g and Z_m shows that acoustic waves can now be considered as a firmly established heating mechanism for chromospheres of stars with large rotation period P_{Rot}.

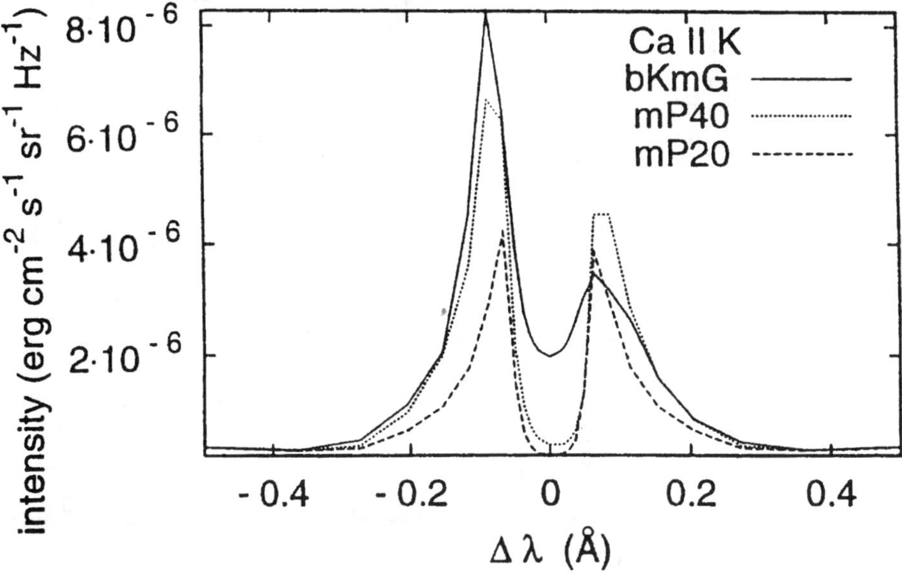

Figure 8. Theoretical time-averaged Ca II K-line profile with partial redistribution (PRD) for calculations using two monochromatic wave cases (P=20, 40 s) and a case with an acoustic spectrum (bKmG).

5. Theoretical Chromospheres

Let us now address the topic of theoretical chromosphere models. Figure 10 (from Theurer 1998) shows theoretical time-averaged temperature profiles for the chromosphere of the Sun. The line marked T_0 shows the radiative equilibrium atmosphere before the start of the wave calculation. The lines marked mP20, mP40 and bKmG indicate the averaged temperatures of two monochromatic wave cases (with period $P = 20,\ 40\ s$) and a calculation with an acoustic Kolmogorov-type spectrum, all with the same initial acoustic energy. It is seen that the acoustic spectra case shows a large temperature depression below the T_0 distribution. This is a well known effect (Ulmschneider et al. 1978), caused by the very strong shock and the nonlinearity of the Planck function and must be considered as an artifact. It is significant that all three wave models above 1400 km height show an outward temperature rise as is present in a classical chromosphere. We are convinced that a better treatment of the single strong shock will bring the acoustic spectrum (bKmG) result much closer to the mP40 case. Comparing Figure 10 to Figure 5 of Carlsson & Stein (1994, also 1995) it is seen that their time-averaged temperature, which decreases in an outward direction almost completely coincides with our radiative equilibrium temperature distribution T_0. These authors conclude that a non-magnetic chromosphere is essentially an outwardly decreasing radiative equilibrium layer on top of which single strong shocks propagate. The emission from such shocks then was wrongly

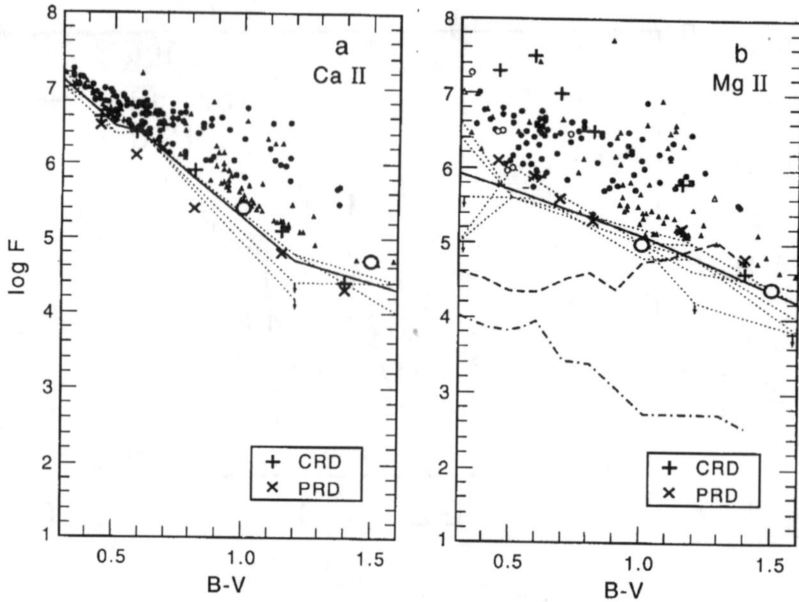

Figure 9. Theoretical and empirical chromospheric emission fluxes in the Ca II and Mg II lines versus colour for late-type stars.

attributed by empirical models to a classical chromosphere with an outwardly rising mean temperature distribution. We believe that some of the conclusions of Carlsson & Stein are a direct consequence of the neglect of the largest fraction of the acoustic spectrum. A treatment of the full spectrum would also produce the little shocks and thus would lead to a remnant of the classical chromosphere, that is, to an outwardly rising mean temperature. Clearly, however, both Carlsson & Stein's and our calculations indicate, that single large shocks are a persistent feature in the chromosphere and that the emission from shock waves is a dominant effect which must be taken into account when constructing empirical models.

Figure 11 (from Buchholz et al. 1998) shows the mean temperature distribution of six theoretical chromosphere models for main sequence stars ranging from spectral type F5V to M0V. Also shown are the corresponding initial radiative equilibrium distributions. Due to the low acoustic energy fluxes of M-stars, their photospheres extend to very low mass column density m, where finally shock formation leads to heating and a chromospheric temperature rise develops. For earlier spectral type the acoustic fluxes are much larger and the location of the temperature minimum and the associated chromospheric temperature rise shifts to greater mass column density. For the earliest spectral types a large temperature depression below the radiative equilibrium values develops. This is the result of large amplitude acoustic waves. These calculated properties of theoretical chromosphere models are in nice qualitative agreement with empirical models by Kelch, Linsky & Worden (1979) based on Ca II line observations.

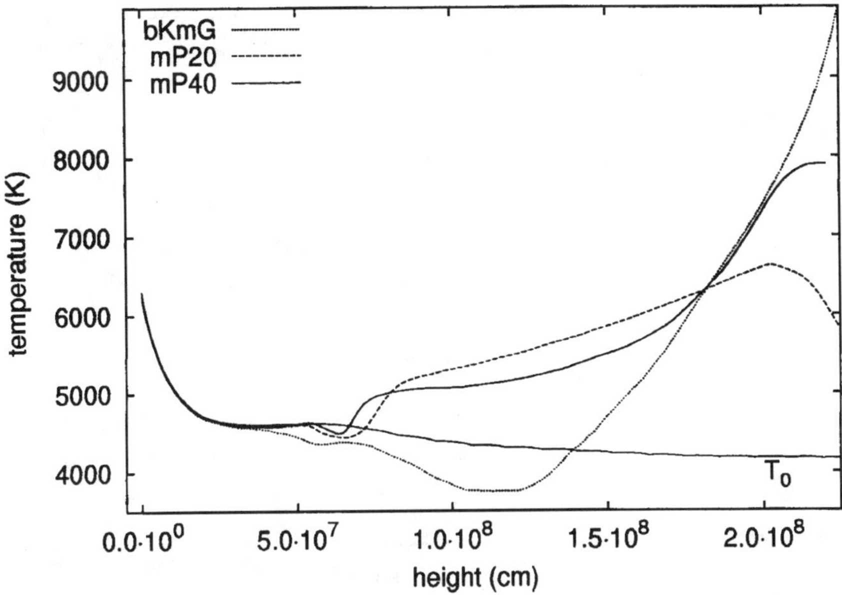

Figure 10. Time averaged solar temperature distributions for two monochromatic wave calculations (mP20, mP40) and a one using an acoustic spectrum (bKmG). Also shown is the initial radiative equilibrium distribution (T_0) before the start of the wave computations.

The large temperature depression in the theoretical models is connected to the large 'photospheric temperature excess' in the Ca II line models. One urgently looks forward to more reliable chromosphere models in the future, based on the full acoustic spectra. Ultimately acoustic chromosphere models as function of T_{eff}, g and Z_m should become as reliable as the present photosphere models for slowly rotating stars. With the identification of the magnetic heating mechanisms, where P_{Rot} comes into play, this may even be true for the chromosphere models of all late-type stars which depend on the full set of four fundamental parameters.

Acknowledgments. I want to thank my collaborators B. Buchholz, M. Cuntz, D. Fawzy, Z.E. Musielak, W. Rammacher and J. Theurer who all contributed to this work, and NATO for grant CGR-910058.

References

Buchholz B., Ulmschneider P. & Cuntz M., 1998, ApJ, 494, 700

Carlsson M. & Stein R.F., 1994, *'Chromospheric Dynamics', mini-workshop*, Univ. of Oslo, M. Carlsson Ed., p. 47

Carlsson M. & Stein R.F., 1995, ApJ, 440, L29

Choudhuri A.R., 1998, The Physics of Fluids and Plasmas, Camb. Univ. Press

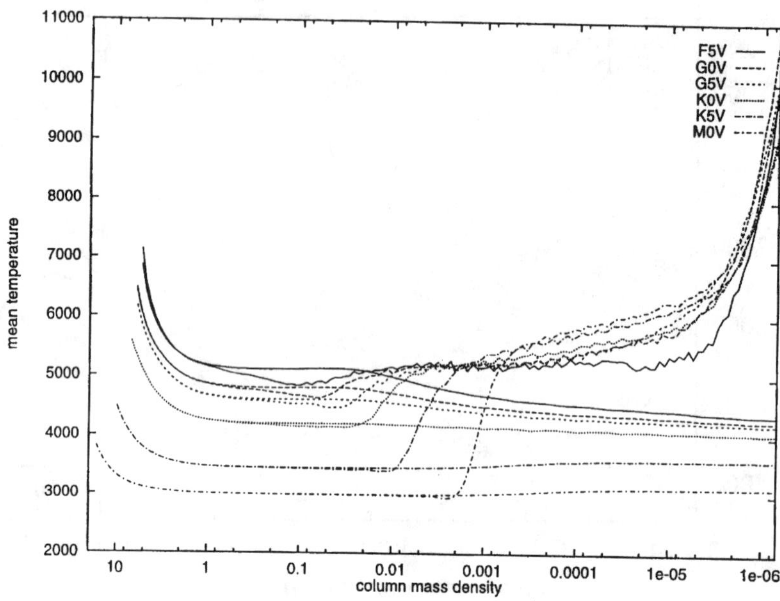

Figure 11. Time-averaged temperature distributions for six main-sequence stars based on monochromatic calculations. Also shown are the corresponding initial radiative equilibrium distributions.

Cuntz M., Rammacher W. & Ulmschneider P., 1994, ApJ, 432, 690

Huang P., Musielak Z.E. & Ulmschneider P., 1995, A&A, 297, 579

Kelch W.L., Linsky J.L. & Worden S.P., 1979, ApJ, 229, 700

Musielak Z.E., Rosner R., Stein R.F. & Ulmschneider P., 1994, ApJ, 423, 474

Narain U. & Ulmschneider P., 1996, Space Sci.Rev., 75, 453

Proctor M.R.E. & Gilbert A.D., 1994, Lectures on Solar and Planetary Dynamos, Cambridge Univ. Press

Schüssler M., Schmitt D. & Ferriz-Mas A.: 1997, in: Advances in Physics of Sunspots, ASP Conf. Ser. 118, B. Schmieder, J.C. del Toro Iniesta & M. Vazquez Eds., p. 39

Theurer J.: 1998, Ph.D. thesis, Univ. Heidelberg

Theurer J., Ulmschneider P., Kalkofen W., 1997, A&A, 324, 717

Ulmschneider P. & Musielak Z.E., 1998, A&A, 338, 311

Ulmschneider P., Musielak Z.E. & Fawzy D.:, 1999, A&A, to be submitted

Ulmschneider P., Schmitz F., Kalkofen W. & Bohn H.U.: 1978, A&A, 70, 487

Ulmschneider P., Theurer J. & Musielak Z.E., 1996, A&A, 315, 212

Ulmschneider P., Theurer J., Musielak Z.E. & Kurucz R., 1998, A&A, to be submitted

Solar and Stellar Activity: Similarities and Differences
ASP Conference Series, Vol. 158, 1999
C.J. Butler and J.G. Doyle, eds.

Two-Component Chromosphere Models: Observations versus Simulations

M. Cuntz

Center for Space Plasma, Aeronomy, and Astrophysics Research, TH 101, University of Alabama in Huntsville, Huntsville, AL 35899

Abstract. We discuss theoretical chromosphere models of single late-type stars of different magnetic activity. The models are assumed to consist of two components: a magnetic component heated by longitudinal flux tube waves and a nonmagnetic component heated by acoustic waves. As an example, we discuss models of K2 V stars with rotation periods between 10 d and 40 d. The rotation periods determine the photospheric and chromospheric magnetic filling factors, which in turn determine the shape of the flux tubes, the propagation and dissipation of the wave energy flux and the spectral line emission. We show that for stars with very slow rotation we are able to reproduce the basal flux limit of chromospheric emission previously identified as due to pure acoustic heating. Most importantly, we find that the relationship between the Ca II H+K emission and the stellar rotation rate deduced from our models is consistent with the empirical relationship given by observations.

1. Introduction

An outstanding problem in stellar astrophysics concerns the identification of the processes responsible for heating stellar chromospheres. P. Ulmschneider (this volume) has outlined the physical logic for calculating theoretical chromosphere models using four basic physical parameters, which are: effective temperature, gravity, metallicity, and rotation period. The stellar rotation period is crucial for describing and modelling the chromospheric heating processes in stars of different magnetic activity. Observationally, it has been found that fast rotating stars have increased chromospheric emission, whereas slow rotating stars have low chromospheric emission, which asymptotically approaches "basal emission" that is now believed to be due to pure acoustic heating (Buchholz, Ulmschneider, & Cuntz 1998). Modelling chromospheric emission in rapidly rotating stars allows to get insight into outer atmospheric heating of stars with increased magnetic activity. In this review, we outline the underlying basic ideas for calculating chromospheric heating models for magnetically active stars with applications to K dwarf stars. The concepts presented are used for explaining observed chromospheric emission — stellar rotation relationships, but can also be utilized for other purposes such as chromospheric variability due to stellar activity cycles or describing rotational modulation of chromospheric emission in stars which are not observed pole-on. For more details see the forthcoming paper by Cuntz, Rammacher, Ulmschneider, Musielak, & Saar (1999) (CRUMS).

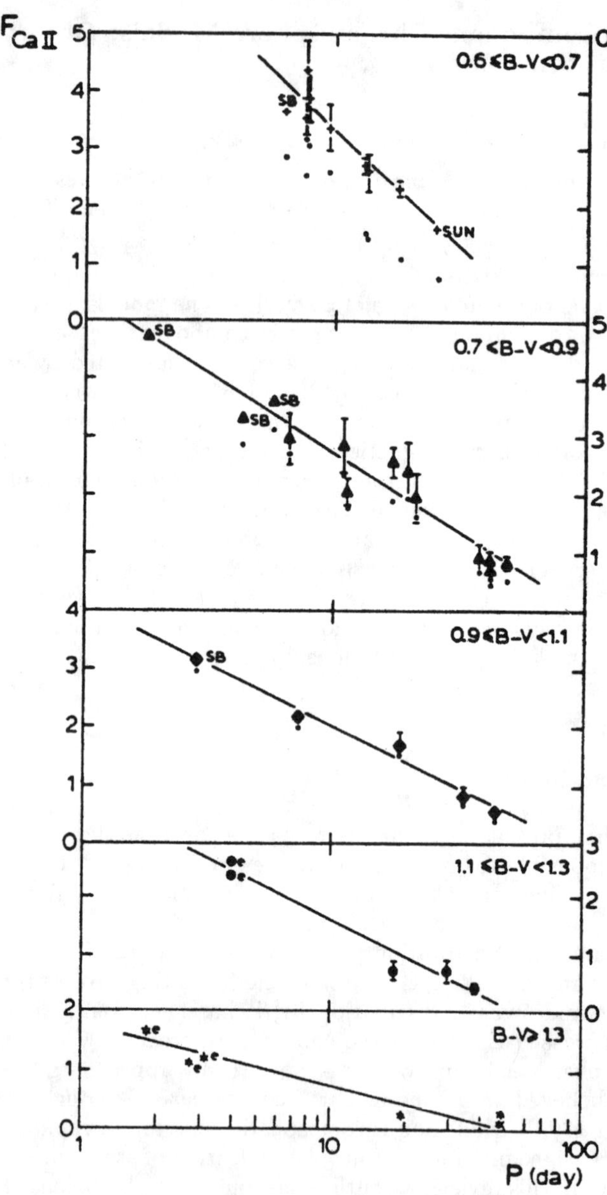

Figure 1. Line core flux in Ca II (log. units) versus rotation period P for late-type main-sequence stars. (adopted from Rutten 1986)

In this paper, we want to discuss the derivation of theoretical chromospheric emission — stellar rotation relationships for single late-type stars. Observationally, the increase of chromospheric emission with decreasing stellar rotation rate has been verified for main-sequence stars (Rutten 1986, 1987; Schrijver 1987; Rutten et al. 1991) and (sub-)giants (Rutten 1987; Rutten & Pylyser 1988; Simon & Drake 1989; Gray 1989; Strassmeier et al. 1994) (see Figure 1). It has also been found that increased stellar rotation rate usually leads to an increased photospheric magnetic filling factor f_o. In fact, it is now possible to link $B_o f_o$ (with B_o being the photospheric magnetic field strength) both to the stellar rotation period P_{rot} (Marcy & Basri 1989; Montesinos & Jordan 1993; see also Cuntz, Ulmschneider, & Musielak 1998) and to the emergent chromospheric emission flux (Saar & Schrijver 1987; Schrijver et al. 1989; Montesinos & Jordan 1993; Jordan 1997).

With respect to Ca II emission, a variety of studies exist. Many years ago, Skumanich (1972) proposed the now famous $t^{-1/2}$ law to describe the decay of the Ca II flux and the stellar spin. The decline in rotation was attributed to a steady loss of angular momentum in the coronal wind, itself a by-product of the activity. The fading of chromospheric tracers was ascribed to a weakening of the spin-catalyzed magnetic "dynamo". Noyes et al. (1984) identified a tight correlation between the Ca II emission and the Rossby number, a key parameter in dynamos theories. Additional observational results were presented by Simon, Herbig, & Boesgaard (1985) and more recently by Soon, Baliunas, & Zhang (1994) and Ayres et al. (1996). The latter authors made use of the Faint Object Spectrograph of the *Hubble Space Telescope* to record the ultraviolet emissions of solar-type stars in galactic clusters of largely different age. The observations lead to unique proof of the decay of chromospheric and transition-layer activity as function of the stellar evolutionary time. Theoretical results about the change of dynamo action on evolutionary time scales have also been given by Schrijver & Pols (1993), Keppens, MacGregor, & Charbonneau (1995), and Charbonneau, Schrijver, & MacGregor (1997). The latter two studies contain a detailed MHD description of the time-development of the internal angular momentum redistribution allowing estimates about the change of surface magnetic field strengths and rotational velocities of main-sequence stars. These studies also provide the theoretical explanation why the stellar rotation rates decrease with time and which changes to expect for the $B_o f_o$ values.

The relative size of the magnetic and nonmagnetic regions of our target stars is determined by the photospheric magnetic filling factor f_o, which is assumed to depend on the stellar rotation period P_{rot}. The rotation periods determine the shape of the flux tube models, which are relevant for the propagation and dissipation of the wave energy flux and the formation of the Ca II emission lines. With respect to theoretical chromospheric heating models for magnetically active stars, serious efforts are made to accomplish three distinctly different steps in the modelling process, which are: (1) the calculation of acoustic and magnetic wave energy generation in the stellar convection zone, (2) the modelling of the propagation and dissipation of acoustic and magnetic wave energy at the different atmospheric heights, and (3) the simulation of specific chromospheric emission lines which are compared with observations. In all these steps, the two-component structure of the stellar photosphere and chromosphere is fully taken into account. These different steps are described in the forthcoming sections.

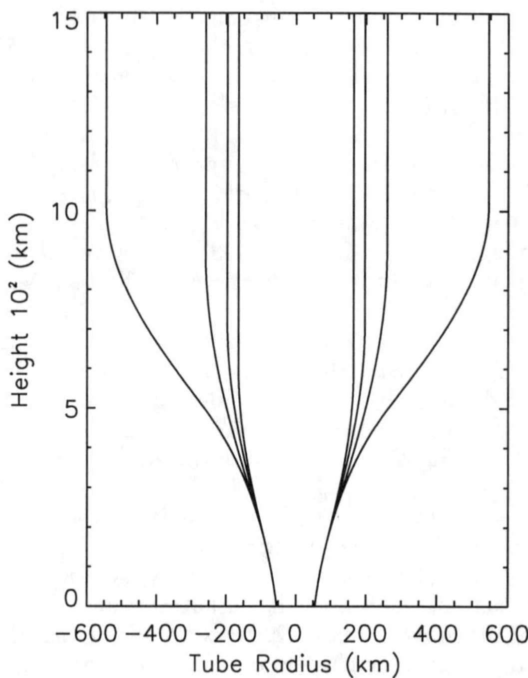

Figure 2. Shape of flux tubes. The flux tube parameters correspond to stars with rotation periods of $P_{\rm rot} = 10, 20, 30$, and 40 d, respectively, with smaller rotation periods corresponding to narrower tubes.

2. Stellar Parameters and Flux Tube Shapes

For our K2 V star models, we take the following stellar parameters: $T_{\rm eff}$ = 4900 K, log $g_* = 4.51$ (cgs), and $R_* = 0.8$ $R_\odot$. Various assumptions must be made to properly select the magnetic parameters of the star, i.e. the photospheric magnetic filling factor and magnetic field strength, the shape of the flux tubes, and distribution of the tubes on the stellar surface.

To calculate the magnetic field strength $B_{\rm o}$ inside the tubes at the photospheric level $\tau_{5000} = 1$, we assume approximate equipartition between the external gas pressure and magnetic pressure (e.g., Solanki 1996; Hasan & van Ballegooijen 1997) assuming that the gas pressure inside and outside the tube (p_i and p_e, respectively) is given by $p_i : p_e = 1 : 4$. This leads to a photospheric magnetic field strength $B_{\rm o} = 2100$ G. As we want to investigate the influence of stellar rotation on our two-component chromosphere models and the resulting Ca II emission, we have to relate the photospheric magnetic filling factor $f_{\rm o}$ to the stellar rotation rate $P_{\rm rot}$. Here we use the $B_{\rm o} f_{\rm o}$ − $P_{\rm rot}$ relation given by

Cuntz et al. (1998) based on recent observational results of Rüedi et al. (1997). An analysis of these data yields $B_o f_o = 238 - 5.51 \cdot P_{rot}$ (cgs; P_{rot} in days).

With the quoted value of B_o, we can link the photospheric filling factor f_o directly to P_{rot}. For P_{rot} = 10, 20, 30, and 40 d, we obtain photospheric magnetic filling factors of 0.087, 0.061, 0.035, 0.008, respectively. Fast rotating stars are thus expected to have higher coverage of flux tubes than slow rotating stars leading to increased photospheric magnetic filling factors. The above given $B_o f_o - P_{rot}$ relationship can also be used to constrain the shape of the flux tubes. As fast rotating stars are expected to have higher coverage of flux tubes than slow rotating stars, the tube spreading in fast rotating stars should remain smaller, resulting in smaller tube opening radii. For P_{rot} = 10, 20, 30, and 40 d, we obtain tube opening radii of 165, 197, 260, and 545 km, respectively. These values are calculated based on the approximation that the tubes are uniformly distributed on the stellar surface, which uniquely constrains the top radius of the tubes for each star. The diameter of the tubes at the bottom are assumed to be equal to the local density scale height, which is 110 km. The shape of the tubes for the different stars is depicted in Figure 2.

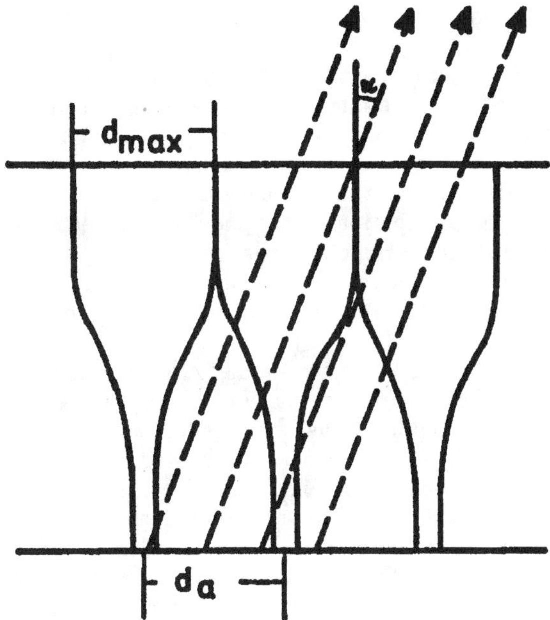

Figure 3. Array of flux tubes as used for the radiative transfer computations.

3. Methods and Model Computations

A pivotal task of this project is to provide heating calculations both for the acoustically heated and magnetically heated chromospheric components. Therefore, we have to provide initial wave energy fluxes and wave periods that are used for the heating models. For the acoustic heating model, we take $F_{\rm w,ac} = 8.0 \cdot 10^6$ ergs cm^{-2} s^{-1} and for the magnetic heating model, we take $F_{\rm w,m} = 2.2 \cdot 10^8$ ergs cm^{-2} s^{-1}. For the magnetic and acoustic wave periods, we take identical values of 60 s. These values are based on recent time-dependent simulations of the acoustic and magnetic energy generation given by Ulmschneider, Theurer, & Musielak (1996) and Ulmschneider & Musielak (1998), respectively. The magnetic wave energy flux is based on the study of longitudinal tube waves, but also considers contributions by transverse tube waves through the process of mode coupling. These initial wave energy fluxes and wave periods are used to compute wave models for the acoustically heated and magnetically heated chromospheric components.

The method used for calculating the acoustically heated chromosphere component has been described in detail by Buchholz et al. (1998). For the magnetic chromospheric component, we apply a wave code suitable for treating radiatively-damped, nonlinear longitudinal flux tube waves. This wave code has already been successfully used for the case of the Sun (e.g., Herbold et al. 1985; Rammacher & Ulmschneider 1989; Fawzy, Ulmschneider, & Cuntz 1998). In this code the set of the governing MHD equations is solved in the thin flux tube approximation using the modified method of characteristics. Similarly to the acoustic models, we introduce the wave energy by an oscillating piston at the bottom of the tubes. The waves are followed to the point of shock formation and beyond. The energy dissipation by the shocks is calculated self-consistently by solving the MHD Rankine-Hugoniot relations that also consider the distensibility of the flux tubes. After the insertion of approximately 20 wave periods, a dynamical steady state has been reached both in the acoustic and magnetic wave models.

The most important step is then to simulate the chromospheric emission for our two-component chromosphere models. For the flux tubes we adopt solutions of our longitudinal wave computations, whereas for the regions surrounding the flux tubes we take solutions of the acoustically heated models. Here we simulate the chromospheric emission using detailed radiation transfer along path rays for the two-component chromospheric flux tube forest (see Figure 3). For the radiative transfer computation, we consider 5 different angles. Figure 4 represents the Ca II K line profiles for the different angles for the model with $P_{\rm rot} = 10$ d and 40 d, respectively. Here we see that the line emission is highest in the case that the flux tubes are viewed under a shallow angle. In this case, the rays intersect a very large number of flux tubes assembling a relatively high contribution of the magnetically heated regions to the total emission. As expected, the Ca II emission is highest for narrow tubes related to fast rotating stars due to the relatively small voids ("canopies") between the tubes. It is found that the Ca II K line emission is about a factor of six greater in the case of $P_{\rm rot} = 10$ d than for $P_{\rm rot} = 40$ d. Due to our 2-level approach, we have calculated the line transfer

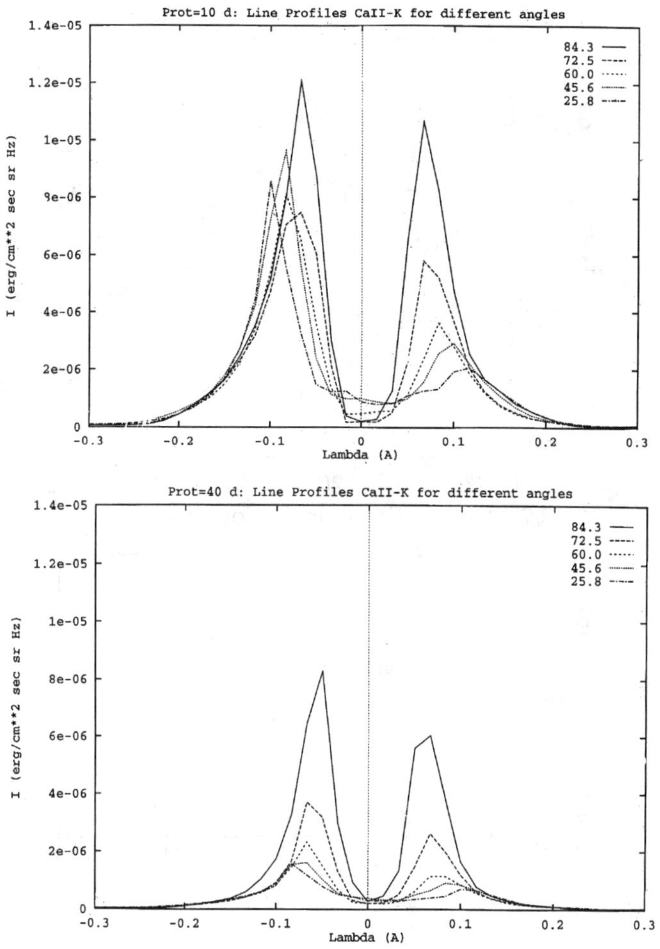

Figure 4. Ca II K line profiles for different angles of view for the model with $P_{\rm rot} = 10$ d (*upper figure*) and $P_{\rm rot} = 40$ d (*lower figure*).

for the Ca II K line only. The total Ca II H+K, line flux, also referred to as $F_{\rm HK}$, is calculated by scaling the line flux of Ca II K.

The fact that the Ca II H+K emission is significantly lower in the star with $P_{\rm rot} = 40$ d is fully consistent with the behavior of the wave energy flux as a function of atmospheric height. Due to the fact that in the star with lower rotation the flux tubes have a larger spreading as a function of atmospheric height, a lesser amount of energy is available in the Ca II formation region. In addition, as shown by Cuntz et al. (1998) and Fawzy et al. (1998), tubes with wider spreading usually lead to reduced shock strength and lower energy dissipation

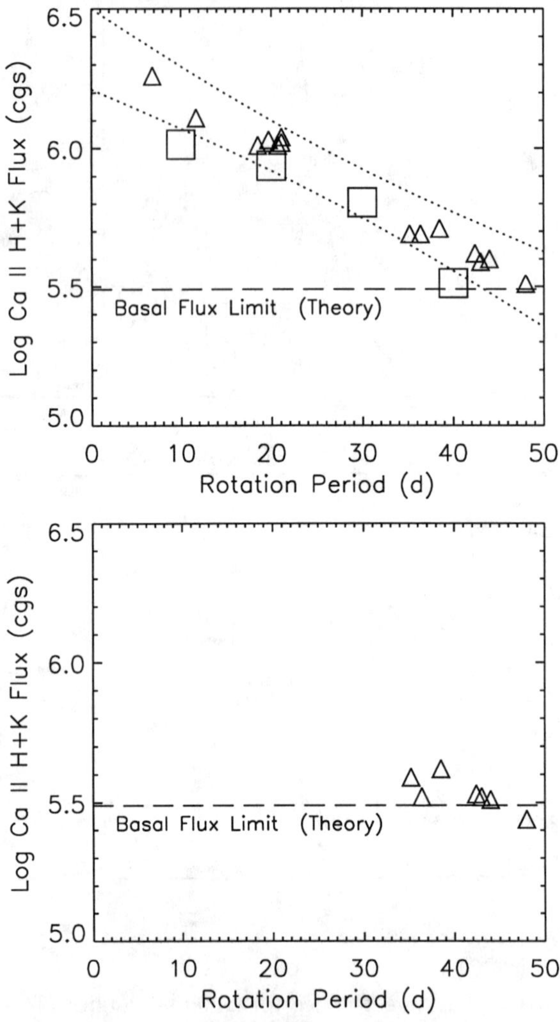

Figure 5. *Top:* Ca II H+K emission flux as function of stellar rotation. The triangles give the observational results for a set of stars with spectral type between K0 V and K3 V The squares give the results from two-component theoretical chromosphere models (magnetic and acoustic heating; see text) for K2 V stars of different rotation periods. The dotted lines indicate the 3σ flux limits permissible by the observations. In this respect, only the K2 V stars have been considered. The dashed line indicates the theoretical Ca II H+K basal flux limit given by pure acoustic heating (Buchholz et al. 1998). *Bottom:* Ca II H+K emission flux as function of stellar rotation for a selected number of stars for which a minimum activity level has been identified.

rates, which also reduces the chromospheric line cooling. The models calculated also show a remarkable behavior concerning the efficiency of converting the initial wave energy flux into Ca II radiation. Although the initial magnetic wave energy flux is greater by a factor of 27 than the initial acoustic wave energy flux, the $F_{\rm HK}$ emission by the two-component chromosphere models is increased only by a factor of 1.5 to 4.5 depending on the model relative to the nonmagnetic (i.e., acoustically heated) component. There are two reasons for this behavior: First, in magnetic flux tubes the wave energy flux is considerably reduced with atmospheric height due to the increase of the cross section as discussed before. Second, in magnetic flux tubes the density inside the tubes is considerably lower than outside of the tubes, which results in reduced Ca II radiative emission.

4. Comparison with Observations

In § 3, we found that the heating and chromospheric emission in magnetic flux tubes is significantly increased relative to the nonmagnetic component, as expected. The heating and chromospheric emission is found to be strongest in flux tubes with small spreading factors which correspond to fast rotating stars. We also calculated the theoretical Ca II H+K flux emission using multi-ray radiative transfer. These fluxes can now be compared with observations. In order to consider a decent number of stars, we have chosen a set of stars ranging from spectral type K0 V to K3 V with $B - V$ between 0.78 and 0.98. The stars have been taken from data sets given by Noyes et al. (1984), Rutten (1986, 1987), Baliunas et al. (1995), and Donahue, Saar, & Baliunas (1996). The calibration of the observed fluxes $F'_{\rm HK}$ was done with the calibration method by Noyes et al. (1984). Note that only stars could be considered that have known rotational periods and reliable Ca II measurements. As we want to concentrate on single stars, we disregard binary stars with significant binary interaction. In addition, we did not include stars with stellar rotation periods smaller than 5 d, which allows us to focus on the linear part of the Ca II emission — stellar rotation relationship. We therefore considered a total of 14 stars.

The comparison between the theoretical results and the observations is given in Figure 5 (top). Here we see that our theoretical results from our models nicely agree with the observations considering a 3σ standard deviation for the best fit. As an underlying function for the best fit, we assumed a linear relationship given by $\log F_{\rm HK} = A_{\rm HK} + B_{\rm HK} \cdot P_{\rm rot}$ where $A_{\rm HK}$ and $B_{\rm HK}$ are constants. To the best of our knowledge, this is the first time that an empirical chromospheric emission — stellar rotation relationship could be matched by theoretical models by starting from first principles. As discussed by CRUMS, the quality of that comparison remains unchanged if we restrict our sample to K2 V stars. All stars studied in this paper show to some degree variable Ca II H+K emission (e.g., Baliunas et al. 1995). The Ca II emission can thus be viewed as due to a variable magnetic dynamo, which is expected to also invoke changes of the photospheric $f_{\rm o}B_{\rm o}$ values. The only situation where a (almost) purely acoustically heated chromosphere might be approached would be at the magnetic cycle minima of the least active stars. We have estimated the minimum seasonal average of the Ca II fluxes from the plots in Baliunas et al. (1995) for selected low activity stars of our sample and converted them to $F'_{\rm HK}$(min). Those minima were found in 7 out of the 14 stars. Figure 5 (bottom) gives the comparison of these "minimum activity fluxes" to the theoretical Ca II H+K basal flux limit for pure acoustic heating as obtained by Buchholz et al. (1998). The observational results show nice agreement with the theoretical models which strongly supports the conjecture that inactive chromospheres of late-type stars are dominantly heated by acoustic waves, which is also consistent with the findings by Schrijver (1995).

We caution, however, that even in these cases, there is still likely magnetic activity remaining from both residual dynamo activity at cycle minimum and from a background, non-variable turbulent dynamo (e.g., Saar 1998). This magnetic dynamo activity is expected to also provide magnetic heating, which may or may not be relevant for the formation of the Ca II H and K lines. Our present study does not provide serious evidence of such relevance, but further studies are needed. $F'_{\rm HK}({\rm min})$ is a good empirical estimate of $F_{\rm ac}$ only if magnetic activity at cycle minimum does not seriously influence the Ca II H and K line formation region. The presence of magnetic heating itself in low-activity stars is undisputed as it is implied by the presence of a coronae. If magnetic contributions to $F'_{\rm HK}({\rm min})$ exist, the model acoustic fluxes $F_{\rm ac}$ will show $F_{\rm ac} < F'_{\rm HK}({\rm min})$ even for very inactive stars.

5. Conclusions and Outlook

In this paper, we discussed the derivation of a theoretical chromospheric emission — stellar rotation relationship using K2 V stars as an example. Therefore, we computed time-dependent two-component (acoustic and MHD) chromospheric heating models assuming different levels of magnetic activity as implied by the stellar rotation period $P_{\rm rot}$. The photospheric magnetic filling factors were chosen in agreement with recent observations, which allowed us to deduce a semiempirical relationship between the stellar rotation period $P_{\rm rot}$ and $B_{\rm o} f_{\rm o}$ with $B_{\rm o}$ and $f_{\rm o}$ being the photospheric magnetic field strength and the magnetic filling factor, respectively. This concept allowed us to construct flux tube models for different values of $P_{\rm rot}$.

The chromosphere models were constructed by performing state-of-the-art calculations of the generation of acoustic and magnetic energy in stellar convection zones, the propagation and dissipation of this energy at the different atmospheric heights, and the formation of specific chromospheric emission lines, which are then compared to the observational data. In all these steps, the two-component structure of stellar photospheres and chromospheres was fully taken into account. We find that due to the presence of magnetic flux tubes, the heating and chromospheric emission is significantly increased in the magnetic component. The heating and chromospheric emission is found to be the strongest in flux tubes with small spreading factors which are expected to be present in fast rotating stars. For stars with very slow rotation we were able to reproduce the basal flux limit of chromospheric emission previously attributed to pure acoustic heating. Most importantly, however, we find that the relationship between the Ca II H+K emission and the stellar rotation rate deduced from our models is consistent with the empirical relationship given by observations.

We consider these results as an important step toward a consistent theoretical explanation of chromospheric emission in different types of stars with different levels of magnetic activity. We anticipate the following future projects: First, we aim to expand our two-component chromospheric heating models to other types of stars, including subgiants, giants, and supergiants. In the past, an extremely large amount of observational data has been collected which deserve to be evaluated by theoretical models. We thus should be able to gain insight into heating of multi-component atmospheres, the evolution of magnetic activity in the HR diagram, the role of magnetic braking in chromosphere heating, and the appearance of HRD dividing lines. In addition, we also plan to target transition layer and coronal emission of magnetically active stars as indicated by *HST*, *ROSAT* and *AXAF*.

Acknowledgments. This work has been supported by the NASA Astrophysical Theory Program under NAG5-3027 to the University of Alabama in Huntsville. I also want to thank my collaborators B. Buchholz, D. E. Fawzy, Z. E. Musielak, W. Rammacher, S. H. Saar, and P. Ulmschneider who all contributed to this work.

References

Ayres, T. R., et al. 1996, ApJ, 473, 279

Baliunas, S. L., et al. 1995, ApJ, 438, 269

Buchholz, B., Ulmschneider, P., & Cuntz, M. 1998, ApJ, 494, 700

Charbonneau, P., Schrijver, C. J., & MacGregor, K. B. 1997, in Cosmic Winds and the Heliosphere, J. R. Jokipii, C. P. Sonett, & M. S. Giampapa, Space Science Series, Tucson: University of Arizona Press, 677

Cuntz, M., Rammacher, W., Ulmschneider, P., Musielak, Z. E., & Saar S. H. 1999, ApJ, submitted (CRUMS)

Cuntz, M., Ulmschneider, P., & Musielak, Z. E. 1998, ApJ, 493, L117

Donahue, R. A., Saar, S. H., & Baliunas, S. L. 1996, ApJ, 466, 384

Fawzy, D. E., Ulmschneider, P., & Cuntz, M. 1998, A&A, 336, 1029

Gray, D. F. 1989, ApJ, 347, 1021

Hasan, S. S., & van Ballegooijen, A. A. 1997, in Cool Stars, Stellar Systems, and the Sun X, R. A. Donahue & J. A. Bookbinder, poster edition

Herbold, G., Ulmschneider, P., Spruit, H. C., & Rosner, R. 1985, A&A, 145, 157

Jordan, C. 1997, Astron. Geophys., 38 (2), 10

Keppens, R., MacGregor, K. B., & Charbonneau, P. 1995, A&A, 294, 469

Marcy, G. W., & Basri, G. 1989, ApJ, 345, 480

Montesinos, B., & Jordan, C. 1993, MNRAS, 264, 900

Noyes, R. W, Hartmann, L. W., Baliunas, S. L., Duncan, D. K., & Vaughan, A. H. 1984, ApJ, 279, 763

Rammacher, W., & Ulmschneider, P. 1989, in Solar and Stellar Granulation, R. J. Rutten & G. Severino, Dordrecht: Kluwer, 589

Rüedi, I., Solanki, S. K., Mathys, G., & Saar, S. H. 1997, A&A, 318, 429

Rutten, R. G. M. 1986, A&A, 159, 291

Rutten, R. G. M. 1987, A&A, 177, 131

Rutten, R. G. M., & Pylyser, E. 1988, A&A, 191, 227
Rutten, R. G. M., Schrijver, C. J., Lemmens, A. F. P., & Zwaan, C. 1991, A&A, 252, 203
Saar, S. H. 1998, in Cool Stars, Stellar Systems, and the Sun X, R. A. Donahue & J. A. Bookbinder, San Francisco: ASP, Vol. 154, in press
Saar, S. H., & Schrijver, C. J. 1987, in Cool Stars, Stellar System, and the Sun V, ed. J. L. Linsky & R. E. Stencel, Berlin: Springer, 38
Schrijver, C. J. 1987, A&A, 172, 111
Schrijver, C. J. 1995, A&ARev, 6, 181
Schrijver, C. J., Coté, J., Zwaan, C., & Saar, S. H. 1989, ApJ, 337, 964
Schrijver, C. J., & Pols O. R. 1993, A&A, 278, 51
Simon, T., & Drake, S. A. 1989, ApJ, 346, 303
Simon, T., Herbig, G., & Boesgaard, A. M. 1985, ApJ, 293, 551
Skumanich, A. 1972, ApJ, 171, 565
Solanki, S. K. 1996, in Stellar Surface Structure, IAU Symp. 176, K. G. Strassmeier & J. L. Linsky, Dordrecht: Kluwer, 201
Soon, W. H., Baliunas, S. L., & Zhang, Q. 1994, Sol. Phys., 154, 385
Strassmeier, K. G., Handler G., Paunzen, E., & Rauth, M. 1994, A&A, 281, 855
Ulmschneider, P., & Musielak, Z. E. 1998, A&A, 338, 311
Ulmschneider, P., Theurer, J., & Musielak, Z. E. 1996, A&A, 315, 212

Solar and Stellar Activity: Similarities and Differences
ASP Conference Series, Vol. 158, 1999
C.J. Butler and J.G. Doyle, eds.

Diagnostics of Stellar Chromospheres and Transition Regions

A.C. Lanzafame

Institute of Astronomy, University of Catania, Viale A. Doria 6, I-95125 Catania, Italy

Abstract. Diagnostics of stellar chromospheres and transition regions are discussed. Their validity is controversial. Mathematically, the (underlying) problem is ill-posed, the existing solutions being neither unique nor stable. Because of inhomogeneity, non-uniformity and possible non-equilibrium conditions of the plasma, some of the physical assumptions on which such methods rely may be inaccurate or invalid. Notwithstanding these comments, the gross picture obtained is valuable in establishing common trends, systematic behaviour and the relevant physics involved. The possibility that plasma dynamics may invalidate current diagnostic techniques is examined, together with possible methods of investigation. It is suggested that some uncertainties may be smaller for the most active main sequence stars of the latest spectral type.

1. Introduction

Recent numerical simulations have shown that, in dynamical conditions, semi-empirical non-local thermodynamic equilibrium (NLTE) chromospheric models may have little or no relationship with the "mean" chromosphere, thus questioning the physical significance of such models (Carlsson & Stein 1995). Similarly, in the optically thin regime, the evolution of transition region and coronal plasma may have time-scales shorter that the ionization/recombination time-scales. In general, therefore, detailed atmospheric *inverse* models may not provide reliable constraints on real atmospheres and, to date, only a few results can be considered trustworthy.

The thermal interface that exists between the corona and the chromosphere can be referred to as the "classical transition region" (CTR). Feldman (1983) originally developed a physical picture of the transition region in which "unresolved fine structures" (UFS) are superimposed on the CTR and are physically unconnected with the corona. A fundamental problem in our current understanding of the outer atmosphere of the Sun and solar-like stars is whether the line emission from the plasma at temperatures $1 \cdot 10^4 \leq T_e \leq 7 \cdot 10^5$ is dominated by CTR structure or UFS.

The consequences of a possible dominance of UFS in the transition region and of the absence of a relationship between forward "mean" and inverse "semi-empirical" model chromospheres (Carlsson & Stein 1995) have profound implications on our ability to diagnose the outer atmospheres of late-type dwarfs, as outlined in the following sections.

2. Limitations

In this paper semi-empirical NLTE modelling, emission measure and line-ratios analysis are all considered *inverse* methods for the determination of the stellar atmospheric physical parameters and structure. Consideration of geometrical factors for the emission measure are not relevant to the present discussion and are therefore ignored. Furthermore, for the purpose of this paper, line ratio analysis is viewed as a special case of the emission measure method, in which the plasma distribution in temperature and density is approximated as a Dirac delta function in the (T_e, N_e) domain, where T_e is the electron kinetic temperature and N_e the electron number density. The limitations considered here are essentially twofold: on the one hand to what extent is the problem ill-posed mathematically and on the other hand to what extent does the underlying physical theory break-down due to plasma dynamics.

2.1. Integral mapping

In the optically thick regime the time-independent radiative transfer equation in plane parallel geometry has the *formal* solution (see, e.g., Mihalas 1978)

$$\left[I(\mu,\nu,t)e^{-t}\right]_{\tau_1}^{\tau_2} = -\int_{\tau_1}^{\tau_2} S(\mu,\nu,t)e^{-t}\,dt \tag{1}$$

where $d\tau \equiv -\chi(z,\mu,\nu)dz/\mu$, $S \equiv \eta/\mu$, z is the spatial dimension, $I(\mu,\nu,\tau)$ the specific intensity at τ having frequency ν and spatial direction $\mu = \cos\theta$ with θ the polar angle, χ the total opacity and η the total emissivity. In the optically thin regime ($\chi \sim 0$) the radiative transfer equation can be reduced to the well known expression for the integrated line intensity (for a column of plasma of cross-sectional area A)

$$I_{j\to k} = \frac{1}{4\pi A}\int A_{j\to k} N_j\,dz \tag{2}$$

where N_j denotes the population density of ions in the upper state j and $A_{j\to k}$ is the radiative transition probability for the transition $j \to k$. The integrand in Eq. (2) can be written as a product of the abundance of the emitting element (of nuclear charge Z) relative to hydrogen ($\mathcal{A}(Z)$, assumed constant in the region of formation of the line), a function $G(N_e, T_e)$ describing the plasma ionization and excitation and a function $\phi(N_e, T_e)$ (the differential emission measure DEM) giving the plasma distribution in density and temperature (see, e.g., Judge et al. 1997 and references therein). Two common assumptions are that the plasma electron density or pressure is constant over the relatively small range of temperature where $G(T_e, N_e)$ has significant values. That is, a model assumption is important even though the G function is usually written as a function of temperature alone. Also, we introduce the concept of surfaces of constant temperature of which there may be more than one in the viewed volume (Craig & Brown

1976). The intensity becomes

$$I_{j\to k} = \frac{\mathcal{A}(Z)}{4\pi} \int_{T_a}^{T_b} G_{j\to k}(T_e)\phi(T_e)dT_e \,. \tag{3}$$

To our purposes, therefore, we can identify *inverse* methods as those which attempt to derive models for the atmospheric plasma by *inverting* Eq. (1) (optically thick case) or Eq. (3) (optically thin case). It is important to stress here that, while in the optically thin case it is possible to perform a proper inversion of Eq. (3), in the optically thick case a proper inversion of Eq. (1) is not possible, Eq. (1) being the *formal* solution of a non-local and non-linear problem. Note also that, in this latter case, the atmospheric structure is determined through its effects on the line source function. That is why, in the optically thick NLTE regime, a semi-empirical method is used, which consists in manually adjusting the atmospheric structure in an attempt to reproduce some spectroscopic features. In either cases, the aim is to derive models for the atmospheric plasma directly from the observations using the least physical assumptions.

Unfortunately, the integral operators in Eqs. (1) and (3) have the property of mapping a wide domain of models onto a small domain of data. In the worst possible case this would mean that no model of interest would be falsifiable, all models being compatible with the data, within its errors. For what matters here, it is important to stress that solutions of Eqs. (1) and (3) are neither unique nor stable. For the inversion of Eq. (3), regularization techniques exist in which the extra information required to stabilise the inversion is introduced by way of a smoothness condition on the source function (see, e.g., Craig & Brown 1986). For the *inversion* of Eq. (1), it is mandatory to use as many lines and continua as possible in order to put as much constraint as possible on the solution (see also below).

2.2. Rate Equations

Considering the life-times of the various states of atoms, ions and electrons, it may be assumed that in most circumstances the free electrons have a Maxwellian distribution (or quasi-Maxwellian) and that the dominant populations of impurities in the plasma are those of the ground and metastable states of ions. When the dominant populations evolve on time-scales of the order of plasma diffusion time-scales, they should be modelled dynamically, that is in the particle number continuity equations, along with the momentum and energy equations of plasma transport theory. Even in such cases, however, the excited populations of impurities may be assumed relaxed with respect to the instantaneous dominant populations, that is they are in *quasi-equilibrium.* In the optically thin case, the quasi-equilibrium is determined by local conditions of electron temperature and electron density.

Suppose the dominant populations are a recombined ion ground state ($i = 1$) and a recombining ion ground state denoted by $+$. These states alone are assumed significantly populated that excited level populations (N_i for $i > 1$) are small in comparison. The set of radiative and collisional couplings between levels is denoted by C_{ij} (an element of the "collisional-radiative matrix" representing transitions from j to i), which has different forms in the optically thin and thick cases (see, e.g., Mihalas 1978 and McWirther & Summers 1984). To C_{ij}, the direct ionization from each level of the ion to the next ionization stage (coefficient q_i) and direct recombinations to each level of the ion from the next ionization stage (coefficient r_i) are added. Then the quasi-equilibrium statistical balance is

$$\frac{d}{dt}N_1 = \sum_{j\neq 1} C_{1j}N_j + C_{11}N_1 + N_e N_+ r_1 \quad (4)$$

$$0 = \sum_{j\neq 1} C_{ij}N_j + C_{i1}N_1 + N_e N_+ r_i \quad , \quad i = 2, ... \quad (5)$$

where

$$-C_{11} = \sum_{j\neq 1} C_{j1} + N_e q_1 \quad (6)$$

is the total loss rate coefficient for the ground state population and N_e the electron density. Here the important point is that the theory underlying diagnostic methods breaks down when the term dN_1/dt in Eq. (5) is not negligible with respect to the other terms, i.e. when the plasma evolution takes place on time-scales comparable with the ionization and recombination time-scales.

The atomic relaxation time-scales for the ground and metastable states must therefore be short enough to allow the plasma to reach equilibrium. In the optically thin approximation, it is possible to make a simple estimate of this time-scale ($\tau_{\rm transient}$) by looking at the summation, over all ionization stages of an element, of the reciprocal of the sum of the collisional dielectronic recombination and ionization coefficients (Brooks et al. 1998). In this way, we can obtain an estimate for the minimum amount of time required for a particular species to relax into equilibrium in a plasma of given electron temperature and density. Transient conditions for the plasma hold if it is subject to evolution with a time-scale $t < \tau_{\rm transient}$. In such cases the Li-like and Na-like ions are expected to be most affected because of their long recombination time-scales from He-like and Ne-like ionization stages (see below).

3. Methods

The diagnostic techniques considered in this paper are the (differential) emission measure and line ratios in the optically thin regime (transition region) and NLTE semi-empirical modelling in the optically thick regime (chromosphere).

3.1. Semi-empirical NLTE Modelling

In general, radiation emitted from the more dense chromosphere cannot be treated in the optically thin approximation. Furthermore, since the photon mean free path is larger than the typical scale-height of temperature and/or density variation, the non local thermodynamic equilibrium (NLTE) approximation must be employed. This consists in the simultaneous solution of radiation transfer and the statistical equilibrium equation for the atomic population densities. Given the complexity of the problem, a physical description of the chromosphere, both in the Sun and in late type stars, is still very approximate. The role of one-dimensional NLTE semi-empirical models is to provide some kind of average information about the chromosphere and its energy budget. However, since the averaging process is non-linear and non-resolved dynamics are likely to have a major role in the atmospheric state, such models can provide only a rough indication of the physical conditions of the atmosphere and they could fail in reproducing spectroscopic signatures formed in overlapping regions.

Semi-empirical NLTE chromospheric modelling remains the only method for inferring the chromospheric structure with the least physical assumptions. However, its weaknesses are such that only a few results could be considered trustworthy and, for instance, the energy constraints derived from them may not be realistic. The method does not (and cannot) represent a proper mathematical inversion of the relevant integral equation and is prone to subjective choices. Filling factors are sometimes used as free parameters making the determination of the chromospheric structure even more uncertain. The physics involved is also somewhat uncertain (e.g. dynamics, PRD, diffusion, turbulence). But, foremost, the physical meaning of resulting models is uncertain, e.g. whether they can be considered *distribution / state functions* for the chromosphere as suggested by Athay (1981).

A few results for dwarf stars of spectral type M and K that establish systematic trends, common behaviour and the relevant physics involved - and therefore considered here as trustworthy - are the following:

- Even Balmer absorption is indicative of the presence of a chromosphere (Cram & Mullan 1979; Cram & Giampapa 1987);
- A marked chromosphere is present even for stars with the lowest level of activity (see, e.g., Mauas et al. 1997 and references therein);
- Stars with Hα in emission have both a chromospheric temperature rise at larger column mass and a higher chromospheric temperature (see, e.g., Mauas et al. 1997 and references therein);
- Significant non-radiative heating is required in the upper chromosphere and lower transition region (Lanzafame 1995; Mauas et al. 1997);
- In active dMe stars (dense chromosphere) the chromosphere may become an efficient radiator at continuum wavelengths thus producing a UV excess due to the star's level of activity (Houdebine et al. 1996);
- The treatment of line blanketing in NLTE is critical for the evaluation of the continuum and, therefore, for line intensities. This assumption affects more strongly less active stars (Falchi & Mauas 1998);
- The Lyα must be modelled in partial redistribution (PRD). The treatment of Lyα in PRD or CRD can affect the statistical equilibrium and the formation of other lines, in particular Hα. This assumption affects more strongly less active stars (Falchi & Mauas 1998 and references therein).

The impact of some not fully justified assumptions commonly used in semi-empirical NLTE modelling have been recently studied by Falchi & Mauas (1998). They showed, for instance, how the comparison with observations is to be done on the flux spectrum not normalised to the continuum. In fact, the modifications induced by the spectral line depends on the underlying continuum, which must be matched with the observations as well.

Comparing Lyα and Hα surface fluxes, Doyle et al. (1990) notice that, for dMe stars, the fluxes are equal to within a factor of two. This has been considered an important constraint in the model calculations of the Lyman and Balmer series fluxes by Houdebine & Panagi (1990) and Houdebine & Doyle (1994). However, the observational data for Lyα from late type stars must be treated with caution. Woodgate et al. (1991) obtained rapid time sequence of spectra in the Lyα region of the dMe star AU Mic with the Goddard High Resolution Spectrograph on the Hubble Space Telescope, which showed a strong interstellar absorption which was not taken into account in the measurement of Doyle et al. (1990). On the theoretical side, plane-parallel semi-empirical modelling for the Lyα on the Sun has not yet proven to give a satisfactory description for the formation of this line (see, e.g., Fontenla et al. 1991). More uncertainties rest on the unknown role of UFS and/or up- and down-flows in the transition region. PRD effects on the Lyα and the super-imposition of the region of formation with Hα depends critically on the activity level (see, e.g., Falchi and Mauas 1998). In view of such uncertainties, the use of model atmospheres for calculating the Lyα attenuation factor due to interstellar medium (Doyle et al. 1997) must also be viewed with caution.

Many authors have argued that, in the case of late-type stellar atmospheres where surface inhomogeneities are present, a one-component, plane-parallel model atmosphere would fail to reproduce spectral lines whose formation takes place at depths having overlapping values of pressure and temperature. Heretofore, the extent to which such inhomogeneities would invalidate single-component models is very largely uncertain. The use of multi-component atmospheric models has been suggested, which make use of a *filling factor* to take into account the different contribution to the line formation from different features on the stellar surface. Houdebine & Doyle (1994), for instance, argued that in order to reproduce the Lyα/Hα ratio, the Hα central reversal and the Balmer decrement in dMe stars, a very high-pressure, thin transition region was required with a *filling factor* of the order of 30% of the stellar surface. The validity of such an approach is doubtful, because of the aforementioned theoretical difficulties in reproducing the Lyα flux, the major difficulties in reproducing line central reversals (see below) and the lack of independent estimates on such filling factors. Other factors such as dynamics and partial redistribution effects have not been fully explored and therefore, before a premature introduction of the filling factor as a free parameter, an exhaustive search of the physics of line formation is required. Recent progresses in imaging techniques (see, e. g., Walter et al. 1987; Neff et al. 1989) offer some hope to derive, in some cases, independent information on the coverage of active regions.

Consider now the use of line central reversal (like in the Hα emission) as diagnostics. NLTE semi-empirical modelling for the solar atmosphere has so far failed to reproduce emission line profiles with central reversal. Lanzafame (1994b), for instance, investigating the formation of the Si II resonance multiplet in the Sun, found that the theoretical lines show a central reversal which is not present in the observations. This may be due to our poor understanding of micro- and macro-turbulence, inhomogeneities or PRD effects. A similar problem is found in other lines, e.g. the Mg II h&k, for which Uitenbroek (1989) computed theoretical line profiles under PRD. In his calculations the theoretical profiles agree reasonably with the observations, but the computed Mg II h&k emission peaks are much narrower and the computed central reversal is much deeper than observed. Note, however, that NRL observations of the Si II (1533 Å) at 12" from the limb above a solar coronal hole (Feldman et al. 1976) show a pronounced central reversal and that the quiet sun HRTS observations of Si II (1526 Å) and Si II (1533 Å) at $\mu = 0.65 - 0.70$ and $\mu = 0.17 - 0.20$ also show a weak central reversal, which confirm the presence of multiple scattering of photons near line centre. A pronounced central reversal in these lines was also observed by Boland et al. (1975). Line central reversal tends to disappear when the intensities are averaged over a larger area on the solar surface and this has profound consequences in the modelling of stellar chromospheres.

Stationary, plane-parallel NLTE models show that in active M and K dwarfs the Hα central reversal *increases* with the temperature gradient (e.g. Lanzafame 1995). The reason for this can be ascribed to the provision of a geometrically thicker Hα emitting plasma at lower transition region temperatures in the models with a smaller temperature gradient. Surface inhomogeneities, however, can cause a smearing of the profile and therefore a less pronounced central reversal. Intuitively, the existence of a spicule-type pattern of velocities on the surface would imply the existence of Hα emitting elements with different velocities with respect to the observer; then, the observed line would be the sum of the emission fluxes from these elements or, in other words, of Hα emission-elements with a certain distribution of wavelength shifts due to the Doppler effect. This scenario is indeed similar to what is observed in the solar case for optically thick emission lines. It is concluded here that the line central reversal cannot be used as a stringent constraint in stellar diagnostics. but only as a broad criterion to select possible models. Lanzafame (1995) argued that when the discrepancies between the Hα profile produced by some model and the observation are very large, so that it a much greater "smearing" by inhomogeneities than other models is required, this model should be disregarded on the basis that it is difficult to imagine a distribution of line shifts that would bring such lines in better agreement with the observations. Using such arguments, Lanzafame (1995) showed that high temperature gradients are unrealistic and the lower transition region in dMe stars turns out to have a smaller temperature gradient than the solar one.

Lanzafame (1994a; 1995) and Lanzafame & Byrne (1995) have studied the upper chromosphere and lower transition region of active stars by means of multi-line semi-empirical modelling based on the hydrogen Lyα and Hα lines, the Si II multiplets near 1814 Å and 1262 Å and the Mg II h&k multiplet. Under the assumption that a CTR in hydrostatic equilibrium dominates the emission from the plasma at temperatures $1 \cdot 10^4 \leq T_e \leq 7 \cdot 10^5$ such models can be effectively interfaced with the results from emission measure analysis. Lanzafame & Byrne (1995) showed that for ϵ Eri it is possible to match satisfactorily the transition region structure derived from the emission measure distribution and semi-empirical chromospheric NLTE models, satisfactorily reproducing, therefore, a number of spectroscopic signatures in the visible and UV range. An interesting result is that, unlike the solar case, for the dK star ϵ Eri and the dMe star V 1005 Ori it is possible to reproduce the observed helium lines with a simple model compatible with other observational constraints. Over-ionization by coronal EUV radiation is found to play a minor role in the formation of the triplet helium lines. Should these results extend to a wide range of active stars, it would imply significant differences with the solar case such as a higher role of electron-atomic collisions in establishing the atomic population densities and in the energy deposition and balance. Note also that since ionization/recombination time-scales depend significantly on density, being smaller at higher densities, ionization equilibrium is more likely to occur in active stellar atmospheres than on the Sun.

Comparison of NLTE models with observation has also outlined the evidence of a mass circulation in the outer atmosphere similar to what is observed in the solar case. The Hα emission profiles observed in active late type stars show systematic asymmetries in the core central reversal which are consistent with the presence of a neutral hydrogen down-flow in the lower transition region. The presence of possible down-flows similar to what are observed in the solar case is confirmed by high-resolution UV GHRS spectroscopy of active late-type stars (e.g. AU Mic and HR 1099 - Wood 1996). Byrne et al. (1995) have also discussed the asymmetries observed in the emission Hα of the RS CVn II Peg. They suggested that material is accelerated outward in the deep chromosphere as revealed by the predominantly blue wing emission, while the small blue shift of the bulk of the quasi-Gaussian Hα emission arises in decelerated, but still rising upper chromospheric material, and the slightly red-shifted absorption reversal arises in the plasma during the early phases of in-fall at the base of the transition region.

A number of long-standing problems with chromospheric NLTE modelling remain, whose solution probably relies on radiation hydrodynamic modelling such that of Carlsson & Stein (1997). Amongst these are the difficulties in reproducing the spectroscopic signature formed in overlapping regions, such as Ca II and Mg II (Giampapa et al. 1982); Na I and Hα (Andretta et al. 1997; Short & Doyle 1998); the conflict between IR CO lines and UV emission lines (see, e.g., Avrett 1997); the formation Ca II bright grains (Carlsson & Stein 1997).

3.2. Emission Measure Analysis

The degree of equilibrium of the outer atmospheric plasma has profound consequences on the diagnostic methods and on our capability of describing the processes occurring in stellar atmospheres. It has long been recognised that dynamic evolution of the TR and coronal plasma can take place on time-scales shorter than ionization/recombination time-scales. Solar and stellar TR plasma diagnostics rely on several simplifying assumptions the most critical of which is the ionization balance. It must be stressed that no "diagnostic" as such can be reliable if the physical assumption of ionization equilibrium does not hold. Clearly, this would be the case if UFS evolving on time-scales shorter than the ionization/recombination time-scales dominate the radiative emission. Current estimates of the radiative output of stellar outer atmospheres (e.g. Doyle 1996 and references therein) critically depend on such assumption. Judge et al. (1995) found highly significant and systematic discrepancies in the emission measure analysis of transition region lines observed with SOLSTICE on the *UARS* spacecraft. These authors suggest that the most likely explanation for such discrepancies is the breakdown of the equilibrium ionization balance by dynamic and diffusive effects. Such highly significant and systematic discrepancies diminish toward coronal temperatures. These results, however, critically depend on the accuracy of the atomic data, as systematic discrepancies amongst iso-electronic sequences can possibly indicate errors in the atomic data (Lang et al. 1990).

Beside the limitations discussed in § 2., large errors may arise from noise in the observational data and from uncertainties in the fundamental theoretical atomic data and their derived quantities which enter the kernel of the integral equation. Other important issues in such analyses concern the density dependence of atomic populations. In conditions typical of the solar corona, both the ionization balance and excitation rates depend significantly on the electron density. The former is principally because dielectronic recombination is sensitive to electron density. Considering excitation, several iso-electronic sequences allow the presence of metastable levels, which can have population densities comparable to the ground level. The population distribution amongst ground and metastable levels is sensitive to electron density. In general, at coronal densities, levels of an LS term are not populated according to their statistical weight, so that LSJ resolution in the kernel calculations becomes mandatory.

A sample case: SERTS observations Analysis of coronal spectra by means of the DEM method (e.g. Brosius et al. 1996 for SERTS-91, 93 data) has led to the suggestion that the distribution of plasma in temperature might exhibit a double peak between $\log T_e = 6.1$ and 6.7. According to Brosius et al. (1996), this could indicate distinct contributions from both the quiet Sun and active regions, the higher temperature peak being due to flaring or to some other enhanced level of activity. Analyses on the SERTS-89 observations have either suggested a DEM structure with a single high temperature peak around $\log T_e \sim 6.5$ (Brickhouse et al. 1995) or with a triple peak between $\log T_e \sim 6$ and 6.7 (Landi & Landini 1997). The existence of such structures in the SERTS-89 observations remains unproven, since an inaccurate treatment of atomic level populations or an integral inversion technique with arbitrary smoothing can also lead to a similar but spurious multiple peak in the DEM.

In Figure 1, the SERTS-89 DEM is shown, assuming a grid of uniform P_e models in the evaluation of kernels. A similar analysis has been performed assuming a grid of uniform N_e models. Full details of the analysis will be given elsewhere. At low P_e, i.e. approaching the zero density limit, the DEM structure shows a well defined double peak, which disappears for $P_e \geq 3 \cdot 10^{14}$ cm^{-3} K. Density-sensitive line ratios suggest an electron density $N_e \geq 5\ 10^9$ cm^{-3} at $T_e \sim 10^6$ K, from which it is estimated, independently of the DEM analysis, $P_e \sim 5\ 10^{15}$ cm^{-3} K. The double peak is therefore spurious and arises from an incorrect treatment of the atomic coefficients, i.e. those approaching the zero density limit. For $3\ 10^{14} \leq P_e \leq 10^{18}$ cm^{-3} K in the uniform pressure case or $10^9 \leq N_e \leq 10^{12}$ cm^{-3} in the uniform density case, the DEM does not show marked variations. A similar behaviour, with more substantial double peaks having different heights or appearing for different values of uniform P_e or N_e, is found when lines having collisional data of different accuracy are selected for the integral inversion or when lines more sensitive to N_e are included. In this latter

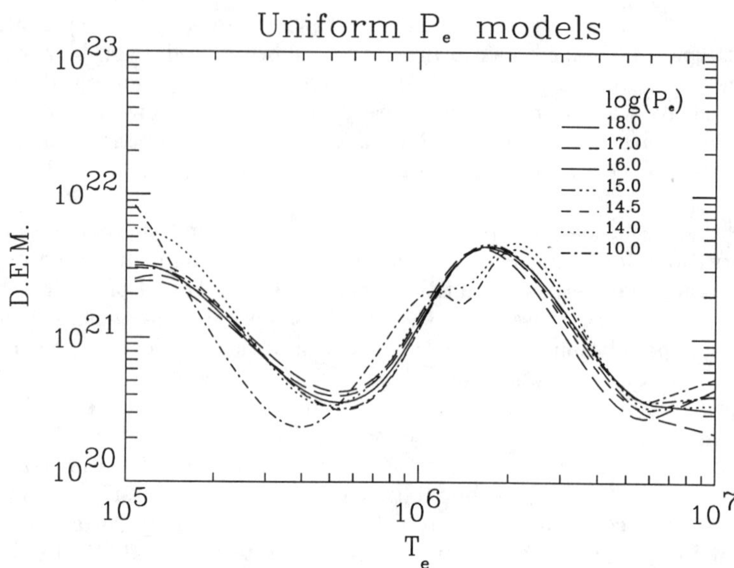

Figure 1. Differential emission measure for an averaged solar active region using EUV line intensities observed by SERTS-89. The different curves are obtained from different sets of G functions. Each set is derived from a *model* of uniform electron pressure whose value is indicated in the figure. Note the spurious double peak for low P_e.

case the DEM shows large and more severe variations with P_e or N_e. These results, together with the well known limitations of inversion techniques with arbitrary smoothing (that is when the smoothing parameter is chosen arbitrarily to balance a minimum deviation with the appearance of structures sharper than the width of the G functions) show that the uncertainties in the atomic physics, in the observations and in the degree of equilibrium of the plasma pose severe limitations on the possibility of resolving real multiple peaks at $T_e \sim 10^6$ K.

Consider now the issue of ionization non-equilibrium which could be revealed by significant and systematic discrepancies between iso-electronic sequences in the emission measure analysis. The reason for this is that some iso-electronic sequences are more susceptible than others to departures from equilibrium because of their long time-scales for recombination from the adjacent ionization stage (see § 2.2.).

For the SERTS-89 observations, lines with relatively larger deviations in the DEM analysis belong to the B-like, C-like and Na-like iso-electronic sequences. I^{theo} for lines of the Na-like sequence, unlike the others, have values systematically *smaller* than I^{obs}, a fact that could indicate ionization non-equilibrium. However, deviations of B-like and C-like sequences have both small and large values and no significant trend with respect to temperature. This leads to the suggestion that large deviations could be mainly due to uncertainties in the atomic collision data and in the observations. Any possible evidence for ionization non-equilibrium is probably overwhelmed by these uncertainties. This analysis could confirm the absence of a discrepancy in the corona or the incapability of unambiguously revealing it, which may reflect the absence of steep temperature gradients in the corona and/or different energy balance regimes then in TR.

4. Conclusions

The degree to which the mathematical problem underlying chromospheric and transition region diagnostics is ill posed have been discussed together with possible breakdowns of physical assumptions. While in the optically thin case it is possible to stabilise the inversion by way of a smoothness condition on the source function, for the *inversion* in the optically thick regime it is mandatory to use as many lines and continua as possible in order to put as much constraints as possible on the solutions. Dynamic conditions are likely to invalidate the ionization equilibrium assumption and cause semi-empirical NLTE chromospheric models to have no relationship to the "mean" chromosphere. A study of the systematic behaviour by iso-electronic sequences can help in investigating on such possible breakdowns. Trustworthy results and controversial points have been discussed. Many results point to the idea that some of the uncertainties in the physical assumptions underlying current diagnostic techniques may be smaller for the chromosphere and the transition region of the most active main sequence stars of the latest spectral type. These facts are likely to become of foremost importance in establishing the limitations and validity of current diagnostic techniques and, therefore, in our understanding of solar and solar-like outer atmospheres.

Acknowledgments. This paper is my personal tribute to Brendan who, as my Ph.D. supervisor, introduced me to this field in 1990, supporting and encouraging me ever since. His friendship has been a great help in all these years. Thanks also to Hugh Summers and Jim Lang for giving me the opportunity of extending my interests to optically thin plasma spectroscopy. For introducing me to Brendan Byrne and Armagh Observatory and for his encouragement and support, I thank Marcello Rodonò.

References

Andretta V., Doyle J.G. & Byrne P.B., 1997, A&A, 322, 266

Athay R.G., 1981, in S. Jordan, The Sun as a Star, Washington DC: NASA, 84

Avrett E.H., 1997, in K.S. Balasubramaniam & J. Harvey, D. Rabin, Synoptic Solar Physics, 27

Brooks D.H., Summers H.P., Harrison R.A., Lang J. & Lanzafame A.C. 1998, in M.T.V.T. Lago &, A.Blanchard, The Non-Sleeping Universe, Astrophysics & Space Science, Kluwer, in press.

Brosius J.W., Davila J.M., Thomas R.J. & Monsignori-Fossi B.C., 1996, ApJS, 97, 551

Booland B.C., et al. 1975, MNRAS, 171, 697

Brickhouse N.S., Raymond J.C. & Smith B.W., 1995, ApJS, 97, 551

Byrne P.B., Panagi P.M., Lanzafame A.C., Avgoloupis, S., Huenemoerder D.P., Kilkenny D., Marang F., Panov K.P., Roberts G., Seiradakis J.H. & Van Wyk F., 1995, A&A, 299, 115

Carlsson M. & Stein R. F., 1995, ApJ, 440, 29

Carlsson M. & Stein R. F., 1997, ApJ, 481, 500

Craig I.J.D. & Brown J.C., 1976

Craig I.J.D. & Brown J.C., 1986, Inverse Problems in Astronomy, Bristol: Adam Hilger

Cram L.E. & Giampapa M.S., 1987, ApJ, 323, 316

Cram L.E., Mullan D.J., 1979, ApJ, 294, 626

Doyle J.G., Panagi P.M. & Byrne P.B., 1990, A&A, 228, 443

Doyle J.G., 1996, A&A, 307, 162

Doyle J.G., Mathioudakis M., Andretta V., Short C.I. & Jelinsky P., 1997, A&A 318, 835

Falchi A. & Mauas P.J.D., 1998, A&A, 336, 281

Feldman U., Doshek G.A. & Patterson N.P., 1976, ApJ, 209, 279

Feldman U., 1983, ApJ, 275, 367

Fontenla J.M., Avrett E.H. & Loeser R., 1991, ApJ, 377, 712

Giampapa M.S., Worden S.P. & Linsky J.L., 1982, ApJ,258, 740

Houdebine E.R. & Panagi P.M., 1990, A&A, 231, 459

Houdebine E.R. & Doyle J.G., 1994, A&A, 289, 185

Houdebine E.R., Mathioudakis M., Doyle J.G. & Foing B.H., 1996, A&A, 305, 209

Judge P.G., Woods T.N., Brekke P. & Rottman G.J., 1995, ApJ 455L, 85

Judge P.G., Hubeny V. & Brown J.C., 1997, ApJ, 475, 275

Landi E., Landini M., 1997, A&A, 327, 1230

Lang J., Mason H.E. & McWhirter R.W.P., 1990, Solar Physics, 129, 31

Lanzafame A.C., 1994a, Ph.D. Thesis, Queen's University of Belfast

Lanzafame A.C., 1994b, A&A, 287, 972

Lanzafame A.C., 1995, A&A, 303, 839

Lanzafame A.C. & Byrne P.B., 1995, A&A, 303, 155

Mauas P.J.D., Falchi, A., Pasquini L. & Pallavicini R., 1997, A&A, 326, 249

McWhirter R.W.P. & Summers H. P., 1984, in Applied Atomic Collision Physics Vol. 2, C. F. Barnett & M. F. A. Harrison eds., Academic Press, 51

Mihalas D., 1978, Stellar Atmospheres, W. H. Freeman and co.

Short C.I. & Doyle J.G., 1998, A&A, 336, 613

Neff J.E., Walter F.M., Rodonò M. & Linsky J.L., 1989, A&A, 215, 79

Uitebroek H., 1989, A&A, 213, 360

Vernazza J.E., Avrett E.H. & Loeser R., 1981, ApJS, 45, 635

Walter F.M., Neff J.E., Gibson D.M., Linsky J.L. & Rodonò M., 1987, A&A, 186, 241

Wood B.E., 1996, Bull. American Astron. Soc., 188, 8702

Solar and Stellar Activity: Similarities and Differences
ASP Conference Series, Vol. 158, 1999
C.J. Butler and J.G. Doyle, eds.

Oscillations in Chromospheric Network Bright Points

D. Banerjee & J.G. Doyle

Armagh Observatory, College Hill, Armagh, BT61 9DG, N. Ireland

E. O'Shea

Dept. of Pure & Applied Phys., Queens University Belfast, Belfast BT7 1NN, N. Ireland

Abstract. We examine spectral properties of the solar chromospheric and transition region network from the SUMER instrument onboard *SOHO*. Observations were obtained for a number of lines ranging in temperature from Log T=4.1 K to 5.5 K. We concentrate on a network region and study intensity and velocity oscillations in two chromospheric spectral lines, observed simultaneously at O VI 1037.6Å and C II 1037.2Å. The 3.7 mHz intensity oscillations observed in O VI are often accompanied with a blue-shifted line profile of 3–5 km s^{-1}. These oscillations can be interpreted in terms of magneto-acoustic waves propagating upwards along thin magnetic flux tubes. Observations are compared with numerical simulations of waves propagating in a thin flux tube.

1. Introduction

Using SUMER, Doyle et al. (1999) have studied the intensity oscillations of two chromospheric lines N I 1319 Åand C II 1335Å. They have looked for differences between the power spectra of network and internetwork regions, and searched for evidence of propagating waves by considering phase differences in the emission of these lines which are formed at different temperatures. O'Shea et al. (1999) have enlarged the scope of this type of analysis using different lines formed at different heights (from Log T= 4.1 K to Log T = 5.5 K) in the outer solar atmosphere. In this short presentation we concentrate on a typical network region. The results are compared with magneto-acoustic waves.

2. Observation

The observations were made for a quiet region, typical of both network and internetwork(cell center). The dataset analysed here was acquired from 15:47 UT to 16:25 UT on 30 July '96 at 6 arcsec West of disk center, zero arcsec North, using SUMER on *SOHO*. Two spectral lines were observed simultaneously, O VI 1037.6Å and C II 1037.2Å, covering 50 wavelength pixels (~2.5Å) and 360 arcsec in the North-South direction. A full description of the Fourier technique can be found in Doyle et al. (1997). For line shifts, an average wavelength was derived

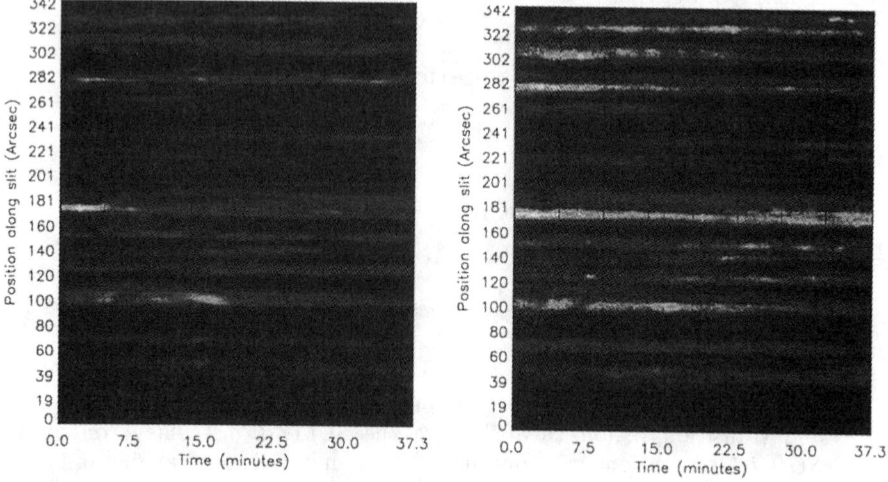

Figure 1. Surface plots for the summed counts (intensity map) for O VI 1037.6Å (left panel) and C II 1037.2Å (right panel) in the dataset starting at 15:47 UT on 30 July 1996.

via summing the data along the entire slit (20 to 320 arc sec) for each data time-point. The spectral data were then fitted with a Gaussian for each spatial pixel at each data time-point.

3. Results

Oscillations were found at several locations along the slit and are not confined to the brightest areas (see Figure 1). Network regions can clearly be seen in Figure 1, at positions along the slit that remain persistently bright, showing (almost) as horizontal streaks. Oscillations were seen to occur in packets of about 10-40 minutes duration. The best example of an oscillation group is plotted in Figure 2. for the region 175-180 arc sec along the slit corresponding to a network. In this 'quiet' region dataset, the frequency of intensity oscillations is typically around 3.7 mHz, which corresponds to an oscillation period of $\sim$ 4.8 minutes. These lines show velocity variations of 2-5 km/s amplitude with a spatial scale of several arc seconds. Furthermore we see clear evidence of blue shifts of $\sim$ 5 km/s for the first 25 minutes period of observation.

4. Numerical Model

We assume that a pulse is generated by the foot-point motion of the magnetic flux tubes in the network boundary, and has a Gaussian velocity profile, and this is consistent with the observations of photospheric bright points associated with magnetic elements, Berger et al. (1998). We study the propagation of transverse magnetoacoustic (also called kink) waves within a thin flux tube

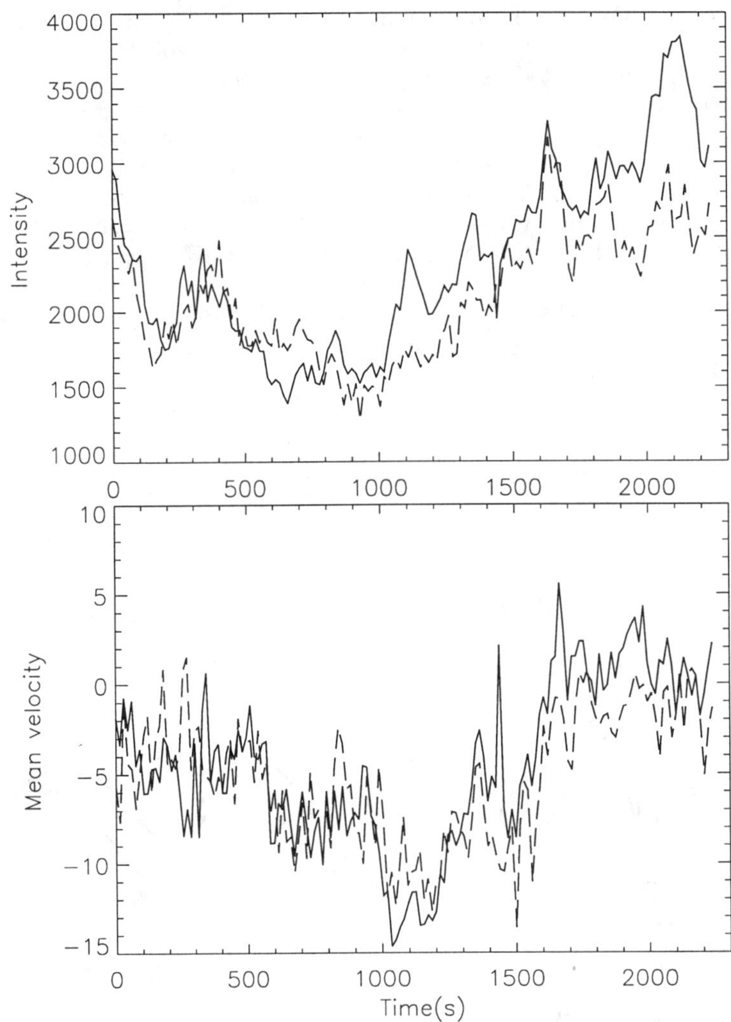

Figure 2. The behaviour with time of the intensity and Doppler shift of the O VI (solid line) and C II (dashed line) lines. The plots are summed over spatial pixels 175-200.

embedded in a two-layer isothermal atmosphere. The displacements of the flux tubes at different heights for a particular maximum velocity of the foot-point motion, can be derived (see Choudhuri et al. [1993] & Banerjee et al. [1999]). Figure 3 shows the displacements of the flux tubes at the formation heights of

O VI and C II as labeled. The parameter space is characterized by dimensionless variables, $\alpha = h/4H_1$, the measure of the thickness of the first layer, where h is the height of the layer in kilometers and $r = \sqrt{T_1/T_2}$, the measure of the temperature contrast between the two layers. The cut-off frequencies of the two layers are also related as $\omega_{c_2}/\omega_{c_1} = r$. We place the temperature jump around 2000 km ($\alpha = 2$) above the photosphere and the temperature contrast corresponds to the temperature jump in the transition layer. Whenever a pulse propagates through a stratified medium, it is known to leave a wake behind it oscillating with the cut-off frequency of the atmosphere. Note that the wake oscillates with a frequency which is neither the cut-off frequency of the lower layer ($\omega_{c_1} = 12.6$mHz) nor the upper layer ($\omega_{c_2} = 10.4$mHz). Instead it oscillates with a frequency $\omega = \omega_{c_1} + \omega_{c_2} = 23.0$ mHz. Thus this layer is oscillating with a cyclic frequency ($\omega/2\pi$) of 3.66 mHz, which is similar to the observed frequency 3.7 mHz (see Banerjee et al. [1999] for details).

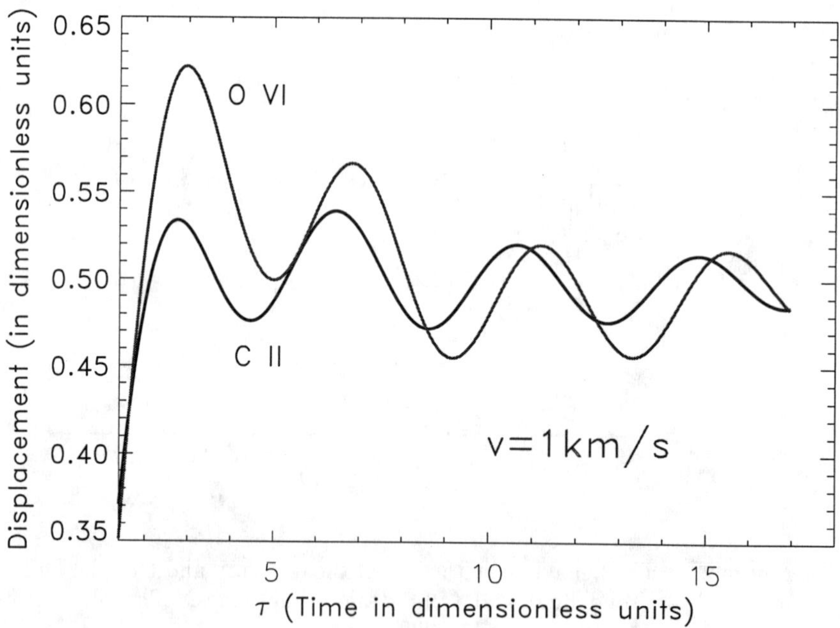

Figure 3. Displacement of the flux tube, at rough formation heights of C II and O VI lines as a function of time in units of cut-off frequency.

5. Conclusions

Our network observations suggests that the lower chromospheric intensity oscillations come from small regions (presumably magnetic flux tubes) of at most 5-8 arc sec along the slit and last for 10–30 mins. These intensity oscillations could be related to the impulsive motions at the photospheric level. The clear presence of a blue-shift indicates an upward propagating wave. The observed 4–5 min network oscillations can be interpreted in terms of longitudinal magneto-acoustic sausage waves propagating upwards along thin magnetic flux tubes. Kink waves can be generated by random foot-point motions at the photospheric level. As they propagate within flux tubes, their amplitude grows exponentially with height and become non-linear, thereby undergoing a mode transformation becoming sausage waves, which are detected on the disk.

References

Banerjee, D., Doyle, J.G. & O'Shea, E., 1999, A&A (in preparation)

Berger, T. E., Lofdahl, M. G., Shine, R. S. & Title, A., 1998, ApJ, 495, 973

Choudhuri, A.R., Dikpati, M., & Banerjee, D. 1993, ApJ, 413, 811

Doyle, J.G., van den Oord, G.H.J. & O'Shea, E., 1997, A&A, 327, 365

Doyle, J.G., van den Oord, G.H.J. & O'Shea, E., Banerjee, D. 1999, A&A (in press)

O'Shea, E., Doyle, J.G., Banerjee, D. & van den Oord, G.H.J., 1999, A&A (in preparation)

Solar and Stellar Activity: Similarities and Differences
ASP Conference Series, Vol. 158, 1999
C.J. Butler and J.G. Doyle, eds.

Chromospheric Activity of *ROSAT* discovered Weak-line T Tauri stars

D. Montes[1,2] & L.W. Ramsey[2]

The Pennsylvania State University, Department of Astronomy and Astrophysics, 525 Davey Laboratory, University Park, PA 16802, USA

Abstract.

We have started a high resolution optical observation program dedicated to the study of chromospheric activity in weak-lined T Tauri stars (WTTS) recently discovered by the *ROSAT* All-Sky Survey (RASS). It is our intention to quantify the phenomenology of the chromospheric activity of each star determining stellar surface fluxes in the more important chromospheric activity indicators, as well as to obtain the Li I abundance, a better determination of the stellar parameters, spectral type, and possible binarity. In this contribution we present preliminary results from high resolution echelle spectroscopic observations at the 2.1m telescope of the McDonald Observatory. We have analysed, using the spectral subtraction technique, the Hα and Ca II IRT lines of six WTTS located in and around the Taurus-Auriga molecular clouds.

1. Introduction

Weak-lined T Tauri stars (WTTS) are low mass pre-main sequence stars (PMS) with Hα equivalent widths $\leq$ 10 Å in which no signs of accretion are observed. The emission spectrum of these stars is not affected by the complications of star-disk interaction which often masks the underlying absorption lines as well as obscuring the central stellar object in classical T Tauri stars (CTTS). The WTTS are thus ideal targets to study the behavior of surface activity in the PMS stage of the stellar evolution. While there are a large number of studies in UV, X-ray and radio wavelengths, little research has been directed towards the study of the chromospheric activity using optical observations. Those which have been done are based on low resolution spectroscopic observations. Only some recent higher resolution studies centered in bona-fide WTTS in Taurus are available (see Feigelson et al. 1994; Welty 1995; Welty & Ramsey 1995, 1998; Poncet et al. 1998; Montes & Miranda 1999). In order to improve the knowledge of the WTTS chromospheres high resolution optical observations are needed. The WTTS discovered very recently by the *ROSAT* All-Sky Survey are good targets

[1]Departamento de Astrofísica, Facultad de Físicas, Universidad Complutense de Madrid, E-28040 Madrid, Spain

[2]Guest observer at McDonald Observatory

to accomplish these objectives. A large number of them have been found far away from the star formation clouds (Neuhäuser et al. 1995; Alcalá et al. 1996; Wichmann et al. 1996; Magazzù et al. 1997; Krautter et al. 1997). Whether these stars are really WTTS, or post TTS, or even young main sequence stars is a matter of ongoing debate (Feigelson 1996, Briceño et al. 1997, Favata et al. 1997). However, we will study only those in the Taurus Auriga molecular cloud for which very recent studies clearly confirmed their PMS nature.

In this contribution we present preliminary results of our high resolution echelle spectroscopic observations of RX J0312.8-0414NW, SE; RX J0333.1+1036; RX J0348.5+0832; RX J0512.0+1020; and RX J0444.9+2717.

2. Observations

The spectroscopic observations were obtained during a 10 night run 12-21 January 1998 using the 2.1m telescope at McDonald Observatory and the Sandiford Cassegrain Echelle Spectrograph (McCarthy et al. 1993). The spectrograph setup was chosen to cover the Hα and Ca II IRT lines. The wavelength coverage is about 6400-8800Å and the reciprocal dispersion ranges from 0.06 to 0.08 Å/pixel and the spectral resolution from 0.13 to 0.20 Å in the Hα line region. The spectra have been extracted using the standard reduction procedures in the IRAF package The wavelength calibration was obtained by taking spectra of a Th-Ar lamp. Finally, the spectra have been normalized by a low-order polynomial fit to the observed continuum. The chromospheric contribution in these features is determined using the spectral subtraction technique (Huenemoerder & Ramsey 1987; Montes et al. 1995; 1997). The synthesized spectrum was constructed using the program STARMOD (Barden 1985).

3. Results

We have analysed the Hα and Ca II IRT lines of six WTTS located in and around the Taurus-Auriga molecular clouds. These targets were selected from two sources: (1) From the *ROSAT* detected late-type stars south of the Taurus nebula (Neuhäuser et al. 1995; Magazzù et al. 1997 (hereafter M97)) we selected the stars that the spectroscopic studies of M97 and Neuhäuser et al. (1997, hereafter N97) clearly identified as WTTS from their greater Li abundance than Pleiades of the same spectral type. Some of them have been classified by these authors as single- and double-lined spectroscopic binaries (SB1, and SB2) and others are visual binaries (Sterzik et al. 1997, hereafter S97). (2) From the list of Wichmann et al. (1996) of new WTTS stars in Taurus, we selected the stars in which radio emission was detected by Carkner et al. (1997, hereafter C97) supporting their identification as genuine WTTS rather than ZANS.

RX J0312.8-0414NW, SE: This is a visual binary with components NW and SE separated by 14″ (M97, S97). The NW component is a G0V with $v\sin i$ = 33 km s^{-1} and is a SB2. The SE component is a G8V with $v\sin i$ = 11 km s^{-1}. Both components exhibit Hα absorption with a EW(Hα) of 3.5 and of of 2.5 Å respectively (M97; N97). Our spectra exhibit a strong Li I 6708 Å line confirming the PMS nature of these objects. However, the level of chromospheric activity

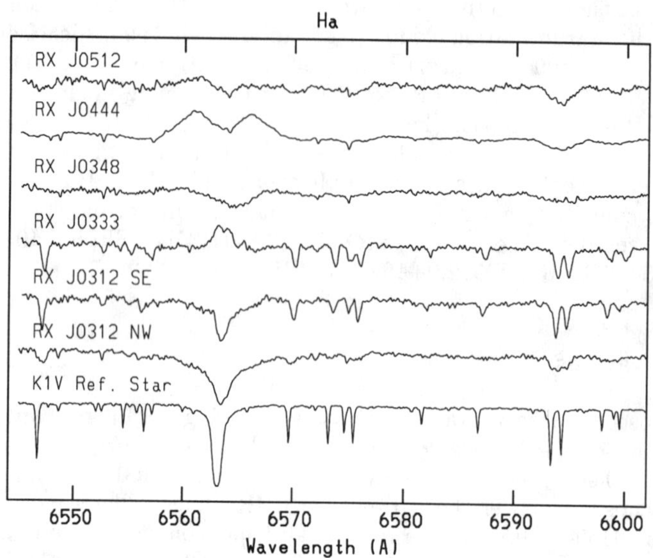

Figure 1. Representative spectra in the Hα line region. A K1V reference star is also plotted for comparison

is very low, only a small filling-in of Hα and the Ca II IRT lines is detected, in agreement with the earlier spectral type of both stars.

RX J0333.1+1036: This star is classified as a confirmed PMS star by M97 and N97 on the basis of its Li I abundance, however, C97 detect no radio emission. M97 give a spectral type of K3 and observed the Hα line in emission with a EW of –0.8 Å. N97 measured a $v\sin i$ of 20 km s^{-1} Emission above the continuum in Hα and Ca II IRT lines is detected in our five spectra from January 14 to January 20 1998 with small variations from night to night.

RX J0348.5+0832: This PMS star is rapidly-rotating ($v\sin i = 127$ km s^{-1}, N97), of spectral type G7 and has a small emission in the Hα line (EW = –0.1 Å, M97). In our five spectra (from 01/12/98 to 01/18/98) we observe the Hα line always in absorption, but filled in. The Ca II IRT lines are also filled in by chromospheric emission.

RX J0444.9+2717: This is a K1 star with Hα emission above the continuum (EW = –2.1 Å) and classified as a PMS star by W96 on the basis of its Li I abundance. The detection of radio emission by C97 confirms its PMS nature. Kohler & Leinert (1998) found a IR companion with a separation of 1.754″ and a brightness ratio at K of 0.102. We have eight spectra of this star available (from 01/12/98 to 01/20/98). The observed spectra are well matched using a K1V reference star with a rotational broadening of 65 $km\ s^{-1}$. Some of the more intense photospheric lines exhibit a flat-bottomed core (i.e. the core is noticeable filled in with respect to the reference profile) as is observed in other rapidly-rotating and spotted stars. A broad and variable double-picked

Hα emission above the continuum is observed. The Hα EW in the observed spectra changes from –1.2 Å to –2.6 Å. The Ca II IRT lines exhibit a strong filling-in.

RX J0512.0+1020: M97 give a spectral type K2 for this star and observed a small emission in the Hα line (EW = –0.1 Å). N97 measured a rotational velocity of 57 km s^{-1}. In our three spectra (from 01/14/98 to 01/17/98) we observe a variable filling-in of the Hα and Ca II IRT lines. The Hα line shows emission in the blue wing in one of the spectra.

Acknowledgments. This work was supported by the Universidad Complutense de Madrid and the Spanish Dirección General de Investigación Científica y Técnica (DGICYT) under grant PB94-0263, and by National Science Foundation (NSF) grant AST 92-18008. We thank the Staff of McDonald Observatory for their allocation of observing time and their assistance with our observations.

References

Alcalá J.M., et al., 1996, A&AS 119, 7

Briceño C., Hartmann, L.W., Stauffer J.R., et al., 1997, AJ 113, 740

Barden S.C., 1985, ApJ 295, 162

Carkner L., Mamajek E., Feigelson E.D., et al., 1997, ApJ 490, 735

Favata F., Micela G. & Sciortino S., 1997, A&A 326, 647

Feigelson E.D., et al., 1994, ApJ 432, 373

Feigelson E.D., 1996, ApJ 468, 306

Huenemoerder D.P. & Ramsey L.W., 1987, ApJ 319, 392

Kohler R. & Leinert C., 1998, A&A 331, 977

Krautter J., 1997, A&AS 123, 329

Magazzù A., Martín E.L., Sterzik M.F., et al., 1997, A&AS 124, 449

McCarthy J.K., Sandiford B.A., Boyd D. & Booth J., 1993, PASP 105, 881

Montes D., Fernández-Figueroa M.J., De Castro E., Cornide M., 1995, A&A 294, 165

Montes D., Fernández-Figueroa M.J., De Castro E. & Sanz-Forcada J., 1997, A&AS 125, 263

Montes D. & Miranda L.F., 1999, A&AS in preparation

Neuhäuser R., Sterzik M.F., Torres G. & Martín E.L., 1995, A&A 299, L13

Neuhäuser R., Torres G., Sterzik M. F. & Randich S., 1997, A&A 325, 647

Poncet A., Montes D., Fernández-Figueroa M.J. & Miranda L.F. 1998, in ASP Conf. Ser. 155, Cool Stars, Stellar Systems, and the Sun, 10th Cambridge Workshop, eds. R.A. Donahue & J.A. Bookbinder, CD-1772

Sterzik M.F., Durisen R.H., Brandner W., et al., 1997, AJ 114, 1673

Welty A.D., 1995, AJ 110, 776

Welty A.D. & Ramsey L.W., 1995, AJ 110, 336

Welty A.D. & Ramsey L.W., 1998, AJ in press

Wichmann R., Krautter J., Schmitt J.H.M.M., et al., 1996, A&A 312, 439

Solar and Stellar Activity: Similarities and Differences
ASP Conference Series, Vol. 158, 1999
C.J. Butler and J.G. Doyle, eds.

Radiative Transfer for Grabs

Robert J. Rutten

Sterrekundig Instituut, Utrecht, The Netherlands

Abstract. This contribution advertises freely available teaching materials on the theory of radiative transfer in stellar atmospheres.

1. Elementary Course

I teach a yearly course on the basics of radiative transfer to second-year students of physics and astronomy at Utrecht. The course is summarized in the second chapter of the advanced course described below and represents an extended version of the first chapter of the book by Rybicki & Lightman. This elementary course is not yet available on the web, awaiting rewriting in English, but the Dutch-language notes have been translated provisionally by Ruth C. Peterson (Santa Cruz) in 1992. You could ask her for a paper copy, or request a slightly improved version from the astronomers at Uppsala.

2. Advanced Course

Utrecht astronomy students also get a more advanced course on radiative transfer in stellar atmospheres from me. The lecture notes are freely available in the form of postscript files[1]. They represent a middle road between Mihalas' *Stellar Atmospheres* and the books by Novotny and Boehm-Vitense.

3. Numerical Exercises

These courses are accompanied by numerical exercises using IDL:

- Stellar Spectra A, *Basic Line Formation*, for first-year students;
- Stellar Spectra B, *LTE Line Formation*, for second-year students;
- Stellar Spectra C, *NLTE Line Formation*, for third-year students.

The first set is complete and on the web. The second set is planned for the spring of 1999. The third set has been developed by Mandy Hagenaar over the past years and should become available on the world-wide-web soon.

[1] http://www.astro.uu.nl/~rutten

Solar and Stellar Activity: Similarities and Differences
ASP Conference Series, Vol. 158, 1999
C.J. Butler and J.G. Doyle, eds.

Multiwavelength Optical Observations of the Chromospherically Active Binary System MS Ser

J. Sanz-Forcada, D. Montes[1], M.J. Fernández-Figueroa, E. De Castro & M. Cornide

Departamento de Astrofísica, Facultad de Físicas, Universidad Complutense de Madrid, E-28040 Madrid, Spain

Abstract. We present here a continuation of our ongoing project of multiwavelength optical observations aimed at studying the chromosphere of active binary systems using the information provided for several optical spectroscopic features that are formed at different heights in the chromosphere. In this contribution we focus our study on the preliminary analysis of the active binary system MS Ser. We have taken Hα and Hβ spectra in 1995 and high resolution echelle spectra (covering Hα, Hβ, Na I D_1, D_2, He I D_3, Ca II H & K, and Ca II λ8662 lines) in 1998. Strong emission in the Ca II H & K and Ca II IRT lines, coming from the primary component (recently classified as K2IV) is observed, and the secondary (G8V) also exhibits a small amount of emission. A near complete and variable filling-in of the Hα and Hβ is obtained after the application of the spectral subtraction technique. We detect also some seasonal variations between these two observing runs in comparison with our previous Ca II H & K observations taken in 1993.

1. Introduction

Griffin (1978) first observed the binary nature of MS Ser (HD 143313) and calculated the orbital elements for the system (see Table 1). Griffin gave its T_0 in MJD, which lead Strassmeier et al. (1993) to a bad calculation of the T_0 in HJD. Griffin proposed also K2 V/K6 V as spectral types for the components, based on photometric arguments for the secondary star. Bopp et al. (1981) observed a variable filling-in of the Hα line, and calculated a photometric period of 9.60 days, slightly different from the orbital period. Miller & Osborn (1996) confirmed the value of the photometric period, and Strassmeier et al. (1990) observed a strong emission in the Ca II H & K composite spectrum. Dempsey et al. (1993) noted some filling-in of Ca IRT lines for MS Ser, but not an emission reversal. Alekseev (1999) made a photometric and polarimetric study of MS Ser, calculating a spot area of 15% of the total stellar surface, and observed some

[1]The Pennsylvania State University, Department of Astronomy and Astrophysics, 525 Davey Laboratory, University Park, PA 16802, USA

variations. Finally, Osten & Saar (1998) revised the stellar parameters for MS Ser, suggesting with their available data K2IV/G8V as a better classification.

Following the series of papers devoted to the study of RS CVn stars through the simultaneous analysis of several optical activity indicators (see Montes et al. 1997, 1998; Sanz-Forcada et al. 1998), we present spectroscopic observations for MS Ser taken in different epochs for the Hα, Hβ, Na I D_1, D_2, He I D_3, Ca II H & K, and Ca II IRT lines. We also revise the luminosity class of the primary star.

2. Observations

Spectroscopic observations of MS Ser in several optical chromospheric activity indicators have been obtained during two observing runs. The first was carried out in 1995 June 12th with the 2.2 m telescope at the German Spanish Observatory (CAHA) in Calar Alto (Almería, Spain), covering two ranges: Hα (from 6510 to 6638 Å), and Hβ (from 4807 to 4926 Å), with a resolution of $\Delta\lambda$ 0.26Å in both cases. We have also made two observations with the 2.56 m *Nordic Optical Telescope* (*NOT*) at the Observatorio del Roque de Los Muchachos (La Palma, Spain), on April 5-14, using the Soviet Finish High Resolution Echelle Spectrograph (SOFIN). The wavelength range covers from 3640 to 10085 Å in a total of 40 echelle orders. The reciprocal dispersion achieved ranges from 0.07 to 0.18 Å/pixel. For the spectral subtraction we use α Boo (K1.5 III) and HD 45410 (K0 IV) as standard stars, applying 0.85/0.15 as proportional intensities for the stars. Table 1 shows the stellar parameters for the system. We give the correct value for T_{conj} in HJD (phase 0 means the primary star behind). Distance is given by the Hipparcos catalogue (ESA, 1997); B-V, T_{conj} and P_{orb} are given by Griffin (1978); T_{sp}, radius and Vsin i are as given by Osten & Saar (1998), and V-R from Alekseev (1999).

Table 1. Stellar parameters

T_{sp}	SB	R (R)	d (pc)	B-V	V-R	T_{conj} (H.J.D.)	P_{orb} (days)	P_{rot} (days)	Vsin$i$ (km s^{-1})
K2IV/G8V	2	3.5/1.0	88	0.94/1.23	0.73	2442616.142	9.01490	9.60	15/7

3. Results

3.1. Luminosity Class

We have revised the luminosity class of the primary star in MS Ser, through the study of the behaviour of Ti I and some other metallic lines. We have compared some of these lines (Ti I 8382, Ti I 6625) with those from α Boo (K1 III), and HD 45410 (K0 IV). These lines are very sensitive to the luminosity class for late type stars (see Montes al. 1998), and so we could better classify the primary component as luminosity class IV-III, or perhaps a giant. According to the data provided by the Hipparcos catalogue (ESA, 1997), we can calculate, through the distance modulus, the absolute visual magnitude, M_V, for MS Ser, which is 3.5. This suggests that the primary star could be a subgiant. By application of the Wilson-Bappu effect (Wilson & Bappu 1957; Montes et al. 1994) to our Ca II K spectra we obtain $M_V = 4.9$. This value is lower than the M_V that corresponds to a K1 V ($M_V(T_{sp}) = 6.6$ from Landolt-Börnstein (Schmidt-Kaler 1982)) but is higher than for a K1 III ($M_V(T_{sp}) = 0.6$). This result indicates that the primary component of MS Ser is of luminosity class IV or higher.

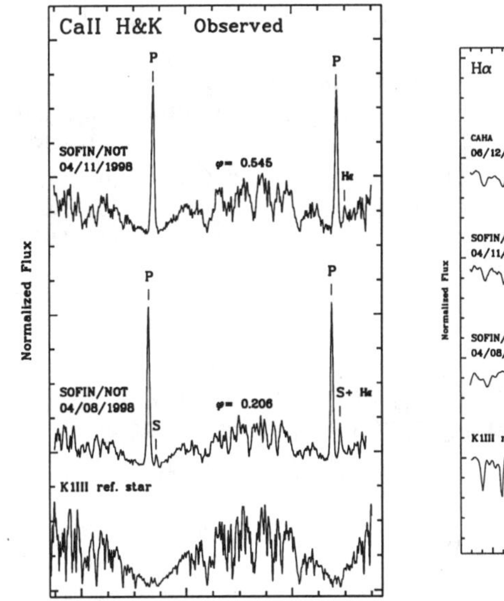

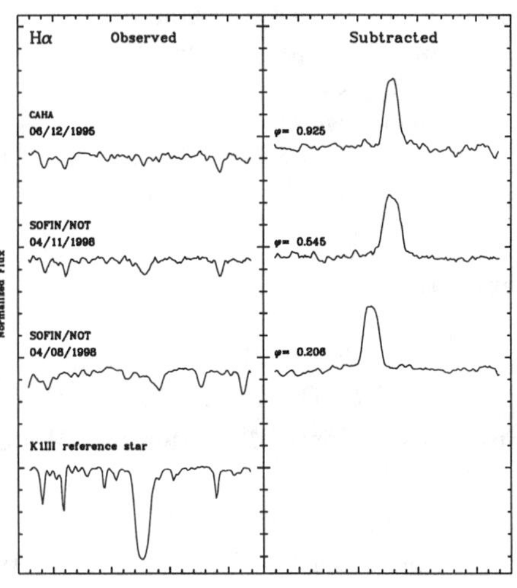

Figure 1. Ca II H & K (left) and Hα (right) spectra. The wavelength position of the emission lines of the primary (P) and secondary (S) components are marked.

3.2. Chromospherical Activity Indicators

The Ca II *H & K and Hϵ lines*: Strong emission in these lines and the Hϵ line also in emission arising from the hot component is observed in our previous observations of this system in the March 1993 at orbital phase 0.10 (Montes et al. 1995). In the present observations (April 1998) we have deblended the emission arising from both components in the spectrum taken at an orbital phase near quadrature ($\varphi = 0.206$). The stronger emission, centered at the absorption line, arise from the hot component, which is the component with the larger contri-

bution to the continuum. The red-shifted and less intense emission corresponds to the cool component (see Figure 1). In the $\varphi = 0.545$ observation we can not separate the contribution from both stars. The Hϵ line appears also in emission in both spectra. The emission intensity observed in our 1993 and 1998 spectra is larger than the emission intensity observed in the 1988 spectrum presented by Strassmeier et al. (1990).
The Ca II IRT lines: A clear emission reversal is observed in the core of the Ca II IRT absorption lines λ8542 and λ8662. After applying the spectral subtraction technique, we can clearly see a small emission arising from the secondary component, in the spectra near quadrature, as in the case of the Ca II H & K lines. The emission reversal observed here clearly contrasts with the filled-in lines reported by Dempsey et al. (1993).
The Hα, and Hβ lines: A nearly complete filling-in of the Hα and Hβ lines is observed in the 1995 spectra, and a small absorption is present in 1998. After applying the spectral subtraction, a clear filling-in of the Hα and Hβ lines is observed in the three spectra (see Figure 1). The small emission, arising from the secondary component, that is observed in the Ca II H & K and IRT lines is not detected in these two Balmer lines.
He I D_3 line: We can distinguish this triplet in absorption from the primary star. This is typical in cases of activity in evolved stars (Montes et al. 1997). In our observations we have measured 35 mÅ in the spectrum at $\varphi = 0.206$, and 17 mÅ at $\varphi = 0.545$ However, the line could be slightly blended with other minor lines, and this could be responsible for its variation.

References

Alekseev I.Y., 1999, A Rep, (in press)

Bopp B.W., Noah P., Klimke A. & Africano J., 1981, ApJ 249, 210

Dempsey R.C., Bopp B.W., Henry G.W. & Hall D.S., 1993, ApJS 86, 293

ESA, 1997, The Hipparcos and Tycho Catalogues, ESA SP-1200

Griffin R.F., 1978, Obs. 98, 257

Miller R. & Osborn W., 1996, Obs. 116, 382

Montes D., et al., 1994, A&A 285, 609

Montes D., et al., 1995, A&AS 114, 287

Montes D., et al., 1997, A&AS 125, 263

Montes D., et al., 1998, A&A 330, 155

Osten R.A. & Saar S.H., 1998, MNRAS 295, 257

Sanz-Forcada J., et al., 1998, in ASP Conf. Ser. 155, The Tenth Cambridge Workshop on Cool Stars, Stellar Systems, and the Sun, eds. R.A. Donahue & J.A. Bookbinder, CD-1450

Schmidt-Kaler T. 1982, in Landolt-Börnstein, Vol. 2b, ed K. Schaifers, H.H. Voig (Heidelberg: Springer)

Strassmeier K.G., et al., 1990, ApJS 72, 191

Strassmeier K.G., Hall D.S., Fekel F.C., Scheck M., 1993, A&AS 100, 173

Wilson O.C. & Bappu M.K.V. 1957, ApJ 125, 661

Solar and Stellar Activity: Similarities and Differences
ASP Conference Series, Vol. 158, 1999
C.J. Butler and J.G. Doyle, eds.

The Chromosphere of II Peg: Multi-line Modelling of an RS Cvn Star

C.I. Short

Department of Physics & Astronomy, University of Georgia, Athens, GA, 30605, USA

P.B. Byrne

Armagh Observatory, College Hill, Armagh, BT61 9DG, N. Ireland

Abstract. We present the first multi-line fitting of a semi-empirical atmospheric model to the H I and Ca II spectra of an RS CVn star (II Peg).

1. Introduction

We carry out detailed NLTE radiative transfer calculations for the first four lines of the H I Balmer series, and the resonance lines (H&K) and the infra-red triplet (*IRT*) lines of Ca II for a variety of reasonable chromospheric models for the RS CVn star II Peg and compare the results with observed profiles. By combining several diagnostic lines which sample different portions of the chromosphere and lower transition region, we can address the uniqueness problems that arise when fitting a single line profile.

2. Computational Method

Figure 1 and Table 1 show the grid of fifteen atmospheric models that were used in the present study. These models have as their photospheric base an RE model that corresponds to a star of $T_{\rm eff} = 4600$ K, log $g = 3.5$, and $[\frac{A}{H}] = 0.0$. Byrne *et al.* found $T_{\rm eff} = 4650 \pm 100$ K. The value of $\log g$ is within the range found by Vogt (1981). The grid explores various functional forms of T versus m in the chromosphere, a range in the thickness of the Transition Region (TR) and the functional form of $T(m)$ in the TR, and a range of values of $T_{\rm min}$ and $m_{\rm TR}$.

We have used the code MULTI, adapted for PRD (Uitenbroek, 1989), to solve the non-LTE problem for atomic models of H I/ II and Ca I/II/III. The synthetic flux profiles were convolved with a function to account for the rotational broadening of $v \sin i = 21$ km s^{-1} (Vogt, 1981) and the instrumental resolution of the corresponding observed profile.

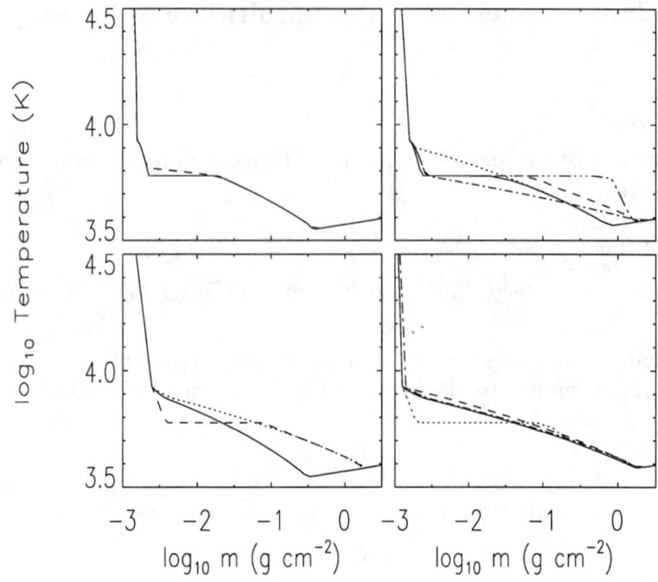

Figure 1. Temperature structure of models in the grid. Upper left: Series 1, upper right: Series 2 and 2A, lower left: Series 3, lower right: Series 4.

Table 1. Parameters of grid models

Model	Series	$m_{\rm TR}$	TR, $\frac{dT}{d\log m}$ or $\frac{d\log T}{d\log m}$	$m_{\rm Tmin}$	$T_{\rm min}$	Chromos. $\frac{dT}{d\log m}$
1	1	-2.8	-6.2×10^5	-0.4	3550	variable
2	1	-2.8	-5.1×10^5	-0.4	3550	variable
3	1	-2.8	-6.2×10^5	-0.4	3590	variable
4	2	-2.8	-5.4	-0.1	3690	variable
5	2	-2.8	-5.4	-0.1	3690	-1600
6	2A	-2.8	-5.4	0.24	3870	variable
7	2A	-2.8	-5.4	0.33	3900	-750
8	2A	-2.8	-5.4	0.2-0.4	3850	variable
9	3	-2.6	-2.7	-0.5	3520	-2100
10	3	-2.6	-2.7	0.1	3830	-1400
11	3	-2.6	-2.7	0.1	3860	variable
12	4	-2.9	-10.7	0.25	3870	-1300
13	4	-2.9	-10.7	0.25	3850	variable
14	4	-2.9	-10.7	0.25	3880	-1500
15	4	-2.85	-10.7	0.25	3870	-1300

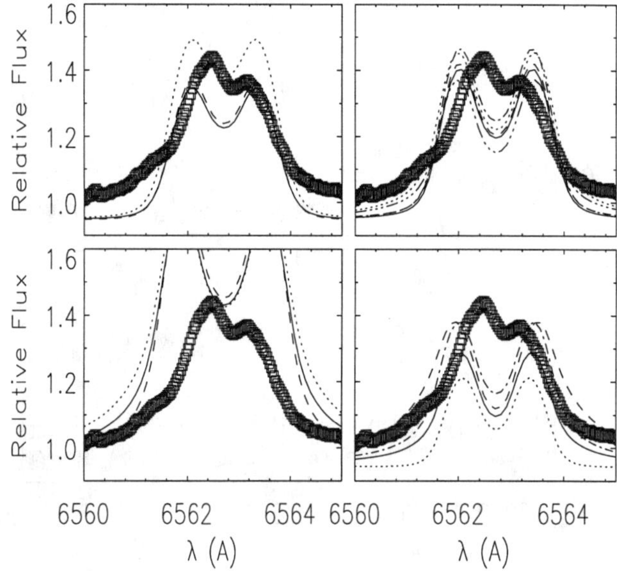

Figure 2. Observed and computed Hα line profiles. Observed profile (squares). Line-styles of the synthetic profiles follow those of the atmospheric models in Figure 1.

3. The Multi-line Fit

Figures 2 and 3 show the line profile fits for the Hα and Ca II K lines. Model 13 in Series 4 provides a relatively close fit to both the H I Balmer spectrum and the Ca II K line. Among the grid models, those of Series 4 have the lowest chromospheric pressure, the deepest $T_{\rm min}$ location, and the thickest chromosphere with the shallowest chromospheric T gradient. By contrast, all the Series 4 models are grossly discrepant with the observed Ca II IRT spectrum. The IR^3 lines are best fitted by models of Series 1, which have intermediate chromospheric pressure, a relatively shallow $T_{\rm min}$ location, and a relatively steep T gradient in the lower chromosphere.

We note that model 13 and the models in Series 1 all have a 6000 K temperature plateau in the upper chromosphere that starts in the middle chromosphere and extends to the TR. Also, the range in chromospheric pressure, specified by the value of $m_{\rm TR}$, spanned by the grid, is relatively narrow, with model 13 differing from the Series 1 models by only 0.1 in $\log m_{\rm TR}$. Models of Series 1 and 4 both have a TR thickness of ≈ 0.15 in log column mass density. However, the two series have different functional forms for $T(m)_{\rm TR}$, with Series 1 having constant $d{\rm T}/d\log m$ and Series 4 having constant $d\log T/d\log m$. Therefore, all five lines in this study are best matched by models having approximately the same upper chromosphere and TR structure.

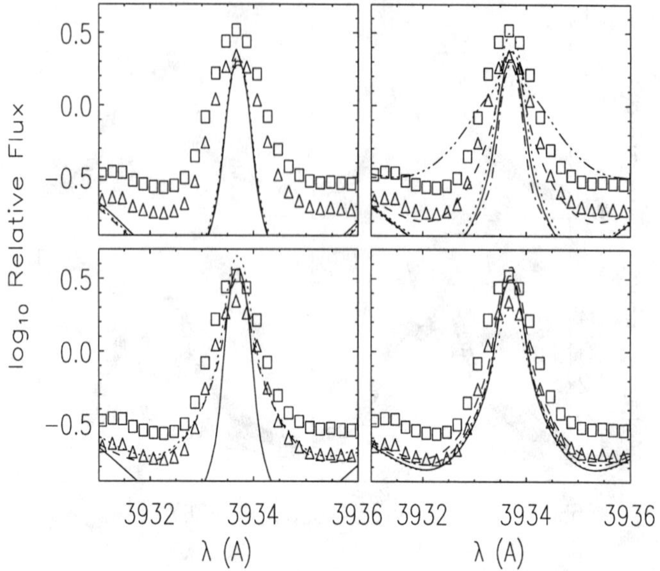

Figure 3. Ca II K: See Figure 1 caption. Squares and triangles: two rectifications of the observed spectrum.

4. Conclusion

To a first approximation, all the lines included in this study are best fitted by the same upper chromospheric structure with a 6000 K plateau that spans the Hα emission core, the Hβ core and the inner Ca II K profile on one hand, and the Ca II IRT lines on the other; the two sets of lines are most sensitive to different parts of the outer atmosphere, and there is no one model in our grid that has both the correct lower and correct upper atmospheric structure.

We note that the unusually steep Balmer decrement, with strong Hα emission and weak Hβ absorption, can be entirely reproduced with a conventional $1D$ hydrostatic atmospheric structure. The only difference between these models and those of active dwarfs in which the entire Balmer series behaves en-mass, is the lower value of $\log g$ that corresponds to a sub-giant.

References

Byrne, P.B., Panagi, P.M., Lanzafame, A.C., Avgol oupis, S., Huenemoerder, D.P., Kilkenny, D., Marang, F., Panov, K.P., Roberts, G., Seiradakis, J.H., & van Wyk, F., 1995a, A&A, 299, 115

Uitenbroek, H., 1989, A&A, 213, 360

Vogt, S.S., 1981, ApJ, 247, 975

Solar and Stellar Activity: Similarities and Differences
ASP Conference Series, Vol. 158, 1999
C.J. Butler and J.G. Doyle, eds.

The Wilson-Bappu Relation for RS CVn Stars

F.F. Özeren[1,2] & J.G. Doyle[1]

[1] *Armagh Observatory, College Hill, Armagh BT61 9DG, N.Ireland*

[2] *Ankara University, Science Faculty, Astronomy and Space Department 06100 Tandoğan-Ankara/Turkey*

Abstract.
We investigate the extent to which the Wilson-Bappu relationship holds for chromospherically active binaries using the Mg II h&k lines of 32 RS CVn stars observed with *IUE*. The resulting fits are different from the relationships obtained for single less active stars. The parallax's used were from the *HIPPARCOS* catalog, which give a much better correlation than those taken from CABS.

1. Introduction

The Wilson-Bappu relationship (hereafter the WB relationship) was first established as a relation between the absolute visual magnitude (M_v) and the line width of Ca II H&K emission lines for late type stars by Wilson & Bappu (1957). Subsequent work, based on COPERNICUS data for a very limited sample of K giants (McClintock et al. 1975) extended it to other lines, notably Mg II h&k and Ly α. Many other authors, (e.g. Kondo et al. 1976; Stencel 1977; Weiler & Oegerle, 1979; Vladilo et al., 1987, Montes et al. 1994) have looked at the WB relationship and attempted to explain the under-lying physics.

There are two plausible basic theories to explain the WB relationship (i) Doppler broadening with stellar absolute magnitude (ii) column density above the temperature minimum. These were discussed in papers by Engovold & Rygh (1978), Ayres (1979), Linsky (1980) and Lutz & Pagel (1982). The dominant parameter seems to be the chromospheric mass column density.

To extend and test the WB relationship for low luminosities, the red dwarf AU Mic and the red dwarf binary AT Mic (Elgoray, 1988) and the M dwarf flare star, AD Leonis (Ambruster et al., 1989) were used. The results were in agreement with those of Vladilo et al. (1987). Going to the more active stars, Gurzadyan (1991) used 10 RS CVn-type stars. He concluded that the observed magnesium emission was generated by ionized gas in the space between the components of the binary systems. Montes et al. (1994) used Ca II H&K for 28 chromospherically active binary systems and 18 single active stars. They noted that there is a systematic difference among the behaviour of stars belonging to different groups. Elgoray et al. (1997) worked with Mg II h&k data for 78 single stars observed with the *IUE*. They also noted that the active stars showed a somewhat larger spread in line widths than the 'quieter' stars, thus giving a different WB relation.

Due to the prolific output from *IUE* during it's 19 years of operation, an excellent dataset of Mg II h&k line profiles has been obtained for a large selection of RS CVn binaries. With the exception of the latter few references, most of the developed WB relationships have excluded RS CVn's, due to their intense chromospheric activity and binary nature. Here, we look again at the WB relationship, confining ourselves to the RS CVn's using the absolute magnitudes based on the new *HIPPARCOS* parallax. The different parallax alone gives substantially different results than those based on Strasmeier et al. (1993) catalogue.

2. Observational Data

We have searched the *IUE* Data Archive for all objects in the CABS (Strasmeier et al. 1993). Mostly LWP data was used, but sometimes data from the LWR camera was also used.

To analyse the *IUE* images we used the STARLINK package DIPSO (Howarth et al. 1996). The widths and the emission line fluxes of the Mg II h&k lines were measured by using a least squares Gaussian fit. All derived line widths are corrected for the instrumental broadening using a square correction.

$$W^2 = W^2_{\textbf{observed}} - W^2_{\textbf{instrumental}} \quad (1)$$

where $W^2_{\textbf{instrumental}} = 21kms^{-1}$ (Turnrose et al., 1984).
M_v is calculated by

$$M_v = V - 5logd(pc) + 5 - A_v \quad (2)$$

A_v is the correction for interstellar absorption expressed in V magnitude. The corrections were obtained using

$$A_v = 3.2 * E(B - V) \quad (3)$$

$$E(B - V) = (B - V)_{\textbf{observed}} - (B - V)_{\textbf{intrinsic}} \quad (4)$$

with the intrinsic (B–V) colors from Fitzgerald (1970). The surface fluxes were obtained from the observed flux via

$$F_{\textbf{surf}} = (d/R)^2 f_{\textbf{observed}} \quad (5)$$

Stellar radii were from CABS, Gunn et al. (1997) and Straizys & Kuriliene (1981).

3. Results

With the exception of a few objects, there is a strong correlation between the $\mathrm{P_{orb}}$ versus the $\mathrm{P_{rot}}$ (r = 0.96) implying tidally locked systems. All fluxes are more than log $\mathrm{F_{surface,Mg\ II\ k}}$ = 5.2. This limit was given as an activity

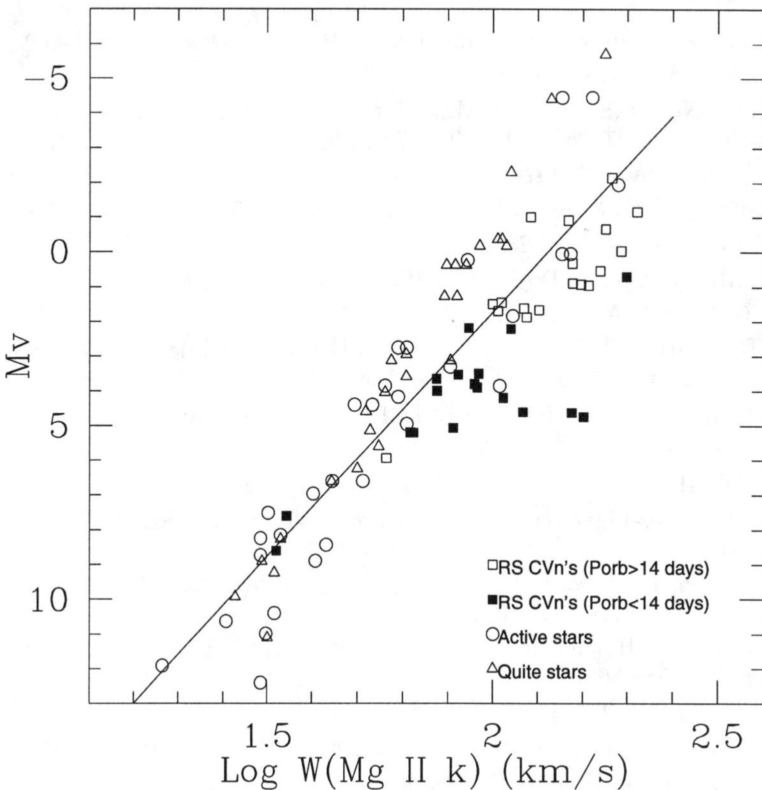

Figure 1. The Wilson-Bappu relation for the Mg II k line. The solid line shows the relation including all stars.

measurement by Elgoray et al., (1997). Figure 1 shows the M_v - log W (km s^{-1}) relation of Mg II k for our sample 32 RS CVn and the active and 'quite' single stars from Elgoray (1997). The least squares fit to all data-points is:

$$M_v = -14.10(\pm 0.71) log\, W_{\mathrm{Mg\ II\ k}} \; + \; 29.92(\pm 1.34) \tag{6}$$

with a correlation coefficient of r=0.82.

This relation is somewhat different from that for single stars although active single stars have also greater variation.

Acknowledgments. Research at Armagh Observatory is grant aided by the Dep. for Educational for N.Ireland. Support for software and hardware is provided by the STARLINK Project which is funded by the UK PPARC. FFO wished to thank the Turkish Scientific and Technical Research Council for a grant which enabled a visit to Armagh.

References

Ambruster, C.W., Pettersen, B.R., Sundaland, S.R. 1989, A&A, 208, 198-200

Ayres, T.R. 1979, ApJ, 228, 509

Dempsey, R.,C., Neff, J.,E., Thorpe, M.,J., Linsky, J.,L., Brown, A., Cutispoto, G. & Rodono M. 1996, ApJ, 470, 1172-1186

Elgaroy, O. 1988, A&A, 204, 147-148

Elgaroy, O., Engvold, O. & Joras, P. 1997, A&A , 326, 165-176

Fitzgerald, M.P. 1970, A&A, 4, 234-243

Gunn, A.G., Mitrou, C.K. & Doyle, J.G. 1998, MNRAS, 296, 150

Gurzadyan, G.A. 1991, ASS, 179,293-302

Howarth, I.D., Murray, J. & Berry, D.S. 1996, DIPSO - A Friendly spectrum Analysis Program, STARLINK User Note 50.19

Kondo, Y., Morgan, T.H. & Modisette, J.L. 1976, ApJ, 207, 167

Linsky, J.L. 1980, Ann.Rev.Astron.Astrophys., 18, 439

McClintock, W., Henry, R.C. & Moos, H.W. 1975, ApJ, 202, 733

Montes, D., Fernandez-Figueroa, M.J., DeCastro, E. & Cornide, M. 1994, A&A, 285, 609

Neff, J.E., Pagano, I., Rodono, M., Brown, A., Dempsey, R.C., Fox, D.C. & Linsky, J.L. 1996, A&A, 310, 173-180

Olah, K., Marik, D., Houdebine, E.R., Dempsey, R.C. & Budding, E. 1998, A&Ap, 330, 559-568.

Stencel, R.E. 1977,ApJ, 215, 176

Strassmeier, K.G., Hall, D.S., Fekel, F.C. & Scheck, M. 1993, ApJS, 72, 191

Straizys, V. & Kuriliene, G. 1981, A&SS, 80, 353-368.

Turnrose, B.E & Thomson R.W. 1984, *IUE* Processing Information Manual Version 2.0, CSC/TM-84/6058

Vladilo, G., Molaro, P., Crivellari, L., Foing, B.H., Beckman, J.E, & Genova R. 1987, A&A, 185, 233-246

Weiler, E.J. & Oegerle, W.R 1979, ApJS, 39, 537

Wilson, O.C. & Bappu, M.K.V. 1957, ApJ, 125, 661

Part 6

CORONAL DYNAMICS

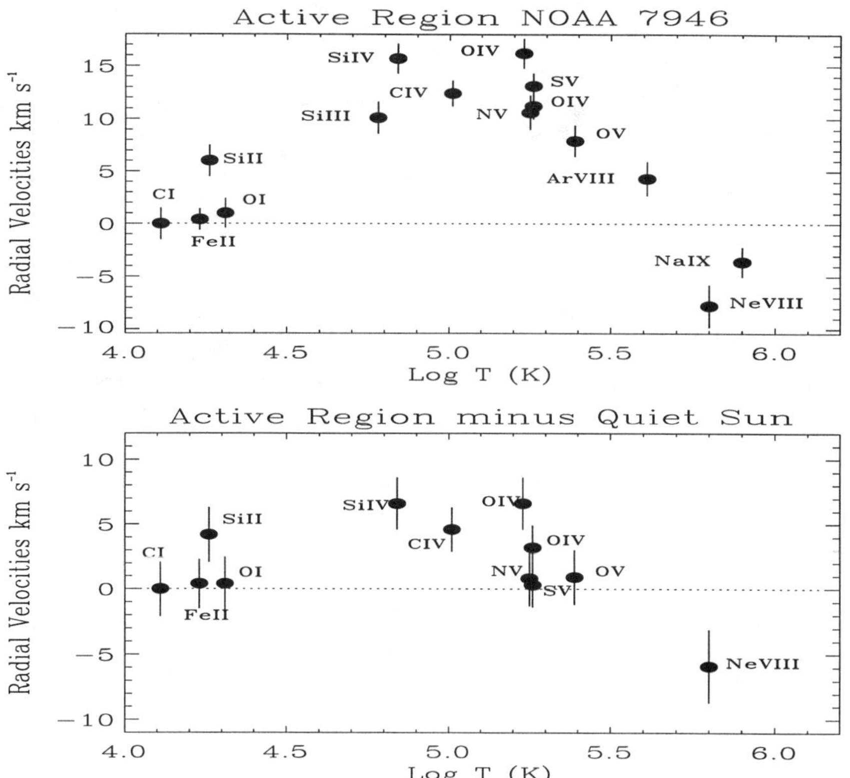

SUMER measurement of Radial Velocities on Active Region NOAA 7946 (top panel), and differences between Active Region and 'quiet' Sun (lower panel).

Solar and Stellar Activity: Similarities and Differences
ASP Conference Series, Vol. 158, 1999
C.J. Butler and J.G. Doyle, eds.

How is the Solar Corona Heated?

E.R. Priest
Mathematical and Computational Sciences Department, St Andrews, KY16 9SS, Scotland

Abstract.
Recent observations from the *Yohkoh* and *SOHO* satellites have produced several advances in tackling the problem of coronal heating. The solar corona has a three-fold structure of coronal loop, coronal hole and X-ray bright point. The question of how the corona is heated is therefore a complex multi-faceted problem, with possible different mechanisms at work in different types of coronal loop, in coronal holes and in bright points. This review summarises some recent advances in understanding from the *Yohkoh* and *SOHO* satellites and focuses on two particular discoveries, namely the mechanism for heating bright points and the nature of the heating of large-scale coronal loops.

1. Introduction

Brendan was one of my favourite people, who I am sure would have wanted us to enjoy ourselves here, and so I look forward to reading the many memories of him recounted over these few days. The last time I met him was at a conference on solar prominences in France. In my office in St Andrews there is a picture taken at the conference which shows me at a table having difficulty entertaining a group of young ladies but being helped out by a gracious Brendan who, with a twinkle in his eye, had just joined us from another table.

During the last ten years there has been a revolution in our understanding of the Sun's corona, which has shown the plasma atmosphere to be highly structured and dynamic, with most of what we see in the corona caused by the magnetic field and its subtle nonlinear interaction with the plasma. At an eclipse you see magnetically open *coronal holes*, along which the fast solar wind escapes and magnetically closed *coronal loops*, which contain the plasma. I was fortunate to witness the eclipse in February in Guadeloupe (Figure 1), which was one of the great experiences of my life – you can find an account on my web page.

In soft X-rays, however, you see the coronal holes and coronal loops direct with an extra third component, namely, *X-ray bright points*, several hundred of which may be present at one time (Figure 2). The Japanese *Yohkoh* satellite has revealed the corona as an MHD world with myriads of loops continually interacting with one another (e.g. Tsuneta, 1993; Culhane, 1997).

The interaction of a magnetic field and plasma can often be modelled by *magnetohydrodynamics* (MHD for short), when the plasma is treated as a con-

Figure 1. White-light image of the corona during the Feb. 26, 1998 eclipse (courtesy High Altitude Observatory, NCAR).

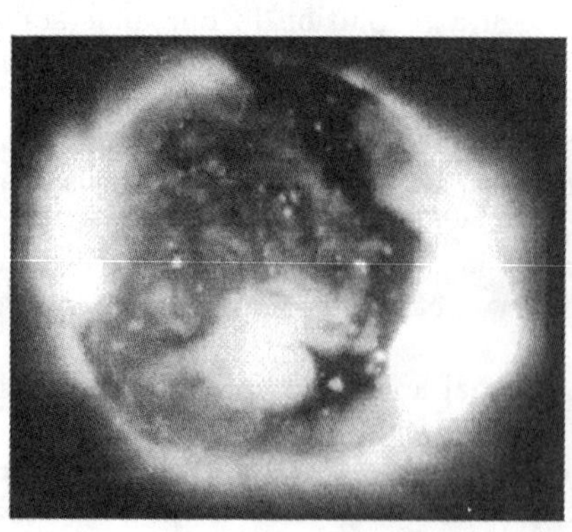

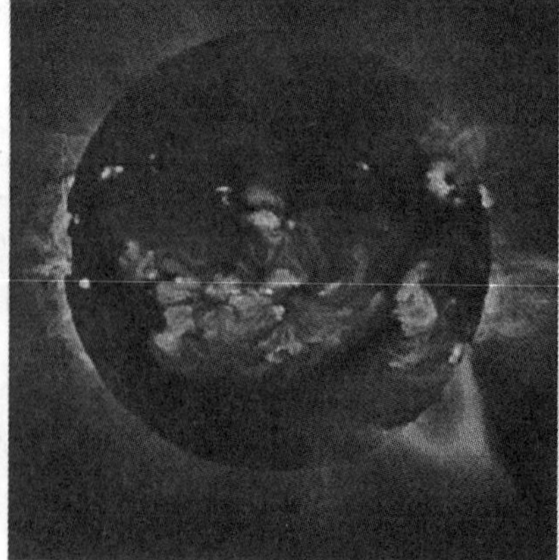

Figure 2. Soft X-ray images of the corona from (a) Skylab (courtesy D. Webb) and (b) *Yohkoh* (courtesy S. Tsuneta).

tinuous medium. The equations are a unification of the equations of slow electromagnetism and fluid mechanics. The two main equations are: the equation of motion

$$\rho\frac{d\mathbf{v}}{dt} = -\nabla p + \mathbf{j}\times\mathbf{B}, \tag{1}$$

where $\mathbf{j}\times\mathbf{B}$ is the force that the magnetic field exerts on a plasma of density ρ, pressure p and velocity $\mathbf{v}$; and the *induction equation*

$$\frac{\partial\mathbf{B}}{\partial t} = \nabla\times(\mathbf{v}\times\mathbf{B}) + \eta\nabla^2\mathbf{B}, \tag{2}$$

which is obtained by eliminating the electric field

$$\mathbf{E} = -\mathbf{v}\times\mathbf{B} + \frac{\mathbf{j}}{\sigma} \tag{3}$$

and electric current

$$\mathbf{j} = \frac{1}{\mu}\nabla\times\mathbf{B} \tag{4}$$

from Maxwell's equations of slow electromagnetism and Ohm's law, where

$$\nabla.\mathbf{B} = 0. \tag{5}$$

Thus, for a given density and pressure, equations (1) and (2) determine the plasma velocity ($\mathbf{v}$) and magnetic field ($\mathbf{B}$), while the electric field and electric current are secondary variables determined by (3) and (4).

Physically, equation (2) implies that the magnetic field changes in time due to transport of the magnetic field with the plasma (the first term on the right) and diffusion of the magnetic field through the plasma (the second term). However, in most of the universe the second term is very much smaller than the first, so the magnetic field is frozen to the plasma. The exception is in singularities called *current sheets*, where the magnetic gradient (and therefore by (4) the electric current) is extremely large. It is in such current sheets that the magnetic field lines can break and *reconnect* by slipping through the plasma, and, in the process, magnetic energy is converted to heat, kinetic energy and fast-particle energy.

Two-dimensional magnetic reconnection is now fairly well understood (e.g. Priest & Forbes, 1999). The early mechanism of Sweet (1958) and Parker (1957) possesses a simple diffusing current sheet, while Petschek's (1964) mechanism has a small central current sheet with most of the energy being released at four slow-mode shock waves that radiate outwards from the central sheet. These classical regimes have now been replaced by a new generation of fast regimes (Priest & Forbes, 1986; Priest & Lee, 1990) which have Petschek's mechanism as a special case.

The emphasis now is on trying to develop an understanding for how reconnection takes place in three dimensions (e.g. Schindler *et al*, 1988; Hornig and Rastätter, 1998; Lau & Finn, 1990, Priest & Forbes, 1999). However, many features of reconnection in three dimensions are quite different from two dimensions, including the nature of the null points, the bifurcations and the topological changes. A typical three-dimensional null point possesses an isolated field line

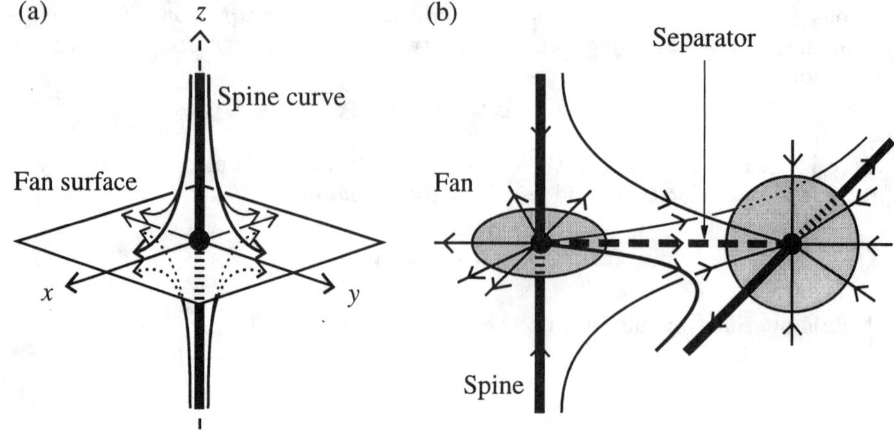

Figure 3. The typical structure of (a) a three-dimensional nullpoint and (b) A pair of null points.

(called a *spine-curve*) that approaches the null and a surface of field lines (called a *fan surface*) that radiates from it (Figure 3).

Reconnection at a null point may occur either by *spine reconnection*, when the current concentrates along the spine, or by *fan reconnection* when it is focussed in the fan, or by *separator reconnection* when the dissipation is located along the special field line (called a separator) that joins one null point to another (Priest & Titov, 1996). It may also take place in the absence of nulls by *singular field-line reconnection* (Priest & Forbes, 1989, Hornig & Rastätter, 1997).

2. The Solar and Heliospheric Observatory (*SOHO*)

SOHO is a collaborative ESA/NASA mission that was launched in December 1995 and reached the first Lagrangian point between the Sun and Earth in February 1996. It is observing the Sun continuously with a suite of 12 instrument packages in unprecedented detail that are giving us the first comprehensive view of the Sun, from the deep interior, through the different layers of the atmosphere and out into the solar wind.

The atmospheric instruments include: EIT (The EUV Imaging Telescope), which is producing global images of the transition region and corona at four wavelengths; SUMER (Solar UV Measurements of Emitted Radiation), which uses emission lines to deduce the physical properties of the transition region and corona; CDS (Coronal Diagnostics Spectrometer), which does the same over a different EUV range with normal and grazing incidence spectrometers; UVCS (Ultra-Violet Coronagraph Spectrometer), which enable UV spectroscopy out to 12 $R_\odot$; and LASCO (Large Angle Spectroscopic Coronagraph), which has three overlapping coronagraphs extending from 1.1 $R_\odot$ out to 30 $R_\odot$.

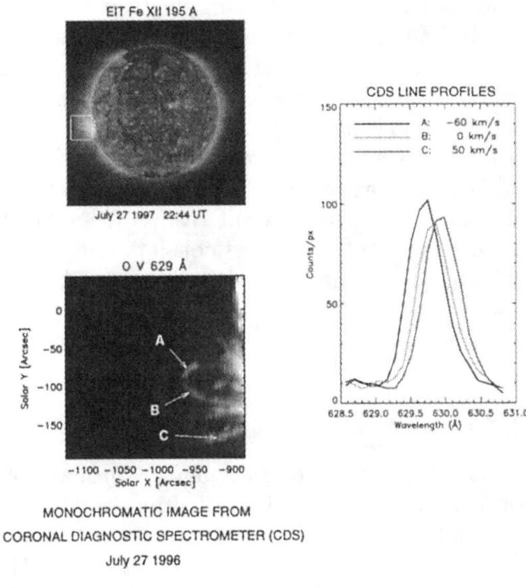

Figure 4. Global image from EIT together with local CDS image in O V and line profiles (courtesy P. Brekke).

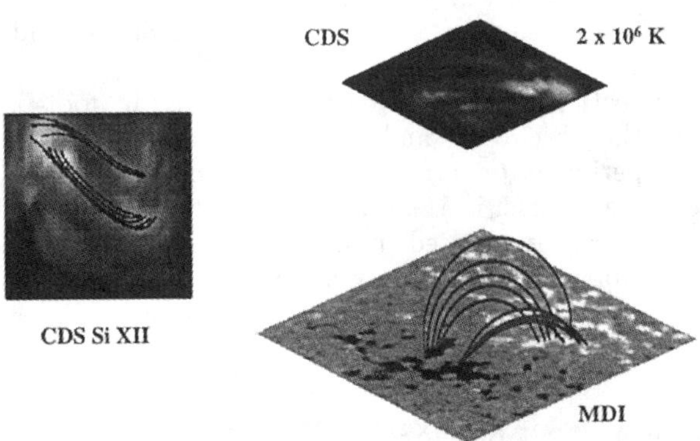

Figure 5. An active region viewed with CDS (top) and the underlying photospheric magnetic field from the MDI instrument (bottom), together with the calculated overlying magnetic field lines (courtesy R. Walsh).

EIT produces images at four different temperatures and has deduced a rough global map of the temperature, which shows that: coronal holes are indeed cool; bright points are hotter; and active regions contain much fine structure with hot and cool loops close to one another.

With CDS we can zoom in on, for example, the active region shown at the limb in Figure 4 and make simultaneous images in many emission lines, literally taking the atmosphere to pieces at many different temperatures. In this example, line profiles in O v at 250,000 K show plasma moving towards us at 60 $km\,s^{-1}$ (A) and away at 50 $km\,s^{-1}$ (C).

What is the nature of the transition region between the chromosphere and corona? The classical picture was of a static plane-parallel layer smoothly linking the chromosphere and corona, but recent results from *SOHO* show instead the plasma at transition-region temperatures consists of a highly dynamic, continually changing set of loops. Perhaps this is because of the dynamic nature of the heating mechanism.

3. Coronal Heating

How is the corona heated? How are bright points, coronal holes and different types of coronal loop heated? Possibly by different mechanisms. The coronal heating problem is multi-faceted, but recently progress has been made in two directions, namely how X-ray bright points are heated and the nature of the heating of large-scale loops.

Two classes of theoretical models have been proposed, namely magnetic waves and magnetic reconnection. Alfvén waves, for instance, may be dissipated by phase mixing or resonant absorption. Recently we have looked for magnetic waves and failed to find them, at least with periods between 30 and 1000 secs (Ireland *et al*, 1997).

We studied the active region shown in Figure 6 and measured the intensity as a function of time at several wavelengths. In the chromosphere and transition region there was substantial power at 300 seconds and 600 seconds, but up in the corona no significant power was seen.

Magnetic reconnection may be driven at null points by foot-point motions or it may occur in the absence of null points by braiding (Parker, 1972). A recent numerical experiment on braiding (Glasgaard & Nordlund, 1996) has shown that current sheets form in response to braiding at sporadic locations in space and that the energy is released impulsively in time (Figure 6). The net effect, however, is to dump the heat fairly uniformly along the loop (Glasgaard *et al*, 1998).

4. The Magnetic Carpet and X-Ray Bright Points

X-ray bright points occur above oppositely directed fragments. They appear to be caused either by emerging flux or more often by the convergence of the photospheric fragments which drives overlying reconnection in the corona (Priest *et al*, 1994).

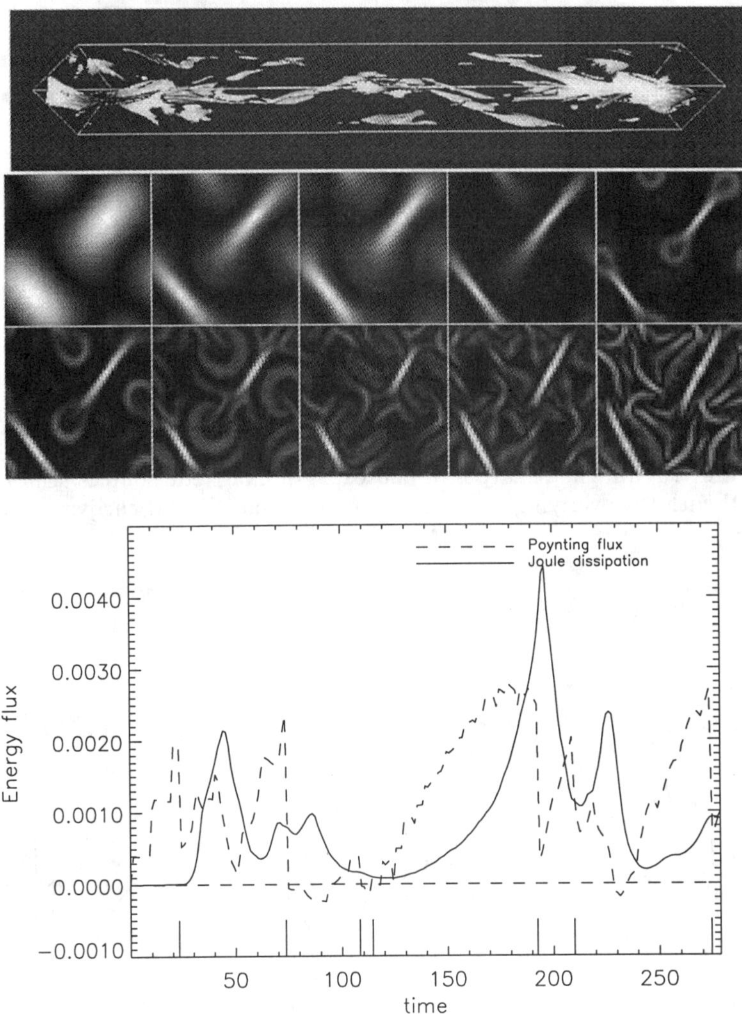

Figure 6. A numerical experiment on braiding showing the current sheets (top) in 3D and in a cross-section (middle). The bottom panel shows heating as a function of time. (Glasgaard & Nordlund, 1996).

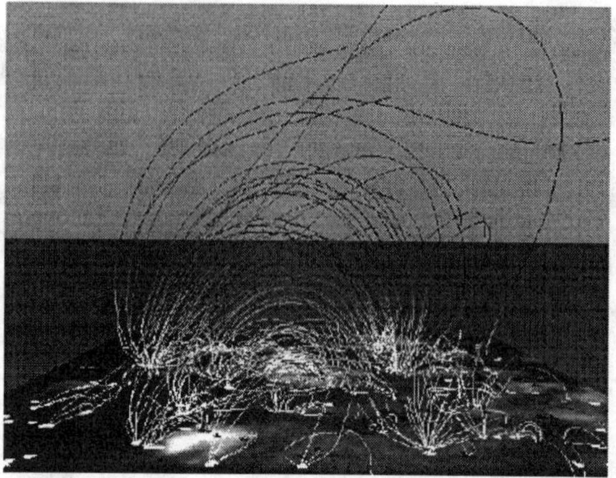

Figure 7. Magnetic carpet of photospheric magnetic sources seen in MDI with the overlying complex magnetic connections (Schrijver *et al*, 1998).

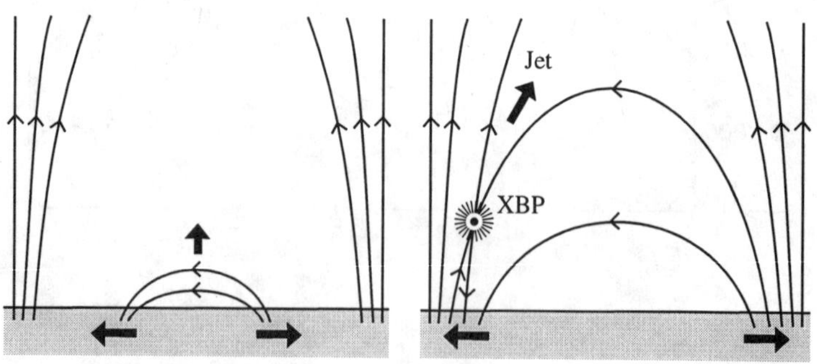

Figure 8. Converging Flux Model for X-Ray Bright Points (Priest *et al*, 1994).

The surface of the Sun is covered with a "magnetic carpet", which may be seen in magnetograms from MDI (Michelson Doppler Imager) as a network of convection cells with magnetic fields concentrated in the boundaries (Figure 7). The magnetic fragments continually emerge, cancel, merge and fragment, with the result that the magnetic flux at the surface is replaced every 40 hours (Schrijver *et al*, 1998).

The magnetic carpet is then the source of the energy, but how is the energy released? What is the mechanism? The basic idea behind the Converging Flux Model (Priest *et al*, 1994) is to suggest that new flux emerges in a supergranule cell, moves to the boundary and there it reconnects, creating a bright point and sometimes accelerating an X-ray jet (Shibata *et al*, 1990). This model (Figure 8) has been given support by high-resolution observations from the *NIXT* telescope, which reveals the internal structure of a bright point in agreement with the predictions of the model (Parnell *et al*, 1994).

5. Microflares and Micro-Bright Points

X-ray bright points are important in their own right but they do not possess enough energy to heat the quiet Sun. However, in an important development Krucker *et al* (1997) have observed micro-flares with EIT. They have an energy of $10^{18} - 10^{19} J (10^{25} - 10^{26} erg)$ and an energy spectrum of $E^{-2.6}$. They contribute about 20% of the quiet-Sun heating.

So what is the nature of micro-flares? It is likely that they represent local reconnections driven by foot-point motions, as described by the Converging Flux Model. An interesting piece of observational support for this suggestion has come from Falconer *et al* (1998), who searched for bright points with EIT. They compared with a magnetogram and found that the bright points always lie over mixed polarity regions, where reconnection is probably being driven. They then applied a filter to remove the background haze and this showed up many much smaller points of emission that they christened *micro-bright points*. Furthermore, these all lie in the boundaries of supergranule cells and most of them are over mixed magnetic polarity.

You can of course never prove a theory with observations, but can only disprove it, which is rather sad. But now there is a real paradigm shift occurring. Whereas previously reconnection was a fascinating concept which exercised the imagination of theorists like myself, now so many *SOHO* observations fit beautifully into place with the eyes of reconnection that it is becoming the natural explanation for many coronal heating phenomena. As well as the EIT bright points and micro-bright points and the MDI carpet, SUMER has observed *explosive events*. For example, Figure 9 shows a series of Si IV spectra with a step-size of 1 arcsec, which reveals the presence of bidirectional jets accelerated to 200 $km\, s^{-1}$ in opposite directions. This has been interpreted as the jets accelerated by a reconnection event (Innes *et al*, 1997).

Furthermore, Harrison (1997) has discovered transition-region heating events in CDS on the network which he calls *blinkers* and which last typically 10 minutes. These are much longer-lasting than explosive events but they both occur over mixed polarity.

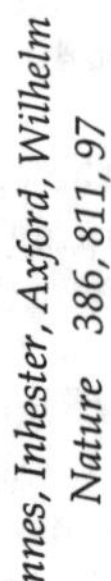

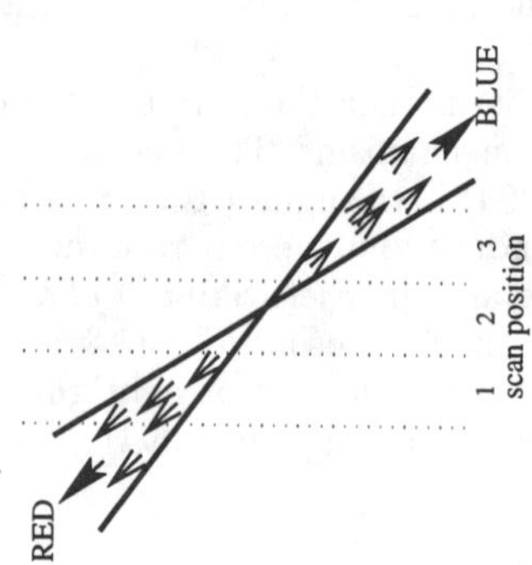

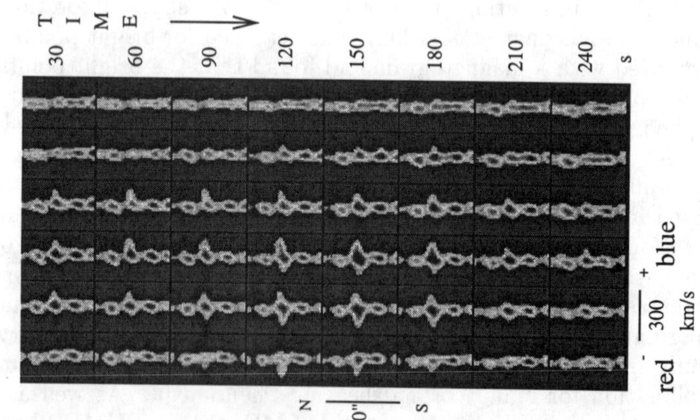

Figure 9. An explosive event from SUMER (courtesy D. Innes).

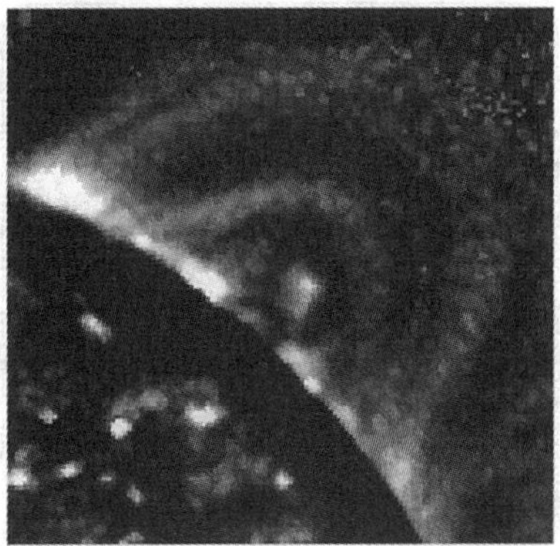

Figure 10. A large-scale loop observed by the *Yohkoh* Soft X-Ray Telescope.

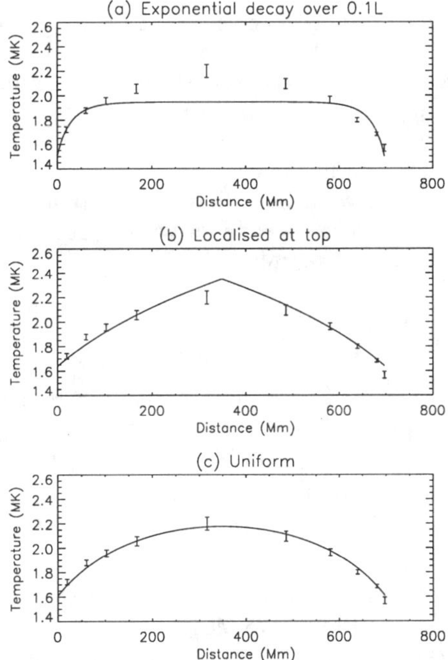

Figure 11. A comparison of the observations of the temperature along the large-scale loop of Figure 10.

6. Large-Scale Coronal Loops

In a letter that has just appeared in Nature, Priest *et al* (1998) building on previous work by *Kano & Tsuneta* (1996) showed that the temperature profile along a loop is highly sensitive to the nature of the heating. Turbulent magnetic dissipation in many small current sheets tends to deposit the heat uniformly along a loop and produce a profile in which $T^{\frac{7}{2}}$ is a quadratic function of distance s along the loop. Heat mainly near the summit, such as may be produced by long-wavelength standing waves, gives a $T^{\frac{7}{2}}(s)$ profile that has constant gradient in the legs of the loop and has strong curvature near the summit.

We used the Japanese *Yohkoh* satellite to measure the temperature along the large loop shown in Figure 10 and compared with a series of models to deduce the likely form of the heating (Figure 11).

The observed temperature increased from about 1.5 MK near the feet up to 2.2 MK at the loop summit. If the heat is localised near the feet, the best fit is rather poor. If the heat is dumped at the summit, the fit is better but still not very good, but a uniform deposition of heat produces an excellent fit with the observations, exactly what turbulent relaxation would tend to give.

7. Conclusion

The primary conclusions are as follows. First of all, the evidence from recent observations is that X-ray bright points and micro-bright points are likely to be heated by driven reconnection. Secondly, large-scale coronal loops appear to be heated rather uniformly, which implies that they are not heated by low-lying micro-bright points but perhaps instead by turbulent reconnection producing micro-flares distributed high in the corona.

Finally, let me re-echo what many have been saying over the past few days, namely that I personally miss Brendan's warm presence and friendship very much – he was a marvellous example to us all.

References

Culhane, J.L. 1997, in Adv. Space. Res., editors E.R. Priest & D. Baker, 19, 1839

Falconer, D.A., Moore, R.L. & Proter, J.G. 1998, in Proc. of High Resolution Solar Atmospheric Dynamics Workshop

Galsgaard, K., Mackay, D., Nordlund, A. & Priest, E.R. 1998, Solar Phys., submitted

Galsgaard, K. & Nordlund, A., 1996, J. Geophys. Res., 101, 13445

Harrison, R.A. 1997, Solar Phys., 175, 467

Hornig, G. & Rastäther. 1997, Adv. Space Res., 19, 1789

Innes, D.E., Inhester, B., Axford, W.I. & Wilhelm, K. 1997, Nature, 386, 811.

Ireland, J., Walsh, R., Harrison, R. & Priest, E.R. 1997, in The Corona and Solar Wind Near Minimum Activity, ESA SP404, 433

Kano, R. & Tsuneta, S. 1996, PASJ, 48, 535

Krucker, S., Benz, A.O., Bastian, T.S. & Acton, L.W. 1998, ApJ, 488, 499
Lau, Y.-T. & Finn, J.M. 1990, ApJ, 350, 672
Parker, E.N. 1957, J. Geophys. Res., 62, 509
Parnell, C.E., Priest, E.R. & Golub, L. 1994, Solar Phys., 151, 57
Petschek, H.E. 1964, AAS-NASA Symp. on Phys. of Solar Flares, NASA SP-50, Washington, D.C., 425
Priest, E.R., Foley, C.R., Heyvaerts, J., Arber, T., Culhane, J.L. & Acton, L.W. 1998, Nature, 393, 545
Priest, E.R. & Forbes, T.G. 1986, J. Geophys. Res., 91, 5579
Priest, E.R. & Forbes, T.G. 1989, Solar Phys., 119, 211
Priest, E.R. & Forbes, T.G. 1999, Magnetic Reconnection, Cambridge University Press, Cambridge
Priest, E.R. & Lee, L.C. 1990, J. Plasma Phys., 44, 337
Priest, E.R., Parnell, C.E. and Martin, S.F. 1994, ApJ, 427, 459
Priest, E.R. & Titov, V.S. 1996, Phil. Trans. Roy. Soc., 354, 2951
Schindler, K., Hesse, M. & Birn, J. 1988, J. Geophys. Res., 93, 5547
Schrijver, C.J., Title, A.M., Harvey, K.L., Sheeley, N.R., Wang, Y-M., van der Oord, G.H.J., Shine, R.A., Tarbell, T.D. & Hurlburt, N.F. 1998, Nature, 487, 424
Shibata, K., Nozawa, S., Matsumoto, R., Sterling, A.C. & Tajima, T. 1990, ApJ, 351, L25
Sweet, P.A. 1958a, IAU Symp. 6, 123
Tsuneta, S. 1993, in The Magnetic and Velocity Fields of Solar Active Regions, ed. H. Zirin, G. Ai, H. Wang, Astron. Soc. Pacific, San Francisco, p 239–248

Solar and Stellar Activity: Similarities and Differences
ASP Conference Series, Vol. 158, 1999
C.J. Butler and J.G. Doyle, eds.

X-ray Coronae of Stars: Some Theoretical Questions

R. Pallavicini

Osservatorio Astronomico di Palermo, Palazzo dei Normanni, I-90134 Palermo, Italy

Abstract. I discuss theoretical implications of X-ray observations of coronal emission in stars, with an emphasis on coronal heating mechanisms and dynamo action, and on the modelling of quiescent and flaring coronal emission with solar-type loop models. I also report on the recent detection by *Beppo*SAX of hard (>20 keV) X-ray emission from stellar flares.

1. Introduction

The subject of stellar coronae and their X-ray emission has been reviewed many times recently (see, e.g., Pallavicini 1998 and references therein). In this volume, stellar coronal observations are reviewed by Stern (stellar magnetic activity and variability), Jeffries (X-ray stars in open clusters) and Walter (pre-main sequence stars), whereas some aspects of stellar coronal observations are also discussed by Linsky (magnetic activity across the HR diagram), Mathioudakis (coronal spectroscopy) and van den Oord (stellar flares). There is no point therefore in trying to summarize again our present observational knowledge of X-ray coronae. Rather what I will do here is to address a few selected theoretical questions in coronal physics that can be effectively tackled with X-ray observations. It is important to stress that our theoretical understanding and modelling of coronal emission in stars is still in a rather rudimentary stage in comparison to the wealth of data that have been gathered over the past twenty years. For late-type stars in particular, our theoretical understanding is largely based on the solar analogy and on the application to stars of models and concepts originally developed for the solar corona. Yet, in spite of the many limitations, we are starting to have a sufficiently accurate description of coronal emission at various stages in the evolution of low-mass stars, from the pre-main sequence contraction phase, to the hydrogen burning main-sequence phase, and to the post-main sequence expansion phase. By comparing spatially-unresolved stellar observations with the detailed knowledge we have of the solar corona, it is possible to investigate how coronal emission depends on parameters like mass, gravity, rotation, magnetic fields and age, i.e. on parameters for which only a single value is accessible in the case of the Sun. In addition, stellar phenomena often occur on a much larger energy scale than on the Sun, thus allowing us to study coronal processes under conditions of extreme activity.

There are many questions in stellar coronal physics that should be addressed. Let me list just a few of them:

- **Coronal heating:** how the coronae of different types of stars (both hot and cool) are heated ? Which is the role of winds (in hot stars) and of magnetic fields (in cool stars) in the heating process ? How important are flares and micro-flares in producing the apparently quiescent X-ray emission of active stars ?

- **Stellar dynamos:** how the coronal activity of low-mass stars depends on rotation and convection ? Are there different types of dynamos for stars of different masses and/or different activity levels ?

- **Stellar evolution:** how coronal activity varies in the course of stellar evolution ? What is the origin of coronal emission in pre-main sequence stars ? Why coronal emission disappears (or is drastically reduced) in red giants with large mass losses ?

- **Coronal structures:** how stellar coronae are spatially and thermally structured ? Are magnetically-confined loops, similar to those observed on the Sun, the coronal building blocks for all types of stars ? How binarity affects the structure of coronae, particularly in close binary systems ?

- **Coronal abundances:** are coronal and photospheric abundances different, as suggested by recent spectroscopic observations ? Does a FIP effect exist for stellar coronal emission ? Do elemental abundances vary during flares ?

- **Coronal variability:** how stellar coronae vary on different time scales ? What powers the very energetic flares (typically much stronger than solar ones) observed in stars ? How realistic are the flare parameters inferred by applying solar-type models to stellar flares ?

In the following, I will elaborate on some of the above problems, with emphasis on coronal heating and dynamo action, and on the modelling of stellar coronae and flares by means of loop-like structures.

2. Coronal Heating and Dynamos

It is generally accepted that coronal heating in late-type stars results from the dissipation of dynamo-generated magnetic fields, as is believed to occur for the Sun. Indirect evidence for this is provided by the observed dependence of coronal emission on rotation rate for F to M stars with outer convective zones (Pallavicini et al. 1981, Hempelmann et al. 1995). The exact functional dependence of this relationship is still a matter of debate, and it is not even clear whether the dependence is simply upon rotation rate or Rossby number, i.e. the ratio of rotation period to the convective turnover time. A dependence on Rossby number is more satisfactory from a theoretical point of view, since it involves both rotation and convection, i.e. the two factors that are most relevant for the efficiency of the dynamo process. However, for field stars there is little observational evidence that a formulation in terms of the Rossby number is better than in terms of rotation periods or $v \sin i$. Much stronger evidence for a dependence on both rotation and convection is provided by observations of stars in clusters. By observing stars in open clusters of different ages, in fact, one can compare homogeneous samples of stars with approximately the same age and chemical composition, but different masses. Thus, by observing stars with the same mass in different clusters, one can investigate the evolution of angular momentum and coronal emission as a function of age; conversely, by comparing stars of different masses in clusters, one can investigate how angular momentum evolution and coronal emission depends on convection zone properties. The extensive observations of open clusters carried out by *ROSAT* (Randich 1997 and references therein; see also Jeffries, this volume) have shown a general trend of decreasing coronal emission with age, as expected by loss of angular momentum on the main-sequence due to the braking action of stellar winds. They have also shown, however, that this overall decrease is itself a function of mass and of convection zone depth. G stars with shallow convective zones are braked much more efficiently than late-K and M dwarfs with deeper convection zones; therefore, coronal emission in G stars decreases more rapidly with age than for lower mass stars, consistently with what expected from simple dynamo models. A further complication is the existence among active stars of a saturation limit (at $L_x/L_{bol} \sim 10^{-3}$), for which stars rotating more rapidly than a given threshold ($\sim 15\ km\ s^{-1}$ for the Pleiades K stars) all have the same X-ray luminosity. The existence of saturation, together with a dependence of coronal emission on Rossby number, explains satisfactorily the observed pattern of coronal emission with age in clusters spanning the age range from $\sim$ 30 Myr (like IC 2602 and IC2391) to $\sim$ 700 Myr (the age of the Hyades).

The existence of a saturation limit is an important and still little understood problem ... see Mathioudakis (this volume). It could be due to a saturation of the dynamo process itself, or to the filling of the whole available coronal volume by active regions with a maximum temperature and pressure. If so, it remains to be determined why there is a maximum in the temperature and density of stellar active regions: it must result from the heating process itself, which however remains elusive. Recently, the phenomenon of "supersaturation" has also been discovered in cluster data, i.e. a turnover of coronal emission at still higher rotation velocities than those typical of the saturation regime (Randich 1998). This supersaturation effect is difficult to understand, unless it is only a selection effect due to higher plasma temperatures attained at very high rotation rates, which shift the observed coronal emission outside of the *ROSAT* passband. If on the contrary supersaturation is an intrinsic property of the dynamo process and/or of the heating mechanism, it provides an important clue, together with saturation, for understanding coronal heating and dynamo action in late-type stars. It should be emphasized that the physical link between coronal activity and the dynamo process is only indirect, and mediated through the efficiency of the coronal heating mechanism for stars of different spectral types and activity levels. The heating mechanism remains unknown (even for the Sun!), although it is generally accepted that it may result from the dissipation of either electric currents or of magneto-acoustic waves. Stellar coronal observations have provided compelling evidence that the heating mechanism must be magnetic in nature (a purely acoustic heating would have produced a strong dependence of coronal emission on spectral type and mass, which is not observed); yet, it has not been possible up to now to distinguish between alternative heating mechanisms.

If the efficiency of coronal heating is not a strong function of mass (which however remains to be proved), X-ray coronal observation can provide information on the dynamo process itself, and therefore on processes which occur in the deep interior of the star. The apparent decrease of coronal emission that has been reported from early *Einstein* observations of very-late M stars suggested that coronal activity could be reduced, or totally suppressed, in fully convective stars. If this is the case, a dynamo model operating in a shallow layer at the interface between the convective zone and the radiative interior (as thought to occur for the Sun) is favoured with respect to a distributed dynamo operating throughout the convective zone. More recent data from *ROSAT* (Fleming et al. 1993) do not support the view that coronal activity is suppressed in fully convective stars, and in fact there is no apparent decrease in coronal efficiency (as measured by the L_x/L_{bol} ratio) for fully-convective stars. The vigorous coronal activity of pre-main sequence (PMS) stars also argues against such a possibility, thus favouring a distributed dynamo rather than a shallow dynamo at the bottom of the convective zone. It must be stressed however that our understanding of the dynamo mechanism is so imperfect that it would be dangerous to use these qualitative arguments, as well as the indirect evidence provided by coronal observations, to make strong statements about stellar dynamos. Similarly, while there is some indication (from both ground-based Ca II data and from coronal X-ray observations) that different types of dynamos (chaotic *vs* cyclic) may be at work in stars with widely different rotation rates and activity levels, the data are still insufficient to constrain dynamo models in an effective way.

An interesting and related question is the origin of coronal emission in PMS stars. It is now well established that PMS stars, both Classical T-Tauri (CTT) and Weak-lined T-Tauri (WTT) stars, are vigorous X-ray emitters. This extends the relationship between X-ray emission and rotation to ages at least as young as $\sim$ 1 to 10 Myr. Extensive observations by *Einstein* and *ROSAT* of PMS stars in star forming regions (SFR) in Taurus-Auriga, Chamaeleon, ρ Oph, Lupus, Sco-Cen and other regions have revealed dozens of X-ray sources which were only partially coincident with previously known PMS stars (Neuhäuser 1997, Feigelson & Montmerle 1998, and references therein). X-ray surveys have proven to be the most effective way to identify young stellar objects and thus to determine in an unbiased way the Initial Mass Function of SFRs. The observed emission most likely originates from magnetic processes at the star surface (as in WTTs) and/or in magnetic structures which connect the central star to the surrounding disk (as may be the case, at least partly, for CTTs). The high level of time variability of these sources, with frequent long-duration flares, suggests that X-ray activity originates via magnetic reconnection as in solar flares. Determination of rotation rates of these stars by means of photometric and spectroscopic observations from the ground has shown, somewhat surprisingly, that PMS stars are in general not very fast rotators, with rotation rates ranging from a minimum detectable $v\ sini \sim 5\ km\ s^{-1}$ up to $\sim 50\ km\ s^{-1}$. On the contrary, stars in young clusters like α Per (with an age of 50 Myr) show a much broader range of rotational velocities, with values of up to $\sim 200\ km\ s^{-1}$. This can easily be understood as a consequence of stellar spin-up as a PMS star approaches the ZAMS along the radiative track (Bouvier 1994). The broad range of rotational values shown by cluster stars is due to the different times scales for the dissipation of the disk during PMS evolution. Shorter time scales imply a larger amount of spin-up and a higher velocity on the ZAMS; on the contrary, a longer time scale for the dissipation of the disk implies a stronger coupling between the star and the disk and a smaller amount of spin-up. The study of angular momentum evolution is thus fundamental for understanding coronal activity in PMS stars as well as in the subsequent evolution on the main-sequence (Bouvier 1997).

3. Modelling of Quiescent Coronae

For only a small number of stars (typically eclipsing binaries) it is possible, at least in principle, to obtain direct information on the spatial structure of their coronae. For all the others, we must rely on indirect evidence such as provided, for instance, by a model-dependent analysis of their temperature structure. X-ray spectra of stellar coronae are usually fitted by one (1-T) or two (2-T) temperature isothermal models, which provide a crude description of the corona in terms of a minimum numbers of parameters: two temperatures T_1 and T_2 and two normalization factors, which are usually expressed in terms of the volume emission measures EM_1 and EM_2 (we assume for simplicity here that the hydrogen column density N_H is fixed and that elemental abundances are solar). Analysis of *Einstein*, *EXOSAT* and *ROSAT* spectra of moderate resolution show that 1-T and 2-T models provide an adequate fit of the data (in terms of χ^2 statistics) and that 2-T models are invariably needed whenever we have data of sufficiently high S/N (e.g. Schmitt et al. 1990, Dempsey et al. 1993, Singh et al. 1996). The low-temperature component typically ranges from a few to several million degrees whereas the high-temperature component is often at temperature higher than $\sim 10^7$ K. More active stars (like RS CVn binaries and young rapidly rotating stars) are usually hotter than inactive stars, and giants tend to be hotter than main-sequence stars of the same spectral type. There is, however, a large range of possible temperature distributions, as indicated by the different emission measure ratios observed in stars of different ages, rotation rates and activity levels (Güdel et al. 1997).

The two well separated temperatures observed in RS CVn binaries and other active stars (which constitute so far the bulk of the available high quality spectra) may suggest at first sight that we are dealing with two different families of coronal loops, one at relatively low temperature and the other at higher temperature. Early observations of eclipsing binary systems, such as those obtained by *EXOSAT*, gave some support to this interpretation showing that the high-temperature component was much more extended than the low-temperature one (e.g. White et al. 1990; we now know, however, that the interpretation of eclipse data is not so straightforward and that unique solutions are usually not obtained, cf. Siarkowski et al. 1996). The two well-separated peaks usually obtained in the derivation of differential emission measure distributions from high-resolution data (Mewe et al. 1996) tend to support this concept. On the other hand, the two temperatures derived by using different instruments were often substantially different, suggesting some dependence on the detector spectral response and the existence of a continuous emission measure distribution. In many cases, the observed spectra could be equally well fitted by either a 2-T model and a power-law differential emission measure model, which mimics the temperature distribution inside a coronal loop (Preibisch 1997). At any rate, the interpretation of the 2-T fits in terms of a physically meaningful model remains problematic.

A more satisfactory approach is to fit the observed spectra with realistic loop models which, if successful, can provide physical information on loop structures and heating mechanisms. Spatially resolved observations of the Sun show that the solar corona consists of an ensemble of magnetically confined loop-like structures, with different lengths, pressures and temperatures. Thus, the relevant questions are: can a single average loop-like structure, or a small numbers of different families of loops, give integrated spectra which are similar to those observed from stellar coronae ? Can 2-T models fit the spectra of coronae formed by only one family of loops, or do we need necessarily two families of loops to reproduce spectra compatible with 2-T fits ? Can we discriminate from high-quality spectra between 2-T isothermal models and loop models ? What physical information can we derive on coronal structures and hence on the coronal heating mechanisms for stars of different ages, spectral types, rotations and ages ? These questions are still far from being solved but are starting to be tackled by current modelling efforts.

We understand quite well the physics of a single magnetically confined loop in hydrostatic equilibrium and energy balance. Its temperature and density structure is determined by a balance between the heating by some magnetic process and the radiative and conductive losses. For the simplified case of constant heating along the loop and constant loop cross-section, this leads to simple scaling laws which relate the maximum temperature T_{max} at the loop apex, the loop semi-length L, the pressure p_o at the loop base and the heating per unit volume E_H (Rosner et al. 1978, Serio et al. 1981). These read (in c.g.s. units):

$$T_{max} \sim 1.4 \times 10^3 (p_o L)^{1/3} \exp\left(-0.04 L/s_p\right) \tag{1}$$

$$E_H \sim 1.0 \times 10^5 (p_o)^{7/6} L^{-5/6} \exp\left(-0.5 L/s_p\right) \tag{2}$$

where $s_p \sim 5 \times 10^3 T_{max} g_\odot / g_\star$ is the pressure scale-height at the loop top. For a loop much shorter than the pressure scale-height, T_{max} depends only on the product $p_o L$ (and not separately on p_o and L), whereas the heating rate E_H is uniquely determined once two of the three quantities L, p_o and T_{max} are fixed.

For a spatially unresolved stellar corona, the total coronal emission will result from the integrated contribution of all loops. In the simplest case that only one type of loops dominates (this is at variance with the solar case, but can be justified if, for instance, the integrated emission is dominated by active regions loops, all with similar lengths and pressures), the total X-ray luminosity will scale as

$$L_X \sim \frac{p_o^2}{T^2} \Lambda(T) L f R_\star^2 \tag{3}$$

where $\Lambda(T)$ is the radiative loss function in the X-ray band, T is an effective temperature averaged along the loop (similar to the one provided by 1-T fits of observed spectra), and f is the fractional area of the stellar surface covered by the loop foot-points. If we combine Eq. (1) and (3) above, taking into account that the "effective" temperature T is related to the maximum temperature T_{max} at the loop apex, it is easy to see that the observed coronal emission (and its

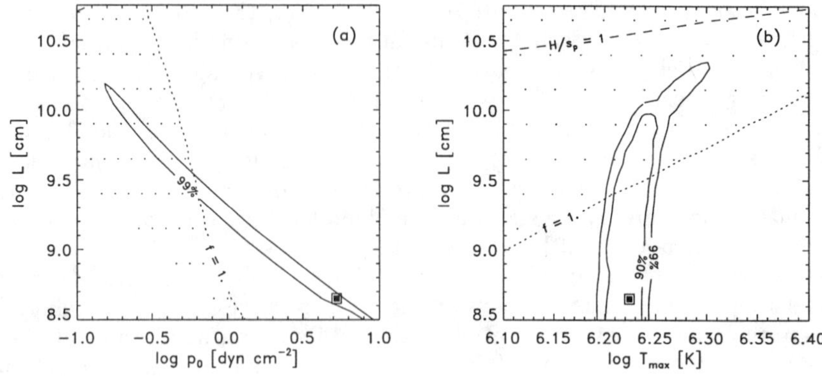

Figure 1. Loop modelling of Procyon (from Maggio & Peres 1997).

spectral shape) will depend on three of the above four parameters, e.g. T_{max}, p_o (or equivalently L), and f (with $f \leq 1$). In fact,

$$L_X \sim T\Lambda(T)p_o f \tag{4}$$

where $\Lambda(T)$ is a known function of T ($\sim T^{-1/2}$ in the range of temperatures typical of stellar coronae). Fits of observed spectral data will allow us to constrain T_{max} (from the shape of the spectrum) and the product $p_o f$ (from the normalization factor (in much the same way as for 1-T models), but they will put little constraint on p_o, L and f separately. This is a fundamental limitation of loop models for loops much smaller than the pressure scale height. Additional independent information are needed in order to determine the individual loop model parameters. One possibility is to have loops comparable to the pressure scale height (for which $L \sim s_p$); the other is to have an independent estimate of density (and hence of p_o) from density sensitive line ratios.

The above considerations are sustained by extensive loop modelling of stellar coronal spectra (Maggio & Peres 1996, 1997; Ciaravella et al. 1996, 1997; Ventura et al. 1998) using both simulated spectra (generated from loop models) and fitting of real spectra (mostly from the *ROSAT* PSPC). 1-loop and 2-loop models have been used, as well the usual 1-T and 2-T isothermal fits. The simulations show that in some cases (typically when the ratio of the hot to cool component $EM_2/EM_1 > 1$) 2-T models can fit spectra generated with a 1-loop model, indicating the 2-T models are not necessarily a proof of the existence in stellar coronae of two distinct families of loops (a similar conclusion is implied by the equivalence in many cases of 2-T fits and of fits obtained by using a $\sim T^{\alpha}$ differential emission measure distribution, Preibisch 1997). On the contrary, when the ratio EM_2/EM_1 given by 2-T fits is less than 1, a 1-loop model is usually unable to fit the observed spectra for realistic values of L, and a 2-loop model is required. The latter model requires 6 free parameters, i.e. two temperatures T, two loop lengths L (or equivalently two base pressure p_o) and two filling factors f, with the constraint that $f_1 + f_2 \leq 1$. The interpretation of the 6-dimensional parameter space in order to determine confidence regions becomes exceedingly complex in this case, even keeping fixed other parameters

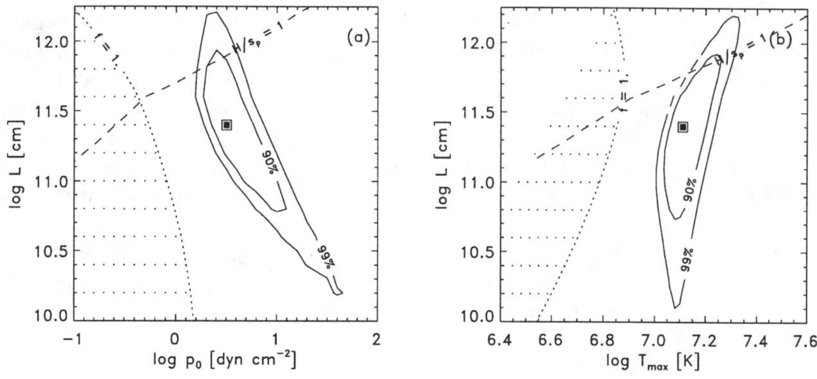

Figure 2. Loop modelling of ι Vir (from Maggio & Peres 1997).

(e.g. elemental abundances). As for 1-loop models, it is usually not possible to constrain effectively all individual parameters, albeit some useful constraints can be obtained in some cases and for high-quality spectra.

When a 1-loop model is able to fit the data, and the loops are shorter than the pressure scale-height, (as occurred for the *ROSAT* spectrum of Procyon, cf. Figure 1 from Maggio & Peres 1997, or for the young stars HD283572 and Hz739, Maggio et al. 1997, and for η Boo, Ventura et al. 1998), T_{max} is well constrained by the shape of the spectrum, but L, p_o and f are not. Only an upper limit to L (and a lower limit to p_o) can be determined by imposing the condition that $f \leq 1$. On the contrary, when the length of the loop is comparable with the pressure scale-height (as occurred for ι Vir in Maggio & Peres 1997, cf. Figure 2), the confidence regions are closed and one can constrain all loop parameters. For loops much larger than the pressure scale height (as for the 1-loop model solutions shown to be unphysical for most G-type stars by Ventura et al. 1998), only p_o is well constrained, while only lower limits can be obtained for L and T_{max}.

When a 1-loop model fails to fit the observed spectrum (as for HD3625 in Maggio & Peres 1997 or for most of the G-type stars in Ventura et al. 1998), a 2-loop model provides an acceptable fit. Two distinct families of loops are definitely needed in this case. Only the temperature is well constrained (particularly the lower one in the case of *ROSAT*), while the loop lengths and pressures are usually not constrained, albeit the range of acceptable 2-loop solutions can be considerably reduced by imposing that the total filling factor $f_1 + f_2 \leq 1$, and thus determining lower limits to L and upper limits to p_o for both families of loops. A more detailed interpretation of the solutions in terms of physical models requires a careful inspection of appropriate multi-dimensional representations of the parameter space (Maggio & Peres 1997, Ventura et al. 1998).

The models run so far suggest the existence in many stars of at least two distinct families of bright loops, one with a temperature of a few million degrees and moderately high pressures (from ~ 1 to 10^2 dyn cm^{-2}), covering a significant fraction of the stellar surface, and the other with both higher temperature ($T_{max} \sim 1-3\ 10^7$ K) and higher pressure ($\sim 10^2$ to 10^4 dyn cm^{-2}), covering only a few percent of the star. These results give us a first glimpse of the diagnostic potentials of loop models and should be considered as preliminary at this stage. They are based on *ROSAT* PSPC spectra of moderate resolution and limited bandwidth, and on the assumption of solar abundances. It would be interesting to see the results of loop modelling for instruments with higher resolution and/or larger bandwidth (such as those on *ASCA* and *Beppo*SAX or to be flown on *AXAF*, *XMM* and *ASTRO-E*), taking also into account the possibility of non-solar abundances. It can be expected that data with better resolution, larger bandwidth and higher S/N may ultimately lead to discrimination, in terms of χ^2 statistics, between 2-T isothermal models and more physically sound loop models. First attempts in this direction using *Beppo*SAX data (e.g. Sciortino et al. 1998) look promising.

4. Modelling of Stellar Flares

Flare-like brightenings similar to those observed from the Sun but on a much larger energy scale are observed from a variety of late-type stars, including UV Ceti flare stars, RS CVn and Algol type binaries, and PMS stars (e.g. Pallavicini et al. 1990; Ottmann & Schmitt 1994, 1996; Stern et al. 1992; Pallavicini & Tagliaferri 1998a,b). These events are believed to occur via magnetic reconnection, either in magnetically-confined loop-like structures (as in solar compact events) or in the relaxation of an open magnetic configuration to a closed one (as in solar two-ribbon flares). X-ray observations of flares have been discussed by a number of authors and the modelling of flares has been reviewed by Schmitt (1994), Pallavicini (1995) and van den Oord (this volume). Here I will consider only recent advances in the hydrodynamic modelling of stellar flares, while referring to the above reviews for a more comprehensive discussion of previous modelling efforts.

The simplest way to model spatially unresolved observations of stellar flares is by comparing the observed decay time to the characteristic times for radiative (τ_R) and conductive (τ_C) cooling. Assuming that the flare occurs in a single loop, and that radiative and conductive cooling times are approximately equal, it is possible to estimate the density n, volume V, loop semi-length L and minimum magnetic field strength B_{min} from the light curve decay time and the observed values of temperature T and volume emission measure EM (derived from spectral fits of X-ray data). This is possible provided there is no appreciable heating during the decay time. Although this may be a plausible assumption for short-lived impulsive flares, it is likely to be completely wrong for intense long-duration events for which the long time scale of the flare suggests prolonged heating in the decay phase. As discussed by Schmitt (1994) and Pallavicini (1995), these cases can be tackled by either a quasi-static cooling model (which approximates the flare decay as a succession of static loop models, van den Oord & Mewe 1989) or by the reconnection model of Poletto et al. (1989), in which heating in the decay phase is provided by the gradual reconnection of an open magnetic field configuration which relaxes to a closed one. Both models are able to reproduce the observed light curve of the flare and can provide plausible values for the relevant physical parameters. Unfortunately, they reproduce the same data sets under completely different assumptions and do not constrain the flare parameters in a unique way. Not only is the reconnection model unable to discriminate between a small region with high magnetic field strength and a large region with a lower value of the field; it cannot also be distinguished (in its ability to fit real data) from the quasi-static cooling model, thus raising strong doubts about the reliability of the derived flare parameters. For instance, a long duration flare observed by *ROSAT* from EV Lac could be modelled equally well with the quasi-static cooling model and with the reconnection model, but the inferred size of the region was orders of magnitude different in the two cases (Schmitt 1994).

For flares which occur in magnetically confined loops, the best way to get information on flare loop sizes and energy release is to use full hydrodynamic models which predict the time behaviour of temperature, density and flow velocity in the loop, under the action of a prescribed heating perturbation (e.g. Reale et al. 1988, Cheng & Pallavicini 1991). This is a complex and rather time consuming job which requires the run of a large number of time-dependent simulations for a variety of loop geometries, preflare conditions and energy release rates. Reale et al. (1997) and Reale & Micela (1998) have shown that this approach can be considerably simplified for the flare decay phase by using a semi-analytical expression which relates the flare loop semi-length, the observed light curve decay time τ_{LC}, the maximum temperature T_{max} at the beginning of the flare decay phase, and the slope ζ of the temperature vs. density (or equivalently of the temperature vs. $\sqrt{EM}$) trajectory during the flare decay. The slope ζ is itself related to the time scale of the heating during the decay phase (Jakimiec et al. 1992), a large value of ζ corresponding to negligible heating during the decay phase, whereas smaller and smaller values of ζ correspond to increasingly longer time scales τ_H for heating during the flare decay. In formulae:

$$L = 2.7 \times 10^3 \frac{\tau_{LC}\sqrt{T_{max}}}{F(\zeta)} \tag{5}$$

where L is the loop semilength and the function $F(\zeta) = \tau_{LC}/\tau_{th}$ is an empirical fitting to the results of flare loop simulations for a variety of different physical conditions (two different fits are required for loops much smaller than the pressure scale height and for loops comparable to, or larger than the pressure scale height). $F(\zeta)$ describes the variation with ζ of the flare light curve decay time τ_{LC} normalized to the spontaneous loop decay time $\tau_{th} = 3.7 \times 10^4 L/\sqrt{T_{max}}$. The latter is a characteristic time for the decay of a loop by radiation and conduction, once the heating term has been switched off (Serio et al. 1991).

The method has been extensively tested on solar flares observed by the SXT on *Yohkoh* (Reale et al. 1997) and has been shown to reproduce quite accurately the lengths of flaring loops (which, in the solar case, can be directly observed). It has also been applied to flares observed by *ROSAT* (Reale & Micela 1998) and *ASCA* (Ortolani et al. 1998), providing estimates of the loop length and of the heating duration during the decay. The advantage of this method is that it uses only observable quantities (the flare decay time and the temperature and volume emission measure at different times during the decay) which can easily be extracted from the data, provided these are of sufficiently good quality. The disadvantage, and limitations, of the method is that it relies on the determination of the function $F(\zeta)$ through a fitting of flare loop model

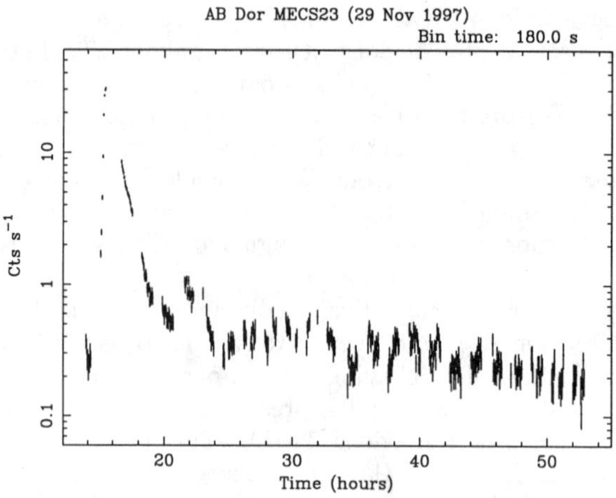

Figure 3. MECS light curve of a *Beppo*SAX observation of AB Doradus on November 29-30, 1997 (from Pallavicini & Tagliaferri 1998b).

simulations folded through the response of the X-ray instrument used. This introduces some scatter, which ultimately limits the accuracy of loop length determinations, and requires that appropriate numerical expressions for $F(\zeta)$ are computed for each X-ray instrument. In addition, the detailed physics of the hydro-simulations remains somewhat hidden in the numerical computations, making the interpretation of the relevant quantities not immediately evident. In spite of these limitations, the method appears promising as a short-cut to detailed modelling of flares with hydrocodes.

Preliminary applications of the method to flares on the dMe stars AD Leo and CN Leo (Reale & Micela 1998) and of the young star AB Dor (Ortolani et al. 1998) show that the derived loop lengths are substantially shorter than inferred by simply equating τ_{LC} to some characteristic decay time like τ_{th} or τ_R. This is due to the fact that allowance is made of possible heating during the decay. The duration of the heating supply determines the slope ζ of the temperature vs. density trajectory as the flare cools down (Jakimiec et al. 1992). In practice, since the density cannot usually be determined from current data, use is made of the temperature T and of the volume emission measure EM, which can easily be determined from fits of spectral data (the slope will be the same in a $logT$ vs. $log\sqrt{EM}$ plot). For spectral data of lower statistics, the observed count rate and a sort of spectral hardness ratio can also be used (Reale & Micela 1998). Note that T_{max} that enters Eq.(5) is the flare maximum temperature at the loop apex, which is somewhat higher (by an amount which can be estimated) than the observed maximum temperature at the flare peak (which is averaged over the loop temperature distribution and the instrument spectral response).

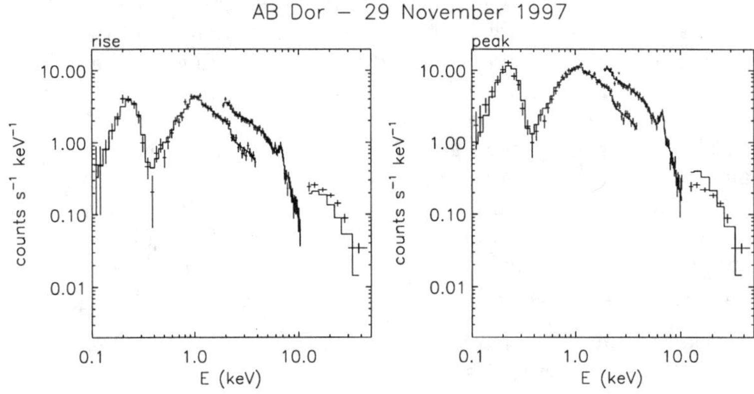

Figure 4. The LECS, MECS and PDS spectra of the Nov 29, 1997 flare on AB Doradus during the rise and peak phase. Also shown is the 2-T models that best fit the LECS and MECS data alone.

5. Hard X-ray Emission from Stellar Flares

I conclude this review of theoretical problems in stellar coronal physics by reporting on a new important observational result obtained recently by the Italian-Dutch satellite *Beppo*SAX. This is the first detection from stars other than the Sun of hard (>20 keV) X-ray emission during large flares (Pallavicini et al. 1998, Favata 1998). This emission, extending up to ~ 50 keV, has been detected by the PDS instrument onbord *Beppo*SAX during the rise phase and at the peak of large flares on Algol (Favata 1998), UX Ari (Pallavicini & Tagliaferri 1998a) and AB Dor (Pallavicini & Tagliaferri 1998b; cf. Figure 3). The emission appears to be thermal and due to a plasma at $\sim 10^8$ K. Such high temperatures are required to fit the low energy part of the spectrum, between 0.1 and 10 keV, observed with the LECS and MECS instruments. If this thermal emission is extrapolated to higher energies it accounts fairly well for the hard tail observed by the PDS (Figure 4). There is no need therefore to invoke a non-thermal power-law component as observed during the impulsive phase of solar flares. Moreover, while in solar flares, the hard X-ray emission is much smaller than the soft one (typically $\sim 10^{-5}$), the hard tail seen by the PDS is about one tenth of the soft X-ray emission observed at the peak of the flare.

Hard X-ray emission from stellar flares is not unexpected on theoretical grounds, albeit realistic estimates based on the solar analogy makes its detection extremely unlikely (a detector several square kilometers in size would be required if this emission were similar to that observed in solar flares!). This emission is expected to originate from non-thermal bremsstrahlung of high-energy electrons accelerated during the impulsive phase of flares. The electrons are channeled by the magnetic fields and lose their energy as they impinge on the denser chromospheric layers at the loop footpoints. Study of this component provides essential information on the flare primary energy release and particle acceleration. So far hard (> 20 keV) X-ray emission had never been observed from stellar flares. The high sensitivity and wide spectral response of *Beppo*SAX has made this detection possible, but the preliminary analysis of the data shows that the observed emission is most likely thermal and much more intense that would be expected from the scaled-up solar case. If this is confirmed, this hard energy tail will provide additional information on the high temperature component of stellar flares, but will give little information on the primary energy release and particle acceleration. For the latter, we must rely on optical continuum emission as a proxy, unless the ratio of non-thermal hard X-ray emission to thermal soft X-ray emission in stellar flares is orders of magnitude larger than in solar flares. There is no indication for this in the recent detection of stellar hard X-rays by *Beppo*SAX.

Acknowledgments. I thank A. Maggio, G. Peres and F. Reale for many useful discussions on these topics.

References

Bouvier, J. 1994, in 8th Cambridge Workshop on Cool Stars, Stellar Systems, and the Sun, ed. J.-P. Caillaut, San Francisco: ASP Conf. Ser., Vol. 64, p. 174

Bouvier, J. 1997, Memorie SAIt, 68, 881

Cheng, C.-C. & Pallavicini, R. 1991, ApJ, 381, 234

Ciaravella, A., Maggio, A. & Peres, G. 1997, A&A, 320, 945

Ciaravella, A., Peres, G., Maggio, A. & Serio, S. 1996, A&A, 306, 553

Dempsey, R.C., Linsky, J.L., Fleming, T.A. & Schmitt, J.H.M.M. 1993, ApJ, 413, 333

Favata, F. 1998, in The Active X-Ray Sky: Results from BeppoSAX and Rossi-XTE, eds. L. Scarsi et al., Amsterdam: North-Holland, p. 23

Feigelson, E.D. & Montmerle 1998, T., ARA&A, in press

Fleming, T.A., Giampapa, M.S., Schmitt, J.H.M.M. & Bookbinder, J.A. 1993, ApJ, 410, 387

Güdel, M., Guinan, E.F. & Skinner, S.L. 1997, ApJ, 483, 947

Hempelmann, A., Schmitt, J.H.M.M., Schultz, M., Rüdiger, G. & Stępień, K. 1995, A&A, 294, 515

Jakimiec, J., Sylwester, B., Sylwester, J., Serio, S., Peres, G. & Reale, F. 1992, A&A, 253, 269

Maggio, A., Micela, G. & Peres, G 1997, Memorie SAIt, 68, 1095

Maggio, A. & Peres, G. 1996, A&A, 306, 563

Maggio, A. & Peres, G. 1997, A&A, 325, 237

Mewe, R., Kaastra, J.S., White, S.M. & Pallavicini, R., 1996, A&A, 315, 170

Neuhäuser, R. 1997, Science, 276, 1363

Ortolani, A., Pallavicini, R., Maggio, A., Reale, F. & White, S.M. 1998, in 10th Cambridge Workshop on Cool Stars, Stellar Systems, and the Sun, eds. R.A. Donahue & J.A. Bookbinder, San Francisco: ASP Conf. Ser., Vol. 154, in press

Ottmann, R. & Schmitt, J.H.M.M. 1994, A&A, 283, 871

Ottmann, R. & Schmitt, J.H.M.M. 1996, A&A, 307, 813

Pallavicini, R. 1995, in Flares and Flashes, J. Greiner et al. eds., Berlin: Springer, p. 148

Pallavicini, R. 1998, Space Sci. Rev., in press

Pallavicini, R., Golub, L., Rosner, R., Vaiana, G.S., Ayres, T. & Linsky, J.L. 1981, ApJ, 248, 279

Pallavicini, R., Stella, L. & Tagliaferri, G. 1990, A&A, 228, 443

Pallavicini, R. & Tagliaferri, G. 1998a, in The Active X-Ray Sky: Results from BeppoSAX and Rossi-XTE, eds. L. Scarsi et al., Amsterdam: North-Holland, p. 29

Pallavicini, R. & Tagliaferri, G. 1998b, in Highlights in X-ray Astronomy, eds. B. Aschenbach & M. Freyberg, Garching: MPE Report, in press

Pallavicini, R., Tagliaferri, G & Maggio, A. 1998, Space Sci. Rev., in press

Preibisch, T. 1997, A&A, 320, 525

Poletto, G., Pallavicini, R. & Kopp, R.A. 1989, A&A, 201, 93

Randich, S. 1997, Memorie SAIt, 68, 971

Randich, S. 1998, in 10th Cambridge Workshop on Cool Stars, Stellar Systems, and the Sun, eds. R.A. Donahue & J.A. Bookbinder, San Francisco: ASP Conf. Ser., Vol. 154, in press

Reale, F., Betta, R., Peres, G., Serio, S. & McTiernan, J. 1997, A&A, 325, 782

Reale, F. & Micela, G. 1998, A&A, in press

Reale, F., Peres, G., Serio, S., Rosner, R. & Schmitt, J.H.M.M. 1988, ApJ328, 256

Rosner, R., Tucker, W.H. & Vaiana, G.S. 1978, ApJ, 220, 643

Schmitt, J.H.M.M. 1994, ApJS, 90, 735

Schmitt, J.H.M.M., Collura, A., Sciortino, S., Vaiana, G.S., Harnden, F.R.Jr. & Rosner, R. 1990, ApJ, 365, 704

Sciortino, S., Maggio, A., Favata, F. & Orlando, S. 1998, A&A, submitted

Serio, S., Peres, G., Vaiana, G.S., Golub, L. & Rosner, R. 1981, ApJ, 243, 288

Serio, S., Reale, F., Jakimiec, J., Sylwester, B. & Sylwester, J. 1991, A&A, 241, 197

Siarkowski, M., Preś, P., Drake, S.A., White, N.E. & Singh, K.P. 1996, ApJ, 473, 470

Singh, K.P., White, N.E., Drake, S.A. 1996, ApJ, 456, 766

Stern, R.A., Uchida, Y., Tsuneta, S. & Nagase, F. 1992, ApJ, 400, 321

van den Oord, G.H.J. & Mewe, R., 1989, A&A, 213, 245

Ventura, R., Maggio, A. & Peres, G. 1998, A&A, 334, 188

White, N.E., Shafer, R.A., Horne, K., Parmar, A.N. & Culhane, J.L. 1990, ApJ, 350, 776

Solar and Stellar Activity: Similarities and Differences
ASP Conference Series, Vol. 158, 1999
C.J. Butler and J.G. Doyle, eds.

SOHO - Where has the Quiet Sun gone?

H.E. Mason

Department of Applied Mathematics and Theoretical Physics, Silver Street, Cambridge CB3 9EW

Abstract. The observations from the *Solar and Heliospheric Observatory* (*SOHO*) have revolutionised our understanding of the Sun. In this review we explore the nature of the quiet Sun with spectroscopic observations from the CDS (Coronal Diagnostic Spectrometer) and SUMER (Solar Ultraviolet Measurement of Emitted Radiation). With CDS and SUMER, simultaneous observations can be obtained over a wide range of temperatures, so that the spatial relationship between the chromospheric, transition region and coronal emission can be unambiguously established. CDS and SUMER also allow non-thermal broadenings and velocity shifts to be determined from the spectral line profiles. The *SOHO* observations have dispelled any notion we might have had of a *quiet* Sun. Even during solar minimum, the solar atmosphere is far from tranquil; it is characterised by transient and dynamic features.

1. Introduction

Faced with the choice between changing one's mind and proving that there is no need to do so, almost everyone gets busy on the proof. Galbraith's Law

Brendan Byrne had a wonderful sense of humour and a deep understanding of human nature. He often sent jokes, like the one above, to brighten a rather dull, grey, winter's day. He was chosen by NASA to participate in the first press conference on *SOHO* observations. He had a way of reaching out to the public and getting across the excitement of scientific research. He loved astronomy and took an interest in the latest developments in solar physics. The Sun is after all our nearest star. If we can't understand the physical processes taking place in the solar atmosphere, what hope is there for stellar atmospheres. This lecture is an overview of what has been learnt from *SOHO* about what we have traditionally called the *Quiet Sun.*

2. Life before *SOHO* - Quiet Sun Models

In the 1970's, the type of model which was prevalent for the solar transition region was that of a plane parallel atmosphere (Gabriel & Mason, 1982). A more realistic model was proposed by Gabriel (1976) to take account of the magnetic field geometry resulting from super-granular convective flow. Observations of the transition region in UV emission from instruments such as the High Resolution Telescope and Spectrometer (*HRTS*) (Breuckner & Bartoe, 1983) showed inhomogeneities. The electron density derived from O IV diagnostic line ratios indicated the existence of sub-resolution filamentary structures, with *fill factors* of 10^{-2}–10^{-5} (Dere *et al*, 1989). These observations led to the suggestion by Feldman (1983) that the transition region is composed of Unresolved Filamentary Structures (UFS's) - one has to surmise that if a more inviting acronym had been chosen, this proposal might have received universal acclaim. Antiochos & Noci (1986) proposed the co-existence of hot and cool loops to explain the upturn in the emission measure towards the lower temperatures ($< 10^5$ K). Dowdy *et al* (1986) went one step further and described the transition region as a *magnetic junkyard*, comprising cool (10^5 K) loops of varying sizes and hotter coronal (10^6 K) funnels.

3. *SOHO* - the 11 year Cycle of Re-discovery?

The European Space Agency/NASA Solar and Heliospheric Observatory spacecraft, *SOHO*, is the most sophisticated solar observatory ever flown. *SOHO* carries a payload of 12 instruments designed to probe the interior of the Sun, the solar atmosphere and interplanetary space. Launched on December 2nd 1995, *SOHO* operated successfully for over two years. Contact was lost with *SOHO* in June 1998, but with an immense amount of patience and tenacity, ESA and NASA space technicians recovered control of *SOHO* in September 1998 and all the instruments are now operating again.

The primary scientific objectives of *SOHO* are to:

- study the structure and dynamics of the solar interior
- study the heating mechanisms of the solar corona
- investigate the solar wind and its acceleration processes.

SOHO has provided some interesting discoveries, including *rivers of plasma* flowing beneath the solar surface, the *magnetic carpet* - a complex web of magnetic fields, *blinkers* which flash on and off in the transition region, *tornadoes* in the polar regions and *coronal mass ejections*, CME's which erupt out into the solar wind. In this review, we concentrate on the spectroscopic observations from the instruments which are looking at the solar atmosphere, that is the transition region and corona. Details of all the *SOHO* observations and operations can be found on the home page (http://sohowww.nascom.nasa.gov/).

The Michelson Doppler Imager (MDI), which provides magnetic data for the photosphere, has revealed a complex network of positive and negative polarity covering the entire quiet Sun. The patterns trace out the super-granular network, 30,000 km wide convection cells, just below the solar surface. The small-scale magnetic flux regions emerge with up-welling flows, and drift to the edges of the cells. The locations where the fields are forced together are sites where magnetic energy is released in the form of thermal and kinetic energy. The magnetic flux on the solar surface is replaced every 40 hours. This re-structuring of the magnetic field seems to be at the heart of the coronal heating puzzle.

The Extreme Ultraviolet Imaging Telescope (EIT) instrument is able to take images of the Sun in four wavelength bands, He II, Fe IX/Fe X, Fe XII and Fe XV, effectively four different temperatures. The clarity of the EIT images is excellent and the EIT movies demonstrate how variable the Sun's atmosphere is - even 'quiet' regions of the Sun's atmosphere are riddled with small transient UV brightenings.

The Coronal Diagnostic Spectrometer (CDS) (160-800 Å) is being used to study the relationship between the transition region and coronal emission - the CDS is designed to determine the physical parameters of the solar atmosphere. The CDS instrument consists of a Wolter II grazing incidence telescope feeding two spectrometers. The Normal Incidence Spectrometer (NIS) covers two wavelength ranges (307-379 and 513-633 Å) using a micro-channel plate/CCD combination detector. The Grazing Incidence Spectrometer (GIS) has four microchannel plate plus spiral anode detectors which cover the wavelength ranges 151-221, 256-341, 393-492 and 659-785 Å. The CDS principal investigator is Dr R.A. Harrison at the Rutherford Appleton Laboratory (RAL). The home page at RAL (http://solg2.bnsc.rl.ac.uk) provides an excellent overview of the CDS observations in the form of an atlas.

The CDS allows us to take slices through the solar atmosphere at different temperatures between 2 10^4 - 3 10^6 K. Simultaneous rasters, up to 4 x 4 arcmin^2 in size, can be obtained at different wavelengths, with a high spatial resolution (approximately 3 arcsec). The NIS channels cover a range of ions, including the low temperature emission from He I at 584Å (2 10^4 K); transition region emission from O V 630 Å (2.5 10^5 K); the Mg IX 368 Å line at coronal temperatures (10^6 K) and the Fe XVI 335 and 361 Å lines observed in active regions (2 10^6 K). A typical quiet Sun raster is shown in Figure 1.

Carrington rotation maps have been compiled from the CDS synoptic observations showing the evolution of coronal holes and active regions spanning several solar rotations. The quiet Sun network has been studied by Gallagher *et al* (1998), who found a sharp change in contrast in the network structure between 2 x 10^5 K and 10^6 K, consistent with the model proposed by Gabriel (1976). The observation of spectral lines from different elements allows the determination of elemental abundances in the solar atmosphere. This can be used as a tracer for the source and acceleration of the corona and solar wind. Reviews of this work are given in Young & Mason (1998) and Mason & Boschler (1998).

The Solar Ultraviolet Measurement of Emitted Radiation (SUMER) instrument can resolve the fine structures in the chromosphere and transition region, flows and wave motions. SUMER is a stigmatic normal-incidence spectrograph operating in the range 450–1610 Å. The detectors cover approximately 40 Å in the first order and the full spectrum is obtained by moving the scan mirror. Different slits can be used, with a spatial resolution around 1 arcsec.

A key feature of the CDS and SUMER instruments is the capability of obtaining simultaneous spatial and spectral information. This allows the determination of temperature and density estimates from diagnostic line ratios, emission measure distributions from sequences of ions, elemental abundance variations and flow patterns from Doppler shifts. Details of the CDS instrument, science objectives and first results are published in Harrison *et al* (1997) with spectroscopic diagnostics discussed in Mason *et al* (1997), Mason & Pike

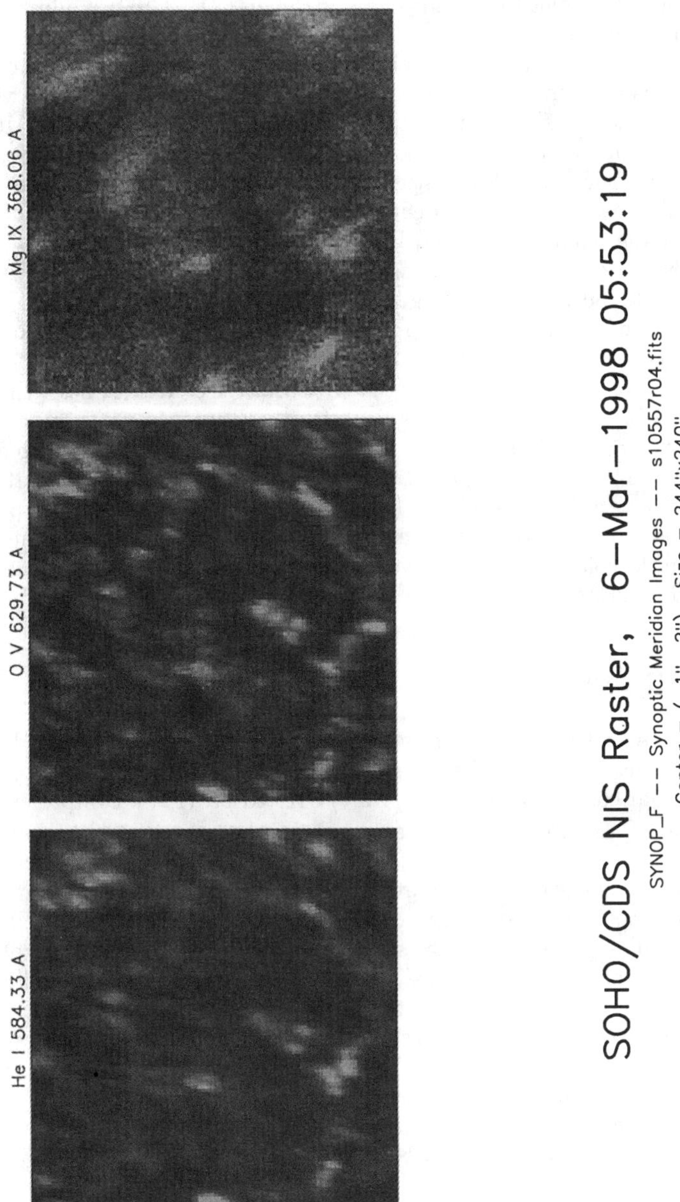

Figure 1. CDS synoptic raster of a quiet Sun region, showing He I, O V and Mg IX emission. Provided by Pike.

(1997). SUMER instrument details and first results are given in Wilhelm *et al* (1997) and Lemaire *et al* (1997).

The Ultraviolet Coronagraph Spectrometer (UVCS) provides measurements of temperature and plasma velocity in the outer corona. UVCS has revealed high particle velocity distributions above the polar regions, possibly non-thermal in origin. UVCS has also provided measurements of the elemental abundance in streamers, which shows gravitational settling of heavy elements. It seems that the slow wind is formed in streamer legs, the boundary between open and closed magnetic structures. Such studies combined with data from MDI, CDS and SUMER are being used to define the boundary conditions for the solar wind, both for the polar regions and from equatorial zones.

The Large Angle Spectroscopic Coronagraph (LASCO) views the corona from 1.1 to 30 solar radii, again with an impressive image quality. LASCO has observed many CMEs erupting into the solar wind, some of which have interacted with the terrestrial environment.

Some of these phenomena studied with SOHO are not new, they have been observed before in a different guise. In fact, there seems to be an eleven year cycle of discovery in solar research - Skylab in the 1970's, the *Solar Maximum Mission* (*SMM*) in the 1980's, now *SOHO* in the 1990's. What is new and remarkable about *SOHO* is that we are able to study these solar phenomena in a more comprehensive way, with co-ordinated observations from the photosphere out into the solar wind. What we had seen previously were tantalising snapshots, which hinted at the importance of different solar features.

4. CHIANTI

The CHIANTI atomic database (Dere *et al*, 1997) and associated software are being extensively used to analyse SOHO observations. It is the result of an international collaboration (USA/Italy/UK) to provide a comprehensive dataset for ions of astrophysical interest seen in the wavelength region $\lambda > 50$ Å. There are three basic sets of files - the observed and theoretical energy levels, the wavelengths and radiative data for each transition in an ion and the electron collisional data for each transition. CHIANTI also includes a set of IDL routines to solve the statistical equilibrium equations; to provide theoretical line intensity ratios; to calculate CDS synthetic spectra and to determine the differential emission measure. CHIANTI can be accessed via the web site: http://wwwsolar.nrl.navy.mil/chianti.html. A detailed comparison of the CHIANTI synthetic spectrum with the SERTS spectrum was carried out by Young *et al* (1997). The second version of CHIANTI has just been released (Landi *et al*, 1998). A comparison of the CHIANTI synthetic spectrum with part of a CDS-NIS quiet Sun spectrum is given in Figure 2.

5. Red-shifts and non-thermal velocities

Line shifts and broadenings give information about the dynamic nature of the solar and stellar atmospheres. The transition region spectra from the solar atmosphere are characterised by broadened line profiles. The nature of this

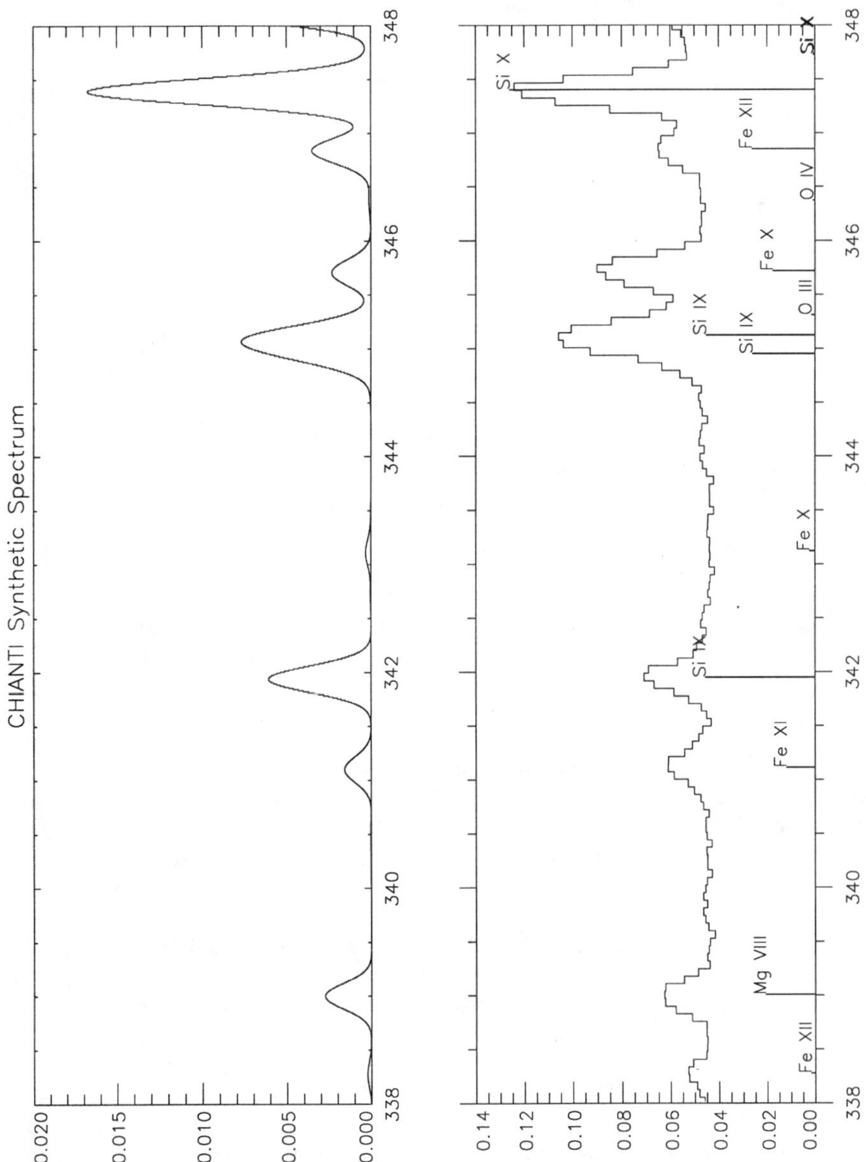

Figure 2. CDS-NIS spectrum (bottom) of a quiet Sun region, compared to a synthetic spectrum from CHIANTI (top), with identifications marked

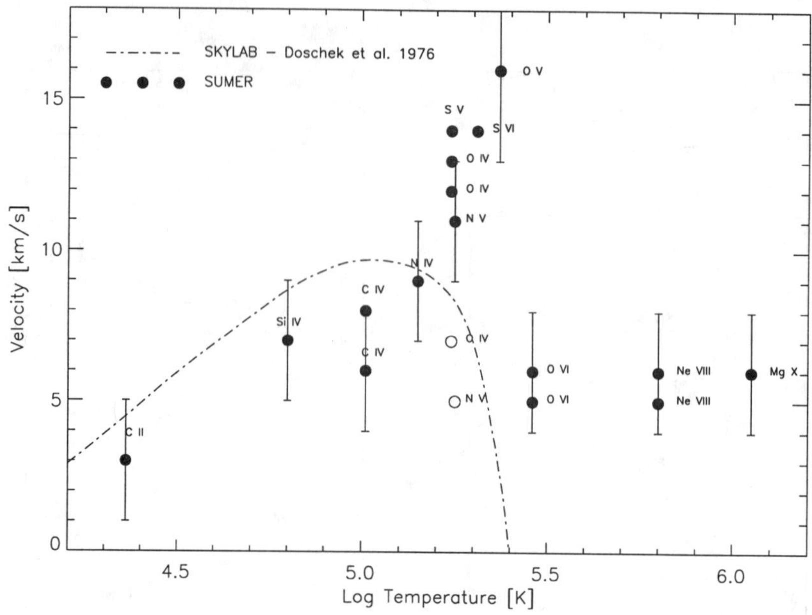

Figure 3. The variation of the redshift with temperature of formation of the quiet Sun as observed by SUMER on August 12 1996. Taken from Brekke *et al* (1997)

excess broadening puts constraints on possible heating processes. Systematic red-shifts in transition region lines have been observed in both solar spectra and stellar spectra of late type stars. On the Sun, outflows of coronal material have been correlated with coronal holes, a probable source of the solar wind. The excess broadening of coronal lines above the limb provides information on wave propagation in the solar wind.

The persistent redshift observed in the quiet Sun transition region lines remains an enigma. It has been observed in the UV lines with many different instruments. A useful summary is given in Brekke *et al* (1997). The net redshift peaks at around 10 km s^{-1} for spectral lines from ions formed at around 10^5 K. This has been confirmed using the SUMER data shown in Figure 3.

The average red-shifted emission observed at 10^5 K is sufficient to empty the corona in a few minutes. Brekke *et al* discuss several possible explanations for the red-shifts. It seems likely that up-flows must also occur (see paper by Teriaca et al. in these proceedings), but these could be less dense or cooler (spicular material). Alternatively, wave motions generated by nano-flares could be causing the apparent red-shifts.

Dere & Mason (1993) used *HRTS* spectra to study the non-thermal velocities (ntv) in the solar transition zone for a range of ions formed between 10^4 K and 10^6 K . They found that the ntv's increased with the temperature of formation of the ion between 10^4 K and 10^5 K, reaching a peak at around

25 km s^{-1}. This confirmed earlier studies with Skylab observations (Mariska *et al*, 1978, 1979). They found that there is enough energy in the acoustic mode waves to supply the coronal energy loss. One limitation on the heating of the corona by Alfvén waves was pointed out by Parker (1987). The observation of coronal line widths (Cheng et al, 1979) places an upper limit of 20 km s^{-1} on the ntv. This severely limits the upward energy flux in waves carried into the corona. Parker proposed the nano-flare concept as an alternative means of heating the corona. Doyle *et al* (1997) have analysed more HRTS data and compared their results with heating models involving acoustic and Alfvén waves. They found that acoustic waves may be able to provide the necessary energy input into the corona assuming constant electron pressure up to 10^5 K, while above this a constant electron density assumption is valid. The observational constraints were not sufficient to give a definitive answer about the validity of different dissipation theories.

Erdélyi *et al* (1998) have studied the spectral line widths from the SUMER observations at the disk centre and at the limb. They find that numerical estimates based on linear MHD, favour the existence of Alfvén wave heating over magneto-acoustic heating. Sheeley *et al* (1997) used SUMER to study the widths of spectral lines from several different elements as a function of height above the limb (30–209 arcsec) in coronal streamers. For the ions which they observed, formed between 5 10^5 K and 1.6 10^6 K, they found that the inferred plasma motion velocity fell from around 22 km s^{-1} at 30 arcsec to less that 10 km s^{-1} for the higher altitudes (see paper by Doyle et al. in these proceedings). They suggest that this result casts doubt on the heating of the corona and acceleration of the solar wind by hydro-magnetic waves in the streamers.

It is interesting to consider the possibility that torsional Alfvén waves may have a role to play in coronal heating. This was suggested by Sterling & Hollweg (1984) and more recently discussed by Kudoh & Shibata (1997). Their models exhibit a levelling out of the ntv's at coronal temperatures.

6. Explosive Events, Jets and Bright Points

The *HRTS* observations showed a dynamic transition region in C IV emission (10^5 K) with explosive or turbulent events. The average extent of these explosive events was 2 arcsec along the slit and Doppler shifts in excess of 100 km s^{-1} in both blue and red-shifted components, with an average lifetime of around 60 s. There was a tendency for these explosive events to be correlated with the weak magnetic network. Porter et al (1987) explored the nature of C IV micro-flares with the UltraViolet Spectrometer and Polarimeter (UVSP) on-board *SMM*. These corresponded to the sites of neutral lines in small magnetic bipoles, possibly smaller scale X-ray bright points.

SUMER has observed bi-directional jets in Si IV emission (Innes *et al*, 1997) which are 2 arcsec in spatial extent, with lifetimes around 60 s. Doppler shifts of 150 km s^{-1}, both red and blue shifted, suggest bi-directional flows, possible sites for energy release (see paper by Perez & Doyle in these proceedings). These jets are distributed over the entire Sun, predominantly located at the cell boundaries.

Rapid imaging with the CDS instrument, at selected wavelengths, has revealed transient bursts in intensity, or flashes, which have become known as 'blinkers' (Harrison, 1997). The O V intensities increase by a factor of 2-4 on time-scales of up to 5 mins, over an area of 20-40 arcsec. These blinkers also seem to be located at network boundaries, with a frequency such that approximately 3000 are in progress on the solar surface at any one time. They could be related to nanoflares (Parker, 1988), but Harrison estimated their energy content to be about two orders of magnitude less than the radiation losses of the upper chromosphere and corona.

The small scale brightenings seen with EIT in the quiet Sun images have been studied by Berghmans *et al* (1998). They find a continuum of sizes and a correlation between the area and duration of the brightening in Fe XII emission. These are possibly the UV extension of X-ray network flares. Falconer *et al* (1998) have carried out a comparison between the EIT brightenings and magnetogram measurements. They find that the coronal bright points cluster along network lanes. Pres & Phillips (1998 and these proceedings) have studied individual brightenings and find a correlation between the change in EIT flux in Fe XII, the soft X-ray flux from *Yohkoh*-Soft X-ray Telescope and the magnetic flux from MDI.

All of these small scale features observed in the quiet Sun network regions show the importance of the magnetic field restructuring. Sometimes the magnetic energy is released as thermal energy, heating the transition region (blinkers) or the corona (EIT brightenings). For other features (jets, explosive events) the energy is released as kinetic energy. The relationship between these small scale features remains to be established.

7. Macrospicules and Tornadoes - another twist to the Story

Using data from the Skylab - NRL slitless spectrometer, Bohlin *et al* (1975) found He II macrospicules to be an important feature in coronal holes. They are spike like features 5-15 arcsec in diameter, 5-50 arcsec in length, with velocities 10-150 $km\ s^{-1}$ moving outward into the corona. The EUV macrospicules could be associated with Hα macrospicules - possibly flaring X-ray bright points.

A macro-spicule was observed by CDS on the southern limb, during a synoptic scan in He I, O V and Mg IX. These data were analysed by Pike and Harrison (1997). The macro-spicule had an extent of more than 40 arcsec (31,000 km), a width of 20 arcsec (14,000 km) and was bright in Mg IX as well as O V and He I. They found strong evidence for acceleration and outflows along the western footpoint, which, depending on the orientation of the macro-spicule, could indicate velocities up to 580 $km\ s^{-1}$, consistent with solar wind flows.

Pike & Mason (1998) have carried out a comprehensive study of dynamic features observed with the CDS synoptic data in the polar regions. They found several examples of features in O V emission, possibly macrospicules, which appear to be spinning with velocities around 20/30 $km\ s^{-1}$. They called these solar tornadoes. Pike & Mason (1998) also found some O V features on the disk with outward flow velocities around 170 $km\ s^{-1}$, which also seem to be spinning. It seems plausible that if the footpoints of magnetic flux tubes are twisted by the convection zone, then when they emerge into the transition region and release their magnetic energy some rotational component is retained. In most theoretical models this aspect of magnetic energy release has received little attention. They suggest that the solar tornado observations lend support to Shibata's (1997) model of a spinning magnetic-twist jet. Dynamic features observed by CDS and SUMER were reviewed in Mason *et al* (1998).

8. Polar exploration - Cool Plumes, the Source of the Fast Solar Wind?

Measurements of electron temperature and electron density above a polar coronal hole have been carried out by Wilhelm *et al* (1998) with *SOHO*/SUMER. These were obtained during the roll manoeuvre of *SOHO* on September 3 1997, with the advantage of placing the SUMER slit in the E-W direction. It was possible to step the slit off limb and to distinguish between a plume and an inter-plume lane. The plumes were found to be significantly cooler than the inter-plume lanes. A controversy rages over whether or not plumes contribute to the fast solar wind. Doppler shifts have not been detected in the plumes with SUMER.

The electron temperature in polar coronal holes as a function of height above the limb has been derived by David *et al* (1998) from CDS and SUMER observations using diagnostic line ratios from O VI. They measured an electron temperature of 850,000 K close to the limb, rising to 10^6 K at 1.15 $R_\odot$, but falling to 400,000 K at 1.3 $R_\odot$.

Fludra *et al* (1998) found from CDS observations that in the coronal hole, the temperature increased from 7.5 10^5 K at the limb to 9 10^5 K at 1.2 $R_\odot$. The electron number density drops from 2 10^8 cm^{-3} to 0.6 10^8 cm^{-3} between the limb and 1.1 $R_\odot$. Del Zanna & Bromage (1998a, 1998b) have provided the differential emission measure analysis of CDS spectra shown in Figure 4.

They found that the emission measure for the quiet Sun peaked at 10^6 K, for the coronal hole at just below 10^6 K and for the plume at 7.6 10^5 K. The coronal hole electron number densities were around 2 10^8 cm^{-3}, a factor of 2–3 lower than the quiet Sun values.

CDS observations of a polar plume have been analysed by Young (1998). The hottest part of the plume, 2 10^6 K, occurs at its base, a height of around 6000 km, directly over one of the O V brightenings. This suggests that the plume base arose through the emergence of a bipolar magnetic flux region underneath the unipolar field of the coronal hole. Higher up, steady conditions prevail with temperatures peaking at around 10^6 K. The electron density at the base of the plume was found to be around 2.5 10^9 cm^{-3}, dropping to $3 - 8$ 10^8 cm^{-3} higher up.

Observations of a polar coronal holes have been carried out with SUMER by Hassler *et al* (1998). They obtained large 9 x 5arcmin^2 rasters, using the 1 x 300 arcsec^2 slit, binned to 3 x 3 arcsec^2. The chromospheric Si II 1533 Å line was used as a wavelength reference to determine the Doppler shift for Ne VIII 770 Å (6 10^5). Out-flowing coronal material (around 10–15 $km\ s^{-1}$) was found to be concentrated at junctions of the chromospheric network. They surmise that these outflows could be the source of the fast solar wind. The underlying magnetic network seems once again to be the dominant driver for the heating and dynamics of the solar atmosphere.

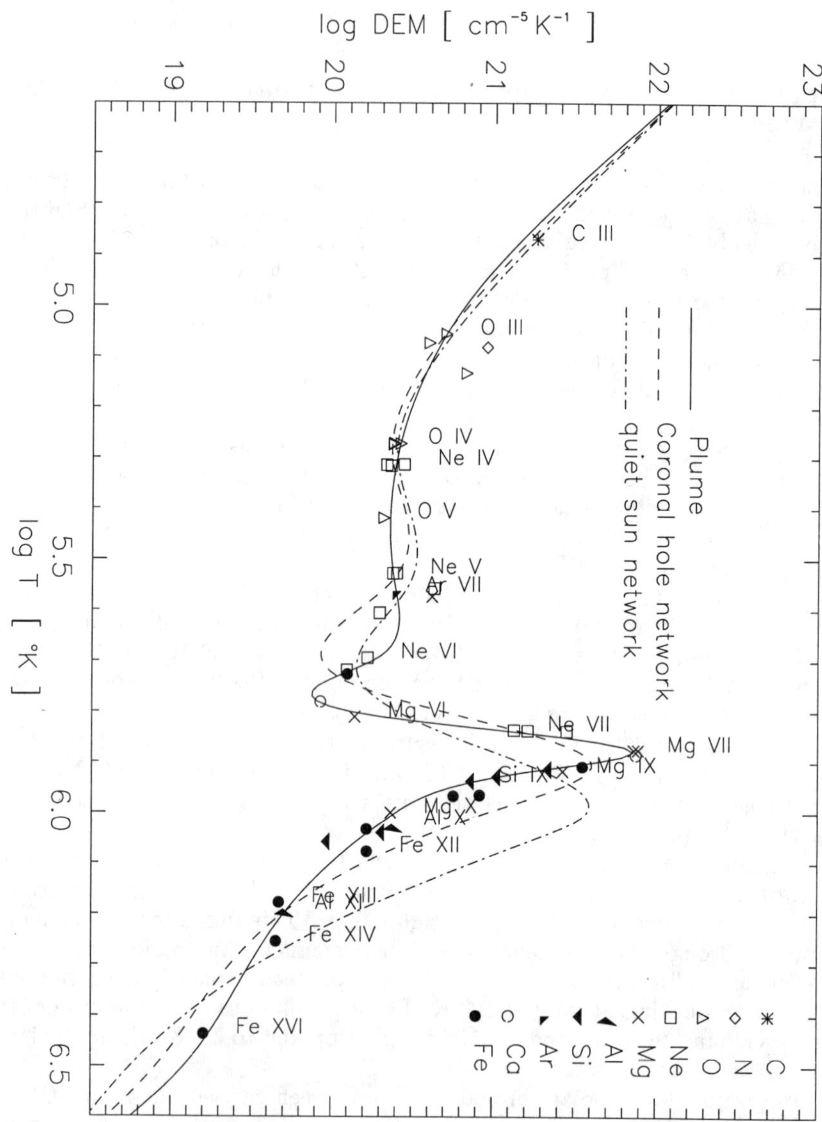

Figure 4. The differential emission measure distribution derived from CDS data using CHIANTI for a polar plume area (with observed points), a coronal hole and quiet Sun network. Taken from Del Zanna & Bromage (1998b)

9. Summary

SOHO has shown us how complex and dynamic the quiet Sun is. In the words of Parker (1997), *"If the pedestrian Sun is so difficult a subject for study, how cautious must the scientist be in extrapolating to unresolved stars with comparable or greater vigor of activity"* .

Acknowledgments. This work has been funded by PPARC. *SOHO* is a joint ESA/NASA project. My thanks go to the CDS team, in particular Richard Harrison and Dave Pike, for their support and encouragement.

This paper is dedicated to my good friend Brendan. There is a time for work and a time for play, both being essential to life.

A grad student, a post-doc, and a professor are walking through a city park and they find an antique oil lamp. They rub it and a Genie comes out in a puff of smoke.

The Genie says, "I usually only grant three wishes, so I'll give each of you just one."

"Me first! Me first!" says the grad student. "I want to be in the Bahamas, driving a speedboat with a gorgeous woman who sunbathes topless." Poof! He's gone.

"Me next! Me next!" says the post-doc. "I want to be in Hawaii, relaxing on the beach with a professional hula dancer on one side and a Mai Tai on the other." Poof! He's gone.

"You're next," the Genie says to the professor.

The professor says, "I want those guys back in the lab after lunch."

Of course, in these politically correct days, the Prof. is a woman!

References

Antiochos, S.K. & Noci, G., 1986, ApJ, 301, 440

Berghmans, D., Clette, F. & Moses, D., 1998, Astron. Astrophys, in press

Bohlin, J.D.. Vogel,S.N., Purcell, J.D., Sheeley, N.R., Tousey, R. & Van Hoosier, M.M., 1975, ApJ, 197, L133.

Brekke, P., Hassler, D.M. & Wilhelm, K., 1997, Solar Phys., 175, 349

Breuckner, G.E. & Bartoe, J.-D.F., 1983, ApJ, 272, 329.

Cheng, C.C., Doschek, G.A. & Feldman, U., 1979, ApJ, 227, 1037

David, C., Gabriel, A.H., Bely-Dubau, F., Fludra, A., Lemaire, P. & Wilhelm, K., 1998, A&A, 336, L90.

Del Zanna, G. & Bromage, B.J.I., 1998a, J. Geophys. Res., in press.

Del Zanna, G. & Bromage, B.J.I., 1998b, The Eigth *SOHO* Workshop, in press.

Dere, K.P., Bartoe, J.-D.F., Breuckner, G.E., Cook, J.W. & Socker, D.G., 1989, Solar Phys., 114, 223.

Dere, K.P., Landi, E., Mason, H.E., Monsignori Foss, B.C. & Young, P.R., 1997, A&AS, 125, 149

Dowdy, J.F., Rabin, D. & Moore, R.L., 1986, Solar Phys., 105, 35

Doyle, J.G., O'Shea, E., Erdélyi, R., Dere, K.P., Socker, D.G. & Keenan, F.P., 1997, Solar Phys., 173, 243

Erdélyi, R., Doyle, J.G., Perez, M.E. & Wilhelm, K., 1998, A&A, 337, 287

Falconer, D.A., Moore, R.L., Porter, J.G. & Hathaway, D.H., 1998, ApJ, 501, 386

Feldman, U., 1983, ApJ, 275, 367

Feldman, U., Schuhle, U., Widing, K.G. & Laming, J.M., 1998, ApJ, 505, 999.

Fludra, A., Del Zanna, G., Alexander, D. & Bromage, B.J.I., 1998, J. Geophys. Res., in press.

Gabriel, A.H. & Mason, H.E., 1982, Solar Phys., Ch 10, in Applied Atomic Collision Physics, Vol 1, eds. H.S.W. Massey, B. Benderson and E.W. McDaniel (Academic Press)

Gabriel, A.H., 1976, Phil. Trans. R. Soc. Lond. A., 281, 339

Gallagher, P.T., Phillips, K.J.H., Harr-Murnion, L.K. & Keenan, F.P., 1998, Solar Jets and Coronal Plumes, ESA SP-421, p365

Harrison, R.A., 1998, Solar Phys., 175, 467

Harrison, R.A., Fludra, A., Pike, C.D., Payne, J., Thompson, W.T., Poland, A.I., Breeveld, E.R., Breeveld, A.A., Culhane, j.L., Kjeldseth-Moe, O. & Aschenbach, B., 1997, Solar Phys., 170, 123

Hassler, D.M., Dammasch, I.E., Lemaires, P., Brekke, P., Curdt, W., Mason, H.E., Vial, J.-C. & Wilhelm, K., 1998, Science, in press

Kudoh, T. & Shibata, K.: 1997, Proc. of the Fifth SOHO Workshop, ESA SP-404, p477.

Innes, D.E., Inhester, B., Axford, W.I. & Wilhelm, K., 1997, Nature, 386, 811.

Landi, E., Landinin, M., Dere, K.P., Young, P.R. & Mason, H.E., 1998, A&A, in press

Lemaire, P. & SUMER team, 1997, Solar Phys., 105

Mariska, J.T., Feldman, U. & Doschek, G.A., 1978, ApJ, 226, 698

Mariska, J.T., Feldman, U. & Doschek, G.A., 1979, A&A, 73, 361

Mason, H.E. & Boschler, P. 1998, The Eigth SOHO Workshop, in press

Mason, H.E. & Pike, C.D., 1997, The Tenth Cambridge Workshop on Cool Stars, Stellar Systems and the Sun, held at Cambridge, USA July 15-19, 1997, (eds R.A. Donahue and J.A. Bookbinder), in press.

Mason, H.E., Pike, C.D. & Young, P.R., 1998, Solar Jets and Plumes, ESA SP-421, p95

Mason, H.E., Young, P.R., Pike, C.D., Harrison, R.A., Fludra, A., Bromage, B.J.I. & Del Zanna G., 1997, Solar Phys., 170, 143

Parker, E,N., 1987, Cambridge Cool Stars Worshop, Springer-Verlag, p341.

Parker, E.N., 1988, ApJ, 330, 474

Perez, M.E. & Doyle, J.G., 1999, these proceedings

Pike, C.D. & Harrison, R.A., 1997, Solar Phys., 175, 457.

Pike, C.D. & Mason, H.E., 1998, Solar Phys., in press

Porter, J.G., Moore, R.L., Reichmann, E.J., Engvold, O. & Harvey, K.L.: 1987, ApJ, 323, 380.

Pres, P. & Phillips, K.J.H., 1998, ApJ, in press

Sheely, J.F., Feldman, U., Schule, U., Wilhelm, K., Curdt, W. & Lemaire, P., 1997, ApJ, 484, L87

Shibata, K.: 1997, Proc. of the Fifth *SOHO* Workshop, ESA SP-404, p103.

Sterling, A.C. & Hollweg, J.V. 1984, ApJ, 285, 843.

Teriaca, L., Doyle, J.G. & Banerjee, D., 1999, these proceedings

Wilhelm, K. & SUMER team, 1997, Solar Phys., 170, 75

Wilhelm, K., Marsch, E., Dwivedi, B.D., Hassler, D.M., Lemaire, P., Gabriel, A.H. & Huber, M.C.E., 1998, ApJ, 500, 1023.

Young, P.R. 1998, *PhD Thesis*, Cambridge University.

Young, P.R. & Mason, H.E., 1998, Solar Composition and its evolution - From Core to Corona : Proceedings of the ISSI Workshop, 26-30 January 1998, Bern, eds. C. Frohlich, M.C.E. Huber, S.K. Solanki, R. von Steiger, in press.

Young P.R., Landi E. & Thomas R.J., 1997, A&A 329, 291

Solar and Stellar Activity: Similarities and Differences
ASP Conference Series, Vol. 158, 1999
C.J. Butler and J.G. Doyle, eds.

SUMER Observations of Doppler Shifts in the Quiet Sun and an Active Region

L. Teriaca, J.G. Doyle & D. Banerjee

Armagh Observatory, College Hill, Armagh, BT61 9DG, N. Ireland

Abstract. The UV spectral lines formed at transition region temperatures in the solar atmosphere, shows a prevailing red-shifted emission. Using the Solar Ultraviolet Measurements of Emitted Radiation spectrometer flown on the *Solar and Heliospheric Observatory* spacecraft, we measure the amount of line-shift as a function of the temperature for several spectral lines formed in the range between 10^4 and 10^6 K. We analyze spectrograms relative to the 'quiet' Sun and to the active region NOAA 7946. The velocities derived are increasing from $\sim$ 1 km s^{-1} at $\sim 20,000$ K to 10 km s^{-1} at $160,000$ K for the 'quiet' Sun, and to $13 - 15$ km s^{-1} at $125,000$ K for the active region. At higher temperature a different behaviour is observed. In the active region a *blue-shift* of 8 km s^{-1} is observed at the Ne VIII formation temperature ($580,000$ K) while in the 'quiet' Sun we have a blueshift of 2 km s^{-1}.

1. Introduction

One of the most interesting problems in solar physics is the observed red-shifted emission of lines formed at transition region(TR) and coronal temperature. During the last two decades, observations of this phenomenon has been reported by many authors, observed with several UV instruments with different spatial resolution. In the earlier investigations the magnitude of the redshift has been found to increase with temperature, reaching a maximum (around 8 km s^{-1} in the 'quiet' Sun) at $T = 10^5$ K, and then to decrease towards higher temperatures. Doschek et al. (1976) found no significant shift in the O V line at 1218 Å at disk center and the commonly quoted average velocity variation with temperature above 10^5 K depends to a large extent on this particular observation of the 1218 Å line. Chae et al. (1998), have shown that, for the 'quiet' Sun, the redshift is peaked around 1.5 10^5 K with a value around 11 km s^{-1} but it is also present at higher temperatures with a value around 5 km s^{-1} for Ne VIII 770.409 Å in the 'quiet' Sun. More recently Peter & Judge (1999) (hereafter, PG) have found *blue-shifts* in disk center for three coronal lines in the dataset (Ne VIII at 770 Å and 780 Å and Mg X at 625 Å, in contradiction to Chae et al. (1998). PG have suggested a rest wavelength of 770.428 Å for Ne VIII, which we have used for this paper. In this short contribution we also focus our attention to the difference in line of sight velocity between an active region and a 'quiet' sun region.

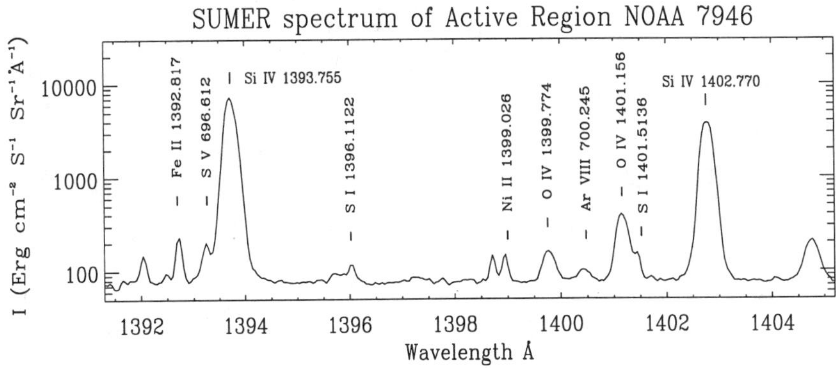

Figure 1. SUMER spectrum of Active Region NOAA 7946.

2. Observations

SUMER is a normal incidence spectrograph operating over the wavelength range 450 Å to 1610 Å. It is a powerful UV instrument capable of making reliable measurements of bulk motions in the chromosphere, TR and low corona with an error better than 1 km s^{-1} (Chae et al. 1998). Observations consist of a series of spectral images covering the wavelength range between 800 and 1590 Å. Every spectrum is partially overlapping the previous and the following, in order to ensure that the spectral lines are recorded on both the bare and the KBr parts of the detector. This allows us to recognize the second order lines from the first order ones using the different wavelength-dependent sensitivity of KBr compared to the bare part. Every single spectrum was exposed for 100 seconds using the 1×300 slit in 'quiet' Sun and in Active Region NOAA 7946. Reduction of SUMER raw images follow several stages, *i.e.* flat field subtraction, correction of geometrical distortion and radiometric calibration (in order to pass from count px^{-1} s^{-1} to erg cm^{-2} s^{-1} Sr^{-1} Å). Particular attention needs to be paid to the problem of the wavelength calibration. For SUMER there is no on-board calibration source, so the wavelength calibration is done using some chromospheric lines of neutral atoms. These lines are formed in the chromosphere at temperatures around 6500 K (*e.g.* Si I and S I, Chae et al. 1998) and are supposed to be at rest (Samain, 1991). These should therefore allow the determination of an absolute wavelength scale. In any case it is important to remember that all the absolute velocity measurement made with SUMER will be relative to the chromospheric reference lines. The necessity to have chromospheric reference lines reduces the wavelength range in which it is possible to make an absolute measurement of velocity to 900 – 1600 Å (practically only the first order). In fact the chromospheric line disappear below 900 Å. In Figure 1 an example of the first order spectra of NOAA 7946 is shown. It is possible to see some second order lines superimposed as well as the chromospheric lines used for wavelength calibration. The measurement of the central wavelength was per-

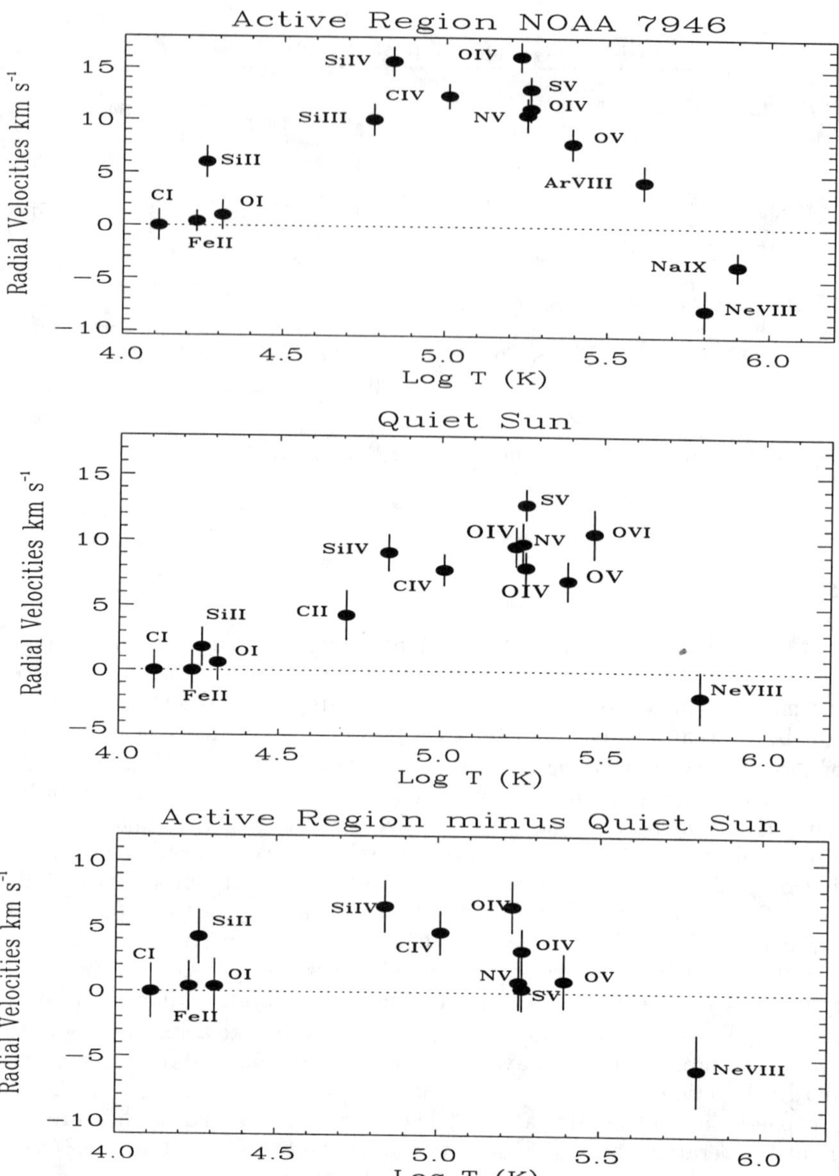

Figure 2. SUMER measurement of Radial Velocities on Active Region NOAA 7946 (top panel), 'quiet' Sun (middle panel) and differences between Active Region and 'quiet' Sun (lower panel).

formed using a multi-Gaussian fit technique. For each spectral line of interest we have calculated the local spectral dispersion using at least two chromospheric lines or two lines of the same ion.

3. Discussion

For active Region NOAA 7946 (top panel, Figure 2), we find that the radial velocity is increasing from ~ 1 km s^{-1} at $T \sim 2\ 10^4$ K to $\sim 13 - 15$ km s^{-1} at $T \sim 1.25\ 10^5$ K. At higher temperature the velocity decreases, leading to a blue-shift at $T \sim 5\ 10^5$ K. The behaviour at high temperature is well represented from the measurement of 3 different spectral lines (*i.e.* Ar VIII, Ne VIII and Na IX). PG have also reported red to blue-shift transition at $T \sim 5\ 10^5$ K for 'quiet' Sun (from roll data of SUMER). In our 'quiet' Sun data (see middle panel of Figure 2) we observe a similar trend; a maximum velocity of ~ 10 km s^{-1} is reached at $T \sim 1.6\ 10^5$ K and drops down to -1.9 km s^{-1} for Ne VIII at 5.8 10^5 K.

Achour et al. (1995), introduced the idea of 'differential redshift measurement', *i.e.* to compare 'quiet' Sun and active region Doppler shift measurement. This method has the advantage of being independent of laboratory wavelengths. They found that the differential velocity increases with increasing temperature, reaching a maximum of 7 km s^{-1} at a temperature of 1.0 – 1.35 10^5 K where the N IV 1486 Å and the O IV 1401 Å lines are formed. Above this temperature the velocity difference decreases abruptly with increasing temperature. It should be noted that they did not use measurements of lines formed above 2.5 10^5 K, so they inferred the disappearance of the differential redshift in the high transition region by extrapolating the result from the O V line. Our results show (see lower panel of Figure 2) a differential velocity of $4 - 5$ km s^{-1} around 10^5 K, smaller than the one observed by Achour et al. Furthermore we find a differential blue-shift after $T \sim 5\ 10^5$ K. We hope to pursue the origin of the observed blue-shift in a subsequent study.

Acknowledgments. Research at Armagh Observatory is grant-aided by the Dept. of Education for N. Ireland while partial support for software and hardware is provided by the STARLINK Project which is funded by the UK PPARC. This work was partly supported by PPARC grant GR/K43315.

References

Achour, H., Brekke, P., Kjeldseth-Moe, O. & Maltby, P., 1995, ApJ 453, 945

Chae J., Yun H. S., Poland A. I., 1998, ApJs 114, 151

Doschek G. A., Feldman, U & Bohlin, J. D., 1976, ApJ 205, L177

Peter, H., & Judge, P., 1999, ApJ (Submitted) (PG)

Samain D., 1991, A&A 244, 217

Solar and Stellar Activity: Similarities and Differences
ASP Conference Series, Vol. 158, 1999
C.J. Butler and J.G. Doyle, eds.

Line Width Variations Above a Coronal Hole: Implications for Heating

J.G. Doyle, L. Teriaca & D. Banerjee

Armagh Observatory, College Hill, Armagh BT61 9DG, N. Ireland

Abstract.
Using data taken above the Sun's northern coronal hole we identify a line due to Fe VIII. The line width variations of this feature are in excellent agreement with that derived from the Si VIII lines confirming that Si VIII it is indeed a good coronal diagnostic. The data confirms that the non-thermal line-of-sight velocity increases from $\sim$ 27 km s^{-1} at 27 arc sec above the limb to $\sim$ 46 km s^{-1} some 250 arc sec ($\sim$180,000 km) above the limb. The electron density shows a decrease from 1.1 10^8 cm^{-3} to 1.6 10^7 cm^{-3} over the same distance. These results argue strongly in favour of hydromagnetic waves with the derived non-thermal velocities and measured electron densities at different heights corresponding to the behaviour of upward propagating Alfvén waves. These waves have sufficient energy to heat the corona and accelerate the solar wind for magnetic field strengths of $\sim$5 G.

1. Introduction

One method to draw conclusions as to coronal heating mechanism(s) (e.g. hydromagnetic waves), is to observe the line widths of high temperature coronal lines above the limb. With the launch of *SOHO*, and in particular with the, SUMER instrument (Wilhelm et al. 1995), line width measurements of coronal lines as a function of position above the limb are possible. Doyle et al. (1998) and Banerjee et al. (1998) reported on observational sequences involving the Si VIII line pair (1445.75Å & 1440.49Å). Observations of this line pair allows us to determine simultaneously the electron density and the line width.

The analysis showed that the non-thermal line-of-sight velocity as derived from the Si VIII line increased above the limb while the electron density decreased. On a closer analysis, the observations revealed that the non-thermal velocity were inversely proportional to the quadratic root of the electron density, in excellent agreement with that predicted for undamped radially propagating Alfvén waves. This raises the question of how representative of coronal conditions are the Si VIII lines. We attempt to answer this question by looking for other high temperature lines which have been observed off-limb, in particular, whether there were additional lines observed close to the Si VIII lines, thus strengthing the usefulness of the existing datasets.

2. Observational Data

SUMER is a normal incidence spectrograph operating over the wavelength range 450 Å to 1610 Å, details which can be obtained from Wilhelm et al. (1995). The dates of the observations discussed here, their locations, pointing, slit sizes and exposure times can be obtained from Doyle et al. (1998) and Banerjee et al. (1998).

Briefly, the data were acquired above a north polar coronal hole on 4 November 1996 and 10 December 1996. Sequences in all the locations were comprised of a temporal series of spectra taken in the same pointing position but at successive times. The objective of the observing programme was to obtain high S/N spectra of the Si VIII 1440.49Å and 1445.75Å lines as a function of position above the limb. The images were acquired with the 1x300 and 4x300 arc sec slits. Using the individual spectral scans a summed spectrum was generated. Standard data reduction procedures were followed as given in Doyle et al. (1998) and Banerjee et al. (1998).

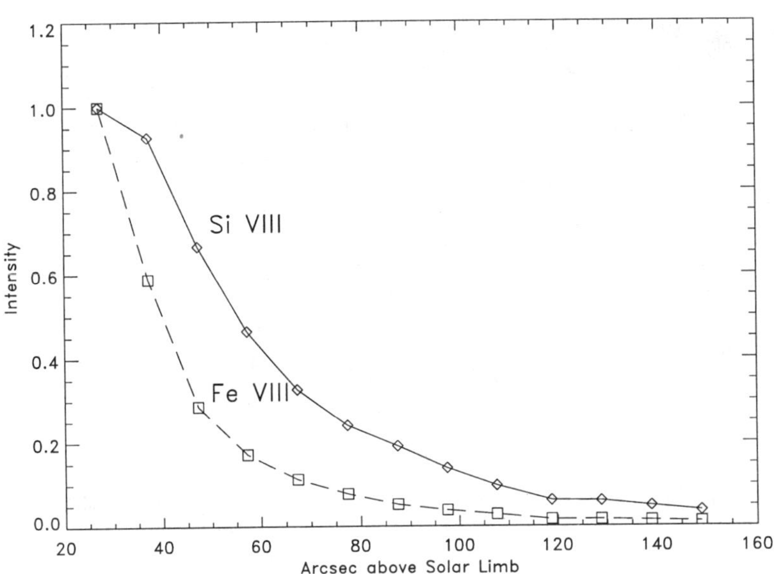

Figure 1. The intensity variation (normalized to 1) for Si VIII 1445.75Å compared to the feature at 1442.47Å.

3. Results & Discussion

Inspection of the spectral region around the Si VIII revealed a strong feature at 1442.47Å. A plot of the intensity variation of this line compared to Si VIII 1445.75Å is given in Figure 1. No identification was given by Feldman et al. (1997) of this line, although they did suggest it to be a hot coronal feature. This is verified by Figure 1 which would suggest it has a temperature lower

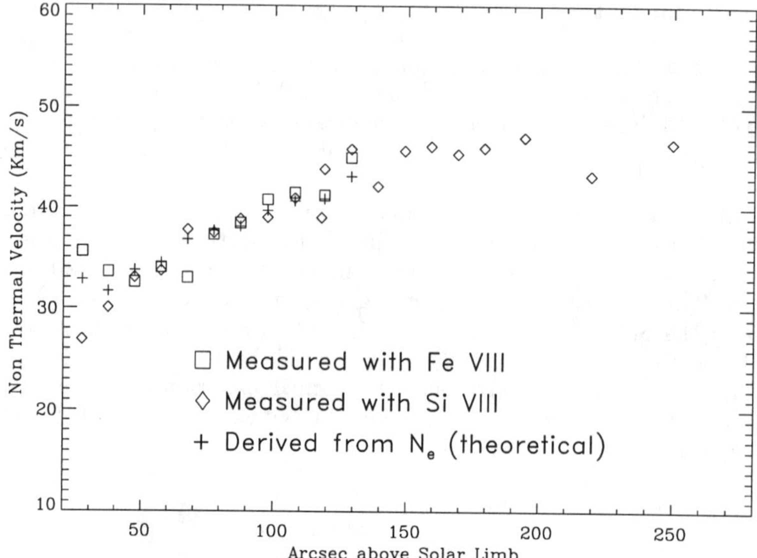

Figure 2. Variation of the non-thermal velocity for the coronal hole, the diamond those measured with Si VIII, the squares those measured with Fe VIII and the + symbols correspond to theoretical values.

than the Si VIII ion temperature. Using the non-thermal velocities as derived from Si VIII and the standard equation for an optically thin line broadened by thermal broadening caused by the ion temperature T_i and broadening caused by non-thermal motions we have,

$$FWHM = \left[4ln2 \left(\frac{\lambda}{c}\right)^2 \left(\frac{2k_BT_i}{M} + \xi^2\right)\right]^{1/2} , \tag{1}$$

where M is the ion mass. ξ is the non-thermal speed, related to the wave amplitude by $\xi^2 = \frac{1}{2} < \delta v^2 >$, where the factor of 2 accounts for the polarization and direction of propagation of a wave relative to the line of sight. If we adopt an ion temperature slightly lower than that of Si VIII (e.g. $T_i = 6\ 10^6 K$), the Si VIII non-thermal widths implies an ion mass of ~53. This imply that the spectral feature is due to Fe, and more specifically, probably Fe VIII.

Following Doyle et al. (1998) it can be shown that the rms wave velocity amplitude and density are related by,

$$< \delta v^2 >^{1/2} \propto \rho^{-1/4} \tag{2}$$

In Figure 2, the (+) represents the velocities calculated on the basis of the above equation, using the measured N_e. The proportionality constant was chosen to match the calculated energy flux (see later). As can be seen, there is excellent agreement, thus the observed non-thermal velocities at different heights are consistent with upward propagating Alfvén waves.

The energy flux density in the corona due to Alfvén waves is given by,

$$F = \sqrt{\frac{\rho}{4\pi}} < \delta v^2 > B \tag{3}$$

where ρ is the plasma mass density (related to N_e as $\rho = m_p N_e$, m_p is the proton mass), $< \delta v^2 >$, the mean square velocity is given previously, and B, is the magnetic field strength. From our dataset at 120 arc sec above the limb using the measured values (i.e. $N_e = 4.8\ 10^7$ cm^{-3}, $< \delta v^2 >= 2 \times (43.9$ km s$^{-1})^2$), we find $F = 4.9\ 10^5$ erg cm^{-2} s^{-1} for $B = 5$ G, which is only sightly lower than the requirements for a coronal hole with a high speed solar wind flow (Withbroe & Noyes 1977), which is estimated to be 8 10^5 erg cm^{-2} s^{-1}. These waves have thus sufficient energy to heat the corona and accelerate the solar wind for magnetic field strengths of ~5 G.

Acknowledgments. Research at Armagh Observatory is grant-aided by the Dept. of Education for N. Ireland while partial support for software and hardware is provided by the STARLINK Project which is funded by the UK PPARC. This work was supported by PPARC grant GR/K43315. We would like to thank the SUMER team at Goddard Space Flight Center for their help in obtaining the data. The SUMER project is financially supported by DLR, CNES, NASA, and PRODEX. SUMER is part of *SOHO*, the Solar and Heliospheric Observatory of ESA and NASA.

References

Banerjee, D, Teriaca, L., & Doyle, J.G., 1998, A&A 339, 208

Doyle, J.G., , Banerjee, D. & Perez, M.E., 1998, Sol Phys 181, 91

Feldman, U., Behring, W.E., Curdt, W., Schuehle, U., Wilhelm, K., Lemaire, P., Moran, T.M., 1997, ApJS 113, 195

Wilhelm, K., Curdt, W., Marsch, E., Schuhle, U., Lemaire, P., Gabriel, A., Vial, J.-C., Grewing, M., Huber, M.C.E., Jordan, S.D., Poland, A.I., Thomas, R.J., Kühne, M., Timothy, J.G., Hassler, D.M. and Siegmund, O.H.W., 1995, Sol Phys 162, 189

Withbroe, G.L. and Noyes, R.W., 1977, ARA&A 15, 363

Solar and Stellar Activity: Similarities and Differences
ASP Conference Series, Vol. 158, 1999
C.J. Butler and J.G. Doyle, eds.

Tales of an Elephant's Trunk

B.J.I. Bromage, J.R. Clegg & G. Del Zanna

Centre for Astrophysics, University of Central Lancashire, Preston, UK

B. Thompson

Space Applications Corp. / NASA GSFC, Greenbelt, MD, USA

Abstract. In August 1996, at solar minimum, a large coronal hole appeared on the Sun, extending from the north polar hole, across the equator to a large active region in the southern hemisphere. It was named "The Elephant's Trunk" because of its characteristic shape. The earlier development of this feature is described here and some of the results obtained from detailed observations made with *SOHO* instruments are summarised.

1. Introduction

The "Elephant's Trunk" coronal hole which appeared around the east limb of the Sun on the 21st August 1996, approached the meridian on the 26th August (see right most image in Figure 1).

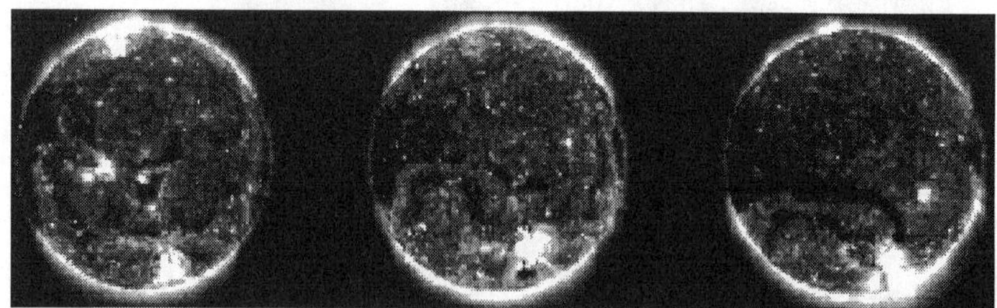

Figure 1. These three *SOHO* EIT images (left to right) are separated in time by one solar rotation (North to the left). They show the development of a prominence channel looping down from the north polar crown arcade system to the large active region in the south-east.

The tip of the 'trunk', which extended down across the equator from the north polar coronal hole, reached across to the edge of a large active region which had been rotating around the Sun for several months. To the east (left), a dark prominence channel can be seen to run parallel to the boundary of the coronal hole. Apart from these dramatic features, the Sun was very quiet, being at solar minimum at this time. Many features of this coronal hole have been reported in Bromage et al (1998) and Del Zanna & Bromage (1999). Some of the main points of interest are described here.

Below, § 2 describes the development of this striking feature from the end of June to its meridian passage in August. § 3 summarises the plasma characteristics of the coronal hole and the final section discusses the characteristics of a plume seen near disc centre within this feature, the first plume to be identified in a low-latitude coronal hole on the solar disc.

2. Development of the Coronal Hole

In many respects this coronal hole is similar to the famous 'CH1' hole seen by Skylab in 1973. Timothy et al (1975) describe the hole as having developed from the merging of the diffuse bipolar field associated with two old active regions. The development of this 1996 hole shares this characteristic of merging of diffuse bipolar regions, but its connection to the north polar flux interestingly appears to involve the emergence of an early new-cycle active region.

Early activity in April and May in the leading parts of a complex active region (located just to the south of the equator) died down, but the following parts developed large loop systems which extended a long way to the north, almost to the arcade which surrounded the north polar hole. At this time of minimum solar activity there were two large polar coronal holes with some equator-ward extensions which were producing fast solar wind streams at the Earth and geomagnetic activity there whenever they crossed the meridian. In May, the first, small new-cycle regions began appearing.

On 28 June, the return of the large active region was preceded by the appearance of a bright facular region at N24E63. This can be seen crossing the disc as a bright, dynamic region near the meridian, in the top image in Figure 1. Reconnection is occurring in this bright region between the arcade system (polar crown) around the north polar hole and the loop systems to the north of the large active region. This is taking place at a point where an equator-ward extension of the polar hole boundary has turned down towards the active region. It may have been triggered by the slower rotation of the polar hole (compared to the equatorial active region) bringing the two loop systems together. The emergence of new cycle flux at this point may also have helped the process.

Shortly afterwards, a 17-spot group re-emerged from the plage in the old active region at S10W11. It was numbered 7978 and rapidly increased in area until on 9 July it generated an X2.6 flare with a coronal mass ejection (the first X-flare since 1992).

Region 7978 reappeared as 7981 towards the end of July as a stable region comprising two bipoles. The remnants of the new-cycle active region can just be seen in the EIT image of July 30, in the middle of a large prominence channel resembling a "question mark" which has formed within the arcade structure. The extension of the polar hole was elongated and the "Elephant's Trunk" had already begun to form just to the west of the prominence channel.

By the end of August, the "question mark" has been straightened a little, probably because of differential rotation, and the arcade system appears to have

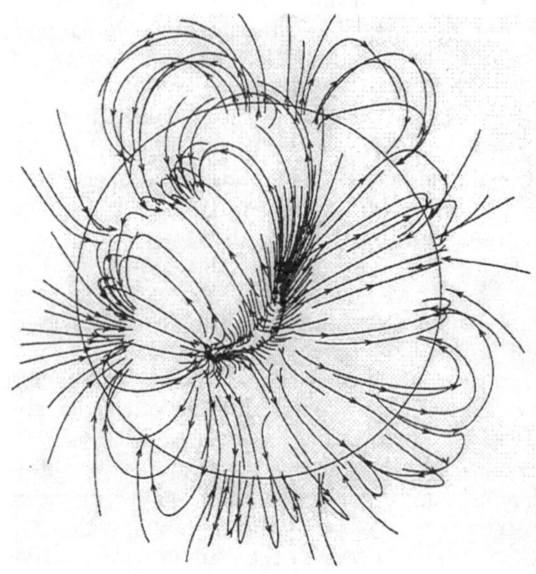

Figure 2. A potential field model of the solar magnetic field associated with the coronal hole and prominence channel seen at the end of August 1996.

stabilised, now connecting through to the active region remnants south of the equator.

The arcade structure overlying the prominence channel can be seen in a potential field model of the solar magnetic field at this time (Figure 2).

The potential field model shown was derived from a series of *SOHO* MDI magnetograms using a new technique which combines one month of daily data to obtain the initial photospheric values over the whole solar surface. This technique is described in more detail in Clegg et al (1999).

3. Characteristics of the Coronal Hole

The plasma characteristics of the coronal hole were studied using *SOHO* CDS spectroscopic data (see Del Zanna & Bromage, 1999). Electron density in the hole was found to be $\simeq 2\ 10^8\ cm^{-3}$, about one-third of the value found in the quiet sun boundary region. The temperature in the hole was typically about $8\ 10^5$K, while the temperature in the boundary was about $1.1\ 10^6$K.

4. Identification of a Low-latitude Plume

Del Zanna & Bromage also reported the results of spectroscopic analysis of plumes. It was found that they are typically cooler than the surrounding coronal

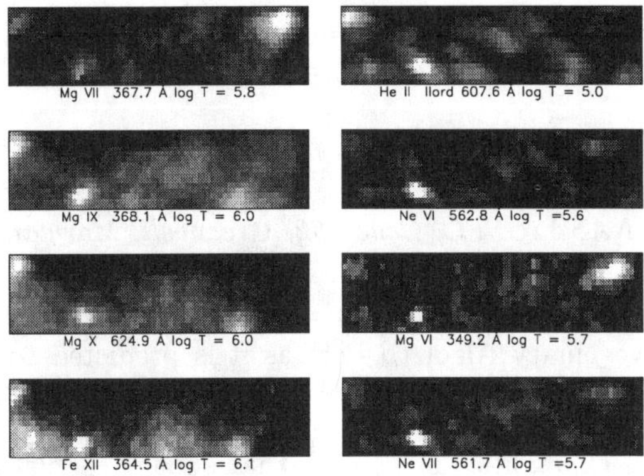

Figure 3. Monochromatic images of a 4 arcmin x 1 arcmin region of the coronal hole and its western boundary near the centre of the Sun. The boundary is seen clearly in Mg X and Fe XII. The plume can be seen as a bright feature at the upper left of the Mg VI and Mg VII images, within the coronal hole.

hole, with peak emission occurring at about 7.4 10^5K, where it appeared brighter than the coronal hole (in lines such as Mg VI and Mg VII). The emission measure profile of plumes was also found to have a narrower temperature peak than the coronal hole. This difference in shape of the emission profile also resulted in a strong enhancement of the Mg/Ne intensity ratio. These characteristics were identified in a small diffuse region in the Elephant's Trunk, near the centre of the disc, revealing it to be a plume. Its structure appears clearly in the upper left part of Mg VI and Mg VII images shown in Figure 3.

It appears as a faint feature in Mg@IX but is absent in the hotter lines of Mg X and Fe XII.

References

Bromage, B.J.I., Alexander, D., Breen, A., Clegg, J. R., Del Zanna, G., DeForest, C., Dobrzycka, D., Gopalswamy, N. & Thompson, B., 1998, *Science* (submitted).

Clegg, J.R., Bromage, B.J.I. & Browning, P., 1999, *J. Geophys. Res.* (in press).

Del Zanna, G. & Bromage, B.J.I., 1999, *J. Geophys. Res.* (in press).

Timothy, A.F., Krieger, A.S. & Vaiana, G.S. 1975, *Sol. Phys.* **42**, 135

Solar and Stellar Activity: Similarities and Differences
ASP Conference Series, Vol. 158, 1999
C.J. Butler and J.G. Doyle, eds.

High-Resolution X-Ray Spectroscopy: What will *AXAF* Tell Us about Stellar Coronae?

Stephen A. Drake

USRA and NASA/GSFC, Code 662, Greenbelt, Maryland, 20771, USA

Abstract. I present a simulation of the soft X-ray spectrum of the well-known active binary Algol (β Per) as it is predicted to appear in a 50 kilosecond exposure using the *AXAF* CCD Imaging Spectrometer (ACIS) detector and High Energy Transmission Grating (HETG) combination. I discuss the capabilities of *AXAF* that should enable it to resolve the ongoing debate about the non-solar coronal abundances inferred from analyses of much lower resolution *ASCA* spectra of Algol and other active stars. I also give examples of the types of analyses that may be usefully applied to *AXAF* grating spectra, and discuss the known limitations of the present atomic data and coronal plasma emission codes.

1. Introduction

One of the most interesting findings in the field of stellar coronae in recent years is that their low- (e.g. *ROSAT* PSPC) and moderate-resolution (e.g., *ASCA* SIS) X-ray spectra cannot in general be well-fit by coronal plasma models having solar photospheric (hereafter solar) abundances (cf. Drake 1996 for a brief review). In many cases, particularly for the most active stars such as RS CVn binaries, the most striking anomaly is that Iron appears to be deficient by a factor of 3-10 relative to the solar value (e.g., Ortolani *et al.* 1997, Mewe *et al.* 1997). Other heavy elements also are generally apparently under-abundant relative to their solar values, by amounts that vary from star to star. These apparent anomalous abundances were initially suspected by some in the field to be artifacts. Various explanations were put forward, including: (i) they are due to an inadequate treatment of the temperature distribution of the coronal plasma; (ii) they are due to opacity effects in the strong resonance lines of the abundant elements such as Fe, Si and Mg; (iii) they are due to an additional power-law continuum 'diluting' the lines; (iv) they are due to inadequacies in the atomic physics contained in the standard coronal plasma models of Raymond & Smith (RS) or Mewe & Kaastra (MEKA); or (v) (in some cases) they are due to instrumental effects, e.g., the response matrices not being completely accurate in certain spectral regions.

The consensus of recent work is that, although there are specific cases in which some of the non-intrinsic explanations may apply, the general conclusion that the abundances in many coronae are really non-solar is hard to avoid. In retrospect, this probably should not have been a big surprise since (i) not all stars have photospheric abundances identical to the Sun's, and, in particular, the Sun appears to be metal-rich by about 0.2 dex compared to the average of various types of stars in the solar neighborhood (even stranger, many active binaries, although they have Pop I-type kinematics, appear to be metal deficient by 0.3-1.0 dex compared to the Sun, although the reality of these photospheric abundance anomalies is still being debated), and (ii) most high-resolution EUV and soft X-ray studies of the **solar** corona have derived coronal abundances that differ significantly from the solar photospheric values, with the most common finding being an overabundance of low-FIP (first-ionization potential) elements such as Fe and Mg relative to high-FIP elements such as Ne (see reviews by Feldman 1992 and Meyer 1996).

2. The Capabilities of *AXAF*

The SIS detectors on *ASCA* have a spectral band-pass from 0.5 to 10 keV, roughly, with a spectral resolution increasing from 50 to 150 eV over this range. With this resolution the Fe L-shell complex is completely unresolved and blended with, for example, the He-like and H-like complexes of Ne and Na, and the L-shell complex of Ni. In my opinion, the elements for which *ASCA*-derived abundances are reliable are O (0.57 & 0.65 keV), Mg (1.35 & 1.47 keV), Si (1.86 & 2.01 keV), S (2.46 & 2.62 keV), Ar (3.14 & 3.32 keV), Ca (3.90 & 4.10 keV), and Fe (6.70 & 6.97 keV), where I have listed the energies of the resonance line of the He- and H-like ions of these elements, and assuming that there is sufficient signal-to-noise in the spectrum at these energies to significantly detect these lines. Although the He-like and H-like resonance lines of N at 0.43 keV and 0.50 keV lie on the edge of the SIS bandpass, they do not appear to yield reliable abundances due to the difficulty of getting an accurate calibration in this region where the effective area is rapidly decreasing.

In contrast, *AXAF* will have much improved capabilities compared to previous X-ray telescopes in almost every instrumental parameter. Its exquisite spatial resolution ($\leq$ 1") compared to previous X-ray missions, combined with its large effective area and bandpass ($\geq$ 100 square cm in direct imaging mode using ACIS over a broad bandpass from 0.15 to 8 keV or 1.6 to 80 Å), will prove very useful for many classes of observations, e.g., those of extended objects or in crowded fields, where CCD-type spectral resolution ($E/\Delta E \sim 10 - 60$) is adequate. For observations in which imaging capabilities are of secondary importance, the insertion into the light path of transmission gratings enables a spectral resolution $E/\Delta E \sim 100 - 1500$ to be achievable over the above-quoted bandpass: this translates to a velocity resolution of 200 km/sec at the lowest energies of 0.15 keV, notice. A more typical spectral resolution of $\sim$ 300 is reached by the HETG gratings over the 0.6 - 1.5 keV (8 - 20 Å) range occupied by the Fe L-shell complex (this instrument combination has 10 to 100 square cm effective area, notice): thus at 1 KeV, the energy resolution will be about 3 eV, sufficient to resolve much although not all of the severe line blending in this spectral region that has not hitherto been resolved by non-solar X-ray telescopes.

AXAF has effective area in the soft (0.1 to 0.5 keV) X-ray band, and this will mean that, not only for N but also for C, the He-like and H-like features (at 0.31 & 0.37 keV for C) will be easily detectable (and resolvable using the LETG). Using the gratings will enable the Ne He-like and H-like features at

0.92 & 1.02 keV to be readily resolved. Other elements with lines in the *AXAF* bandpass which may be isolated for the first time in stellar X-ray spectra include Ni, Na, and Al, although the low cosmic abundances of the latter 2 elements would require very deep exposures for them to be detected. Thus, in the best *AXAF* spectra, we may be able to detect spectral lines and infer abundances for 10 even-Z and 3 odd-Z elements, compared to the 7 even-Z elements accessible to *ASCA*. Furthermore, these 13 elements are about equally divided between those having low First Ionization Potential (FIP), (Na, Mg, Al, Si, Ca, Fe and Ni), and those having high FIP (C, N, O, Ne, and Ar) or intermediate FIP (S).

3. Sample Atomic Physics Problems Addressable Using *AXAF*

The He-like resonance line complexes of the abundant ions such as O, Ne, Mg, and Si are in the spectral region accessible to the ACIS HETG and should also be detectable in a reasonable exposure time. The structure of these line complexes has been studied for at least 3 decades (e.g., Gabriel 1972), and the basic atomic physics parameters such as f-values and collisional strengths are well-determined. The strongest features in a He-like multiplet are the resonance (R) and the forbidden line (F) components, with the inter-combination line (I) several times weaker: there are also a score of even weaker satellite lines scattered around (and blending with) the stronger lines.

What can be deduced from such complexes when the individual line components are spectrally resolved?

• One can derive accurate differential emission measures (DEMs) at the peak formation temperatures of these ions from the overall line fluxes with much less confusion than in lower-dispersion spectra where, in some cases (e.g., Mg XI), blending with other ion lines is a problem;

• From the ratio F/I, one can derive electron densities (e.g., Gabriel & Jordan 1969) provided that the density is above the critical low-density limit N_C for that particular ion (e.g., for O VII and Mg XI, $N_C \sim 3\ 10^{10}$ and 10^{12} cm^{-3}, respectively);

• From the ratio (F+I)/R or F/R, one can derive information about possible resonance-line optical depth (τ) effects and/or (possible in flares, probably not in quiescent coronal conditions) non-ionization equilibrium effects;

• By combining the previous items, one can derive constraints on the spatial scale of the emission region.

(Notice that the above information will enable one to confirm or refute several of the proposed non-intrinsic explanations for the anomalous coronal abundances.)

Another important set of lines to study are the Fe L-shell lines in the 9 – 18Å range. It has been realized, partially based on the failure of current plasma codes to fit parts of this spectral range, that there is something missing or erroneous in the line lists in this spectral region that are incorporated in these codes. Some updating of the plasma codes has already occurred (e.g., the 'new' MEKAL code includes updated L-shell spectra calculated by Liedahl *et al.* 1995), but more still remains to be done: e.g., Liedahl & Brickhouse (1998) and Brown *et al.* (1998) have convincingly argued the omission from the plasma codes of the ensemble of high-n transitions of the various Fe ions in this range is one of the causes of the mismatch between observed and predicted spectra. New laboratory data and theoretical calculations are now available for some of the Fe ions (e.g., Fe XVII by Brown *et al.* 1998 and Fe XXIV by Savin *et al.* 1996) and data for more ions should become available in the next year or so. Comparison of the lines in this region of coronal spectra with laboratory plasma spectra and theoretical predictions will be an important test of this new atomic physics, particularly for the ions Fe XXI - XXIV that are predicted to have strong lines in the relatively high-temperature coronae of these two active binaries. For those L-shell lines for which reliable atomic parameters are available and for which sufficient signal-to-noise can be obtained, DEM's, temperatures, optical depths and electron densities can be derived in an analogous manner to the He-like ions.

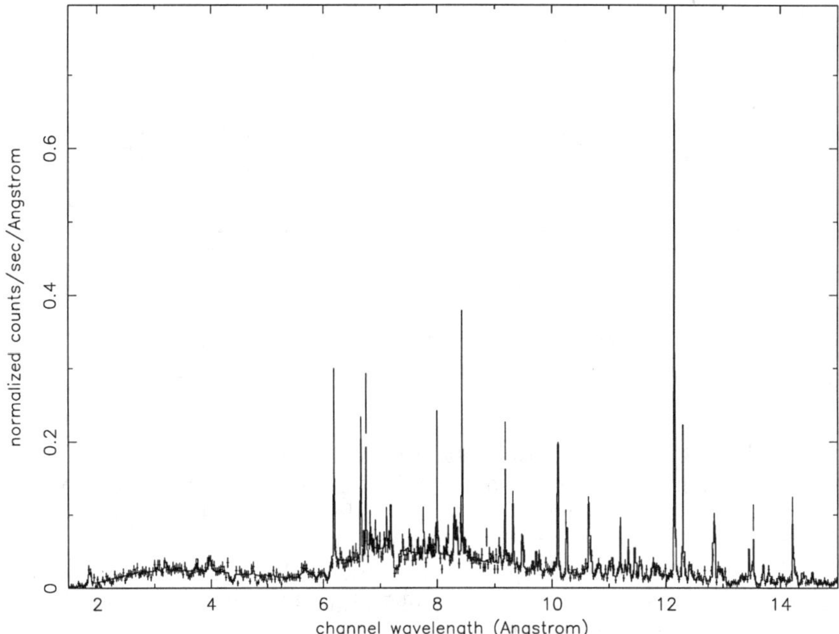

Figure 1. The complete 1.5 to 15$\mathring{A}$ ACIS-S/HEG spectrum of Algol

In Figures 1 & 2, I show simulated Algol ACIS-S/HEG spectra integrating over 50 ksec of exposure time; I have done simulations in XSPEC using the standard coronal codes, which of course include assumptions (e.g., optically thin lines, low-densities, etc.) that such observations are in fact going to confirm or refute. The input model was based on the best-fit 2-T model to the *ASCA* spectrum of Algol presented by Singh *et al.* (1995). In Figure 1, I show an overview of the 1.5 to 15Å spectral range. Notice the wealth of emission lines (the strongest predicted line is the Ne X Lyman α line and the shortest-wavelength detected line is the Fe XXV He-like multiplet at 1.85Å), as well as the well-exposed continuum. In Figure 2 I show a blow-up of the predicted HEG spectrum

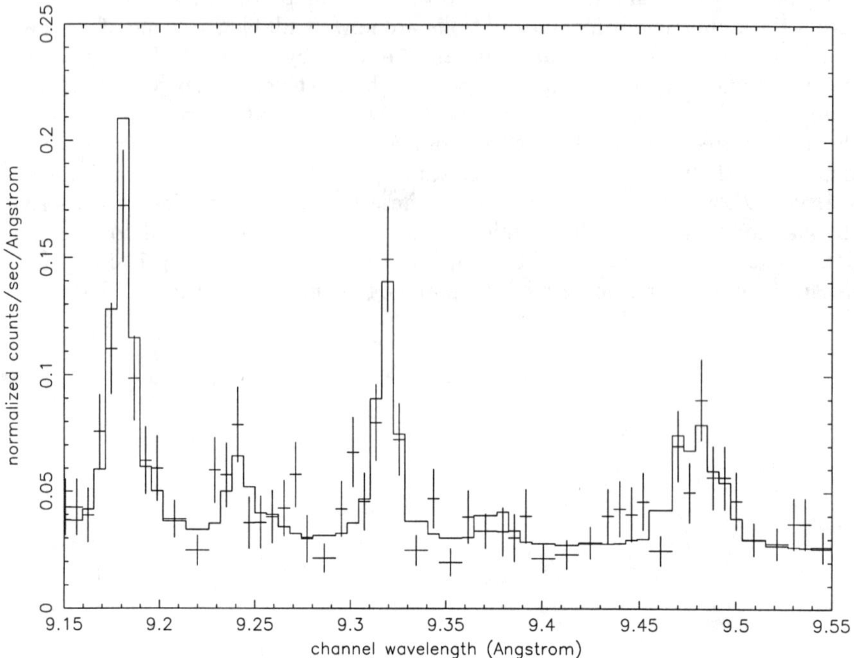

Figure 2. The Mg XI R (9.18Å), I (9.24Å), and F (9.32Å) He-like resonance multiplet region of the ACIS-S/HEG spectrum of Algol

of the Mg XI He-like complex: notice that the R, I and F components are all well-resolved and significantly detected (as they also are in the Ne IX and Si XIII He-like multiplets). The prominence of the other He-like multiplets is predicted to be less, being dependent on the existence of a low-temperature tail to the DEM (for O VII 21.6Å) and on the particular values of the abundances adopted for this simulation (for S XV 5.04Å, Ar XVII 3.95Å, and Ca XIX 3.18Å), and only the R components of these multiplets may be detectable.

4. The Crystal Ball

I expect that initial analyses of the *AXAF* grating spectra will be on a line-by-line basis and that a global fitting approach will not be used to model them, given that both the data and the models are being tested. Ultimately, when the model plasma codes are both complete and reliable enough, at least for the stronger lines, it may be possible to perform XSPEC- and SPEX-type global fitting approaches; this will be of great assistance in identifying the weaker lines in the grating spectra.

In a more speculative vein, I offer some predictions on some of the major things that *AXAF* will reveal about stellar coronae: (i) the non-solar abundances inferred for most stellar coronae by *ASCA* and *EUVE* will be confirmed;

(ii) most coronae will exhibit complex temperature and density structure; (iii) particularly during flares, coronal lines will exhibit detectable velocity shifts and turbulent broadenings; (iv) some strong resonance lines will show evidence for high optical depths; and (v) *AXAF* and the (shortly to follow) X-Ray Multi-Mirror Mission (*XMM*) will revolutionize the field of stellar coronae in the same way that *IUE* had revolutionizing effects on the field of stellar chromospheres and transition regions. I hope to verify some of the above predictions in my forthcoming *AXAF* observations of the active binary systems Algol and UX Ari.

5. Acknowledgements

I would like to thank the staff of the Armagh Observatory, particularly John Butler and Gerry Doyle, for their efficient planning and organization of this meeting which made it so memorable; it was a wonderful tribute to our most esteemed friend and colleague, Brendan Byrne.

References

Brown, G.V. *et al.* 1998, ApJ, 502, 1015
Drake, S.A. 1996, in 'Cosmic Abundances', p 215
Feldman, U. 1992, Phys. Scripta., 46, 202
Gabriel, A.H. 1972, MNRAS, 160, 99
Gabriel, A.H. & Jordan, C. 1969, MNRAS, 145, 241
Liedahl, D.A. & Brickhouse, N. 1998, in preparation
Liedahl, D.A. *et al.* 1995, ApJL, 438, L115
Mewe, R. *et al.* 1997, A&A, 320, 147
Meyer, J.-P. 1996, in 'Cosmic Abundances', p 127
Ortolani, A. *et al.* 1997, A&A, 325, 664
Savin, D.W. *et al.* 1996, ApJL, 470, L73
Singh, K.P., Drake, S.A. & White, N.E. 1995, ApJ, 445, 840

Solar and Stellar Activity: Similarities and Differences
ASP Conference Series, Vol. 158, 1999
C.J. Butler and J.G. Doyle, eds.

Coronal Bright Points and the Magnetic Field Association

Paweł Preś[1,2] & Kenneth J.H. Phillips[1]

[1]*Space Science Department, Rutherford Appleton Laboratory, Chilton, Didcot, Oxon OX11 0QX, U.K.*
[2]*Astronomical Institute, Wrocław University, Kopernika 11, 51-622 Wrocław, Poland*

Abstract. We have studied the time evolution of the magnetic flux and Fe XII EUV emission from several coronal bright points and found them to be extremely well correlated. The energy contained in the magnetic field, estimated from a simple model, is approximately equal to that lost by conduction and radiation from the bright points thermal plasma.

1. Introduction

Coronal X-ray bright points (XBPs) are known to be associated with ephemeral bipolar magnetic regions (Golub et al., 1977), but the connection can now be examined with data from the Michelson Doppler Imager (MDI) on board the Solar and Heliospheric Observatory (*SOHO*) as well as coronal images from the *SOHO* Extreme Ultraviolet Imaging Telescope (EIT) and the Soft X-ray Telescope (SXT) on the *Yohkoh* spacecraft.

2. Observations

The period chosen was 1997 July 31 (0:00 UT) to August 2 (0:00 UT) when solar activity was at almost the lowest level during the recent solar minimum. The X-ray emission as detected by the *GOES* satellites was at the A1 (1×10^{-8} W/m^2) level. The Soft X-ray Telescope (SXT) on *Yohkoh* observed the quiet region at the Sun center with a resolution of 4.9″ and a field of view 10′ × 10′, using Al 0.1 and Al/Mg filters. The EIT instrument obtained full-Sun images at 9 to 20 minute intervals in the Fe XII (195 Å) channel (peak emissivity at $T_e \sim 1.4$ MK). We also used MDI full-disk images averaged over five-minute intervals and available every 90 minutes. The images were co-aligned.

3. Magnetic Flux and Coronal Emission

Inspection of the MDI magnetograms shows many changes in the form of flux-concentration motions which give rise to diverging and converging bipoles and disappearance of magnetic elements. Coronal bright points are observed in SXT and EIT images, with lifetimes of several hours, which are co-spatial and simultaneous with either diverging or converging bipoles in the MDI images. For the

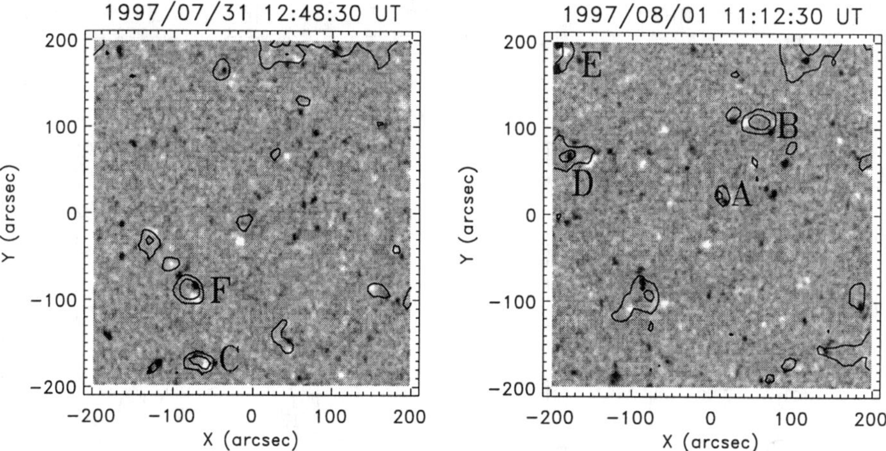

Figure 1. Magnetograms of a quiet-Sun at disk center at two times separated by about a day in the period discussed in the paper. North is at the top, east on the left. White areas in the magnetogram correspond to positive polarity, black to negative. The contours show Fe XII emission and correspond to levels of 120 and 200 DN/pixel/s. Letters A to F denote six bright points discussed in the text.

regions marked A to F in Figure 1, from the MDI data, we calculated values of magnetic flux $\Phi = \int \mid B_l \mid dS$ where B_l is the magnetic field strength measured along the line of sight and the integral is over a small area $\int dS$ which includes the region of significant Fe XII emission. We found that the Fe XII emission within each area is highly time-correlated with Φ, illustrated in Figure 2 for 3 of the regions. Best-fit power-law relations between the two, of the form $F_{\mathrm{Fe\ XII}} = F_0\ (\Phi/10^{18})^{\gamma}$ DN/s, showed that γ was close to unity. A similar value of γ was found for active regions by Fisher et al. (1998).

We estimated the energy contained by the magnetic field, by taking its configuration to be a semicircular loop with footpoints marked by the two monopoles and uniform cross section. We found this energy to be up to $\sim 10^{29}$ erg for all analyzed regions. From the measured SXT flux in the Al 0.1 and Al/Mg filters we calculated plasma temperatures and emission measures of each bright point. Due to the weak SXT signal these values are well determined for regions B, C and F only. From SXT images we estimated the volume V of each bright point and calculated the plasma thermal energy $3N_ekTV$. These values are $\sim 10^{27}$ erg, two orders of magnitude less than the magnetic field energy. We also calculated bright point energy losses due to radiation and thermal conduction. Typically, the radiative energy loss is $\sim 10^{28}$ erg during the bright point lifetime, i.e. only 10% of the magnetic energy. For conductive energy losses we assumed Spitzer conductivity, acting along field lines with simple loop geometry. The values of conductive losses are at least a factor ten more than radiative losses. The total energy losses are within a factor of 2 of the calculated magnetic energy, implying that the conductive and radiative losses of the coronal bright point are completely accounted for by the energy contained in the magnetic field.

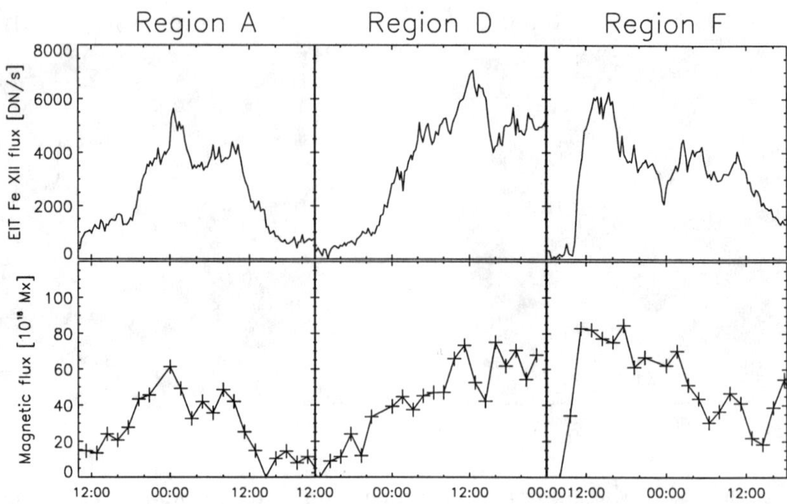

Figure 2. The EIT Fe XII (*top*) light-curves (in DN/s) of three of coronal brights shown in Figure 1, together with (*bottom*) time variations in the local magnetic flux Φ (10^{18} Mx) from the MDI instrument.

4. Summary and Conclusions

We find that the magnetic flux Φ as determined from the photospheric magnetic field is highly time-correlated with the Fe XII line emission from the six coronal bright points studied here. For three of the bright points for which temperature and emission measure are reasonably well determined, the amount of magnetic energy is within a factor of two of the total energy loss of the plasma over its lifetime. In view of the uncertainties of the magnetic energy calculation and the determination of the energy losses, we consider that the difference between magnetic energy and total energy loss is insignificant, i.e. that the magnetic energy supplies practically all the energy lost by the bright point plasma.

Acknowledgments. P.P. acknowledges financial support through the UK Royal Society/NATO Postdoctoral Fellowship Program. We thank the *SOHO* team for use of the data used in this paper, notably the MDI team (P.I. P. H. Scherrer) and the EIT team (P.I. J.-P. Delaboudinière), and the *Yohkoh* team for use of SXT images. Hugh Hudson is thanked for helpful discussions.

References

Fisher, G. H., Longcope, D. W., Metcalf, T. W. & Pevtsov, A. A. 1998, ApJ, submitted

Golub, L., Krieger, A. S., Harvey, J. W. & Vaiana, G. S. 1977, Sol. Phys., 53, 111

Solar and Stellar Activity: Similarities and Differences
ASP Conference Series, Vol. 158, 1999
C.J. Butler and J.G. Doyle, eds.

Turbulent Bursts in the Solar Transition Region

E. O'Shea

Dept. of Pure & Applied Phys., Queens University Belfast, Belfast BT7 1NN, N.Ireland

J.G. Doyle

Armagh Observatory, College Hill, Armagh, BT61 9DG, N. Ireland

Abstract. Using observations taken with SUMER on board *SOHO* we present evidence of turbulent bursts occurring in both 'quiet' and active regions on the Sun. Observations were taken in lines of C IV1548 Å and N V 1242 Å. These bursts manifest themselves as a brightening and a broadening of the line profile, indicating large turbulent velocities. In both the quiet and active regions these bursts were observed to occur in bright network regions, close to explosive event sites, and were found to have a measured periodicity of 5-16 minutes. They may be echoes of explosive event bursts and represent cooling evaporated plasma from the site of these explosive events.

1. Introduction

Explosive events are now generally recognised as being due to magnetic reconnection (Innes et al., 1997a). A characteristic noted by Dere et al. (1991), Dere (1994), Chae et al. (1998) and also by Innes et al. (1997b) is that explosive events tend to occur repeatedly in bursts .. see also paper by Perez & Doyle (these proceedings). It is proposed that the bursts of explosive events found by these authors are related to the turbulent bursts found here. It is likely that the turbulent bursts represent cooling evaporated plasma from the sites of these explosive events.

2. Observations

The Solar Ultraviolet Measurements of Emitted Radiation (SUMER) instrument is a stigmatic normal-incidence spectrograph operating in the range 450 to 1610 Å. For these observations the instrument was operated in a sit-and-stare mode with the SUMER standard rotation compensation switched off. For disk center pointing, this implies a new 1 arc sec region every 377 seconds. The N V 1242 Å dataset (sum_960710_002802) was obtained between 00:28 and 01:01 UT on the 10th July 1996. It is an observation of an active region, at coordinates X=581, Y=–200. The C IV 1548 Å dataset (sum_960710_182013) was obtained between the times of 18:20 and 19:26 UT on the 10th July 1996. This is an observation of the quiet Sun, at coordinates X=0, Y=0.

3. Results

The results have been analysed in terms of moments of the line profiles. The line width was used subsequently to obtain non-thermal velocity estimates. The data has in addition been analysed using a power spectrum analysis, identical to that used by Doyle et al. (1997).

3.1. N V 1242 Å

In Figure 1(a) we show the results of a moments-type analysis for a region at 19-21 arc sec. The line profiles were summed over this 19-21 arc sec region to increase the signal-to-noise. Explosive events during the first 3 minutes are apparent from the large non-thermal velocities and the distinctive asymmetric shape of the line profiles during this period. Comparing the intensity light-curve to that of the line-shift velocity it is apparent that the peaks in intensity at 14.5, 19.5 and 25 minutes correspond to blue-shifts of a few $km\ s^{-1}$. Therefore these turbulent bursts may correspond to upwardly moving material, perhaps away from the source of the explosive event. Note that the light curve, after the explosive event has finished, contains a number of peaks at times of 8, 14.5, 19.5 and 25 minutes. The peaks at these times were found to correspond to very broadened line profiles, indicating non-thermal velocities up to 70 $km\ s^{-1}$. These profiles were found to be well fitted by a single Gaussian fits and are clearly not characteristic of explosive events. The light curve, line-shifts and non-thermal velocities were examined using a power spectrum analysis. All the power spectra were normalised with respect to their maximum values above 2 mHz. In Figure 1(b) we plot the power spectra of the light curve, line-shift and non-thermal velocity for the 19-21 arc sec region. The intensity has a clear periodicity of 2 mHz (~8 minutes), the line-shift velocity has a periodicity of 3 mHz (~5 minutes) while the non-thermal velocity has no distinct peak in its power spectrum, but instead has a spread of frequencies below 2 mHz.

3.2. C IV 1548 Å

The C IV 1548 Å line was observed in a quiet Sun region at Sun centre, X=0, Y=0. Even though it is a 'quiet' Sun region, evidence was found of explosive events at different regions along the slit. A small region from 67-73 arc sec showed evidence of large increases in intensity coupled with a large increase in line width (non-thermal velocity), see Figure 2.

The last three peaks in intensity after a time of 40 minutes have a period of oscillation of ~9 minutes, while the full observation has a frequency of oscillation of 1 mHz (~16 minutes) with a weaker oscillation frequency of 3 mhz (~5 minutes). The line-shift velocity on-the-other-hand has a it's main frequency of oscillation at 2 mHz, while the non-thermal velocity has it's peak at 1.5 mHz. In this case there are no explosive event near the site of the turbulent bursts. The nearest explosive event takes place at region 60-62 arc sec, at least 5 arc sec away.

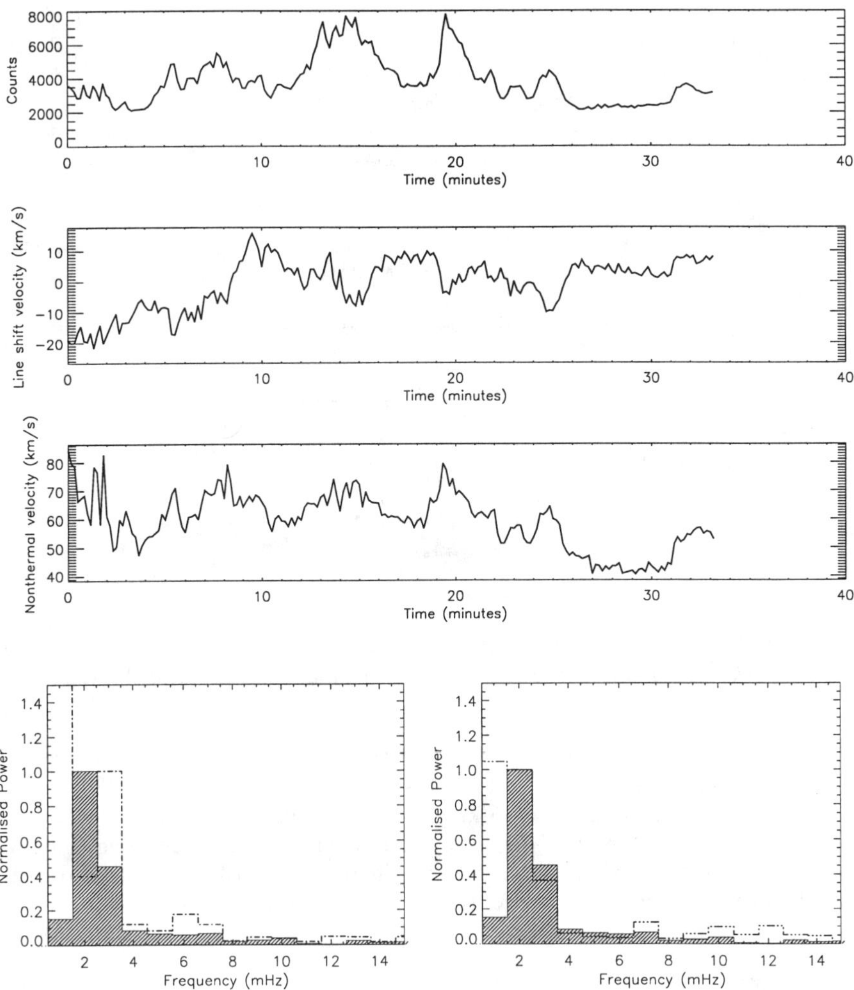

Figure 1. (a) Intensity, line-shift velocity and non-thermal velocity calculated from region 19-21 arc sec of dataset 002802, using the moments of the line profiles. (b) The power spectra of the intensity (shaded area) and line-shift velocity (dot-dash line) and the intensity (dashed area) and non-thermal velocity (double dot-dash line) for the same region.

The light curves from region 60-62 and 67-73 arc sec have been measured and found to have a time delay of 90 seconds with respect to one another, i.e. the light curve from region 60-62 arc sec lags that of region 67-73 arc sec by 90 seconds. An explosive event measured at 53 minutes in region 60-62 arc sec showed a blue-shifted velocity of $\sim$48 $km\ s^{-1}$. If we assume that the plasma

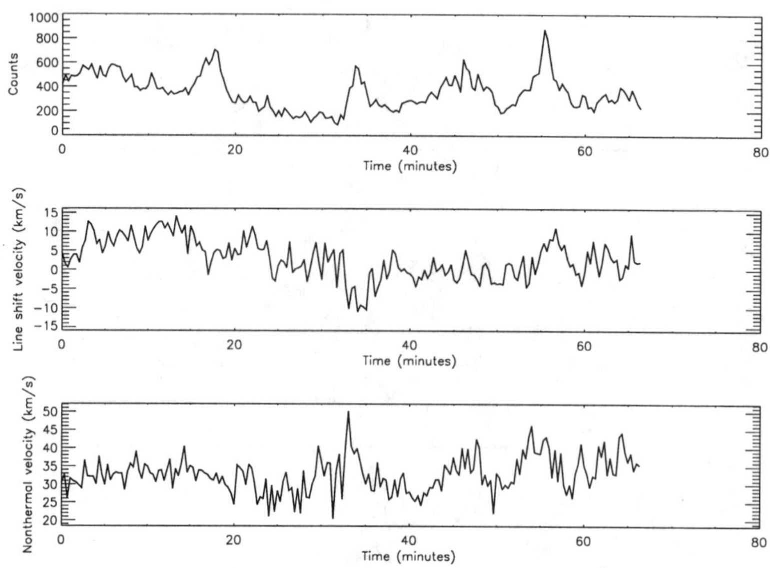

Figure 2. Intensity, line-shift velocity ($km\ s^{-1}$) and non-thermal velocity ($km\ s^{-1}$) calculated from region 67-73 arc sec of dataset 182013

from the explosive event travels at this speed, then over 90 seconds (i.e. the time delay between the regions) the plasma will have travelled ~4300 km or roughly 6 arc sec (1 arc sec = 715 km). This is roughly the difference between the two regions (60-62 and 67-73) being examined here. It is possible therefore to suggest that the turbulent bursts seen in region 67-73 arc sec are directly related to the explosive events occurring at region 60-62 arc sec.

References

Chae, J., Wang, Haimin, Lee, Chik-Yin & Gooder, P.R., 1998, Ap. J, 497, L109-L112

Doyle, J.G., van den Oord, G.H.J. & O'Shea, E, 1997, A&A, 327, 365

Dere, K.P., Bartoe, J.-D.F., Brueckner, G.E., Ewing, J. & Lund, P, 1991, J. Geophys. Res., 96, 9399

Dere, K.P., 1994, Adv. Space Res., 14, 13

Innes, D.E., Inhester, B., Axford, W.I., & Wilhelm, K, 1997a, Nature, 386, 811

Solar and Stellar Activity: Similarities and Differences
ASP Conference Series, Vol. 158, 1999
C.J. Butler and J.G. Doyle, eds.

Observations of Explosive Events in the Solar Atmosphere

M.E. Pérez & J.G. Doyle

Armagh Observatory, College Hill, Armagh BT61 9DG, N. Ireland

Abstract. Two examples of explosive events observed with SUMER in transition region spectral lines are reported here; one detected in C IV 1548 Å, in a region within the northern polar coronal hole, and the other in O VI 1032 Å, in an active region located on the solar disk. In the northern coronal hole we may have observed a bi-directional jet with an observed velocity ranging between 150 km s^{-1} in the blue wing to 150 km s^{-1} in the red wing. It extends approximately over a region of 8 arcsec along the slit (N-S) and 8 arcsec in the E-W direction. The life time for this event is approximately 160 s with a length approximately 35 arcsec. This extension implies that the jet reaches the corona and travels down along most of the transition region. In the events observed in the active region the apparent maximum velocities range between 250 km s^{-1} in the blue wing to 215 km s^{-1} in the red wing. The life times for these events ranges between 60 to 90 s. These explosive events seen in O VI showed a very complex structure of subsonic and supersonic velocity flows, both red-shifted and blue-shifted.

1. Introduction

With the launch of *SOHO* new opportunities have become available for studying short-time scale variability phenomena, such as explosive events. The instrument that allows us to do so is SUMER (Solar Ultraviolet Measurements of Emitted Radiation), a stigmatic normal incidence spectrograph operating in the wavelength range 450 to 1610 Å (Wilhelm et al. 1997). In July 1996, we obtained data with SUMER at several locations on the solar disk. The observing sequences involved the resonance lines C IV 1548Å and O VI 1032Å formed in the transition region between 10^5 and 3 10^5 K. The purpose of obtaining this data was to provide input for an ongoing modelling programme (Erdélyi et al. 1998, 1999). The scope of this programme is to study the relevance of the explosive events phenomena in the process of heating the solar corona.

2. Observational Data

The first observational data set reported here was obtained with SUMER onboard SOHO on 10 July 1996 for O VI 1032Å from an active region SW on the solar disk. The second data set for C IV 1548Å was obtained from a region in the northern polar coronal hole on 14 July 1996. The spatial resolution of SUMER is approximately 1 arcsec (E-W) and 2 arcsec (N-S). We used the 1×120 arcsec^2 slit on 14 July data set and the 0.3×120 arcsec^2 slit on 10 July.

3. Results

3.1. Coronal hole: C IV 1548Å

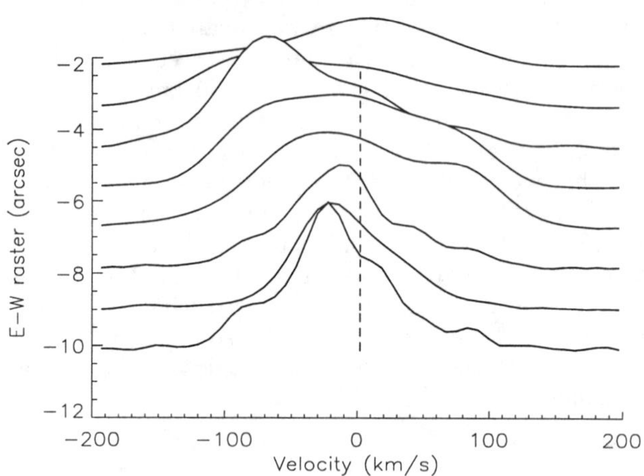

Figure 1. The C IV 1548Å line profile averaged over the whole explosive event as a function of time.

The sequence in Figure 1 lasts 200 s and covers a solar area of 10×20 arcsec^2. In the first time-frame we see a broadening in the C IV line centered at 904 arcsec north of disk center. By the second time-frame (06:31:40 UT) we see a blue-shifted component. For the next 40 s, the line is mostly blue-shifted although there is a weak red-shifted feature. At 06:32:41UT we see another injection of energy resulting in blue and red-shifted plasma. By this stage the center of the feature has drifted southward by three to four arcsec. The latter four raster positions show mostly a blue shifted plasma. The size of the explosive event in the north-south direction had a maximum extent of $\sim$ 6 arcsec. The time-frames in Figure 1 are separated by 1 arcsec (moving eastward), thus the feature is visible over an area of 6×8 arcsec^2. The maximum velocity reached in the blue wing was 150 km s^{-1} and 100 $km\ s^{-1}$ in the red wing.

An estimate of the characteristic sound speed, c, in a region is given by the relation $c \sim 0.17\ T_e^{0.5}$ km s^{-1}. This implies that for the C IV 1548Å line ($T_e = 10^5$ K) a value for c around 50 km s^{-1} can be estimated. Therefore, we are observing a global presence of supersonic up-flows and down-flows all along the sequence shown in Figure 1.

3.2. Active region: O VI 1032Å

The first explosive sequence in Figure 2 lasts for over 4 min. In the first time-frame (07:25:37 UT) we see a brightening to the blue at approximately –180 arcsec in the N-S direction, this slowly fades until 07:26:38 UT where we see a broadened blue-shifted line profile with a maximum velocity of approximately 120 km s^{-1}. The mass motion increases in the red wing to velocities of 100 - 150 km s^{-1} by 07:26:53 UT, remaining at these supersonic velocities for 45 s. Similar velocities are seen in the blue wing. In the frames between 07:27:54 UT and 07:28:39 UT the velocities fall below the sound speed ($c \sim 95$ km s^{-1}) in the red wing, disappearing at 07:28:39 UT, while they stay very close to the sound speed value for the blue wing over a region of 2 - 4 arcsec along the slit. At 07:28:55 UT there is another injection of energy, with plasma again moving both red-ward and blue-ward at a velocity up to 160 km s^{-1}. By 07:29:40 UT, the major component is blue-shifted at close to 240–260 km s^{-1}. Fifteen seconds later this has almost decayed. The later time-frames also showed apparent motions, now southward by 3 - 4 arcsec. At 07:30:26 UT two explosive events take place simultaneously at –202 and –209 arcsec in the N-S direction, with the maximum velocity being 200 km s^{-1} in the blue and 180 km s^{-1} in the red wing. By 07:31:11 UT, both events are gone.

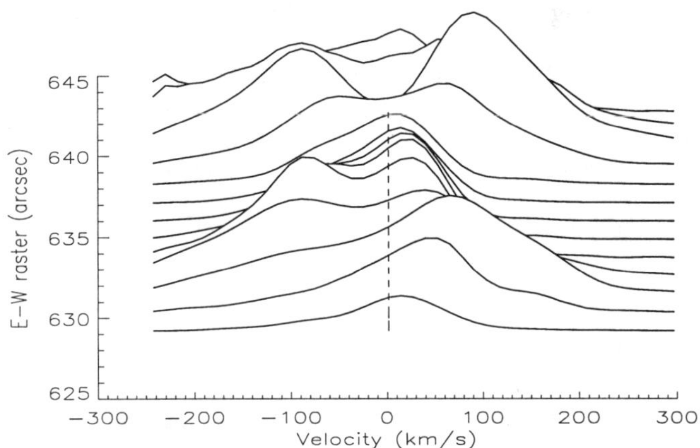

Figure 2. O VI 1032Å: For each time frame in the raster series we show here an averaged profile over the whole explosive event with the vertical line showing the zero velocity reference for the first two events between 07:28:39 UT and 07:29:40 UT (629–643 arcsec E-W)

4. Discussion

The persistence of supersonic flows along our raster series in the active region (i.e., the O VI data) could be due to bursts of explosive events (Chae et al.

1998) in the area observed, which are not connected physically with each other. However, the picture is quite complex and not easy to analyse. Our analysis has shown the complex structure of the velocity fields produced by these events. There are clearly asymmetric profiles with significant fluctuations in intensity. Also the presence of overlapping velocity fields with a consistent structure along the slit is apparent, often leading to supersonic flows.

For the C IV data in Figure 1, it is not possible to identify the signature as a burst of explosive events. The characteristics of this sequence of events can, instead, be compared with those discussed by Innes et al. (1997). If we suppose this sequence is produced by only one jet propagating away from its source, the area in which it is visible (6×8 arcsec^2) and the observed lifetime ($\sim 160\ s$) would be in coincidence with those previous results. The maximum velocities, though, correspond to blue-shifts and globally the blue wing is more intense than the red one. The C IV event can perhaps be explained as a jet of bi-directional nature if we consider a high latitude as a possible explanation for the apparent confusion between the two opposite flows. The jet can be within a plane that forms a short angle with that formed by the line of sight and the E-W direction, and the axis of this jet forms a relatively small angle with the line of sight. In these conditions we can explain why the first raster position shows the coincidence of blue-shifts and red-shifts, while the later ones are mainly blue-shifted. If this is correct, we can assume that the distance that this event covers in our raster (8 arcsec in 160 s) corresponds with its increase of size, and if that increase is due to the propagation of the head of the jet at a velocity equal to its maximum Doppler velocity (150 km s^{-1}), then we can estimate its length . We calculate an inclination angle for the axis of our jet of $\sim 13°$. Then the actual jet length is approximately 35 arc sec or 2.5 10^4 km, estimated from the apparent length of 8 arcsec ($\sim 6\ 10^3$ km). This extension implies that the jet reaches the corona and travels down through most of the transition region.

References

Chae, J., Wang, H., Lee, C., Goode, P.R. & Schühle, U., 1998, ApJ, 497, L109

Dere, K.P., Bartoe, J.-D.F. & Brueckner, G.E., 1991, J. Geophysical Res., 96, No. A6, 9399

Erdélyi, R., Sarro, L.M. & Doyle, J.G., 1998, in *Solar Jets and Coronal Plumes*, ESA SP-421, 207, 1998

Erdélyi, R., Sarro, L.M., Doyle, J.G. & Pérez, M.E., 1999 in progress

Innes, D.E., Inhester, B., Axford, W.I. & Wilhelm, K. 1997, Nature, 386, 811

Pérez, M.E., Doyle, J.G., Erdélyi, R. & Sarro, L.M. 1999, A&A(submitted)

Sterling, A.C., Mariska, J.T., Shibata, K. & Suematu, Y., 1991, ApJ, 381, 131

Wilhelm, K., Lemaire, P.,Curdt, W., Schúhle, U., Marsch, E., Poland, A.I., Jordan, S.D., Thomas, R.J., Hassler, D.M., Huber, M.C.E., Vial, J.-C., Kúhne, M., Siegmund, O.H.W., Gabriel, A., Timothy, J.G., Grewing, M., Feldman, U., Hollandt, J. & Brekke, P., 1997, Solar Phys., 170, 75

Solar and Stellar Activity: Similarities and Differences
ASP Conference Series, Vol. 158, 1999
C.J. Butler and J.G. Doyle, eds.

The Solar-Stellar Connection in X-rays: How to take advantage of *Yohkoh* data

G. Peres[1], S. Orlando[2] & F. Reale

Dipartimento di Scienze Fisiche ed Astronomiche, Univ. di Palermo, Piazza del Parlamento 1, 90134 Palermo, Italy

R. Rosner

Dept. of Astronomy and Astrophysics, University of Chicago; Enrico Fermi Institute, Chicago, Illinois, USA

H. Hudson

Solar Physics Research Corp., 4720 Calle Desecada, Tucson, Arizona 85718, USA

Abstract. We use solar *Yohkoh*/SXT data as a template for observations of stellar coronae. To circumvent the large differences between solar and stellar X-ray studies we have developed a method to put *Yohkoh*/SXT data in the same band, format and context as stellar X-ray observations. From the *Yohkoh*/SXT full-disk images we derive the whole–Sun X-ray emission measure vs. temperature (EM(T)), in the range $10^{5.5}K - 10^{8}K$; then, from the EM(T) we compute the whole Sun X-ray spectrum; folding the spectrum through the instrumental response of non-solar X-ray observatories, for instance *ROSAT*/PSPC and *ASCA*/SIS, we synthesize observations of the Sun similar to the stellar ones and we can analyze them with the same tools. We present the synthesized *ROSAT*/PSPC- and *ASCA*/SIS-like spectra of the Sun as a star at the maximum of the solar cycle with the relevant fits in terms of thermal components.

1. Introduction

The driving idea of this project is to explore the solar-stellar analogy in the X-ray band, in order to check the presumption that the coronae of solar-like stars behave in a way similar to that of our Sun. Solar observations give us a unique opportunity to study in detail an example of a late type star and, in particular, to examine the spatial structure of the source regions of its X-ray emission. Linking the characteristics observed in the spectra of late type stars with phenomena resolved on the Sun can be crucial for our understanding of

---

[1]e-mail: peres@oapa.astropa.unipa.it

[2]Solar System Division – ESA Space Science Dept., ESTEC, Postbus 299, NL–2200 AG Noordwijk, The Netherlands

the physical mechanisms at work in the coronae. Beyond the fact that solar observations allow for spatial resolution and the stellar observations do not, the instruments used to observe the solar corona in general differ substantially from those used for stellar coronal observations; *this results in the additional obstacle that measured physical properties for the solar corona do not correspond, and therefore cannot be easily compared, to those of stellar coronae.*

We present a method which, using the *Yohkoh/SXT* data, allows us to find how the X-ray Sun would be detected if observed as a star with a non-solar instrument, for instance *ROSAT/PSPC* or ASCA/*SIS*. The homogeneous comparison of solar and stellar observations should allow us to study the solar-stellar connection in a direct way avoiding, to some extent, extrapolations and approximate guesses based on very different and not directly comparable data; *a common framework to compare solar and stellar X-ray observations is an important tool for the interpretation of solar and stellar X-ray data.*

Our approach is based on deriving the Emission Measure vs. Temperature (EM(T)) of the whole solar corona from *Yohkoh*/SXT images and on synthesizing from the EM(T) the stellar-like spectra. The spectra are folded with the effective area vs. photon energy of the non-solar instruments, and taking into account their response matrix for the energy channels. The application of the standard analysis used for stellar data to stellar-like solar spectra derived as above is straightforward.

Since we have at the same time detailed information on the morphology, temperature and emission measure of the regions from which the X-ray emission originate, our study can help identify the structures on the solar disk responsible of various features in the spectra. If similar features are then recognized in stellar spectra, one can hypothesize that the star hosts structures or phenomena alike the solar one. At the other extreme this method may show directly any limit in the solar paradigm when applied to stars, particularly the most active ones.

This project opens the possibility to use directly the solar X-ray data to provide phenomenological models for many events and characteristics observed on stars. Such a direct way of bridging the gap between the solar and the stellar coronal physics helps generalize finely resolved solar observations to the stellar environment.

2. Results

Our approach allows us to compute accurately the solar coronal luminosity over various bands of the X-ray spectrum. It is worth noting that, just from the global solar X-ray fluxes collected in the *Yohkoh*/SXT bands (or equivalently over the bands of instruments such as *GOES* or *SOLRAD*), one cannot derive the overall solar spectrum, nor the luminosity over other X-ray bands, and cannot synthesize the data collected by other instruments, including non-solar ones; such fluxes are therefore of no help for our approach.

The distribution EM(T), namely the emission measure vs. temperature for the whole solar corona, is a result of interest by itself and can be useful for solar coronal studies. Our first analysis of solar stellar-like data yields results compatible with those found for other stars: the temperatures of the Sun during maximum of the cycle, obtained with the multiple thermal component fitting of

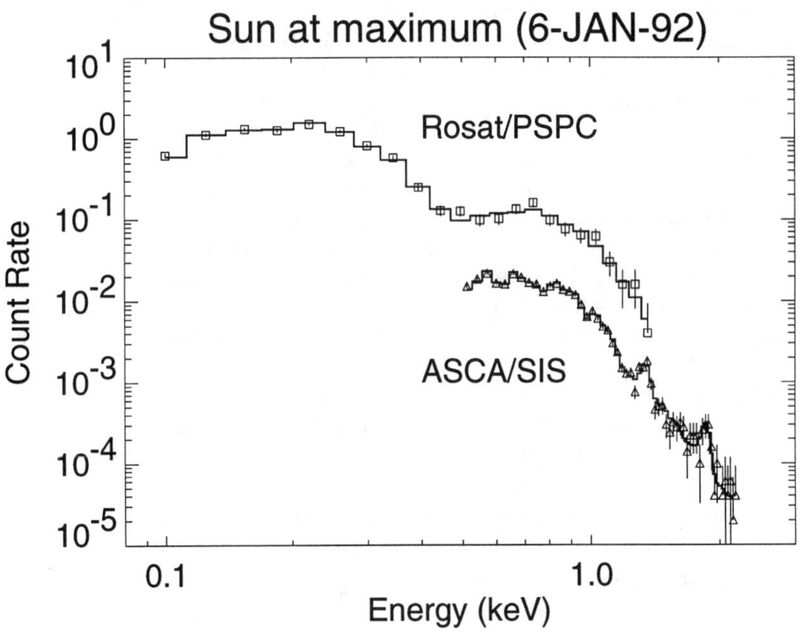

Figure 1. Spectra, in units of photons/s in the focal plane, of the Sun close to the maximum of the cycle (January 6, 1992), after folding through the *ROSAT*/PSPC (squares) and *ASCA*/SIS (triangles). We assume to observe the solar corona from a distance of 1 parsec. The solid lines are the best fitting model spectra.

the *ROSAT*/PSPC-like and *ASCA*/SIS-like spectra (Figure 1), are typical of a moderately active star. The fitting of *ROSAT*/PSPC data yields two components with temperature of $\approx 1.1\ 10^6$ K and $\approx 2.8\ 10^6$ K, emission measure values of $\approx 1.3\ 10^{50}$ cm^{-3} and $\approx 0.7\ 10^{50}$ cm^{-3}, respectively; their total luminosity in the PSPC band is $4.5\ 10^{27}$ erg s^{-1}. The same analysis for the Sun at minimum of the solar cycle yields a luminosity approximately 20 times smaller.

The corresponding fitting of *ASCA*/SIS data yields two components with temperature of $\approx 2.1\ 10^6$ K and $\approx 5.8\ 10^6$ K, emission measure values of $\approx 10^{50}$ cm^{-3} and $\approx 0.1\ 10^{50}$ cm^{-3}, respectively; their total luminosity in the SIS band is $1.5\ 10^{27}$ erg s^{-1} and $\sim$ 50 times smaller at the minimum of the solar cycle. Notice the difference in the luminosity values obtained for *ROSAT* and *ASCA*, due to the different energy bands of the two instruments. The results for the Sun during the minimum of the cycle are alike those of a very quiet star.

The X-ray luminosity of the solar corona undergoes changes, during the cycle, by large factors. Equivalent variability has not been reported (so far) in stellar coronae; it still remains to be explored whether a large cyclic variability is a unusual or relatively common among coronae. Peres et al. (1998) discuss the method in more details and show some applications to a few observations of the solar corona during the solar cycle, and to a large solar flare; the data are compared with typical characteristics of stellar coronae. Orlando et al. (1998), provide an extensive set of simulations dedicated to investigate the features of the *Yohkoh*/SXT full disk data even at low photon counts and over the entire nominal range of temperature sensitivity $10^{5.5} - 10^8$ K and are fundamental to devise appropriately our analysis method.

We can now tap the large amount of information contained in the *Yohkoh* data set for a variety of projects, including studies of variability on various time scales and the analysis of the role of various structures of the corona (active regions, older regions, large scale structures, etc.) in determining the EM(T) and the emitted spectrum.

Acknowledgments. This work has been partially supported by Ministero della Pubblica Istruzione Ricerca Scientifica e Tecnologica and by Agenzia Spaziale Italiana.

References

Orlando et al., 1998, ApJ (submitted)
Peres et al., 1998, ApJ (submitted)

Solar and Stellar Activity: Similarities and Differences
ASP Conference Series, Vol. 158, 1999
C.J. Butler and J.G. Doyle, eds.

Super-Saturation: The Myth Exploded!

David J. James & Andrew C.Cameron

School of Physics and Astronomy, University of St Andrews, St Andrews, KY16 9SS, Scotland

Robin D. Jeffries

Department of Physics, Keele University, Keele, ST5 5BG, UK.

Moira M. Jardine & Joao M. Ferreira

School of Physics and Astronomy, University of St Andrews, St Andrews, KY16 9SS, Scotland

Abstract. We have unearthed from the literature and the SIMBAD database a sample of both single and binary M-dwarfs, with known rotation periods and ROSAT PSPC X-ray detections. An analysis of their X-ray behaviour is presented, to search for evidence of **super-saturation** - a state of decreased X-ray emission observed in some ultra-fast rotating G & K-dwarfs - and to probe their magnetic nature.

1. Introduction

Optical, UV, EUV and X-ray observations of solar-type stars (spectral types late-F→M) both in the field and young ($\lesssim$ 1 Gyr) open clusters have provided evidence for a correlation of increased magnetic activity manifestations (eg, Hα, X-ray fluxes) with increasing rotation rate (cf. Vilhu, 1984; Doyle 1987; Soderblom et al. 1993; Stauffer et al. 1994; Randich et al. 1996). However, especially at X-ray energies, a *saturation* (or emission plateau) in the magnetic-heating induced emissions of more rapidly rotating ($v \sin i \gtrsim$ 15-20 km s^{-1}) solar-type stars occurs. At X-ray luminosities, this saturation occurs at about 10^{-3} of the bolometric luminosity. Furthermore, in the IC 2391 and α Per open clusters, a downward trend from the saturated X-ray emission plateau - termed **super-saturation** - is observed for a few G and K stars with extremely high rotation rates ($\sim$ 150 km s^{-1}), or Rossby numbers, R$_o$, ($\equiv P/\tau_c$), $\lesssim$ -1.6, Unfortunately, there are too few **super-saturated** stars to say whether this occurs at a given rotation rate or Rossby number.

The physical cause of **super-saturation** (and indeed saturation !) of stellar X-ray emission is far from being fully understood. It is possible that the dynamo itself is self-limiting, maybe via a Lorentz back-reaction, or that a saturation of the magnetic heating processes which energizes coronal plasma, trapped in confining magnetic structures, is occurring in the most rapidly rotating stars. An alternative hypothesis is that the X-ray emitting coronal volume of rapid rotators is reduced via centrifugal stripping (*e.g.*, Jardine & Unruh 1998). The models of these authors also show that for increasing rotation rate in G dwarfs (their Figure 1), coronal temperatures increase quite considerably (up to an order of magnitude or more). If their models mirror some degree of physical reality, we must then also consider the possibility that rapid rotators may sustain sufficiently hotter coronae such that the ROSAT PSPC instrument can no longer measure their emission distributions in its relatively narrow 0.1-2.4 keV passband (see also Randich, 1998). This possibility is also backed up by Pleiades observations. Gagné, Caillault & Stauffer (1995) provided analyses of X-ray observations of G & K-dwarf Pleiads which show that single and two temperature plasma models both yielded moderately hotter coronal temperatures for the more rapidly rotating stars in their sample. Our *two cents worth* concerns the behaviour of the dynamo, and the observed X-ray emission, in stars with deep convection zones - approaching the fully convective state where the nature of a solar-like *shell dynamo* may change possibly into a more *distributed* state.

2. The Sample

A sample of rapidly rotating single and binary M-dwarfs has been assembled from a variety of literature searches and the SIMBAD database. All single stars were chosen such that their photometric rotation periods were known, and a *ROSAT* PSPC X-ray observation existed for each. Binary stars were chosen rather more selectively. Their rotation periods are inferred from spectroscopic orbital motions of the binary under the assumption that each component of the binary system is tidally-locked to its companion through a tidal coupling of the orbital and rotational motion of each (for orbital periods of $\sim$ 10 days or less - see Zahn, 1977, 1989). Binary M-dwarf systems were therefore chosen having periods of $\sim$ 10 days or less, and if a PSPC X-ray observation existed for each system. A complete identification and discussion of the sample is deferred to a forthcoming paper.

3. Discussion

We present X-ray luminosities, as a fraction of bolometric luminosities, versus Rossby number, for our sample of single and binary M-dwarfs in Figure 1. An initial warning to the reader is warranted at this stage to avoid over-interpretation of these results. Our primary concerns are centered on the narrow passband of the *ROSAT* observatory and the small number statistics of our sample. Note: the extremely high datum at $L_{\rm x}/L_{\rm bol} = -2.25$ is for Gl 873, and is considerably higher that the expected saturated level of 10^{-3} seen in G, K and M-dwarfs both in the field and young open clusters. We have extracted the PSPC data from the public archive (RP 200984n00 PSPC) and preliminary results indicate a large flare in the X-ray count rates during the exposure of this star.

It is clear from Figure 1 that X-ray emission is saturated in both single and binary M-dwarf systems with rotational/orbital periods less than 6 days ($R_o \lesssim -0.5$). Equally clear is the lack of evidence for **super-saturation** seen in some young G & K-dwarfs. We believe there are several hypotheses for the apparent lack of **super-saturation** observed in the systems in our sample:

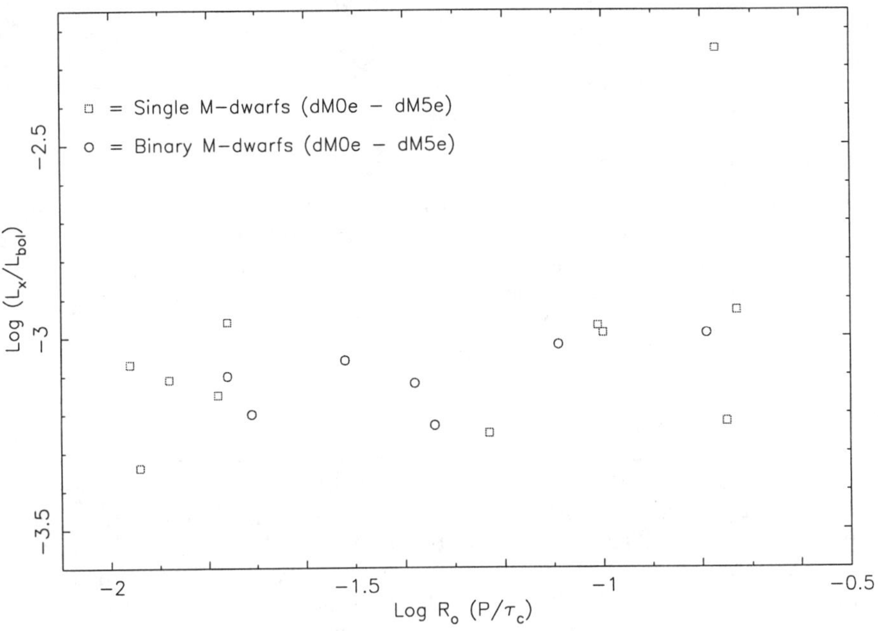

Figure 1. X-ray luminosity, as a fraction of bolometric luminosity, is plotted against Rossby number R_o ($\equiv P_{rot}/\tau_c$) for our sample of single and binary M-dwarfs. There is little evidence of the **super-saturation** effect (at R_o between ~ -1.6 and -2.1) seen in some extremely rapidly rotating IC 2391 and α-Per late-type stars (Prosser et al. 1996; Randich 1998).

- **Centrifugal stripping:**

  As the stellar rotation rate increases, centrifugal forces cause a rise in the pressure and density in the outer parts of the largest magnetic loops (Jardine & Unruh 1998). Such stresses and distortions may break open previously closed field lines, destroying X-ray emitting regions and leading naturally to the saturation of the observed X-ray emission. If, **in addition**, dynamo saturation is invoked, the X-ray emission will decrease further with increasing rotation rate, leading to **super-saturation**. Centrifugal stripping occurs when the co-rotation radius moves inside the closed corona and would be indicated by the presence of slingshot prominences. The rotation rate at which this occurs is determined by the field topology, which is currently an unknown factor in M-dwarfs but one that could be resolved by Doppler imaging.

- **Nature of the Dynamo:**

  M-dwarfs have extremely deep convective zones, $\gtrsim 0.5$ - by fractional radius. Does the nature of the stellar dynamo change as the convection zone deepens in progressively lower mass stars ? The shell dynamo, presumably at work in the Sun, may be reduced in efficiency as a star becomes nearly or fully convective. It is plausible that a variant of the dynamo (shell or distributed) at work in very low mass stars may still exhibit saturation-like phenomena (such as $L_x/L_{bol}=10^{-3}$) whilst not exhibiting the **super-saturation** effect seen in higher mass stars with more shallow convection zones, and presumably a solar-like shell dynamo.

- **Faster rotators, hotter coronae:**

  A set of X-ray observations, with a greater spectral energy coverage (e.g. *XMM*), of rapidly rotating late-type stars of all spectral types is desirable to verify that the temperature(s) of X-ray emitting plasma has/have not climbed out of the narrow passband of the *ROSAT* observatory.

4. Conclusions

Our analyses of rotation and X-ray data show that single and binary M-dwarfs have indistinguishable levels of saturated X-ray emission. There is no evidence of the **super-saturation** effect seen for a couple of G & K stars in IC 2391 and the Alpha Persei cluster. A fuller sample is presently being considered to include stars with slightly longer rotation periods to discern at what point X-ray saturation occurs in M-dwarfs. Dynamic spectral analysis and Doppler imaging of some of the more rapidly rotating (and bright) stars is required in order to better understand the field topology of these systems.

Acknowledgments. DJJ would like to thank María Pilar Escribano Benito for much love and support during this work, and the continued positive influences of Mrs J Pryer. This research has made use of the SIMBAD database, operated at CDS, Strasbourg, France, and the Leicester Database and Archive Service at the Department of Physics and Astronomy, Leicester University, UK. DJJ also thanks PPARC (and ACC) for a post-doctoral research fellowship.

References

Doyle, J. G., 1987, MNRAS, 224, 1P

Gagné, M., Caillault, J.-P., & Stauffer, J. R., 1995, ApJ, 450, 217

Jardine, M., & Unruh, Y. C., 1998, A&A, submitted

Prosser, C. F., et al. 1996, AJ, 112, 1570

Randich, S., et al. 1996, A&A, 305, 785

Randich, S., 1998, in Donahue, D., & Bookbinder, J., eds., 10th Cambridge Workshop on Cool Stars, Stellar Systems and the Sun.

Soderblom, D. R., et al. 1993, ApJS, 85, 315

Stauffer, J. R., et al. 1994, ApJS, 91, 625

Vilhu, O., 1984, A&A, 133, 117

Zahn, J.-P., 1977, A&A, 57, 383

Zahn, J.-P., 1989, A&A220, 112

Part 7

SUMMARY

“Blind belief in authority is the greatest enemy of truth”

Solar and Stellar Activity: Similarities and Differences
ASP Conference Series, Vol. 158, 1999
C.J. Butler and J.G. Doyle, eds.

Activity in the Sun and Late-type Stars – What Have we Learned so Far

Jeffrey L. Linsky

JILA/University of Colorado and NIST
Boulder, CO 80309-0440 USA

Abstract. In my summary of this symposium on "Solar and Stellar Activity: Similarities and Differences" I will highlight those new observations and theory that provide answers to long-standing problems or provide some insight into their essential physics. I also call attention to the new phenomena revealed by the beautiful *TRACE*, *SoHO*, *Yohkoh*, and ground-based data presented at the meeting and suggest some major questions that should be addressed using the upcoming *AXAF* and *XMM* X-ray spectra and images. One example of important recent progress in the field of solar/stellar activity is the detailed calculations of wave propagation and heating in magnetic flux tubes that predict the observed dependence of chromospheric Ca II emission on stellar rotation from basal heating rates up to saturation. Other important accomplishments are the first detailed maps of the magnetic field structure on active stars like AB Dor revealed by Zeeman-Doppler imaging and the drift of Hα transient absorption structures with orbital phase.

"Blind belief in authority is the greatest enemy of truth."
– Albert Einstein.

1. Introduction

I am certain that Brendan Byrne would be proud of what we have accomplished. The organizers of this symposium have properly honored his memory by scheduling talks on the most rapidly advancing topics of solar and stellar activity, which inevitably are the most controversial. The invited speakers and poster presenters have risen to the challenge by highlighting many critical issues and discussing the alternative ways in which they should be addressed. I am pleased that many of the speakers have called attention to the inadequacies of their data sets and theoretical models. Honest assessments of the shortcomings of both observations and theory is required for a scientific discipline to make real progress.

There is likely universal agreement that the underlying cause of the diverse phenomena that we call "stellar activity" lies in the interactions between magnetic fields and turbulent plasmas. Although one can write down the physical equations that describe these interactions, our understanding of how to apply these equations, even at an intuitive level, remains far from complete. For example, is the α-Ω dynamo operating at the interface between the convective envelope and the radiative core of the Sun responsible for generating magnetic fields and the solar cycle, or is the magnetic field regenerated close to the surface as suggested by the recent *TRACE*[1] images. *SUMER*[2] and *Yohkoh* X-ray data may well have settled the long standing controversy as to the nature of the heating process in the solar transition region and corona. It now appears that magnetic reconnection events are the dominant heating mechanism, but we have no detailed physical models that can explain how this really happens or predict the heating rate based on observable quantities.

For many years people have argued that the Sun can serve as our "rosette stone" in which the secrets of stellar active processes can be revealed by observing our nearby bright star with very high spatial, spectral, and temporal resolution. Thanks to *SoHO*, *Yohkoh*, *TRACE*, and very high resolution observations from the ground, we now have better insight into these questions, but observations of stars, even as unresolved point sources, are providing important new information not readily obtained from the Sun. For example, the dependence of presumably non-magnetic basal heating on stellar properties (e.g., $T_{\rm eff}$, g, and chemical composition), saturated heating in rapidly-rotating young stars, flares occurring in stellar-size magnetic loops, polar and near-polar spots (Brendan take note), and the very unsolar-like magnetic field distributions on stars in RS CVn systems revealed by Zeeman Doppler images cannot by learned by studying only one middle-aged, slowly rotating star. The cross-fertilization and synergy between solar and stellar studies can be breathtaking at times. Conferences like this one that bring together solar and stellar astronomers stimulate this cross-fertilization, whereas meetings covering topics only in solar or in stellar astronomy do not benefit from different perspectives. I will mention other examples of this cross-fertilization later.

2. Some Illustrious Quotes

The true flavor of a meeting cannot be appreciated from a bland summary of the conclusions of each speaker. To appreciate this flavor one should hear what, in their unguarded moments, they actually said. So I begin my summary with some direct quotes, ripped out of context and reassembled into a more logical order. Here is a sample of the more enticing quotes:

Suzanne Hawley: [3] "It is strange to be in Armagh without Brendan as my host."

Rob Jeffries: "Brendan was my thesis examiner, and boy did he give me a hard time."

Sami Solanki: (referring to a very complex line profile) "I don't know where the flow is, but Brendan would find it."

---

[1] The Transition Region and Coronal Explorer.

[2] The Solar Ultraviolet Measurements of Emitted Radiation instrument on the Solar and Heliospheric Observatory (*SoHO*).

[3] In this paper I identify the symposium speakers by printing their names in **bold face**. Some of their comments were indeed bold.

Suzanne Hawley: (referring to a scatter diagram) "There is a slight correlation here."

Sami Solanki: (well after his allotted time had ended) "Since the chairman has not barked, I will now misuse this opportunity and speak for another 10 minutes."

Antonio Lanza: (before showing another 20 viewgraphs) "I have only one viewgraph of interest to show."

Sami Solanki: "I hate to criticize my own work."

Fred Walter: (at the beginning of his talk on pre-main sequence stars) "In order to learn about the stars, don't start with the Sun. That is hopeless. Active stars are not scaled-up Suns."

Andrew Collier-Cameron: "When in doubt, you might as well invoke a magnetic field to solve the problem."

Karel Schrijver: (referring to the statistical connectivity of surface and deep magnetic fields, he coined the term) "magneto-chemistry."

Helen Mason: (referring to the present picture of the solar corona with hot and cool magnetic loops interleaved, she used the term) "magnetic junkyard."

Peter Ulmschneider: (referring to the inadequate state of solar chromospheric models) "The truth lies somewhere between the Carlsson-Stein and Avrett calculations."

Peter Ulmschneider: "It is very strange that we are learning more about physical processes at meter scales by studying point sources than by studying the Sun with high spatial resolution."

Peter Ulmschneider: (referring to the inability of present hydrodynamic codes to properly treat heating by strong shocks) "There are presently no good calculations on the market."

Alesandro Lanzafame: (after saying that systematic trends provide more trustworthy results than comparisons of spectra with semi-empirical NLTE models) "A good fit does not mean that you have a good model."

Alesandro Lanzafame: (concerning the reliability of present generation spectroscopic diagnostics of chromosphere and transition region plasmas) "Today I will provide more doubts than answers."

Andrew Collier-Cameron: (referring to whether velocity drifts in Hα absorption features seen in spectra of AB Dor are long-lived prominences or ejected clouds) "I am uncharacteristically attempting to find a middle ground."

Bert van den Oord: (referring to the lack of spatial resolution and inadequate time resolution of stellar flare data) "This is about the situation in solar physics in the 1960s."

Eric Priest: (referring to *TRACE* movies showing that the solar transition region is highly dynamic) "The old model looks a bit outdated."

Eric Priest: "*Yohkoh* has revealed a whole new MHD world."

Brendan Byrne: (sage advice from the past) "May the road rise to meet you."

3. Some Important Results and Interpretations concerning Solar/Stellar Activity

The rich variety of interesting new observations from instruments in space and on the ground, and the insightful interpretations of these observations presented during the meeting were a great delight to all of us. A portion of this material struck me as especially interesting either because it broke new ground, strengthened or demolished some of our preconceptions, or will likely provide the basis for future models. With apologies to those people whose ideas did not strike me quite this way, here is my summary:

3.1. New Perspectives on the Solar Magnetic Field

Long uninterrupted sequences of magnetograms obtained with MDI[4] have changed some previously accepted ideas about the solar magnetic field. **Schrijver** described how the field arises at the surface in bipolar regions with a power law distribution of sizes and fluxes and then diffuses and disappears when it encounters fields of opposite polarity. Since on large scales the photospheric field is replaced in 4 months (at sunspot maximum) or 10 months (at sunspot minimum), the photospheric field forgets its past on these time scales. Thus the connectivity between surface and deep magnetic field lines must be statistical (a concept that he called "magneto-chemistry") rather than deterministic as assumed in the classical Babcock–Leighton dynamo model. Does this mean that the widely-accepted $\alpha\Omega$ dynamo mechanism operating at the interface between the convective envelope and the radiative core must be replaced by a dynamo operating close to the solar surface?

On small scales the replacement of photospheric magnetic flux is far more rapid. **Schrijver** estimates that in very quiet regions on the solar surface the magnetic field is replaced about every 40 hours. Magnetic fields disappear from the photosphere mainly by subduction rather than cancellation, a process that produces no heating. However, the continual replacement of magnetic flux in the photosphere leads to field-line braiding, reconnection, and thus heating in the corona, likely along the lines proposed by Parker (1993). I encourage theoreticians to compute coronal heating rates from the evolution of the photospheric field as seen by MDI to see if the coronal energy budget can be explained by this process.

Mason pointed out that EIT[5] images have completely demolished the old view that 10^5 K (transition region) plasma exists only at the footpoints of magnetic loops between coronal plasma and the chromosphere. Instead, the EIT

[4] The Michelson Doppler Imager instrument on *SoHO*.

[5] The Extreme-Ultraviolet Imaging Telescope instrument on *SoHO*.

images show that hot ($T > 1\ 10^6$ K) and cool ($T \sim 10^5$ K) loops are interspersed in a model that has been called the "magnetic junkyard" (e.g., Dowdy, Rabin & Moore 1986). Thus the corona is magnetically disconnected from the transition region, and typical pressures of transition region and coronal plasma are not be related by hydrostatic equilibrium. This situation provides guidance to those who wish to model stellar atmosphere structure from UV, EUV, and X-ray fluxes. One can infer an emission measure distribution from the data, but it is highly unrealistic to then infer the properties of a one component model structure in hydrostatic equilibrium. The "magnetic junkyard" model simply explains why stellar coronal pressures inferred from density sensitive line ratios are typically much larger than pressures of the 10^5 K plasma. The SUMER instrument on *SoHO* has discovered that macro-spicules at the limb appear to be spinning, leading to their being called "tornados". Are the spinning macrospicules flux tubes filled with chromospheric material?

From their analysis of *HRTS*[6] ultraviolet spectra, Dere et al (1987) published the surprising result that the cross-sectional area of bright features in the transition region are very small (< 70 km) and the filling factors also very small (< 0.01). **Mason** noted that filling factors of 10^{-2} down to perhaps 10^{-5} are implied by *SoHO* ultraviolet spectra, confirming the *HRTS* results. If the bright emission points in the transition region are confined by the magnetic field, then the magnetic flux loops are very thin (as is shown in the spectacular *TRACE* images) and the filling factors measure the cross-sectional area of strong fields on the Sun. **Schmeider** showed theoretical models of solar magnetic field evolution that can explain erupting prominence observations.

3.2. New Observations of Stellar Magnetic Fields

Linsky provided us with a tour through the HR Diagram, identifying the types of stars for which magnetic fields are measured or inferred from reliable proxies and summarizing the many roles that magnetic fields can play in stellar atmospheres. For late-type stars, photospheric magnetic fields can now be measured with a variety of techniques using both unpolarized and polarized light. Beginning with the pioneering work of Robinson, Worden & Harvey (1980), a very successful technique has been to measure the Zeeman broadening of optical lines and splitting of near-infrared lines in unpolarized light. Using this technique, Saar (1990) showed that photospheric field strengths increase with decreasing T_{eff} in main sequence stars consistent with the equipartition of gas and magnetic pressure, and that the magnetic flux and filling factors increase with stellar rotation rate and Rossby number. This method does not, however, provide information on the distribution or three-dimensional structure of the magnetic field across the stellar surface.

Donati and **Collier-Cameron** described the Zeeman Doppler imaging (ZDI) technique that can map the radial, meridional, and azimuthal components of the magnetic field for rapidly-rotating stars using profiles of some 2000 spectral lines in circularly polarized light. Zeeman-Doppler images are now available for six active stars (including AB Dor and HR 1099) using the maximum

---

[6]The High Resolution Telescope Spectrograph rocket experiment.

entropy or optimal reconstruction techniques. The ZDI technique is described by Donati & Brown (1997) and by Donati et al. (1997). Unlike the Sun, the field lines for these active stars are azimuthal in rings (3 for AB Dor). The technique now allows one to follow the magnetic field evolution in active stars and to monitor stellar magnetic cycles directly rather than through a proxy like the Ca II H+K flux. **Collier-Cameron** said that for AB Dor the azimuthal field predicted from potential field extrapolations of the observed radial magnetic field agrees well with the observed azimuthal field. This gives the technique additional credibility. However, **Solanki** pointed out that the ZDI technique likely misses most of the magnetic field because it does not see dark spots of weak field regions.

Recent studies have provided new information on the geometry of stellar magnetic fields in addition to field strengths and magnetic fluxes. **Walter** showed that very large prominence-like magnetic loops are inferred for naked T Tauri stars (NTTS) and other active stars. **Collier-Cameron** showed that for AB Dor the rotation periods of Hα transient absorption features are the same as spots at latitudes of 60°–70°. Since the Hα absorbing clouds/prominences with lifetimes of about 1 week often lie outside the co-rotation radius (about 3 stellar radii), this active star appears to have giant magnetic loops that are anchored at high latitudes on opposite hemispheres.

The evidence for spots lying near the poles of active stars (unlike the case for the slowly rotating Sun) is increasing since Byrne (1996) and Strassmeier (1996) last debated the topic, and the picture is becoming more complex. As you recall, the main evidence for polar spots comes from Doppler imaging of active stars where the filling-in of the cores of absorption lines with no change with rotational phase is usually interpreted as indicating the absence of continuum emission at the rotational poles. Both authors identified other mechanisms that could make absorption lines less deep in active stars than in quiescent stars including greater heating of the upper atmosphere, different micro-turbulence, and magnetic splitting. **Solanki** called attention to the Bruls, Solanki & Schüssler (1998) plage models that show that some lines used for Doppler imaging have filled-in cores but not of the type observed in the spectra of active stars. **Lanza & Rodonò** showed Doppler images indicating that rapidly-rotating stars have polar spots, while more slowly-rotating stars have high latitude spots not at the poles. They also reviewed the evidence for very large spots or spot groups (spot filling factors as large as 0.60 for II Peg) and preferred longitudes for spots on stars in RS CVn systems.

3.3. New Studies of Stellar Flares and Coronae

Although the optical and near-UV continua contain most of the radiative energy loss from flares on M dwarf stars, realistic models that can explain the time dependence of this energy loss are in short supply. For this reason the dynamic flare model atmospheres presented by **Abbett** and **Hawley** were a highlight of the meeting. Using the Carlsson & Stein (1997) radiative hydrodynamic code, **Abbett** and **Hawley** solved for the radiative and dynamic response of the photosphere and chromosphere to a beam of non-thermal particles propagating down along a flux tube from the magnetic energy release site in the corona. The initial response of the atmosphere is increased photoionization that increases the electron density and H^- opacity, thereby decreasing the continuum intensity. This is probably the explanation for the "pre-flare dips" seen in the near infrared continuum before Hα brightenings in some flares. Next is a "gentle" phase in which the photosphere can radiate all of the input energy from the beam in the hydrogen and H^- continua with an increase in temperature. In this stage the emission lines are symmetric as there are no mass motions. When hydrogen becomes highly ionized and cannot radiate the input energy, the atmosphere explodes with a large shock wave removing the input energy largely as kinetic energy. The code predicts shifted, distorted, and doubled emission lines similar to what is observed during flares. I consider this work to be a highlight of the meeting because it demonstrates that models that properly include the essential physics can explain many of the diverse flare phenomena that have puzzled observers and theoreticians alike for many years.

X-ray emission from flares on M dwarf stars and close binaries (RS CVns, Algols, and W UMas) are often explained by analogy with solar compact flares (short time-scales) or two-ribbon flares (longer time-scales). The usual explanation for the factor of 10^3–10^4 larger energy release in large stellar flares compared to large solar flares is that the flare volumes in active stars are much larger than for the Sun. This explanation should have been scrutinized at the meeting. Even if it is right, this explanation begs the question of why the flare volumes are so much larger in active stars. Is this because individual flux loops are larger (as in the case of AB Dor), or that much larger arcades of loops can be flare sites in stars with much larger magnetic fluxes than the Sun?

An important result obtained from the analysis of *ROSAT* and *ASCA* X-ray energy distributions and *EUVE* spectra is that the inferred abundances of metals with low first ionization potentials in the coronae of Algols and RS CVns are typically 1/3 to 1/10 of the abundances measured in the solar photosphere, but are similar to what is measured in the solar corona (the so-called FIP effect). **Mathioudakis** reminded us that the inferred metal abundances are sensitive to the amount of high temperature plasma in the assumed coronal model. **Van den Oord** explained that the low coronal abundances are due to settling in hydrostatic equilibrium. During flares one observes a return to normal abundances most likely due to the evaporation of photospheric material into the corona. An example is the 1994 August 29 flare on UX Ari (Güdel et al. 1998). However, *SoHO* and *TRACE* observations reveal that the solar atmosphere is highly dynamic and thus mixing processes must be included in any diffusion calculations for the Sun and presumably other stars. Realistic calculations are needed in order to better understand the FIP effect.

3.4. New Models of Heating Mechanisms

The question of whether the outer atmospheres of stars with convective zones are heated primarily by the dissipation of acoustic waves, MHD waves, or magnetic reconnection events (now often called "microflaring") has been a major theme of solar/stellar physics for more than half a century. What new results concerning this question were presented at this meeting?

The Ca II H_{2V} grains, the transient bright features that appear in the violet peaks of the Ca II emission on small-scales have often been cited as evidence for localized magnetic heating events. The new *TRACE* observations described by

Rutten now show that the grains are not spatially correlated with the magnetic field, but instead are naturally produced by upwardly propagating acoustic waves generated by pistons in the photosphere. The *TRACE* data also suggest that the bright internetwork grains are produced by weak shocks in the chromosphere. These weak shocks likely account for the basal flux chromospheric heating rate seen in the internetwork regions and in very slowly-rotating stars (Buchholz, Ulmschneider & Cuntz 1998). The enhanced chromospheric emission in the network and active regions on the Sun and in active stars requires some type of magnetic heating process.

Ulmschmeider and **Cuntz** summarized some major advances that are being made in computing the heating rates of MHD waves in stellar chromospheres. **Ulmschneider** showed that convective buffeting of magnetic flux tubes in the solar photosphere can generate longitudinal MHD wave fluxes with 2 10^8 erg cm^{-2} s^{-1} and transverse MHD wave fluxes 15 times larger. In principle, these wave modes contain sufficient energy in these modes to heat stellar chromospheres, especially when one includes a spectrum of wave periods and allows for different period waves to overtake slower waves to form super-shocks. The critical issue is how to properly treat strong shocks where "there is presently no good calculation on the market."

Cuntz presented an insightful set of calculations of the propagation of MHD waves in K2 V stars. He computed the MHD waves fluxes for atmospheres with photospheric field strengths $B_o = 2100$ G, roughly the equipartition value, and magnetic fluxes given by an empirical relation of $B_o f_o$ with $P_{\rm rot}$. In his models the flux tube cross-sectional areas increase with height until the upper chromosphere is completely filled. For slowly rotating stars, the small magnetic filling factors in the photosphere imply that the flux tube cross sections increase rapidly with height so that the upward propagating MHD waves are spread out over a large area and therefore shock high in the atmosphere where the density is low. The computed Ca II H+K flux is therefore small. For rapidly rotating stars the opposite is true, so that the MHD waves spread out slowly with height and therefore shock low in the chromosphere where the densities are large. As a consequence, the chromospheric heating rate is large (c.f. Fawzy, Ulmschneider & Cuntz 1998). From my perspective the important point is that we saw calculations that incorporate much of the essential physics of MHD wave heating and predict the dependence of the Ca II H+K surface flux on $P_{\rm rot}$ in agreement with observations. In particular, the sum of acoustic and MHD wave heating explains the chromospheric heating rate from the basal flux level (pure acoustic waves in the very slowest rotators) to the saturated heating rate of the most rapid rotators without the need for microflares or other heating mechanisms in the chromosphere.

Above the chromosphere other heating mechanisms become important. In her review of the SUMER observations, **Mason** said that the small-scale bidirectional flows seen in 10^5 K emission lines, which were first noted in *HRTS* data, are correlated with the position of transition region explosive events, which are likely produces by microflares. Wood, Linsky & Ayres (1997) have interpreted the broad wings of transition region lines in active stars as produced by microflare events, as the fraction of the emission line fluxes in the broad wings increases with the X-ray and C IV surface fluxes. The EIT and CDS[7] data show that the sum of many microflare events, which are generally assumed to be magnetic reconnection events, may explain 60% or more of the coronal heating. The microflares typically occur in regions of complex polarity. **Priest** concludes that these important results from *SoHO* are producing a paradigm shift in the question of how the corona is heated. X-ray spectroscopy with the upcoming *AXAF* and *XMM* satellites will provide powerful diagnostics of the coronal plasmas that hopefully will provide constraints on the heating mechanism. In particular, solar coronal loops appear to be isothermal and require uniform heating rates. **Priest** argued that small current sheets may be required to heat these loops. Güdel, Guinan & Skinner (1997) found that the emission measure of high temperature coronal plasma decreases with age. This important result must be explained by models of microflare heating.

Mason told us that the Chianti atomic data base for emission line spectroscopy is being extended to $\lambda < 50$ Å for this purpose. One positive aspect of the delayed launch dates for *AXAF* and *XMM* is that the Chianti code extensions should be available in time to analyze these exciting X-ray spectra.

Pallavicini described the recent SAX satellite observations of extremely hot flares from UX Ari and AB Dor with $kT_2 = 9.6$ keV and 20–50 keV photons. If these flare plasmas are thermal, as typically assumed, then the temperatures are about 10^8 K. Since SAX detected only a small number of hard X-ray photons, it is hard to distinguish between X-ray spectra from thermal and non-thermal electron energy distributions. The detection of hard X-rays from non-thermal distributions of electrons during impulsive solar flares makes it likely that some or all of the hard X-rays from stellar flares are non-thermal. Future observations with *SAX* and *XTE* are needed to answer this question. Then we can address the question of whether the heating of coronae during flares and outside of detected flares is the same phenomena only on different scales.

4. Some important unanswered questions concerning Solar/Stellar Activity

Many solar/stellar activity phenomena do not yet have an adequate explanation. I list here some of the more interesting questions as brought to our attention by the various speakers in the hope that progress will be reported soon:

Mathioudakis: How can one explain densities of 10^{12} cm^{-3} at $T \geq 10^7$ K in "nonflaring" coronae as inferred from EUVE spectra? The implied gas pressures are $\sim 10^4$ that of the solar corona, and the required magnetic fields needed to confine the plasma exceed 200 Gauss. Are coronal fields this strong consistent with measured photospheric fields? During flares the inferred densities are even higher, implying larger magnetic field strengths.

Stern: What is the explanation for supersaturation ($R_x = L_x/L_{bol}$ turns over at low $N_R = P_{rot}/\tau_{conv}$)? This phenomenon is seen in rapidly rotating

[7] The Coronal Diagnostic Spectrometer instrument on *SoHO*.

stars in young clusters like α Per. (**Collier-Cameron** suggested that the co-rotation radius moves in to the corona for such stars.)

Hawley: Does $R_x = L_x/L_{bol}$ turn over in the lowest mass stars in clusters?

Stern: Why are binary stars (including wide binaries) more X-ray luminous than single stars of the same age and spectral type?

Stern: After 17 years of observations, why is there no evidence for stellar X-ray activity cycles? The Mt. Wilson Ca II H+K line program has identified cycles in many G and K stars, but as yet no cycles have been identified in the X-ray data. Is this due to the sparse sampling by X-ray satellites or is there another explanation?

Jeffries: Why is there a large spread in vsini for clusters younger than the Hyades?

Jeffries: Why is $< L_x >$ a factor of 2 different in clusters of the same age (e.g., Hyades and Praesepe)?

Solanki: Why is there a log normal distribution of sunspot sizes? What are the implications for stars with very large spots (e.g., II Peg with spots covering 40% of the visible surface)?

Solanki: In penumbral outflows why is the flow velocity horizontal when the magnetic field is inclined?

Solanki: Can the observed filling in of line cores in rapidly rotating active stars be explained by plages rather than the usual interpretation of polar spots?

Lanza: Why does the amplitude of differential rotation decrease as the rotation velocity of stars increase?

van den Oord: Why are the most intense flares observed on the brightest stars (e.g., RS CVns and Algols)?

van den Oord: What is the physical explanation for the Güdel–Benz law: $L_x = 10^{15.5\pm0.5} L_R$? This relation is unexpected because the X-ray emission is thermal whereas the radio emission from active stars is non-thermal.

Phillips: Why are hot points observed by the *Yohkoh* satellite at the tops of solar coronal loops when conduction should make the loops isothermal in about 2 seconds?

Ulmschneider: What is the correct way to treat strong shocks (e.g., heating and propagation)? Also, it is important to look for short period waves produced by small shocks.

Lanzafame: What are reliable plasma diagnostics given the following problems:

- contribution functions are typically broad, covering a range of plasma temperatures and densities leading to a difficult inversion uniqueness problem;

- ionization can be out of collisional equilibrium, especially for the Li-like and Na-like ions;
- the fraction of the aperture filled with bright emission for solar observations is uncertain and the filling factor for stars is unknown; and
- the essential physical processes are difficult to model given the highly dynamic atmospheres, diffusion, turbulence, and complex frequency redistribution. (He suggested in his talk that a search for trends and common behavior of different lines that may lead to conclusions that do not depend heavily on these problems.)

Priest: How does one construct an accurate theory for 3-D reconnection?

Priest: Why is the transition region so dynamic?

Hawley: Why does the Hα emission from M dwarfs not appear to decay with age as do other activity indicators for F, G, and K stars? This new result was unexpected since previously more limited surveys of M dwarfs indicated a decrease in Hα emission with age.

Pallavicini: If solar coronae consist of many loops at different temperatures, then it will be difficult to derive the temperatures, loop lengths, and filling factors for these different loops uniquely even with high quality X-ray spectra obtained by *AXAF* and *XMM.*

To this list I will add a few more critically important questions:

- What exactly do we mean by the term "activity" and how should it be characterized?
- Are all phenomena that we call "active" predominately magnetic in character?
- Does the full range of stellar active phenomena occur on the Sun, even rarely?
- Are there active phenomena on stars for which solar phenomena are not useful prototypes?
- What mechanical forces other than convection can produce active phenomena?
- Are active phenomena in close binary stars and pre-main sequence stars qualitatively different from phenomena on single stars?
- Why are active phenomena in some stars as much as 10^4 times more energetic than in the Sun?
- Which magnetic heating modes are most important in different types of stars and for different temperatures and densities?
- What role does the geometry of the magnetic field play in heating?

Acknowledgments. I thank NASA for my support through grants to the University of Colorado under the *HST*, *AXAF*, and *FUSE* programs. I also wish to thank for their hospitality the staff of Armagh Observatory, including Gerry Doyle, John Butler, and posthumously Brendan Byrne. We will long remember Brendan's scientific contributions and his good advice and companionship.

References

Bruls, J.H.M.J., Solanki, S.K. & Schüssler, M. 1998, A&A, 336, 231

Buchholz, B., Ulmschneider, P. & Cuntz, M. 1998, ApJ, 494, 700

Byrne, P.B. 1996, in Stellar Surface Structure, ed. K.G. Strassmeier & J.L. Linsky (Dordrecht: Kluwer), 299

Carlsson, M. & Stein, R.F. 1997, ApJ, 481, 500

Dere, K.P., Bartoe, J.-D.F., Brueckner, G.E., Cook, J.W. & Socker, D.G., Science, 238, 1267

Donati, J.-F. & Brown, S.F. 1997, A&A, 326, 1135

Donati, J.-F., Semel, M., Carter, B.D., Rees, D.E. & Cameron, A.C. 1997, MNRAS, 291, 658

Dowdy, J.F., Rabin, D. & Moore, R.L. 1986, Solar Physics, 105, 35

Fawzy, D.E., Ulmschneider, P. & Cuntz, M. 1998, A&A, 336, 1029

Güdel, M., Guinan, E.F. & Skinner, S.L. 1997, ApJ, 483, 947

Güdel, M., Linsky, J.L., Brown, A. & Nagase, F. 1998, ApJ, in press

Parker, E.N. 1993, ApJ, 407, 342

Robinson, R.D., Worden, S.P. & Harvey, J.W. 1980, ApJ, 236, L155

Saar, S.H. 1990, in IAU Symposium 138, The Solar Photosphere: Structure, Convection, and Magnetic Fields, ed. J.O. Stenflo, (Dordrecht: Kluwer), p. 427-441.

Strassmeier, K.G. 1996, in Stellar Surface Structure, ed. K.G. Strassmeier & J.L. Linsky (Dordrecht: Kluwer), 289

Wood, B.E., Linsky, J.L. & Ayres, T.R. 1997, ApJ, 478, 745

Author Index

Author Index